# Magnetic Resonance Spectroscopy and Imaging in Neurochemistry

# Advances in Neurochemistry

**SERIES EDITORS:**

**B. W. Agranoff,** *University of Michigan, Ann Arbor*
**K. Suzuki,** *University of North Carolina, Chapel Hill*

*ADVISORY EDITORS:*

| | | | |
|---|---|---|---|
| J. Axelrod | A. Dahlstrom | B. S. McEwen | E. Roberts |
| S. T. Brady | F. Fonnum | P. Morell | L. Sokoloff |

A Continuation Order Plan is available for this series. A continuation order will bring delivery of each new volume immediately upon publication. Volumes are billed only upon actual shipment. For further information please contact the publisher.

# Magnetic Resonance Spectroscopy and Imaging in Neurochemistry

*Edited by*

## Herman Bachelard

*University of Nottingham*
*Nottingham, England*

SPRINGER SCIENCE+BUSINESS MEDIA, LLC

Library of Congress Cataloging-in-Publication Data

---

Magnetic resonance spectroscopy and imaging in neurochemistry / edited
by Herman Bachelard.
        p.   cm. -- (Advances in neurochemistry ; v. 8)
    Includes bibliographical references and index.
    ISBN 978-0-306-45520-9      ISBN 978-1-4615-5863-7 (eBook)
    DOI 10.1007/978-1-4615-5863-7
    1. Brain--Magnetic resonance imaging.  2. Nuclear magnetic
resonance spectroscopy.  3. Brain--Diseases--Diagnosis.
4. Neurochemistry--Methodology.   I. Bachelard, H. S.  II. Series.
QP356.3.A37  vol. 8
[RC386.6.M34]
573.8'419 s--dc21
[616.8'047548]                                              97-15443
                                                              CIP

---

ISBN 978-0-306-45520-9

© 1997 by Springer Science+Business Media New York
Originally published by Plenum Press New York in 1997

http://www.plenum.com

10 9 8 7 6 5 4 3 2 1

# CONTRIBUTORS

CHRIS M. ANDREW • *Department of Clinical Neurosciences, Institute of Psychiatry, De Crespigny Park, London, SE5 8AF, United Kingdom*

D. L. ARNOLD • *Montreal Neurological Institute, McGill University, Montreal, Quebec, Canada H3A 2B4*

HERMAN BACHELARD • *M. R. Centre, Department of Physics, University of Nottingham, Nottingham, NG7 2RD, United Kingdom*

RONNITTE BADAR-GOFFER • *M. R. Centre, Department of Physics, University of Nottingham, Nottingham, NG7 2RD, United Kingdom. Present address: Elscint Ltd., Haifa 31004, Israel*

ODED BEN-YOSEPH • *Department of Radiology, University of Michigan Medical Center, Ann Arbor, Michigan 48109-0648*

MICK J. BRAMMER • *Departments of Neuroscience and Biostatistics & Computing, Institute of Psychiatry, De Crespigny Park, London, SE5 8AF, United Kingdom*

ED T. BULLMORE • *Departments of Psychological Medicine and Biostatistics & Computing, Institute of Psychiatry, De Crespigny Park, London, SE5 8AF, United Kingdom*

ALBERT BUSZA • *RCS Unit of Biophysics, Institute of Child Health, London, WC1N 1EH, United Kingdom*

ERNEST B. CADY • *Department of Medical Physics and Bio-Engineering, University College London Hospitals NHS Trust, London, WC1E 6JA, United Kingdom*

THOMAS L. CHENEVERT • *Department of Radiology, University of Michigan Medical Center, Ann Arbor, Michigan 48109-0648*

N. DE STEFANO • *Montreal Neurological Institute, McGill University, Montreal, Quebec, Canada H3A 2B4*

JENS FRAHM • *Biomedizinische NMR Forschungs GmbH am Max-Planck-Institut für biophysikalische Chemie, D-37077 Göttingen, Germany*

PENNY GOWLAND • *Magnetic Resonance Centre, Department of Physics, University of Nottingham, Nottingham NG7 2RD, United Kingdom*

FOLKER HANEFELD • *Abteilung Kinderheilkunde, Schwerpunkt Neuropädiatrie, Georg-August-Universität, D-37075 Göttingen, Germany*

KEIKO KANAMORI • *Magnetic Resonance Spectroscopy Laboratory, Huntington Medical Research Institutes, Pasadena, California 91105*

MARTIN KING • *RCS Unit of Biophysics, Institute of Child Health, London, WC1N 1EH, United Kingdom*

TERRI L. C. LUVISOTTO • *Department of Clinical Neuroscience, Division of Neurosurgery, The University of Calgary, Calgary, Alberta, Canada T2N 1N4*

PETER MANSFIELD • *The Magnetic Resonance Centre, Department of Physics, University of Nottingham, Nottingham, NG7 2RD, United Kingdom*

P. M. MATTHEWS • *Montreal Neurological Institute, McGill University, Montreal, Quebec, Canada H3A 2B4*

TRACY K. MCINTOSH • *Division of Neurosurgery, University of Pennsylvania School of Medicine, Philadelphia, Pennsylvania 19104*

PETER MORRIS • *Magnetic Resonance Centre, Department of Physics, University of Nottingham, Nottingham, NG7 2RD, United Kingdom*

SOPHIA RABE-HESKETH • *Department of Biostatistics & Computing, Institute of Psychiatry, De Crespigny Park, London, SE5 8AF, United Kingdom*

BRIAN D. ROSS • *Magnetic Resonance Spectroscopy Laboratory, Huntington Medical Research Institutes, Pasadena, California 91105*

BRIAN D. ROSS • *Department of Radiology, University of Michigan Medical Center, Ann Arbor, Michigan 48109-0648*

ARNE SCHOUSBOE • *PharmaBiotec Research Center, Department of Biological Sciences, Royal Danish School of Pharmacy, DK-2100 Copenhagen, Denmark*

ANDREW SIMMONS • *Department of Clinical Neurosciences, Institute of Psychiatry, De Crespigny Park, London, SE5 8AF, United Kingdom*

URSULA SONNEWALD • *MR Center, SINTEF UNIMED, N-7034 Trondheim, Norway*

GARNETTE R. SUTHERLAND • *Department of Clinical Neurosciences, Division of Neurosurgery, The University of Calgary, Calgary, Alberta, Canada T2N 1N4*

ROBERT TURNER • *RCS Unit of Biophysics, Institute of Child Health, London, WC1N 1EH, United Kingdom*

NICK VAN BRUGGEN • *Department of Neuroscience, Genentech Inc., South San Francisco, California 94080*

ROBERT VINK • *Department of Physiology and Pharmacology, James Cook University of North Queensland, Townsville, Queensland 4811, Australia*

NIELS WESTERGAARD • *PharmaBiotec Research Center, Department of Biological Sciences, Royal Danish School of Pharmacy, DK-2100 Copenhagen, Denmark*

STEVE C. R. WILLIAMS • *Department of Clinical Neurosciences, Institute of Psychiatry, De Crespigny Park, London, SE5 8AF, United Kingdom*

# PREFACE

The *Advances in Neurochemistry* series was initiated for a readership of neuroscientists with a background in biochemistry. True to this concept, the present volume brings together various applications of magnetic resonance technology to advance our knowledge of how the nervous system functions. Whether at the cellular, tissue slice, or intact organism level, magnetic resonance techniques are by their nature noninvasive, and thus provide a window through which biochemical reactions can be viewed without grinding, binding, or otherwise perturbing ongoing physiological processes. As technological improvements in methodology, such as higher and more uniform magnetic fields, novel paradigms for data analysis, etc., are made, we find increased sensitivity and improved temporal and spatial resolution for functional imaging techniques on the one hand, and better separation of signals that identify chemical properties in spectral shift studies, on the other. It is upon knowledge such as is described in the twelve chapters that follow, that further advances in scientific discovery and the biomedical applications of tomorrow will be based. We are grateful to Dr. Bachelard, the Volume Editor, and to the authors of the individual chapters for their efforts. We also note that with this volume Dr. Morris Aprison, a co-founder of the *Advances in Neurochemistry* series has stepped down and acknowledge with thanks his major role in its inception. In addition, we thank our past and present Advisory Editors.

Bernard W. Agranoff
Kunihiko Suzuki
*Series Editors*

# CONTENTS

HERMAN BACHELARD

CHAPTER 1

*¹³C AND ¹H MRS OF CULTURED NEURONS AND GLIA*

URSULA SONNEWALD, ARNE SCHOUSBOE, AND NIELS WESTERGAARD

CHAPTER 2

*MEASUREMENT OF FREE INTRACELLULAR CATIONS*

HERMAN BACHELARD AND RONNITTE BADAR-GOFFER

CHAPTER 3

IN VIVO *NITROGEN MRS STUDIES OF RAT BRAIN METABOLISM*

KEIKO KANAMORI AND BRIAN D. ROSS

CHAPTER 4

*TRAUMATIC BRAIN INJURY*

ROBERT VINK AND TRACY K. MCINTOSH

CHAPTER 5

*ANIMAL MODELS OF STROKE*

TERRI L. C. LUVISOTTO AND GARNETTE R. SUTHERLAND

CHAPTER 6

IN VIVO *MAGNETIC RESONANCE IMAGING AND SPECTROSCOPY: APPLICATION TO BRAIN TUMORS*

BRIAN D. ROSS, ODED BEN-YOSEPH, AND THOMAS L. CHENEVERT

CHAPTER 7

*DIFFUSION-WEIGHTED MAGNETIC RESONANCE IMAGING*

MARTIN KING, NICK VAN BRUGGEN, ALBERT BUSZA, AND ROBERT TURNER

CHAPTER 8

*HIGH-SPEED ECHO-PLANAR IMAGING AND ITS APPLICATION
TO NEUROLOGY*

PENNY GOWLAND AND PETER MANSFIELD

CHAPTER 9

*BRAIN ACTIVATION STUDIES USING MAGNETIC RESONANCE IMAGING*

STEVE C. R. WILLIAMS, ANDREW SIMMONS, CHRIS M. ANDREW, MICK J. BRAMMER, ED T. BULLMORE, AND SOPHIA RABE-HESKETH

CHAPTER 10

*MRI AND PROTON MRS IN THE EVALUATION OF MULTIPLE SCLEROSIS*

D. L. ARNOLD, P. M. MATTHEWS, AND N. DE STEFANO

CHAPTER 11

*PHOSPHORUS AND PROTON MAGNETIC RESONANCE SPECTROSCOPY OF THE BRAIN OF THE NEWBORN HUMAN INFANT*

ERNEST B. CADY

CHAPTER 12

*LOCALIZED PROTON MAGNETIC RESONANCE SPECTROSCOPY*
*OF BRAIN DISORDERS IN CHILDHOOD*

JENS FRAHM AND FOLKER HANEFELD

# SYMBOLS AND GLOSSARY

**Angular frequency ($\omega$)** Frequency of rotation or oscillation ($\omega = 2\pi f$, where $f$ is frequency in hertz).

**$B_0$** Magnetic induction caused by the constant field strength of the magnet.

**Chemical shift ($\delta$)** Change in Larmor frequency (q.v.) of the same nucleus in different parts of the molecule, as parts per million (ppm) of the applied resonance frequency.

**Chemical shift imaging** Technique in which spectra are phase-encoded to allow acquisition of the spectra in a number of spatially localized volumes.

**EPI** Echo-planar imaging.

**EVI** Echo-volumar imaging.

**FLASH** Fast low-angle shot imaging.

**Flip angle** Amount of rotation of the magnetization vector from the static magnetic field, produced by the radio-frequency pulse.

**Fourier transform (FT)** Computerized transformation from the time domain to the frequency domain. In magnetic resonance spectroscopy, the amplitude versus time of the FID (q.v.) is transformed to amplitude versus frequency of the spectra.

**Free Induction Decay (FID)** Decay of the transient magnetic resonance signal resulting from the transverse magnetization of the nucleus produced by the excitant radio-frequency pulse.

**Gauss (G)** Unit of magnetic flux density (the Earth's magnetic field is ca. 0.5–1 G). 10,000 G = 1 tesla (q.v.).

**$G_x$, $G_y$, $G_z$** Gradient of the magnetic field in $x$, $y$, or $z$ direction.

**Gyromagnetic ratio ($\gamma$)** The ratio of the magnetic moment to the angular momentum of the nucleus (constant for each nucleus).

**Hertz (Hz)** SI unit of frequency (cycles/sec).

**ISIS** Image-selected *in vivo* spectroscopy.

***k* Space**   Spatial frequency domain (describes the effects of a pulse sequence in frequency space).

$k_x$, $k_y$, $k_z$   k space in *x*, *y*, or *z* direction.

**Larmor frequency ($\omega_0$)**   Frequency of excitation of the magnetic resonance.

**Pixel**   Smallest picture element in digital image display.

**ppm**   Parts per million of the applied magnetic field.

**RARE**   Rapid acquisition with relaxation enhancement.

**STEAM**   Stimulated-echo acquisition mode.

$T_1$   Longitudinal relaxation time—exchange of energy from nuclei to their environment (the "lattice").

$T_2$   Transverse spin–spin relaxation time.

$T_2^*$   Time constant for loss of phase coherence due to a combination of $T_2$ and field inhomogeneities, especially from variations in magnetic susceptibilities of the tissues.

**TE**   Time to echo (in spin-echo radio-frequency pulse sequences).

**Tesla (T)**   Unit of magnetic field (= 42.58 MHz for protons).

**TI**   Inversion time (in inversion recovery radio-frequency pulse sequence).

**Voxel**   Smallest volume element.

$\alpha$   Flip angle.

$\gamma$   Gyromagnetic ratio.

$\delta$   Chemical shift.

$\omega$   Angular frequency.

$\omega_0$   Larmor frequency.

$\chi$   Magnetic susceptibility.

# INTRODUCTION

## HERMAN BACHELARD

The major excitement of magnetic resonance (MR) lies in its potential to study novel aspects of brain function and metabolism noninvasively, with high resolution and especially with unique chemical specificity. Indeed, there are many intriguing investigations that are only feasible using this technique. The early major impact of MR was in the superb, highly resolved anatomical pictures produced by $^1$H MR imaging (MRI), whereas similar resolution is not normally yet possible from MR spectroscopy (MRS). MRS generally suffers from the major disadvantage of relative insensitivity—molecules that occur in small amounts <0.1 mM) cannot normally be detected unless indirect approaches are used.

However, it is the absolute chemical specificity of MRS that provides fascinating novel opportunities to explore biochemical function in the living intact brain, including that of man. It allows for continuous monitoring of the energy state and of selected metabolites and for following rates of flux through the intermediates of metabolic pathways and affords unique possibilities of studying enzyme kinetics. By the use of interleaved spectra of different nuclei, virtually simultaneous measurements of selected intermediary metabolites with intracellular cations (pH, calcium, magnesium, potassium, and sodium) can be performed. By exploiting differences in emphasis of metabolism between neurons and glial cells, we now have the potential means of investigating metabolic relationships

*HERMAN BACHELARD* • *M. R. Centre, Department of Physics, University of Nottingham, Nottingham, NG7 2RD, United Kingdom.*

*Magnetic Resonance Spectroscopy and Imaging in Neurochemistry,* Volume 8 of *Advances in Neurochemistry,* edited by Bachelard, Plenum Press, New York, 1997.

and interdependencies of these cell types. The future benefits for clinical investigations of numerous cerebral disorders are exciting.

During the 15 years since the application of MRS in biomedical research became a real possibility, MRS studies could be criticized for a tendency to "reinvent the wheel"—i.e., many observations seemed merely to reproduce or confirm the results obtained previously from the use of conventional biochemical techniques. However, a noninvasive *in vivo* metabolic technique requires verification before the results obtained from it can be treated with any confidence, and much of the work until recently can properly be regarded as validation studies. Over the past few years the technique has advanced to the extent that it is breaking new ground by providing information on aspects of intermediary metabolism that cannot be studied by other available techniques. Such recent developments, especially in metabolic studies, formed the basis of a special workshop on "Advances in Physiological Chemistry by In Vivo NMR," held in 1995 in Woods Hole, Massachusetts, under the auspices of the Society of Magnetic Resonance.

The contributions to this book describe applications ranging from fundamental studies on isolated brain preparations (cultured cells and brain slices), through *in vivo* investigations using rodents, as a basis for studies on the conscious human brain. These include examples of the use of all of the nuclei most commonly employed for metabolic studies: $^1H$, $^{13}C$, $^{15}N$, $^{19}F$, $^{23}Na$, $^{31}P$, and $^{39}K$. These are all the naturally abundant forms except for $^{13}C$ and $^{15}N$. The latter are relatively insensitive in MRS terms, which makes it difficult to use them for studies on endogenous metabolites and limits such studies to metabolites that occur in high concentrations. However, the low natural abundance of $^{13}C$ and $^{15}N$ can be an advantage in that labeled precursors can be used for metabolic investigations with little or no interference from endogenous metabolites.

Though relatively insensitive even in MRS terms, a major advantage of $^{13}C$ MRS is its chemical specificity, which allows enrichment of individual atoms within the same molecule to be distinguished and quantified. Further information is available from analysis of multiplets ("isotopomers"), provided that there is sufficient resolution of the spectra. This approach is based on observations that the resonance of a $^{13}C$ atom that is attached to another $^{13}C$ atom will be split as a result of the couplings of the attached protons and displaced from the central single resonance that would be present as a singlet if the $^{13}C$ atom were the only one labeled in the molecule; the resonance will therefore appear as a multiplet. For example, glutamate with $^{13}C$ only on position 4 gives rise to a singlet, whereas glutamate with $^{13}C$ on positions 3 and 4 gives rise to a doublet with a characteristic spin–spin coupling constant, $J_{CC}$. A mixture of these species (i.e., [4-$^{13}C$] glutamate plus [3,4-$^{13}C$] glutamate) gives a pseudotriplet, and the relative intensities of their resonances provide a direct measurement of their relative concentrations. The positions of the adjacent labeled atoms can be determined from the $J_{CC}$ constants; for example, the $J_{CC}$ of [3,4-[13]C] glutamate (34 Hz) will be different from that of [4,5-[13]C] glutamate (52 Hz). "Isotopomer analysis" therefore offers the possibility of following time sequences of labeling of the different C atoms of a metabolite such as glutamate, thus providing unique information on metabolic routes and flux rates. [For detailed reviews, see London (1988), Jeffrey et al., (1991), and Badar-Goffer and Bachelard (1991).] Such analyses using brain, liver, and heart preparations in vivo and in vitro have revealed a considerable amount of novel information (Cerdan et al., 1990; Cohen, 1987; Gopher et al., 1990; Malloy et al., 1987, 1990; Sherry et al., 1985) but have proved extremely difficult to apply to studies on human brain metabolism, owing largely to the relative insensitivity of [13]C MRS. Unfortunately, it did not prove possible to recruit a chapter on this topic, so reference to it is offered here.

The sensitivity of [13]C MRS can be greatly improved using the proton-observe, carbon-edited (POCE) technique. The signal from a proton directly bound to a [13]C (unlike that bound to a [12]C) is normally split into two resonances due to J-coupling. Use of appropriate spectral editing techniques enables detection of the protons bound to [13]C only and eliminates the signals from all other protons, including those bound to [12]C (Alger and Shulman, 1984). It is important to note that while [13]C resonances on —COOH and —CH$_2$ groups can be detected by conventional [13]C MRS, the indirect technique cannot readily detect the protons on —COOH groups, not only because these groups are largely dissociated as —COO$^-$ at physiological pH, but particularly because the proton is not directly attached to a [13]C. This is a minor limitation in studies on, for example, amino acid metabolism. The POCE approach takes advantage of the far greater MRS sensitivity of protons as compared to that of [13]C and so enables a considerable increase in sensitivity. The technique was successfully applied by Shulman's group to the calculation of rates of metabolism of lactate and glutamate in rat brain in vivo, by following rates of flux of [13]C from [1-[13]C]glucose into the C-4 and C-3 positions of glutamate (Rothman et al., 1985; Fitzpatrick et al., 1990). The increased sensitivity of the technique was reflected in the ability of these investigators to acquire data in less than 98 sec. The calculated rate of flux through the cycle was ca. 1.4 μmol min$^{-1}$ g$^{-1}$, which is in good agreement with previous assessments. This work paved the way for quantitative measurements of metabolic flux in vivo in human tissues, as demonstrated in the elegant recent studies by the same group. From the direct [13]C observations of Fig. 1 (Rothman et al., 1992; Gruetter et al., 1994; Mason et al., 1995), the calculated rate of flux through the cycle was 0.72 ± 0.18 μmol min$^{-1}$ g$^{-1}$, about 50% of that which these investigators had observed for rat brain, which again is in line with earlier biochemical knowledge—i.e., that intermediary metabolism is faster in the rodent brain than in humans (McIlwain and Bachelard, 1985). Application of the more sensitive POCE technique allowed observation in 12-ml volumes of human

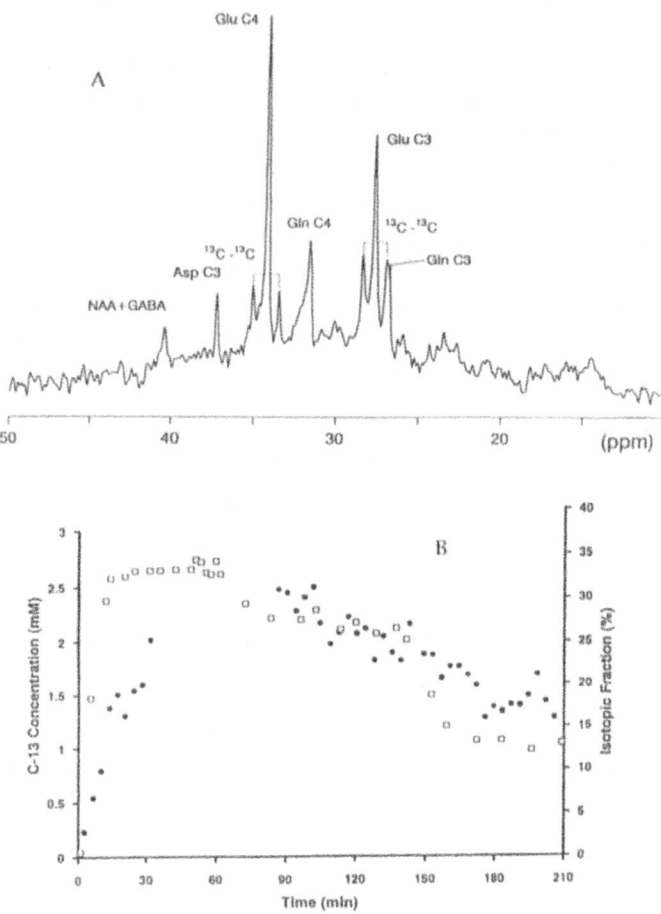

FIGURE 1. (A) $^{13}$C MR spectrum showing amino acid labeling from [1-$^{13}$C] glucose in the human brain. (B) Time course of incorporation of $^{13}$C from plasma glucose (□) into the C-4 position of glutamate (●). [From Gruetter *et al.* (1994), with thanks to E. J. Novotny.]

brain, where similar rates of $0.94 \pm 0.18$ $\mu$mol min$^{-1}$ g$^{-1}$ were calculated (Rothman *et al.*, 1992; Chen *et al.*, 1994; Novotny *et al.*, 1995; Fig. 2).

The first chapter of this book, which concerns studies based on the use of cultured cells, with some comments on brain slices, shows the wealth of metabolic information that can be derived from the use of $^{13}$C-labeled precursors and discusses the interesting information that is emerging on the metabolic relationships between neurons and glial cells. The second chapter describes MRS techniques for measuring a variety of intracellular cations in different cerebral preparations—cultured cells, slices, animal brains *in vivo*—with some comments on

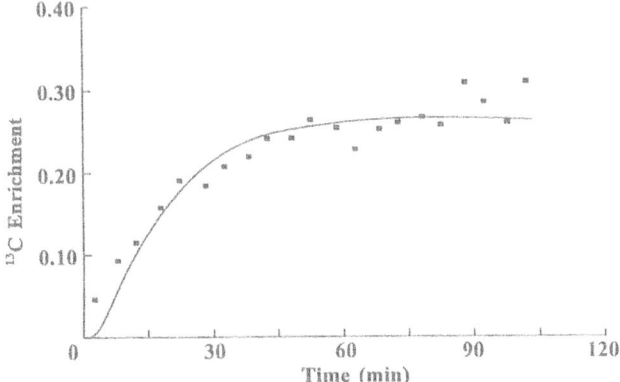

FIGURE 2.   $^1$H-{$^{13}$C} MR measurements of glutamate turnover in human brain. The graph shows rates of incorporation of $^{13}$C from [1-$^{13}$C]glucose into the C-4 position of glutamate. [From Chen *et al.* (1994), with thanks to E. J. Novotny.]

applications to the human brain. Novel opportunities of monitoring amino acid metabolism are now also coming from the use of $^{15}$N-labeled precursors: kinetic measurements of key enzymes (glutamine synthetase, glutaminase) are providing opportunities to follow dynamic aspects in rodent brains *in vivo* as described in Chapter 3. Noninvasive MRS and MRI research on animal models is yielding new information on the biochemical and physiological consequences of head injury, stroke, and brain tumors, as described in Chapters 4–6. Each of these contributions emphasizes the feasibility of applying such studies to the human brain, with the exciting potential of monitoring the effects of therapeutic intervention.

These animal studies lead logically to the chapters devoted to direct imaging and spectroscopy of the human brain. The technique of diffusion-weighted imaging (Chapter 7) exploits the possibility of following the diffusion of water *in vivo,* leading to new perspectives in our understanding of the cerebral pathophysiology associated with such disorders as stroke and spreading depression. The invention of echo-planar imaging (EPI) some years ago at the University of Nottingham (Chapter 8) revolutionized the ability to follow dynamic morphological changes *in vivo.* This remarkable technique allows high-resolution images to be acquired in less than 60 msec ("snapshots"), thus resulting in "movies" of internal organs of the body. Together with other rapid acquisition imaging techniques, it is increasingly being applied to brain activation studies ("functional imaging"; Chapter 9). While the precise biochemical or physiological basis for the "lighting up" of discrete regions of the brain upon sensory or motor stimulation remains to be elucidated, it is understood to be related to the paramagnetic properties of oxygen, such that, on binding of oxygen to hemoglobin, the proton resonances of the protein are affected. It is therefore thought to reflect changes in oxygen delivery or extraction. It provides spatial resolution at least as good as that offered by positron-emission tomography (PET), with the advantage over PET of being essentially noninvasive so that repeated scans can be performed on the same subject or patient.

The chapters which follow are devoted to direct applications to human clinical disorders: to investigations of the pathology of multiple sclerosis, only possible previously by postmortem examination (Chapter 10), to the prognostic evaluations in birth asphyxia pioneered by the authors of the chapter some 15 years ago, when MRS was only beginning to be applied (Chapter 11), and to the exploration of a variety of metabolic disorders of childhood (Chapter 12).

Some of the chapters, particularly those in which novel imaging techniques are discussed, contain a certain amount of physics. While an attempt has been to limit this as much as possible for the neuroscience reader, some treatment of the underlying theory is inevitable. An elementary guide to the principles of magnetic resonance is given in an earlier review (Bachelard and Badar-Goffer, 1993), and for the reader wishing to learn more, detailed treatments are available in the books by Gadian (1982), Morris (1986), and Chakeres and Schmalbrock (1992). General review articles on magnetic resonance in the neurosciences have been published by van Bruggen *et al.* (1994) and Kauppinen and Williams (1993). For those unversed in magnetic resonance, I have compiled a simple glossary (p. xxi). More comprehensive glossaries can be found in the books by Chakeres and Schmalbrock (1992) and Rinck (1993).

After being invited by the series editors to organize this book, I was gratified by the enthusiastic response I received from the contributors and wish to express by appreciation of their efforts and their tolerance in acceding to my editorial requests.

## REFERENCES

Alger, J. R., and Shulman, R. G., 1984, Metabolic applications of high-resolution $^{13}$C nuclear magnetic resonance spectroscopy, *Br. Med. Bull.* **40:**160–164.

Bachelard, H. S., and Badar-Goffer, R. S., 1993, NMR spectroscopy in neurochemistry, *J. Neurochem.* **61:**412–429.

Badar-Goffer, R. S., and Bachelard, H. S., 1991, Metabolic studies using $^{13}$C nuclear magnetic resonance spectroscopy, *Essays Biochem.* **26:**105–119.

Cerdan, S., Kunnecke, B., & Seelig, J., 1990, Cerebral metabolism of [1,2-$^{13}$C$_2$]acetate as detected by *in vivo* and *in vitro* $^{13}$C NMR, *J. Biol. Chem.* **265:**12916–12926.

Chakeres, D. W., and Schmalbrock, P., 1992, "Fundamentals of Magnetic Resonance Imaging," Williams & Wilkins, Baltimore.

Chen, W., Novotny, E. J., Boulware, S. D., Rothman, D. L., Mason, G. F., Zhu, X.-H., Blamire, A., Prichard, J. W., and Shulman, R. G., 1994, Quantitative measurements of regional TCA cycle flux in visual cortex of human brain using $^1$H-{$^{13}$C} NMR spectroscopy, *Abstract, Society of Magnetic Resonance,* p. 63.

Cohen, S. M., 1987, Effects of insulin on perfused liver from streptozotocin-diabetic and untreated rats: $^{13}$C NMR assay of pyruvate kinase flux, *Biochemistry* **26**:573–580.

Fitzpatrick, S. M., Hetherington, H. P., Behar, K. L., and Shulman, R. G., 1990, The flux from glucose to glutamate in the rat brain in vivo as determined by $^1$H-observed, $^{13}$C-edited NMR spectroscopy, *J. Cereb. Blood Flow Metab.* **10**:170–179.

Gadian, D. G., 1982, "Nuclear Magnetic Resonance and Its Application to Living Systems," Clarendon Press, Oxford.

Gopher, A., Vaisman, N., Mandel, H., and Lapidot, A., 1990, Determination of fructose metabolic pathways in normal and fructose-intolerant children: A $^{13}$C NMR study using [U-$^{13}$C]fructose, *Proc. Natl. Acad. Sci. USA* **87**:5449–5453.

Gruetter, R., Novotny, E. J., Boulware, S. D., Mason, G. F., Rothman, D. L., Shulman, G. I., Prichard, J. W., and Shulman, R. G., 1994, Localized $^{13}$C NMR spectroscopy in the human brain of amino acid labeling from D-[1-$^{13}$C] glucose, *J. Neurochem.* **63**:1377–1385.

Jeffrey, F. M. H., Rajagopal, A., Malloy, C. R., and Sherry, A. D., 1991, $^{13}$C-NMR, a simple yet comprehensive method for analysis of intermediary metabolism, *Trends Biochem. Soc.* **16**:5–10.

Kauppinen, R., and Williams, S. R., 1994, Nuclear magnetic resonance spectroscopy studies of the brain, *Prog. Neurobiol.* **44**:87–118.

London, R. E., 1988, $^{13}$C labeling in studies of metabolic regulation, *Prog. NMR Spectrosc.* **20**:337–383.

Malloy, C. R., Sherry, A. D., and Jeffery, M. H., 1987, Carbon flux through citric acid pathways in perfused heart by $^{13}$C-NMR spectroscopy, *FEBS Lett.* 212:58–62.

Malloy, C. R., Thompson, J. R., Jeffery, F. M. H., and Sherry, A. D., 1990, Contribution of exogenous substrates to acetyl coenzyme A: Measurement by carbon-13 NMR under non-steady-state conditions, *Biochemistry* **29**:6756–6761.

Mason, G. F., Gruetter, R., Rothman, D. L., Behar, K. L., Shulman, R. G., and Novotny, E. J., 1995, Simultaneous determination of the rates of the TCA cycle, glucose utilization, and alpha-ketoglutarate/glutamate exchange and glutamine synthesis in human brain by NMR, *J. Cereb. Blood Flow Metab.* **15**:12–25.

McIlwain, H., and Bachelard, H. S., 1985, "Biochemistry and the Central Nervous System," 5th ed., Churchill-Livingstone, Edinburgh.

Morris, P. G., 1986, "Nuclear Magnetic Resonance Imaging in Medicine and Biology," Clarendon Press, Oxford.

Novotny, E. J., Mason, G. F., Gruetter, R., Rothman, D., Chen, W., Behar, K. L., Prichard, J., Boulware, S., Zhu, X.-H., and Shulman, R. G., 1995, Determination of the Krebs cycle, glutamine synthesis and amino acid turnover rates in human brain in vivo by $^{13}$C NMR spectroscopy, *J. Neurochem.* **65**:S206D.

Rinck, P. A. (ed.), 1993, "Magnetic Resonance in Medicine. The Basic Textbook of the European Magnetic Resonance Forum," 3rd ed., Blackwell, Oxford.

Rothman, D. L., Behar, K. L., Hetherington, H. P., den Hollander, J. A., Bendall, M. R., Petroff, O. A. C., and Shulman, R. G., 1985, $^1$H-observe/$^{13}$C-decouple spectroscopic measurements of lactate and glutamate in the rat brain *in vivo, Proc. Natl. Acad. Sci. USA* **82**: 1633–1637.

Rothman, D. L., Novotny, E. J., Shulman, G. I., Howseman, A. M., Petroff, O. A. C., Mason, G., Nixon, T., Hanstock, C. C., Prichard, J. W., and Shulman, R. G., 1992, $^1$H-[$^{13}$C] NMR measurements of [4-$^{13}$C]-glutamate turnover in human brain, *Proc. Natl. Acad. Sci. USA* **89**:9603–9606.

Sherry, A. D., Nunnally, R. L., and Peshock, R. M., 1985, Metabolic studies of pyruvate- and lactate-perfused guinea pig hearts by $^{13}$C NMR: Determination of substrate preference by glutamate isotopomer distribution, *J. Biol. Chem.* **260**:9272–9279.

van Bruggen, N., Roberts, T. L., and Cremer, J. E., 1994, The application of magnetic resonance imaging to the study of experimental cerebral ischaemia, *Cerebrovasc. Brain Metab. Rev.* **6**:180–210.

# 13C AND 1H MRS OF CULTURED NEURONS AND GLIA

## URSULA SONNEWALD, ARNE SCHOUSBOE, and NIELS WESTERGAARD

## SUMMARY

Magnetic resonance spectroscopy (MRS) can be used to probe cellular metabolism. This chapter describes how 1H MRS can be used to detect cellular marker substances and to quantitatively determine amounts of metabolites secreted by different cell types. 13C MRS is a valuable tool for investigation of metabolic pathways. Thus, metabolism of [1-13C]glucose, [2-13C]acetate, [U-13C]glutamate, or [U-13C]glutamine has been studied in cerebral cultures of astrocytes and neurons and in co-cultures thereof. The effects of hypoxia on metabolism are also described.

URSULA SONNEWALD • MR Center, SINTEF UNIMED, N-7034 Trondheim, Norway.   ARNE SCHOUSBOE • PharmaBiotec Research Center, Department of Biological Sciences, Royal Danish School of Pharmacy, DK-2100 Copenhagen, Denmark.   NIELS WESTERGAARD • NOVO-Nordisk A/S, Novo Allè, DK-2880 Bagsværd, Denmark.

Magnetic Resonance Spectroscopy and Imaging in Neurochemistry, Volume 8 of Advances in Neurochemistry, edited by Bachelard, Plenum Press, New York, 1997.

## 1. INTRODUCTION

Magnetic resonance spectroscopy (MRS) has been used extensively for several decades by chemists for the identification of pure compounds, but its application to biological systems is more recent and has led to new insights into cellular metabolism. A large part of the metabolic insight into the biochemistry of the brain gained by MRS so far has been obtained using *in vitro* models, such as tissue extracts, cell cultures, or brain slices. The kind of information gained from MRS experiments is dependent on the system used as well as on how the MRS experiment is conducted. Extracts of cells or tissue provide an excellent basis for metabolic studies and facilitate the interpretation of *in vivo* spectra. *In vivo* spectroscopy, on the other hand, has a much higher clinical relevance and is now approaching the stage where it definitely will find its place in clinical research and, possibly in the not too distant future, in clinical diagnostic studies. Recent reviews in the literature cover various aspects of MRS (Bachelard and Badar-Goffer, 1993; Kauppinen *et al.*, 1993; Sonnewald *et al.*, 1994a). The topic of the present chapter is MRS of primary brain cell cultures from rodent brain and implications of this work for cerebral metabolism in general.

## 2. CELL CULTURES

Cells taken directly from the organism and subsequently grown for at least 24 hr *in vitro* are considered to be primary cultures. Different procedures have been used, often in combination, to enable the establishment of monotypic cultures from mixed cell suspensions. The cell suspensions are obtained after an initial mechanical and/or enzymatic dissociation of the nervous tissue of neonatal or newborn rat, mouse, or chick. Due to the relative abundance of cells, MRS studies are often performed on cultures consisting of excitatory cerebellar granule cells or inhibitory GABAergic cerebral cortex neurons prepared as described by Schousboe *et al.* (1989) and Hertz *et al.* (1989a). The choice of these particular preparations of cultured neurons should be viewed in light of the fact that 90% of the synapses in the brain utilize either glutamate or GABA as the neurotransmitter. As counterparts for the neurons, typically astrocytes from the same brain region may be maintained in culture as detailed by Hertz *et al.* (1989b). Not only may such cultures of neuronal or glial origin be maintained in plastic tissue culture flasks or petri dishes, but alternatively microcarriers (Westergaard *et al.*, 1991a) may be used as the matrix. This technique allows transfer of the cultures directly to the MR spectrometer. Additionally, Westergaard *et al.* (1991b, 1992) have developed a technique to culture neurons on top of preformed layers of astrocytes; these so-called co-cultures have proven useful in studying neuronal/glial interactions. The purity of the neuronal cultures with regard to contamination with astrocytes is over 90% under the prevailing culturing conditions (Messer, 1977; Larsson *et al.*, 1985; Drejer *et al.*, 1985; Westergaard *et al.*, 1991a, 1992). In the cerebellar cultures, more than 95% of the neurons are glutamatergic granule cells, provided the cultures are treated with 50 µM kainic acid (Drejer and Schousboe, 1989). In the case of the cerebral cortical cultures, by far the majority of the neurons stain positively for glutamic acid decarboxylase (Hertz *et al.*, 1989a), confirming the GABAergic nature of the cells (Yu *et al.*, 1984; Westergaard *et al.*, 1992). In case of the astrocytic cultures, no neuronal contamination is present (Hertz *et al.*, 1989b).

Interpretation of results obtained from cell cultures has to be done carefully, since these cultures might not always represent the mature brain but might in some aspects be more representative of fetal stages. Thus, it has been shown that cultured neurons from mice express the M-type lactate dehydrogenase (LDH) isoenzymes necessary for anaerobic glycolysis that are present in the prenatal brain (Schousboe *et al.*, 1993b). Astrocytes, on the other hand, change their LDH isoenzyme pattern during three weeks in culture toward a higher proportion of the H-type, which is what is observed in the brain *in vivo* (Nissen and Schousboe, 1979). The creatine/choline ratio has been shown to increase after birth (Bates *et al.*, 1989), and low amounts of creatine are associated with hypoxic conditions. It has been shown that the creatine/choline ratio does not change in cell cultures to the same extent as *in vivo*, and that the ratio in cell cultures represents the ratio at the time of removal from the animal (Sonnewald *et al.*, 1993a). However, with respect to glutamate-metabolizing enzymes it has been shown that cell cultures follow the same ontogenic development as the living organism (Drejer *et al.*, 1985). Thus, cell cultures are very useful in determining qualitative aspects of metabolism and can reveal pathways that were not previously known to be active (Sonnewald *et al.*, 1993c), but one has to bear in mind the possibility that certain enzymes might be present as the fetal isoenzyme.

## 3. ¹H MRS OF CULTURED NEURONS AND GLIA

### 3.1. Cell-Specific Molecules in the Water-Soluble Phase

The advantage of MRS over other analytical techniques is the fact that mobile components of the sample are detected in a concentration-dependent manner that makes quantification possible in most cases. Derivatization and separation, requirements for other techniques, are not necessary, and chemical specificity is exquisite. A disadvantage of the technique is the low sensitivity as compared to that of techniques like mass spectrometry or radiolabeling techniques. The water-soluble components extracted from brain tissue or cells with perchloric acid (PCA) are typically amino acids, tricarboxylic acid (TCA) cycle intermediates, lactate, glucose, and other small molecules that are present in sufficient quantities. A detailed assignment of $^1$H resonances has been given by Behar and Ogino (1991). Developmental studies on rat brain extracts show a good correlation between $^1$H MRS and chromatographic data (Burri *et al.*, 1990). Bates *et al.* (1989) have shown that the concentration of choline-containing compounds in the rat brain decreases only slightly during the first 21 days after birth, whereas creatine and phosphocreatine concentrations increase by a factor of 2.5. This makes the choline resonance well suited as an internal standard in brain and cell culture spectra.

The brain consists of a complex mixture of different cell types. In order to understand the characteristics of the two most abundant cell types, namely, neurons and astrocytes, it is of importance to analyze them separately, this can be done by using cell culture techniques in combination with extraction or by direct observation in the MR spectrometer (Sonnewald *et al.*, 1992). $^1$H MRS studies have been carried out on PCA extracts of different brain cell types (Gill *et al.*, 1989; Sonnewald *et al.*, 1992, 1993a, 1994b,c; Urenjak *et al.*, 1993; Brand *et al.*, 1993). Characteristic metabolic patterns were identified for different cell types, valuable also for *in vivo* application. *N*-Acetylaspartate has been shown to be a marker for neurons but is also seen in O-2A progenitor cells (Urenjak *et al.*, 1993). In mouse cell cultures, hypotaurine was detected in cerebellar granule neurons as well as in astrocytes (Sonnewald *et al.*, 1994c). Thus, the value of hypotaurine as a marker for oligodendrocytes seems questionable.

With extracts it is possible to analyze compounds that occur in low concentrations, since long accumulation times are possible and metabolites from a large number of cells can be obtained. Thus, it was possible to detect UDP-*N*-acetylglucosamine and UDP-*N*-acetylgalactosamine in cerebellar neurons and cortical astrocytes (Sonnewald *et al.*, 1994b). $^1$H NMR spectra revealed the presence of elevated levels of UDP-*N*-acetylglucosamine and UDP-*N*-acetylgalactosamine in glioblastoma cell line U-87, in comparison with normal brain tissue and primary cell cultures of neurons and astrocytes. UDP-*N*-acetylhexosamines appear to accumulate in cells that are unable to differentiate. Furthermore, it was found that the culture medium had an effect on the concentration of UDP-*N*-acetygalactosamine in the glioblastoma cell line (Sonnewald *et al.*, 1994b) indicating that the effects of components in the incubation medium have to be taken into consideration in the analysis of intracellular metabolites.

## 3.2. The Lipid Extract of Cultured Neurons and Glia

Important information can also be obtained from the lipid extract of cell cultures. The extraction is typically carried out as described by Folch *et al.*

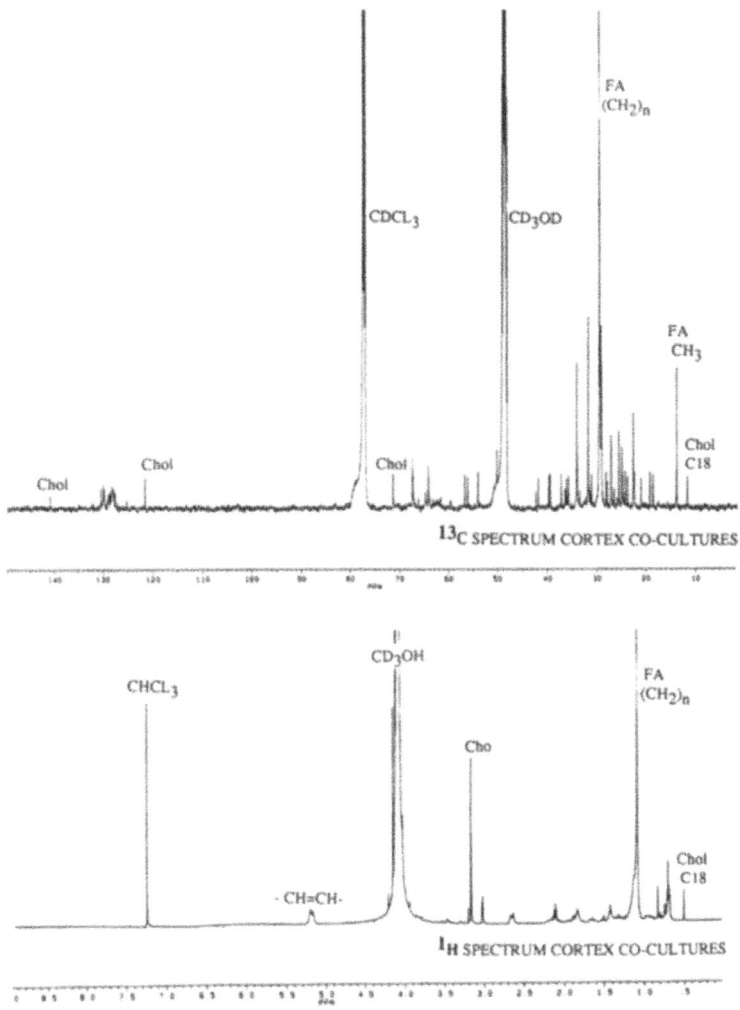

*FIGURE 1.* $^{13}$C (*top*) and $^1$H (*bottom*) MR spectra of the lipid extract of cortical co-cultures in CDCl$_3$ and CD$_3$OD. Abbreviations: Chol, cholesterol; FA, fatty acids; Cho, choline containing compound. [Reproduced, with permission, from Sonnewald *et al.* (1993a).]

(1957). Lipid extracts include glycerophospholipids, lysophospholipids, alkyl and alkenyl ether phospholipids, sphingomyelin and other sphingolipids, acylglycerols, steroids, and steroid esters. The fatty acid group is the major component of most of these lipids, which leads to a considerable overlap in the $^1$H spectrum (Fig. 1, bottom spectrum, from Sonnewald *et al.*, 1993a). A detailed two-dimensional $^1$H spectroscopic analysis of lipid composition in stimulated human lymphocytes has been carried out by Sze and Jardetzky (1990). These authors concluded that the total cholesterol-to-total phospholipid molar ratio and the phosphatidylcholine/phosphatidylethanolamine composition determined by $^1$H MRS agree with those determined by traditional methods. The $^{13}$C (Fig. 1, top) and $^1$H (Fig. 1, bottom) MR spectra of the lipid extract of cortical neurons growing on top of cortical astrocytes (co-cultures) show characteristic peaks for fatty acids and cholesterol (Sonnewald et al., 1993a).

## 3.3. Neurons and Astrocytes during Normoxia and Hypoxia

Proton spectra can be used to determine quantitative changes in metabolite concentrations both intracellularly and in the medium. Furthermore, it is possible to observe differences in metabolism due to changes in the cellular environment, such as hypoxia (Kauppinen and Williams, 1991; Ben-Yoseph et al., 1993; Pirttilä and Kauppinen, 1994; Sonnewald et al., 1994c). The $^1$H MR spectra of PCA extracts from astrocytes and cerebellar granule neurons (Fig. 2) show prominent

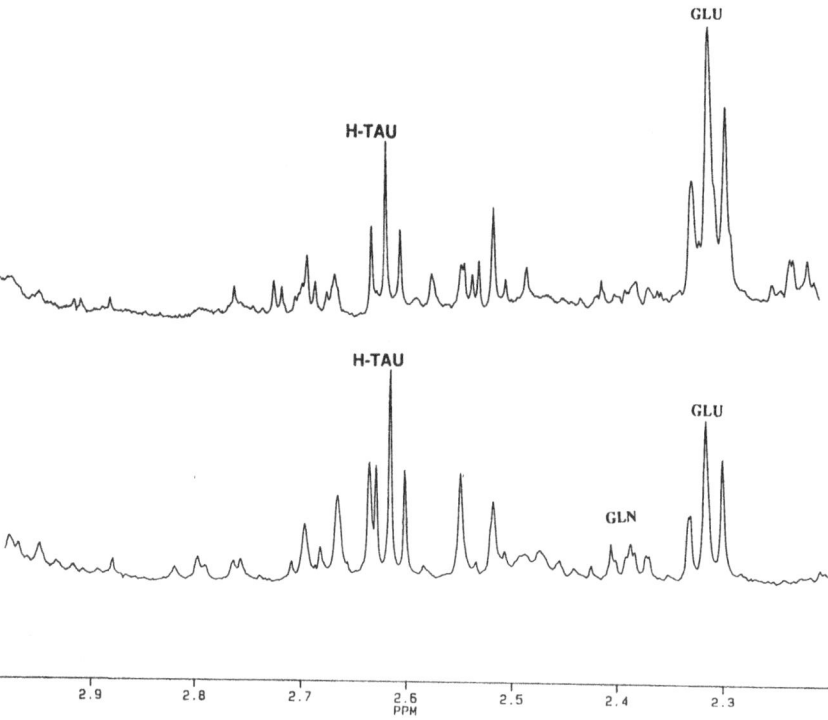

FIGURE 2. Expanded $^1$H MR spectra (3–2.2 ppm) of PCA extracts from cerebral cortical astrocytes (bottom) and cerebellar granule neurons (top). Abbreviations: GLN, glutamine; GLU, glutamate; H-TAU, hypotaurine. [Reproduced, with permission, from Sonnewald et al. (1994c).]

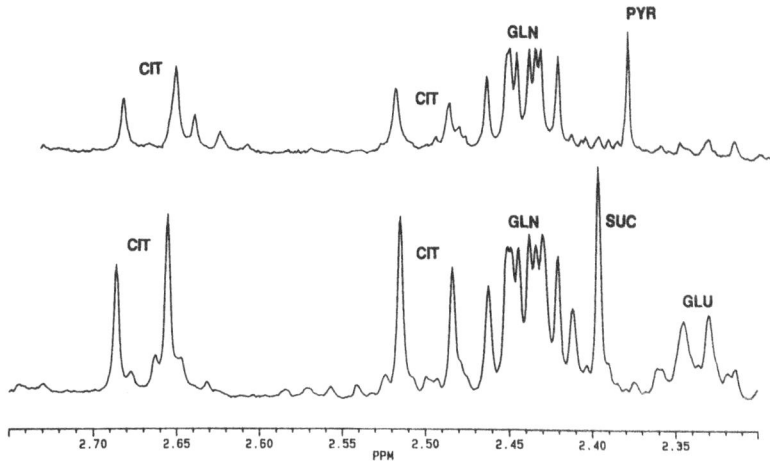

FIGURE 3. Expanded ¹H MR spectra (3–2.2 ppm) of medium (Hanks' balanced salt solution) from astrocytes incubated for 7 hr under hypoxic (*top*) and normoxic (*bottom*) conditions. Abbreviations: CIT, citrate; GLN, glutamine; GLU glutamate; PYR, pyruvate, SUC, succinate. Lactate and alanine resonances are not shown. [Reproduced, with permission, from Sonnewald *et al.* (1994c).]

peaks for hypotaurine and glutamate, whereas peaks representing citrate (for peak asigment see Fig. 3) and glutamine were only visible in the spectra of the extracts from astrocytes. The glutamate peak in these spectra is larger in the case of the granule neurons than in that of the astrocytes, in agreement with the role of granule cells as glutamatergic neurons. Glutamine synthetase is exclusively located in astrocytes (Norenberg and Martinez-Hernandez, 1979), and hence glutamine is observed in larger amounts in astrocytes than in neurons.

Some major metabolites visible in the ¹H spectra of incubation media from astrocytes were glutamine, citrate, and succinate (Fig. 3, bottom). (For detailed release studies, see Westergaard *et al.*, 1993, 1994a,b.) The amount of citrate released during hypoxia is significantly less than during normoxia, which might have implications for metal-ion homeostasis (Westergaard *et al.*, 1994a). Succinate is clearly visible during normoxia, and pyruvate during hypoxia (Fig. 3, top). In astrocytes, lactate production increased slightly during a 7-hr incubation period under hypoxic conditions from 3.3 ± 1 to 4 6 ± 1.3 mmol/mg protein whereas citrate production decreased from 320 ± 1 to 130 ± 1 nmol/mg (Sonnewald *et al.*, 1994c). Since citrate transport from mitochondria proceeds with cotransport of one proton (LaNoue and Schoolwerth, 1979), it is likely that this transport is impaired during hypoxia, which is commonly accompanied by acidosis. Glutamine release was unchanged at 300 ± 1 nmol/mg protein. Succinate release was 39 ± 10 mmol/mg during normoxia and was not detectable during hypoxia, when pyruvate was released (50 ± 20 nmol/mg) (Sonnewald *et al.*, 1994c). Lactate dehydrogenase leakage, a sign of cell damage, was less from cerebellar granule neurons, which were shown to have an increased glycolysis as compared to astrocytes and were capable of increasing their lactate production from $11.2 \pm 1.8$ to $26.7 \pm 4.5$ nmol/mg protein (Sonnewald *et al.,* 1994c).

In order to develop appropriate treatment regimes for the alleviation of damage caused by temporary oxygen deficiency in the brain, it is of importance to distinguish between cause and effect. Thus, establishing which cell type is first affected will have far-reaching implications. Some controversy exists in the literature regarding this question. Neurons have been shown to be more affected by ischemia than astrocytes (Juurlink and Hertz, 1993). The reverse was observed when neurons and astrocytes were exposed to hypoxia as described above (and in Sonnewald *et al.,* 1994a) and by Tholey and Ledig (1990), who showed that oxygen consumption by neurons in the presence of pyruvate or succinate was higher than in astrocytes. In brain slices exposed to hypoxia, an increased vulnerability of astrocytes was also indicated by an increase of glycerol 3-phosphate in glia (Ben-Yoseph *et al.,* 1993). More work is needed to determine the exact sequence of events during hypoxic/ischemic episodes, and cell culture work in combination with MRS can add new aspects to this puzzle.

## 4. METABOLIC PATHWAYS IN ASTROCYTES AND NEURONS

Information about biochemical pathways can be gained using $^{13}$C-labeled substances. These labeling experiments are analogous to conventional $^{14}$C labeling experiments but contain additional information about the location of the label within the molecule and have the potential for *in vivo* applicability. Sensitivity is, however, much lower, and generally only molecules that occur in concentrations above 0.1 mM can be detected. Using MR spectra in a qualitative manner (for assignment of $^{13}$C resonances, see Barany *et al.,* 1985), it is possible to detect unexpected compounds within the tissue (Ben-Yoseph *et al.,* 1993; Badar-Goffer *et al.,* 1990, 1992) or in the medium (Sonnewald *et al.,* 1991). Much additional information is, however, gained by quantitative analysis of $^{13}$C spectra. For an excellent description of the use of $^{13}$C MRS in combination with $^{13}$C-labeled compounds, see Bachelard and Badar-Goffer (1993).

[2-$^{13}$C]Acetate enters the TCA cycle and will label the C-2 position in citrate (Fig. 4a) and the C-4 position in glutamine (which is derived from 2-oxoglutarate through glutamate) during the first turn of the TCA cycle. The second turn will scramble this label to the C-3 and C-4 positions in citrate and the C-2 and C-3 positions in glutamine (Fig. 4a).

Pyruvate is incorporated into the TCA cycle both by pyruvate dehydrogenation, i.e., via acetyl coenzyme A (acetyl-CoA), and by pyruvate carboxylation (Fig. 4a). Exposure to [1-$^{13}$C]glucose will cause similar direct labeling in acetyl-

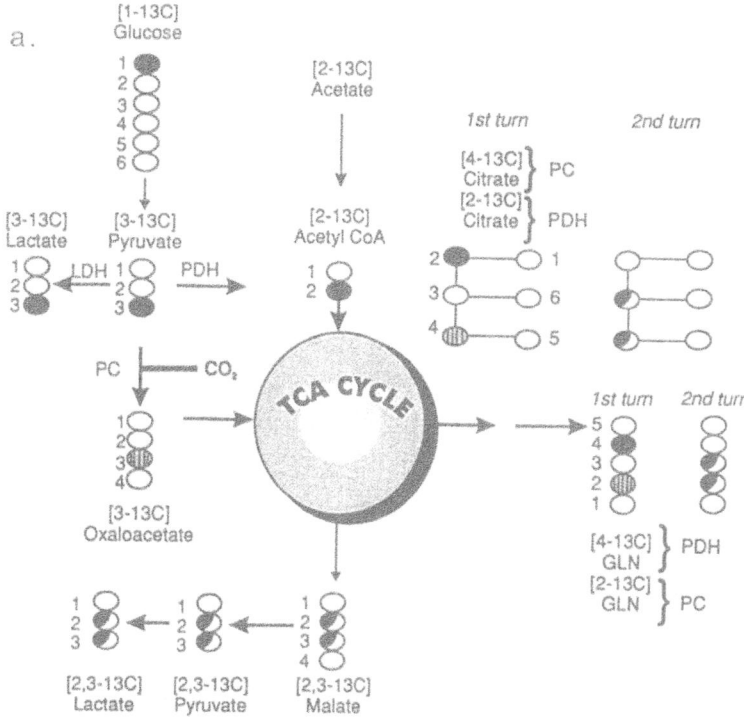

FIGURE 4. (a) Astrocyte TCA cycle, showing ¹³C labeling patterns of TCA cycle intermediates derived from [1-¹³C]glucose. Label in carboxylic acid groups is not shown because it has not been used for calculations. Even though glutamine is synthesized in the cytosol, it is included in the Figure to show the distribution of label resulting from the TCA cycle intermediate 2-oxoglutarate. Abbreviations: LDH, lactate dehydrogenase; PC, pyruvate carboxylase; PDH, pyruvate dehydrogenase. Filled circles represent ¹³C label from glucose through PDH; half-open circles indicate that only one-half of the original label is in this position; stippled circles represent label from PC. [Reproduced, with permission, from Sonnewald *et al.* (1994d; see also Badar-Goffer *et al.*, 1990.]

CoA to that seen during exposure to [2-¹³C]acetate, in astrocytes. It has been shown that pyruvate carboxylase is present exclusively in astrocytes (Yu *et al.*, 1983; Shank *et al.*, 1985). This enzyme condenses pyruvate and carbon dioxide to form oxaloacetate, a truly anaplerotic pathway (Waelsch *et al.*, 1964). Glutamine production is coupled to pyruvate carboxylase activity. Pyruvate carboxylase activity on pyruvate labeled in the C-3 position (i.e., the position labeled from [1-¹³C]glucose) will result in labeling of the C-4 position of citrate and GABA and the C-2 position of glutamine and glutamate in the first turn of the TCA cycle. On the other hand, [2-¹³C]acetyl-CoA (from pyruvate dehy-

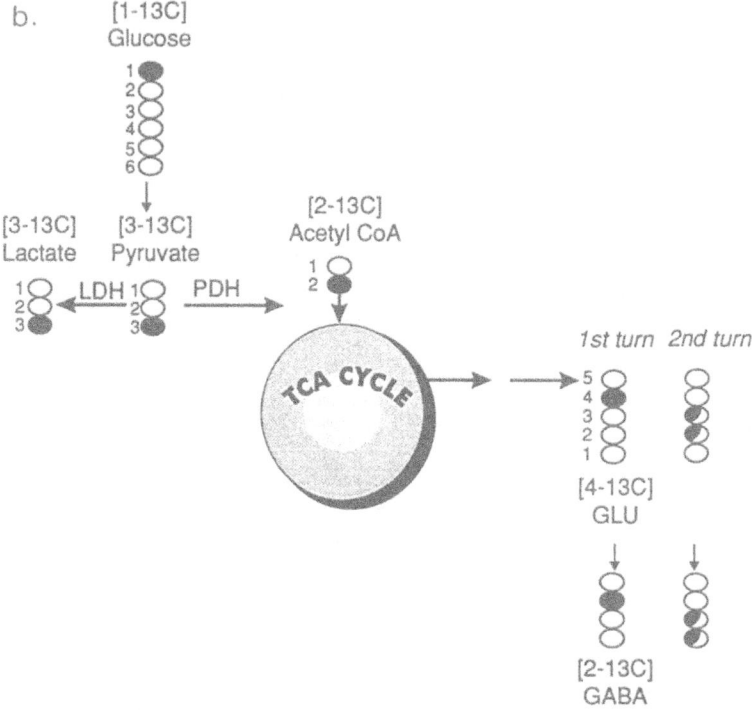

FIGURE 4. (b) Neuronal TCA cycle.

drogenase) will condense with oxaloacetate, giving rise to C-2-labeled citrate, and can after several steps be converted to C-4-labeled glutamine and glutamate and C-2-labeled GABA. If these metabolites stay in the TCA cycle for an additional turn, this label will be scrambled to the C-2 or C-3 positions of glutamine and glutamate and the C-3 or C-4 positions of GABA.

Analysis of the $^{13}C$ contents of the C-2 and C-3 positions of glutamine (Sonnewald *et al.*, 1993d) also gives information about the direction of the TCA cycle. If only the forward direction is operative, then pyruvate carboxylase labeling will lead to a higher incorporation of label into the C-2 position of glutamine than into the C-3 position. However, incorporation of label into the C-2 and C-3 positions was similar (Sonnewald *et al.*, 1994d). This can only be explained by a reversal of the TCA cycle from the oxaloacetate stage to the symmetrical fumarate.

Metabolism of glucose in neurons is schematically described in Fig. 4b. Since neurons lack pyruvate carboxylase (Yu *et al.*, 1983; Shank *et al.*, 1985), entry of label from glucose into the TCA cycle is only possible through pyruvate

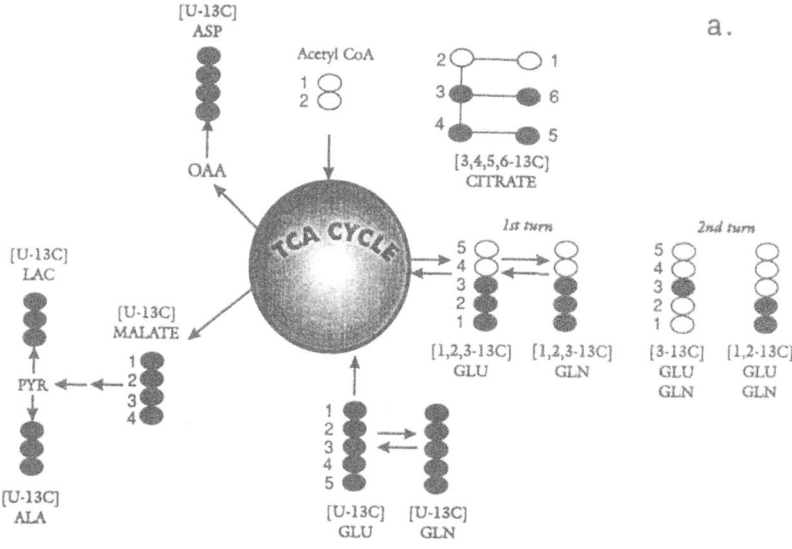

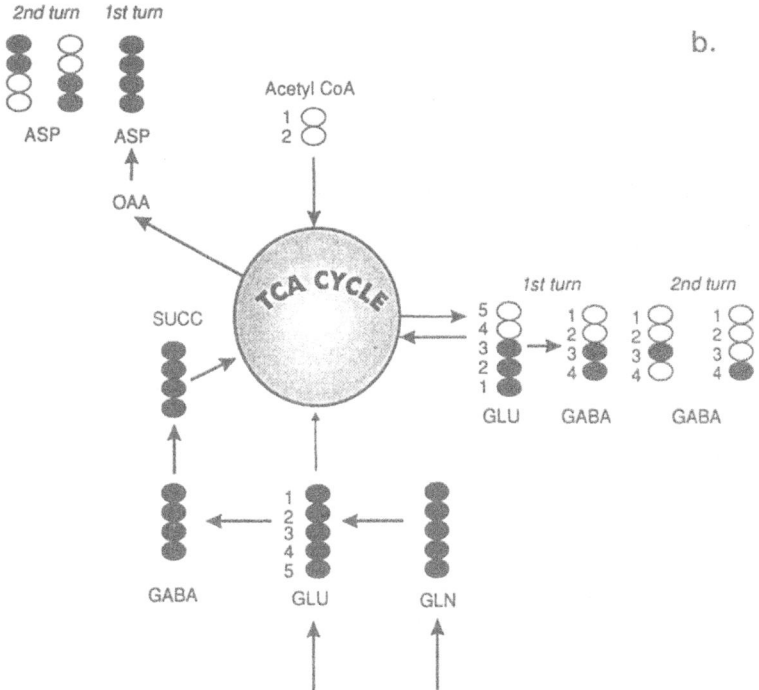

FIGURE 5. (a) Astrocytic TCA cycle, showing $^{13}C$ (●) labeling patterns in a variety of molecules after incubation of cultured mouse cerebral cortical astrocytes with [U-$^{13}$C]glutamate. [Reproduced, with permission, from Schousboe *et al.* (1993a).] (b) Neuronal TCA cycle.

dehydrogenase, and scrambling of label results from repeated turns of the TCA cycle only (Fig. 4b).

Metabolism of uniformly labeled glutamate and glutamine is shown schematically in Fig. 5, for both astrocytes (Fig. 5a) and neurons (Fig. 5b).

## 5. $^{13}$C MRS OF CULTURED ASTROCYTES

Using animal models and brain slices, it was shown that acetate metabolism takes place in the small glutamate compartment, now known to be astrocytes (Berl and Clarke, 1969; Badar-Goffer *et al.*, 1990; Hassel *et al.*, 1992). This was confirmed using cell culture techniques, and it was shown that glucose metabolism proceeds efficiently in both astrocytes and neurons (Sonnewald *et al.*, 1991, 1993b; Portais *et al.*, 1991; Leo *et al.*, 1993). Net release of alanine from astrocytes was reported by Yudkoff *et al.* (1986) using mass spectrometry and was confirmed by Sonnewald *et al.* (1991) using $^{13}$C MRS. Citrate synthesis from acetate, but not from glucose, could be demonstrated in brain slices under resting conditions (Badar-Goffer *et al.*, 1990, 1992), indicating that astrocytes are involved in citrate synthesis. That indeed astrocytes but not neurons are able to carry out a net synthesis and release of citrate was demonstrated directly by MRS of media from the two cell types incubated with [1-$^{13}$C]glucose (Fig. 6; Sonnewald *et al.*, 1991). Subsequently, it was shown that [2-$^{13}$C]acetate also labeled citrate which was released from astrocytes (Sonnewald *et al.*, 1993b). This is illustrated in Fig. 6, which shows typical $^{13}$C MR spectra of media from astrocytes, neurons, and co-cultures, incubated with [1-$^{13}$C]glucose. Citrate resonances are clearly visible at 46.4 ppm (C-2 + C-4) and 76.2 ppm (C-3). The C-4 and C-2 resonances of citrate coincide in the MR spectrum, owing to the symmetry of the molecule. Glutamine resonances appear at 55.6 ppm (C-2), 26.1 ppm (C-3), and 32.2 ppm (C-4). Labeled lactate and alanine are also visible (Fig. 6). It should be noted that the lactate C-3 peak is truncated in the astrocyte and co-culture spectra. There is clearly more release of lactate from astrocytes than from neurons (Sonnewald *et al.*, 1995).

The actual rates for release of alanine and citrate in astrocytes incubated in a medium containing 6 mM glucose and no acetate have been reported to be approximately 25 and 40 nmol hr$^{-1}$ mg$^{-1}$, respectively (Westergaard *et al.*, 1993, 1994b). While the functional role of alanine appears to be mainly that of an amino group donor for neuronal glutamate synthesis (Peng *et al.*, 1993), the role of extracellular citrate is less clear. In cultures of neurons and astrocytes, citrate was not metabolized to any significant extent (Westergaard *et al.*, 1994b). A possible role of extracellular citrate could be in the homeostasis of divalent cations like calcium and magnesium in the extracellular space (Westergaard *et al.*, 1994b).

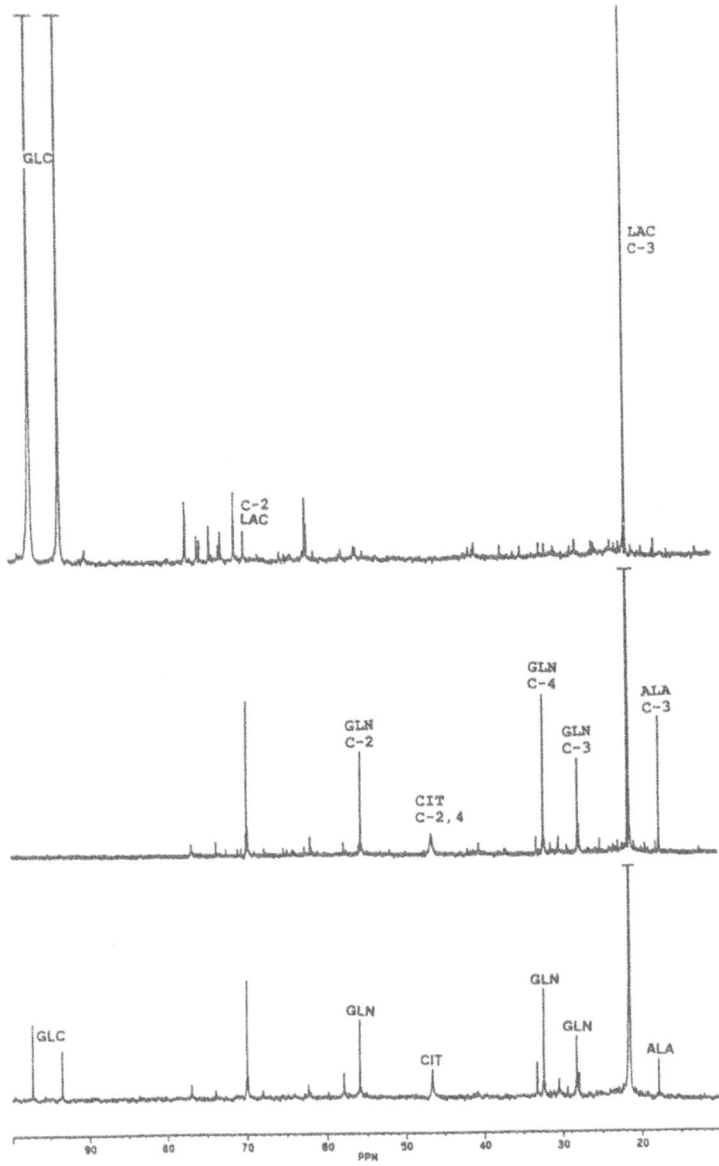

FIGURE 6. 13C MR spectra of lyophilized media collected from cerebral cortical neurons (*top*), astrocytes (*middle*), and co-cultures of neurons on astrocytes (*bottom*). Cells were incubated for 48 hr (*middle, bottom*) or 20 hr (*top*) with [1-13C]glucose. Abbreviations: ALA, alanine; CIT, citrate; GLC, glucose; GLN, glutamine; LAC, lactate. The lactate peak is truncated in the astrocyte and co-culture spectra and glucose in the neuronal spectrum. [Reproduced, with permission, from Sonnewald *et al.* (1991).]

## 5.1. Metabolism of [2-$^{13}$C]Acetate and [1-$^{13}$C]Glucose

### 5.1.1. Labeling of Citrate and Glutamine with [1-$^{13}$C]Glucose or [2-$^{13}$C]Acetate

Glucose is metabolized extensively in brain and brain slices (Berl and Clarke 1969; Badar-Goffer et al., 1990; Hassel et al., 1992, Kauppinen et al., 1994). It can be seen in Fig. 7 that [1-$^{13}$C]glucose is metabolized in neurons, astrocytes, and co-cultures, with label appearing in glutamate, aspartate, GABA (neurons, co-cultures), and glutamine (astrocytes, co-cultures). In the case of intracellular glutamine, it can be seen in Table 1 (from Sonnewald et al., 1994d) that the C-4 position was much more enriched from acetate (22.57 ± 1.13%) than from glucose (9.15 ± 3.64%). Glutamine in the medium, on the other hand, did not exhibit this difference. It should be noted that enrichment in C-4 of glutamine was similar in the medium and in the cells when glucose was used as the precursor but was a factor of 3 higher in the cells when acetate was used as the precursor, lending support to the suggestion of a compartmentation of metabolism in astrocytes (Sonnewald et al., 1993b; Schousboe et al., 1993a). Pyruvate carboxylation is demonstrated by a much higher C-2/C-4 ratio in glutamine from glucose as compared to that in glutamine from acetate. Thus, much higher incorporation was seen in C-2 from glucose as compared to acetate whereas the reverse was true for C-4. It can furthermore be concluded that glucose is a more efficient precursor for glutamine released into the medium. Although the C-4 position of glutamine in the medium was labeled similarly from both precursors, it is evident that the C-2 and C-3 positions are labeled more from glucose than from acetate. Citrate was only measurable in the medium and showed similar enrichment in the C-2 + C-4 positions (the C-2 and C-4 resonances coincide in the MR spectrum and cannot be separated) when glucose or acetate was used as precursor.

$^{13}$C MR studies carried out on medium from cultures of striatal astrocytes incubated with [1-$^{13}$C]glucose showed incorporation of label into the same molecules as described above and additionally into succinate and glycerol. It could be demonstrated that incorporation into succinate, under depolarizing conditions, in the presence of dibutyryl cyclic-AMP (dBcAMP) or glutamate (0.1 mM), was changed but the other resonances were not affected (Leo et al., 1993). Studies of metabolic fluxes have also been carried out using labeled pyruvate on brain cell cultures, and the results have been compared with those for cancer cell lines (Brand et al., 1992).

### 5.1.2. Labeling of Lactate with [1-$^{13}$C]Glucose or [2-$^{13}$C]Acetate

In astrocytes [1-$^{13}$C]glucose labels lactate considerably (32%; Table 1), although not to the theoretical maximum extent (50%), in the C-3 position

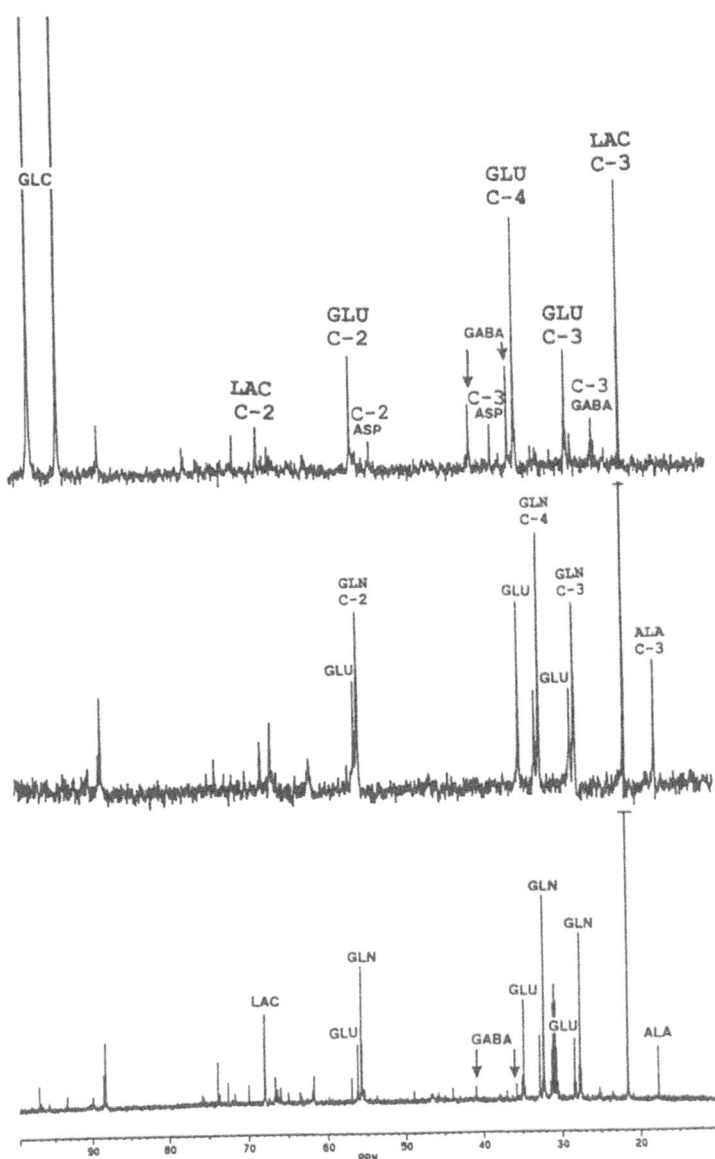

*FIGURE 7.* 13C MR spectra of PCA extracts of cerebral cortical neurons (*top*), astrocytes (*middle*), and co-cultures of neurons on astrocytes (*bottom*). Cells were incubated for 48 hr (*middle, bottom*) or 20 hr (*top*) with [1-13C]glucose. Abbreviations: ASP, aspartate; ALA, alanine; GLC, glucose; GLN, glutamine; GLU, glutamate; LAC, lactate. The lactate peak is truncated in the astrocyte and co-culture spectra. [Reproduced, with permission, from Sonnewald *et al.* (1991).]

TABLE 1. Percentage of $^{13}$C Enrichment Calculated from $^{13}$C MR Spectra of Medium and Cell Extracts from Co-cultures of Astrocytes and Neurons Incubated for 12 hr with [2-$^{13}$C]Acetate or [1-$^{13}$C]Glucose under Normoxic or Hypoxic Conditions[a]

| Analysis of | Condition | Carbon position | Percentage of $^{13}$C enrichment[b] after incubation with: | |
|---|---|---|---|---|
| | | | [1-$^{13}$C]Glucose | [2-$^{13}$C]Acetate |
| Cell extract | Normoxia | Glutamate C-4 | 16.74 ± 0.49 | 14.53 ± 3.72 |
| | | | (0.76 ± 0.07) | (0.45 ± 0.01) |
| | | Glutamine C-4 | 9.15 ± 3.64 | 22.57 ± 1.13 |
| | | | (2.23 ± 0.6) | (0.44 ± 0.01) |
| Medium | Normoxia | Lactate C-2 | 1.16 ± 0.19 | 0.41 ± 0.07 |
| | | Lactate C-3 | 32.70 ± 2.66 | 0.39 ± 0.04 |
| | | Glutamine C-4 | 6.61 ± 0.88 | 7.09 ± 2.27 |
| | | | (1.35 ± 0.11) | (0.30 ± 0.04) |
| | | Citrate C-2 + C-4 | 9.38 ± 1.79 | 10.83 ± 2.16 |
| | Hypoxia | Lactate C-2 | —[c] | 0.23 ± 0.04 |
| | | Lactate C-3 | 31.67 ± 3.6 | 0.19 ± 0.08 |
| | | Glutamine C-4 | —[c] | 1.69 ± 0.71 |
| | | Citrate C-2 + C-4 | —[c] | 13.70 ± 1.77 |

[a]Reproduced, with permission, from Sonnewald *et al.* (1994d).
[b]The percentage enrichment in the different carbon positions was calculated as described by Badar-Goffer *et al.* (1990). Values in parentheses are C-2/C-4 ratios. Results are averages ± SEM of three to four experiments.
[c]The amount was too small for accurate quantification.

(Hassel *et al.*, 1994; Sonnewald *et al.*, 1994c). There are several mechanisms possible for dilution of label in lactate: conversion of amino acids present in the medium to pyruvate; contribution from the pentose phosphate shunt; and, as will be outlined below, production of lactate from the TCA cycle. As observed by Leo *et al.* (1993), Sonnewald *et al.* (1993d, 1995), and Hassel *et al.* (1994), label was also found in the C-2 position of lactate but to a much lesser extent (1.16%; Table 1.) Label incorporation into lactate was also observed with [2-$^{13}$C]acetate as the precursor (Hassel *et al.*, 1994; Sonnewald *et al.*, 1994d). In neurons, however, incorporation into the C-2 position was much less and possibly due to some astrocytic contamination (Sonnewald *et al.*, 1995). Label incorporation into the C-2 position suggests that some glucose, after metabolism through the TCA cycle, is converted to lactate, probably via conversion of malate to pyruvate, catalyzed by malic enzyme, which is illustrated in Fig. 4a. This is in accordance with the previous finding of malic enzyme activity and localization in astrocytes (Kurz *et al.*, 1993). Alternatively, lactate could be produced by the combined effects of phosphoenolpyruvate carboxykinase and pyruvate kinase. This is consistent with results from brain extract studies by Cerdan *et al.* (1990) using uniformly labeled acetate and $^{13}$C MRS. It has been shown that during depolarizing conditions and in hypoxia in brain slices lactate labeling reaches the theoretical maximal amount of 50% (Badar-Goffer *et al.*, 1992, Ben-Yoseph *et al.*, 1993). It has recently been shown that labeling of lactate from mitochondrial precursors also occurs in rodent brain (Hassel and Sonnewald, 1995a). [2-$^{13}$C]Acetate, which is metabolized in astrocytes only (see above), could be shown to label lactate in awake animals whereas labeled glucose failed to do so.

## 5.2. Metabolism of [U-$^{13}$C]Glutamate

It has been demonstrated that astrocytes have an active uptake of glutamate and are responsible for the fast removal of this potentially toxic substance. In order to study the metabolic fate of glutamate, [U-$^{13}$C]glutamate was added to the incubation medium of astrocytes (Sonnewald *et al.*, 1993c). The advantage of uniformly labeled precursors is that detection of incorporation of label into metabolites is unambiguous due to $^{13}$C–$^{13}$C spin–spin coupling patterns. The probability of having two $^{13}$C atoms in the same molecule from the 1.1% naturally abundant $^{13}$C is 0.01%, well below the detection limit for most naturally occurring substances. Thus, using [U-$^{13}$C]glutamate it can be seen from $^{13}$C MR spectra of PCA extracts (Fig. 8, from Sonnewald *et al.*, 1993c) that astrocytes metabolized glutamate extensively, with $^{13}$C label appearing predominantly in aspartate, glutamine and malate. The medium contained glutamine, lactate, alanine, citrate, and glutamate (Sonnewald *et al.*, 1993c) and, in addition, 2-oxoproline (it is not clear whether this compound is a metabolite or a decomposition product of glutamine). The C-3 resonances of uniformly labeled glutamate and glutamine are shown in Fig. 9a and 9b, respectively, and the natural abundance peak of C-3 of lactate is shown in Fig. 9c. In the PCA extract, the glutamate C-3 resonance at 27.5 ppm consisted of a triplet and a doublet (Fig. 9d), indicating the existence of at least two isotopomers. Uniformly labeled glutamate enters the TCA cycle after conversion to uniformly labeled 2-oxoglutarate. After one turn of the cycle, the C-4 and C-5 positions in 2-oxoglutarate will be occupied by $^{12}$C atoms from acetyl-CoA (Fig. 5a). Glutamate derived from this precursor will have $^{12}$C in the C-4 and C-5 positions. The same was found in glutamine (Fig. 9e) derived from this glutamate (Sonnewald *et al.*, 1994c). Incorporation of label into lactate was evidenced by a doublet (Fig. 9f) in the C-3 position at 21 ppm and a multiplet in the C-2 position at 69 ppm (Sonnewald *et al.*, 1994c). In order to accomplish this conversion, glutamate must obviously enter the TCA cycle for conversion to metabolites like citrate, 2-oxoglutarate, and aspartate, which can subsequently leave the mitochondria. This is confirmed by the appearance of label in aspartate (Fig. 8). The further conversion to lactate is described above. It has recently been shown that conversion of glutamate to lactate and aspartate in astrocytes is concentration-dependent. Thus, at concentrations of 0.1 mM and below, no label was detected in lactate or aspartate, whereas at concentrations of

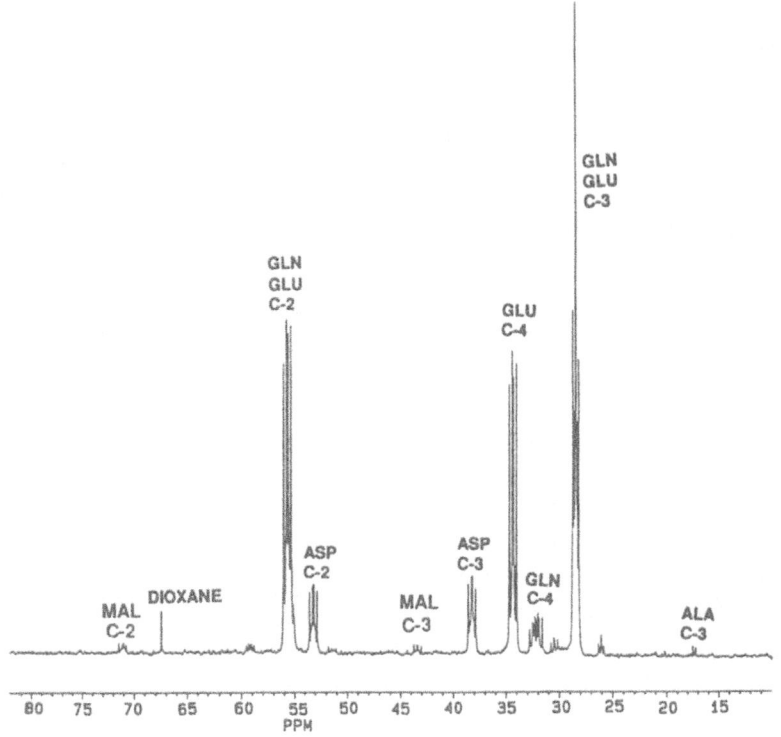

*FIGURE 8.* [13]C MR spectra of metabolites from PCA extracts of cultured cerebral cortical astrocytes incubated for 2 hr in the presence of 0.5 mM [U-[13]C]glutamate. Abbreviations: ASP, aspartate; GLN, glutamine; GLU, glutamate; MAL, malate. [Reproduced, with corrections and permission, from Sonnewald *et al.* (1993c).]

0.2 mM and above, incorporation of label was indeed observed (McKenna *et al.,* 1996). Bachelard *et al.* (1994a,b) have also obtained incorporation of label into lactate using 0.2 mM [U-[13]C]glutamate in the incubation medium of cerebral tissues (Fig. 10). At glutamate concentrations of 0.5 mM, excitotoxicity was observed in brain slices (Badar-Goffer *et al.,* 1994), since they contain neurons as well as astrocytes.

## 6.  [13]C MRS OF CULTURED NEURONS

### 6.1.  Metabolism of [1-[13]C]Glucose

Neurons are not capable of metabolizing acetate, as has been mentioned earlier, but metabolize glucose (Sonnewald *et al.,* 1993b). GABA synthesis could

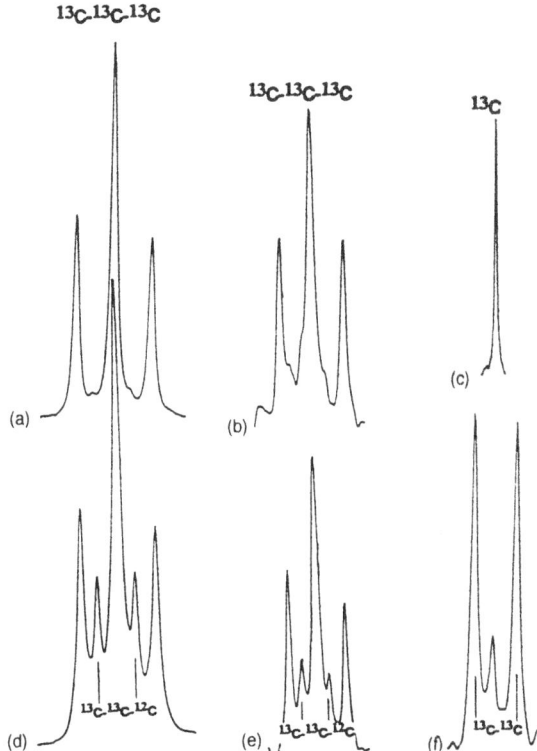

FIGURE 9. ¹³C MR peaks of the C-3 position of uniformly labeled glutamate (a) and glutamine (from PCA extract) (b), unlabeled lactate (natural abundance on a larger scale than the other peaks) (c), and the C-3 position of isotopomer mixtures (triplet ¹³C–¹³C–¹³C, doublet ¹³C–¹³C–¹²C) of metabolites from PCA extracts (d) or medium (e and f) from cultured cerebral cortical astrocytes incubated for 2 hr in the presence of 0.5 mM [U-¹³C]glutamate. [Reproduced, with permission, from Sonnewald *et al.* (1993c).]

be demonstrated in cerebral cortex neurons from [1-¹³C]glucose (Fig. 7). Except for labeled lactate (C-3), there was no measurable incorporation of label into metabolites in the medium (Fig. 7; Sonnewald *et al.*, 1991; Leo *et al.*, 1993).

## 6.2. Metabolism of [U-¹³C]Glutamate and [U-¹³C]Glutamine

As can be seen from ¹³C MR spectra of PCA extracts of the cells (Fig. 11, top), neurons metabolized glutamate extensively, with ¹³C label appearing predominantly in aspartate and GABA (Westergaard *et al.*, 1995). Label from glutamine appeared in the same molecules but with different labeling ratios (Fig. 11, bottom). Incorporation of label into GABA after 3 hr was similar from glutamate and glutamine, 13 and 15%, respectively, implicating the existence of a large

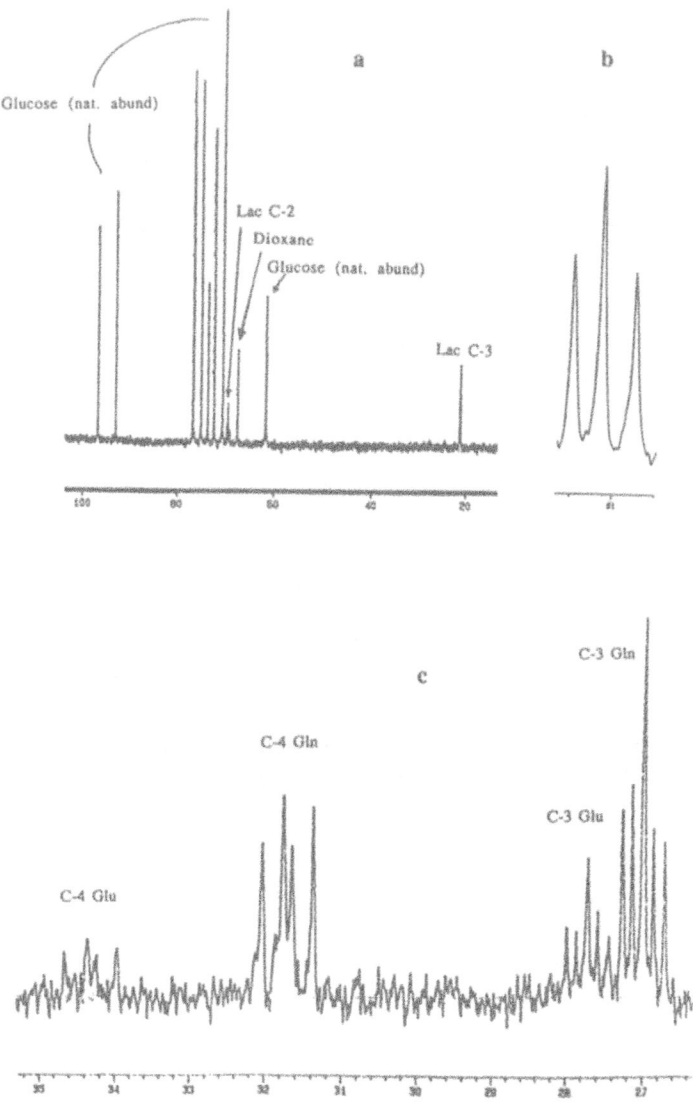

*FIGURE 10.* Isotopomers of $^{13}C$-labeled lactate (a and b) and glutamate (Glu) and glutamine (Gln) (c) released from brain slices incubated with 0.2 mM [U-$^{13}$C]glutamate in the presence of unlabeled 5 mM glucose. The incubation media were desalted on ion-exchange columns used to separate lactate from the amino acids. The C-3 lactate resonance in (a) is expanded in (b) to show the isotopomers. (Bachelard *et al.*, unpublished experiments; see also Bachelard *et al.*, 1994a,b.)

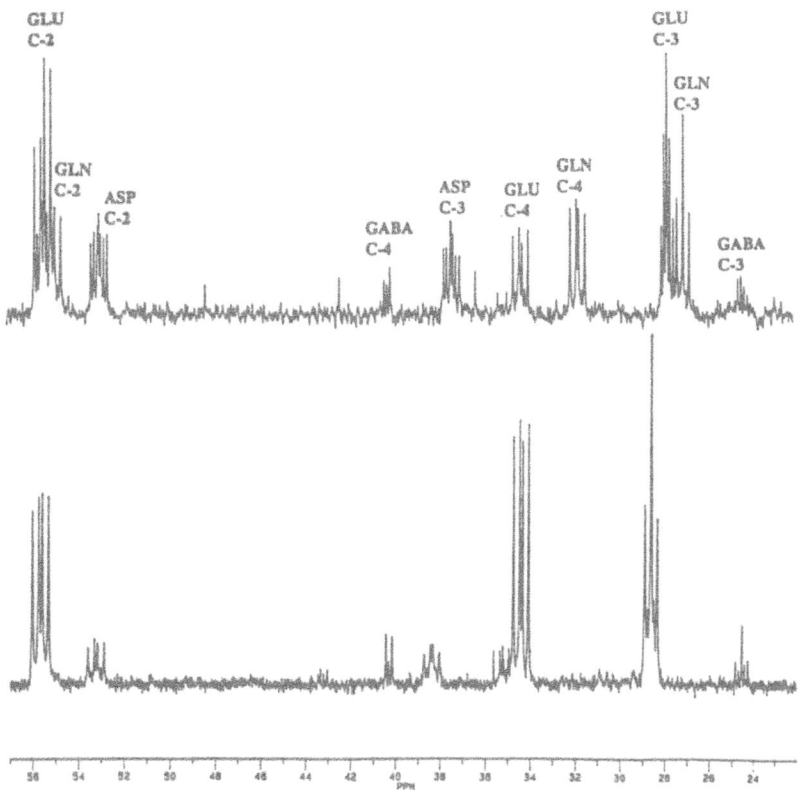

*FIGURE 11.* ¹³C MR spectra of metabolites from PCA extracts of cultured cerebral cortical neurons incubated for 2 hr in the presence of [U-¹³C]glutamate (*bottom*) or [U-¹³C]glutamine (*top*). Abbreviations: ASP, aspartate; GLN, glutamine; GLU, glutamate. [Reproduced, with permission, from Westergaard *et al.* (1995).]

unlabeled GABA pool, due to either a slow turnover of GABA in cell cultures or preferential synthesis from different precursors (Westergaard *et al.,* 1995). Even though incorporation of label from the two precursors was similar, the labeling patterns were different. Of the GABA formed from glutamine as the labeled precursor, 52% was derived from the TCA cycle, whereas this was true for only 30% of the GABA when glutamate was the labeled precursor (Westergaard *et al.,* 1995). TCA cycle involvement in the synthesis of GABA, aspartate, glutamate, and glutamine could be determined by analysis of doublets formed as described for astrocytes and schematically presented for neurons in Fig. 5b. In order to investigate the significance of the succinate shunt for GABA metabolism through the TCA cycle (Fig. 5b), neurons were incubated with the GABA transaminase inhibitor γ-vinyl-GABA (vigabatrin) in the presence of glutamine (Westergaard *et al.*, 1995). Doublet formation was unchanged in GABA, aspartate, and glutamate, but the percent $^{13}C$ incorporation was decreased significantly in all three metabolites. These results demonstrated that entry of label, derived from glutamate or glutamine, into the TCA cycle proceeds via 2-oxoglutarate and only to a minor extent through succinate. Aspartate was labeled almost twice as much from glutamate (37%) as from glutamine (20%). Furthermore, aspartate was more highly labeled than GABA, and when glutamine was the precursor, aspartate was also more highly labeled than glutamate (Westergaard *et al.*, 1995). Generally, glutamine did not give rise to as much incorporation of label as did glutamate, but it gave rise to more extensive metabolism through the TCA cycle, indicating the possibility that glutamine is more efficient as an energy substrate than as a precursor for amino acids.

It has now been demonstrated that glutamate metabolism, as expected, is complex both in neurons and in astrocytes. Unraveling glutamate metabolism in the separate cell types might make it possible to understand the even more complex compartmentation and interdependencies of glutamate metabolism in the brain. Lactate formation through the TCA cycle from glutamate was only observed in brain slices (Bachelard *et al.*, 1994a,b) and astrocytes (Sonnewald *et al.*, 1993c), but not in cultured neurons (Westergaard *et al.*, 1995), and might be a mechanism used to convert the potentially excitotoxic glutamate to a possible fuel for neurons and/or might point toward the importance of lactate for neuronal metabolism.

## 7. NEURONAL–GLIAL INTERACTIONS IN AMINO ACID SYNTHESIS

### 7.1. Synthesis of GABA, Glutamate, and Aspartate from [2-$^{13}$C]Acetate

Using cerebral cortical neurons (GABAergic) growing on a preformed layer of cerebral cortical astrocytes (co-cultures) incubated with labeled acetate, in combination with MRS, it has been shown that precursors produced in astrocytes are able to label GABA in neurons (Schousboe *et al.*, 1992; Sonnewald *et al.*, 1993b). Inhibition of glutamine synthetase with methionine sulfoximine led to a significant decrease in GABA labeling, implicating glutamine as the major labeling substrate (Sonnewald *et al.*, 1993b). Thus, acetate can be used to label metabolites within astrocytes and to determine if such metabolites could be precursors for synthesis of GABA, glutamate, or aspartate in neurons. The scrambling patterns in these amino acids will be dependent on the origin of the precursor. The primary precursor is [2-$^{13}$C]acetyl-CoA, which will enter the TCA cycle and give rise to labeling predominantly in C-4 of glutamate and glutamine, C-2 of GABA, and C-2 or C-3 of aspartate. As the metabolites stay in the TCA cycle during a second turn, there will also be labeling in C-2 and C-3 of glutamate and C-4 and C-3 of GABA (Fig. 46). Glutamate and aspartate produced in neurons with glutamine as a precursor therefore should exhibit the same distribution of label as seen in glutamine. This was indeed observed for glutamate (presumed to be located predominantly in neurons), since the C-2/C-4 ratio (0.45 ± 0.01) was equal to that in glutamine (0.44 ± 0.01) with acetate as the precursor (Table 1; Sonnewald *et al.,* 1994d). Owing to the small amounts present, GABA and aspartate could only be quantified by mass spectrometry.

## 7.2. Synthesis of GABA, Glutamate, and Aspartate from [1-¹³C]Glucose

Analysis of the ¹³C labeling pattern in glutamate shows that pyruvate carboxylation is indeed observable (Table 1), since the C-2/C-4 labeling ratio increased to 0.76 ± 0.07 when glucose was the precursor from 0.45 ± 0.01 with acetate as precursor. With glucose as the precursor, label in GABA can arise both from astrocytic precursors (glutamine) or through neuronal TCA cycle activity. Results on monolabeled precursors (Sonnewald *et al.,* 1994d) in combination with results on neuronal [U-¹³C]glutamine metabolism (Westergaard *et al.,* 1995) could be used to show that GABA was indeed labeled from glutamine. Thus, glutamine in the medium will enter neurons and may be converted to GABA via glutamate. This conversion has been shown to involve mitochondrial activity to a surprising extent (Westergaard *et al.,* 1995). Evidently, label in GABA can be accounted for from glutamine in the medium, and it thus seems likely that in the presence of astrocytes a large fraction of GABA was labeled in this way after 12 hr of incubation (Sonnewald *et al.,* 1994d). Some exclusive neuronal synthesis must, however, take place since GABA shows a higher enrichment (43.6 ± 1.1%) than its precursor glutamine (32.5 ± 3.1%) (Sonnewald *et al.,* 1994d).

In a ¹³C MRS study of the metabolism of [1-¹³C]glucose in rat brain, Shank *et al.* (1993) showed that some precursor from astrocytes, probably glutamine, is supplied to neurons for replenishment of their neurotransmitter pool of GABA. Hassel *et al.* (1995) confirmed this and could further show evidence for the same transfer of precursors for the synthesis of neuronal glutamate. Impairment of GABA synthesis from [1-¹³C]glucose could be demonstrated in the presence of the succinate dehydrogenase inhibitor 3-nitropropionic acid (Hassel and Sonnewald, 1995b). The use of [1-¹³C]glucose in combination with MRS provides the possibility of obtaining *in vivo* spectra of the human brain, as has been demonstrated by Gruetter *et al.* (1994).

## 7.3. Compartmentation of Metabolism

Compartmentation of metabolism in the brain has long been an accepted concept (Berl and Clarke, 1969), but compartmentation within a cell is not well documented. Studies of metabolism of glutamate and glutamine in astrocytes have, however, suggested that a compartmentation of glutamate metabolism exists (Schousboe *et al.*, 1993a). It was shown that the isotopomer distribution in aspartate was different from that in extracellular glutamate or glutamine (Schousboe *et al.*, 1993a). By MRS analysis of $^{13}C$ incorporation into different positions in glutamine, citrate, and lactate, information has been obtained that gives strong support to the concept of mitochondrial heterogeneity (Sonnewald *et al.*, 1993d).

Further support for cellular compartmentation in astrocytes is derived from the following results obtained from co-culture work (Sonnewald *et al.*, 1994d):

1. Glutamine C-4 was labeled differentially intracellularly from acetate to glucose, but extracellular glutamine was labeled similarly, thus indicating the existence of several glutamine pools within astrocytes.

2. The C-2/C-4 ratio for intracellular glutamine was $0.44 \pm 0.01$ but was only $0.30 \pm 0.04$ in the medium, indicating less cycling for exported glutamine.

3. MR data showed clearly a contribution to label incorporation, especially in glutamine, from pyruvate carboxylase activity with glucose as precursor, but this was not reflected in an increase in double labeling (measured by mass spectrometry), which would be expected if two labeling pathways exist simultaneously. One possible explanation is that acetate enters mitochondria which have a higher metabolic rate and will therefore show the same double labeling as metabolites from glucose which enter a compartment with pyruvate carboxylase and a lower metabolic rate. It seems likely that pyruvate carboxylase is located in a compartment optimized for export of glutamine and TCA cycle intermediates such as citrate, 2-oxoglutarate, malate, and oxaloacetate. Acetyl-CoA from acetate, on the other hand, might be more prominently used for energy purposes.

4. During hypoxia, reduction of label in C-4 of glutamine and C-2 + C-4 of citrate in the medium was more pronounced from glucose, though total incorporation of label in intracellular glutamine was the same from both precursors, indicating that pyruvate availability was not the problem, but that different cycling rates might exist.

5. Double labeling in glutamate, in contrast to glutamine, was decreased during hypoxia also when acetate was the precursor and labeled glutamate should be largely localized in astrocytes (Sonnewald *et al.*, 1994d).

Since neurons consist of morphological compartments with very diverse functions, it is not surprising to observe compartmentation of metabolism as well. Several observations described by Westergaard *et al.* (1995) led to this conclusion: (1) aspartate contains more $^{12}C$ when glutamine is the precursor; (2) there is less doublet formation in glutamate than in the product, GABA; and (3) aspartate is more highly labeled than its precursor glutamate (through 2-oxoglutarate and the TCA cycle) when glutamine is the precursor.

Even though heterogeneity of cell cultures cannot be ruled out totally, a growing amount of evidence seems to point strongly toward intracellular compartmentation of metabolism in individual cells.

## 7.4. Effects of Hypoxia

### 7.4.1. Synthesis of GABA, Glutamate, and Aspartate from [1-13C]Glucose or [2-13C]Acetate

The effect of hypoxia on cell cultures and brain tissue has been studied extensively, and glutamate toxicity appears to play a role in hypoxic–ischemic neuronal death (Rothman, 1984; Choi and Rothman, 1990). An impairment of astrocytic glutamate uptake has been reported by Yu *et al.* (1989). Thus, accumulation of extracellular glutamate will occur during hypoxia, as also shown by Sonnewald *et al.* (1994d).

By analyzing label incorporation and pool sizes, it could be shown (Ben-Yoseph *et al.*, 1993; Sonnewald *et al.*, 1994d) that $^{13}C$ labeling of glutamate and aspartate both from acetate and from glucose decreased during hypoxia. Total amounts of these amino acids decreased by a factor of 2 (Sonnewald *et al.*, 1994d) in comparison with a 3.5-fold decrease in the amount of glutamine and a 40% decrease in labeling. Considering the astrocytic location of glutamine and assuming a predominantly neuronal location of glutamate and aspartate, it can be concluded that neuronal metabolism seems to be less impaired. GABA may be localized almost exclusively in neurons and shows a similar behavior to aspartate and glutamate in terms of total amounts, but labeling is reduced to a larger extent both from glucose and from acetate (Sonnewald *et al.*, 1994d). Since glutamine labeling in the medium is almost absent with glucose as the precursor (Table 1), GABA labeling must proceed in this case via the neuronal TCA cycle, as has been observed by Müller *et al.* (1994) by analysis of labeling patterns both in GABA and in glutamine. When acetate was the precursor, glutamine in the medium was still labeled, and thus GABA labeling was observed.

### 7.4.2. Synthesis of Glutamine and Citrate from [1-13C]Glucose or [2-13C]Acetate

During hypoxia the amount of citrate was decreased and the amount of glutamine unchanged in the medium of co-cultures (Sonnewald *et al.*, 1994d), as reported earlier in astrocytes (Sonnewald *et al.*, 1994a). Label incorporation into the C-4 of glutamine in the medium was, however, reduced when acetate was the precursor, and not measurable from glucose (Table 1). Since $^{13}C$ incorporation in metabolites in the medium must reflect intracellular processes, the labeling data obtained by MRS of extracellular glutamine can be compared with data obtained for intracellular glutamine from mass spectrometry (Sonnewald *et al.*, 1994d). Thus, the mass spectrometry results also showed a reduction in label incorporation during hypoxia, but no difference was observed between the two precursors, indicating that a difference in label distribution is responsible for the difference observed by MRS. In contrast to glutamate double labeling, an increase was observed in double labeling of both glutamine and GABA from glucose and acetate, pointing toward an increased number of TCA cycle turns for the precursors of the latter amino acids. Citrate labeling was unchanged during hypoxia when acetate was the precursor, as also observed earlier in astrocytes (Sonnewald *et al.*, 1994a), but it was considerably reduced when glucose was the precursor.

## 8. CONCLUSION

Tholey and Ledig (1990) showed that oxygen consumption by neurons, in the presence of pyruvate or succinate, was higher than in astrocytes. During hypoxia, neuronal pyruvate oxidation increased whereas a decrease was observed in astrocytes. Succinate oxidation was only slightly reduced in neurons but decreased more than 50% in astrocytes during hypoxia. These observations as well as the results on labeling patterns in GABA and glutamine (Müller *et al.*, 1994; Sonnewald *et al.*, 1994d) suggest that neuronal mitochondria are more resistant to the effects of hypoxia than mitochondria in astrocytes.

Neurons in culture appear to have at least two pathways that make them functionally more adaptable to hypoxic conditions than astrocytes during hypoxia. Both glycolysis and TCA cycle activity seem to be less impaired in neurons than in astrocytes. With the advent of adequate MR spectrometers for *in vivo* spectroscopy, it should soon be possible to assess the relevance of the above observations to the *in vivo* situation.

Finally, it appears clear that metabolism is more compartmentalized in individual cells than previously assumed.

ACKNOWLEDGMENTS

One of the authors (U. Sonnewald) would like to thank the following people for valuable discussions: H. S. Bachelard, B. Hassel, M. McKenna, T. B. Müller, and R. P. Shank. The excellent assistance provided by Ms. Mona Eidem in the preparation of the figures is cordially acknowledged. Parts of the text are from Sonnewald *et al.* (1994d), with permission. This research was supported by the Research Council of Norway, the Danish State Biotechnology Program (1991–

95), the NOVO and Lundbeck foundations, and the CEC Biomed I Program (BMH1-CT94-1248).

# REFERENCES

Bachelard, H. S., and Badar-Goffer, R., 1993, NMR spectroscopy in neurochemistry, *J. Neurochem.* **61:**412–430.

Bachelard, H. S., Badar-Goffer, R. S., Ben-Yoseph, O., Morris, P-G., Taylor, A., and Thatcher, N. M., 1994a, Metabolism of [U-13C]glutamate to glutamine and lactate in cortical slices: An NMR study, *J. Neurochem.* **63:**48B.

Bachelard, H. S., Badar-Goffer, R. S., Ben-Yoseph, O., Morris, P.-G., Taylor, A., and Thatcher, N. M., 1994b, 13C-MRS studies on cerebral metabolism, *MAGMA* **2:**284–289.

Badar-Goffer, R. S., Bachelard, H. S., and Morris, P. G., 1990, Cerebral metabolism of acetate and glucose studied by 13C NMR spectroscopy, *Biochem. J.* **266:**133–139.

Badar-Goffer, R. S., Ben-Yoseph, O., Bachelard, H. S., and Morris, P. G., 1992, Neuronal–glial metabolism under depolarizing conditions: a 13C NMR study, *Biochem. J.* **282:**225–230.

Badar-Goffer, R. S., Morris, P. G., Thatcher, N. M., and Bachelard, H. S., 1994, Excitotoxic amino acids cause appearance of MRS-observable zinc in superfused cortical slices, *J. Neurochem.* **62:**2488–2491.

Barany, M., Arus, C., and Chang, Y. C., 1985, Natural abundance 13C NMR of brain, *Magn. Reson. Med.* **2:**289–295.

Bates, T. E., Williams, S. R., Gadian, D. G., Bell, J. D., Small, R. K., and Iles, R. A., 1989, 1H NMR study of cerebral development in the rat, *NMR Biomed.* **2:**225–229.

Behar, K. L., and Ogino, T., 1991, Assignment of resonances in the 1H spectrum of rat brain by two-dimensional shift correlated and J-resolved NMR spectroscopy, *Magn. Reson. Med.* **17:**285–303.

Ben-Yoseph, O., Badar-Goffer, R., Morris, P. G., and Bachelard, H. S., 1993, Glycerol 3-phosphate and lactate as indicators of the cerebral cytoplasmic redox state in severe and mild hypoxia respectively: A 13C- and 31P-NMR study, *Biochem. J.* **291:**915–919.

Berl, S., and Clarke, D. D., 1969, Metabolic compartmentalization of glutamate in the CNS, in "Handbook of Neurochemistry," Vol. 1 (A. Lajtha, ed.), pp. 447–472, Plenum Press, New York.

Brand, A., Engelmann, J., and Leibfritz, D., 1992, A 13C NMR study on fluxes into the TCA cycle of neuronal and glial tumor cell lines and primary cells, *Biochimie* **74:**941–948.

Brand, A., Richter-Landsberg, C., and Leibfritz, D., 1993, Multinuclear NMR studies on the energy metabolism of glial and neuronal cells, *Dev. Neurosci.* **15:**289–298.

Burri, R., Bigler, P., Straehl, P., Posse, S., Colombo, J.-P., and Herschkowitz, N., 1990, Brain development: 1H magnetic resonance spectroscopy of rat brain extracts compared with chromatographic methods, *Neurochem. Res.* **15:**009–1016.

Cerdan, S., Kunnecke, B., and Seelig, J., 1990, Cerebral metabolism of [1,2-13C2]acetate as detected by in vivo and in vitro 13C NMR, *J. Biol. Chem.* **265:**12916–12926.

Choi, D. W., and Rothman, S. M., 1990, The role of glutamate neurotoxicity in hypoxic–ischemic neuronal death, *Annu. Rev. Neurosci.* **13:**171–182.

Drejer, J., and Schousboe, A. 1989, Selection of a pure cerebellar granule cell culture by kainate treatment, *Neurochem. Res.* **14:**751-754.

Drejer, J., Larsson, O. M., Kvamme, E., Svenneby, G., Hertz, L., and Schousboe, A., 1985, Ontogenic development of glutamate metabolizing enzymes in cultured cerebellar granule cells and cerebellum in vivo, *Neurochem. Res.* **10:**49–62.

Folch, J., Lees, M., and Sloane-Stanley, G. H., 1957, A simple method for the isolation and purification of total lipids from animal tissues, *J. Biol. Chem.* **226:**497–509.

Gill, S. S., Small, R. K., Thomas, D. G. T., Patel, P., Porteous, R., Van Bruggen, N., Gadian, D. G., Kauppinen, R. A., and Williams, S. R., 1989, Brain metabolites as ¹H NMR markers of neuronal and glial disorders, *NMR Biomed.* **2:**196–200.

Gruetter, R., Novotny, E. J., Boulware, S. D., Mason, G. F., Rothman, D. L., Shulman, G. I., Prichard, J. W., and Shulman, R. G., 1994, Localized ¹³C NMR spectroscopy in the human brain of amino acid labeling from [1-¹³C]glucose, *J. Neurochem.* **63:**1377–1385.

Hassel, B., and Sonnewald, U., 1995a, Glial formation of pyruvate and lactate from TCA cycle intermediates, *J. Neurochem.* **65:**2227–2234.

Hassel, B., and Sonnewald, U., 1995b, Selective inhibition of the TCA cycle of GABAergic neurons with 3-nitropropionic acid in vivo, *J. Neurochem.* **65:**1184–1191.

Hassel, B., Paulsen, R. E., and Fonnum, F., 1992, Selective inhibition of glial cell metabolism in vivo by fluorocitrate, *Brain Res.* **576:**120–124.

Hassel, B., Sonnewald, U., Unsgård, G., and Fonnum, F., 1994, NMR spectroscopy of cultured astrocytes. Effects of glutamine and the gliotoxin fluorocitrate, *J. Neurochem.* **62:**2187–2194.

Hassel, B., Sonnewald, U., and Fonnum, F., 1995, Glial–neuronal interactions as studied by cerebral metabolism of [2-¹³C]acetate or [10¹³C]glucose. An *ex vivo* ¹³C NMR spectroscopy study, *J. Neurochem.* **64:**2773–2782.

Hertz, E., Yu, A. C. H., Hertz, L., Juurlink, B. H. J., and Schousboe, A., 1989a, Preparation of primary cultures of mouse (rat) neurons, *in* "A Dissection and Tissue Culture Manual of the Nervous System" (A. Shahar, J. De Vellis, A. Vernadakis, and B. Haber, eds.), pp. 183–186, Alan R. Liss, New York.

Hertz, L., Juurlink, B. H. J., Hertz, E., Fosmark, H., and Schousboe, A., 1989b, Preparation of primary cultures of mouse (rat) astrocytes, *in* "A Dissection and Tissue Culture Manual of the Nervous System" (A. Shahar, J. De Vellis, A. Vernadakis, and B. Haber, eds.), pp. 105–108, Alan R. Liss, New York.

Juurlink, B. K., and Hertz, L., 1993, Ischemia-induced death of astrocytes and neurons in primary culture: Pitfalls in quantifying neuronal cell death, *Brain Res. Dev. Brain Res.* **71:**239–246.

Kauppinen, R. A., and Williams, S. R., 1991, Nondestructive detection of glutamate by ¹H NMR spectroscopy in cortical brain slices from guinea pig: Evidence for changes in detectability during severe anoxic insults, *J. Neurochem.* **57:**1136–1144.

Kauppinen, R. A., Williams, S. R., Busza, A. L., and van Bruggen, N., 1993, Application of magnetic resonance spectroscopy and diffusion-weighted imaging to the study of brain biochemistry and pathology, *Trends Neurosci.* **16:**88–95.

Kauppinen, R. A., Pirttilä, T.-R. M., Auriola, S. O. K., and Williams, S. R., 1994, Compartmentation of cerebral glutamate in situ as detected by ¹H/¹³C NMR, *Biochem. J.* **298:**121–127.

Kurz, G. M., Wiesinger, H., and Hamprecht, B., 1993, Purification of cytosolic malic enzyme from bovine brain, generation of monoclonal antibodies, and immunocytochemical localization of the enzyme in glial cells of neural primary cultures, *J. Neurochem.* **60:**1467–1474.

LaNoue, F. K., and Schoolwerth, A. C., 1979, Metabolite transport in mitochondria, *Annu. Rev. Biochem.* **48:**871–922.

Larsson, O. M., Drejer, J., Kvamme, E., Svenneby, G., Hertz, L., and Schousboe, A., 1985, Ontogenic development of glutamate and GABA metabolizing enzymes in cultured cerebral cortex interneurons and in cerebral cortex *in vivo, Int. J. Dev. Neurosci.* **3:**177–185.

Leo, G. C., Driscoll, B. F., Shank, R. P., and Kaufman, E., 1993, Analysis of [1-¹³C]D-glucose metabolism in cultured astrocytes and neurons using nuclear magnetic resonance spectroscopy, *Dev. Neurosci.* **15:**282–288.

McKenna, M. C., Sonnewald, U., Huang, X., Stephenson, J., and Zielke, H. R., 1996, Exogenous glutamate concentration regulates the fate of glutamate in astrocytes, *J. Neurochem.* **66**:386–393.

Messer, A., 1977, The maintenance and identification of mouse cerebellar granule cells in monolayer culture, *Brain Res.* **130**:546–553.

Müller, B. T., Sonnewald, U., Westergaard, N., Schousboe, A., Petersen, B. S., and Unsgård, G., 1994, 13C NMR spectroscopy study of cortical nerve cell cultures exposed to hypoxia, *J. Neurosci. Res.* **38**:319–326.

Nissen, C., and Schousboe, A., 1979, Activity and isoenzyme pattern of lactate dehydrogenase in astroblasts cultured from brains of newborn mice, *J. Neurochem.* **32**:1787–1792.

Norenberg, M. D., and Martinez-Hernandez, A., 1979, Fine structural localization of glutamine synthetase in astrocytes of rat brain, *Brain Res.* **161**:303–310.

Peng, L., Hertz, L., Huang, R., Sonnewald, U., Petersen, S. B., Westergaard, N., Larsson, O., Schousboe, A., 1993, Utilization of glutamine and of TCA cycle constituents as precursors for transmitter glutamate and GABA, *Dev. Neurobiol.* **15**:367–377.

Pirttilä, T.-R. M., and Kauppinen, R. A., 1994, Regulation of intracellular pH in guinea pig cerebral cortex ex vivo studied by 31P and 1H NMR spectroscopy: Role of extracellular bicarbonate and chloride, *J. Neurochem.* **62**:656–664.

Portais, J. C., Pianet, I., Merle, M., Raffard, G., Kien, P., Brian, M., Labouesse, J., Caille, J. M., and Canioni, P., 1991, Magnetic resonance spectroscopy and metabolism. Application of proton and 13C NMR to the study of glutamate metabolism in cultured glial cells and human brain in vivo, *Biochimie* **73**:93–97.

Rothman, S., 1984, Synaptic release of excitatory amino acid neurotransmitter mediates anoxic neuronal death, *J. Neurosci.* **4**:1884–1891.

Schousboe, A., Meier, E., Drejer, J., and Hertz, L., 1989, Preparation of primary cultures of mouse (rat) cerebellar granule cells, in "A Dissection and Tissue Culture Manual of the Nervous System" (A. Shahar, J. De Vellis, A. Vernadakis, and B. Haber, eds.), pp. 183–186, Alan R. Liss, New York.

Schousboe, A., Westergaard, N., Sonnewald, U., Petersen, S. B., Yu, A. C. H., and Hertz, L., 1992, Regulatory role of astrocytes for neuronal biosynthesis and homeostasis of glutamate and GABA, *Prog. Brain Res.* **94**:199–211.

Schousboe, A., Westergaard, N., Sonnewald, U., Petersen, S. B., Huang, R., Peng, L., and Hertz, L., 1993a, Glutamate and glutamine metabolism and compartmentation in astrocytes, *Dev. Neurosci.* **15**:359–366.

Schousboe, I., Tønder, N., Zimmer, J., and Schousboe, A., 1993b, A developmental study of lactate dehydrogenase activity in organotypic rat hippocampal slice cultures and primary cultures of mouse neocortical and cerebellar neurons, *Int. J. Dev. Neurosci.* **11**:765–772.

Shank, R. P., Bennett, G. S., Freytag, S. O., and Campbell, G. L., 1985, Pyruvate carboxylase: Astrocytic-specific enzyme implicated in replenishment of amino acid neurotransmitter pools, *Brain Res.* **329**:364–367.

Shank, R. P., Leo, G. C., and Zielke, H. R., 1993, Cerebral metabolic compartmentation as revealed by nuclear magnetic resonance analysis of D-[1-13C]glucose metabolism, *J. Neurochem.* **61**:315–323.

Sonnewald, U., Westergaard, N., Krane, J., Unsgård, G., Petersen, S. B., and Schousboe, A., 1991, First direct demonstration of preferential release of citrate from astrocytes using 13C NMR spectroscopy of cultured neurons and astrocytes, *Neurosci. Lett.* **128**:235–239.

Sonnewald, U., Petersen, S. B., Krane, J., Westergaard, N., and Schousboe, A., 1992, 1H NMR study of cortex neurons and cerebellar granule cells on microcarriers and their PCA extracts: Lactate production under hypoxia, *Magn. Reson. Med.* **23**:166–171.

Sonnewald, U., Westergaard, N., Isern, E., Müller, T. B., Schousboe, A., Petersen, S. B., and Unsgård, G., 1993a, MRS evaluation of brain metabolites in tissue and cell culture extracts: Cholesterol ester, choline containing compounds and creatine are markers for brain development and pathology, *Int. J. Oncol.* **2**:545–555.

Sonnewald, U., Westergaard, N., Schousboe, A., Svendsen, J. S., Unsgård, G., and Petersen, S. B., 1993b, Direct demonstration by $^{13}$C NMR that glutamine from astrocytes is a precursor for GABA synthesis in neurons, *Neurochem. Int.* **22**:19–29.

Sonnewald, U., Westergaard, N., Petersen, S. B., Unsgård, G., and Schousboe, A., 1993c, Metabolism of [U-$^{13}$C]glutamate in astrocytes studied by $^{13}$C NMR spectroscopy: Incorporation of more label into lactate than into glutamine demonstrates the importance of the TCA cycle, *J. Neurochem.* **61**:1179–1182.

Sonnewald, U., Westergaard, N., Hassel, B., Müller, T. B., Unsgård, G., Fonnum, F., Hertz, L., Schousboe, A., and Petersen, S. B., 1993d, NMR spectroscopic studies of $^{13}$C acetate and $^{13}$C glucose metabolism in neocortical astrocytes: Evidence for mitochondrial heterogeneity, *Dev. Neurosci.* **15**:351–358.

Sonnewald, U., Gribbestad, I. S., Westergaard, N., Nilsen, G., Unsgård, G., Schousboe, A., and Petersen, S. B., 1994a, Nuclear magnetic resonance spectroscopy—biochemical evaluation of brain function in vivo and in vitro, *NeuroToxicol.* **15**:579–590.

Sonnewald, U., Isern, E., Gribbestad, I. S., and Unsgård, G., 1994b, UDP-*N*-acetylhexosamines and hypotaurine in human glioblastoma, normal brain tissue and cell cultures: $^1$H NMR spectroscopy study, *Anticancer Res.* **14**:793–798.

Sonnewald, U., Müller, T. B., Westergaard, N., Schousboe, A., Petersen, S. B., and Unsgård, G., 1994c, NMR spectroscopy study of cell cultures of astrocytes and neurons exposed to hypoxia, *Neurochem. Int.* **24**:473–483.

Sonnewald, U., Müller, T. B., Westergaard, N., Svendsen, J. S., Unsgård, G., and Schousboe, A., 1994d, Neuronal–glial interactions in glutamate and GABA homeostasis: Effects of hypoxia, *in* "Pharmacology of Cerebral Ischemia 1994" (J. Kriegelstein and H. Oberpichler-Schwenk, eds.), pp 1–14, Medpharm Scientific Publishers, Stuttgart.

Sonnewald, U., Wang, A. Y., Petersen, S. B., Westergaard, N., Schousboe, A., Erikson, R., and Skottner, A., 1995, $^{13}$C NMR study of IGF-I- and insulin-effects on mitochondrial function in cultured brain cells, *NeuroReport* **6**:878–880.

Sze, D. Y., and Jardetzky, O., 1990, Characterization of lipid composition in stimulated human lymphocytes by $^1$H NMR, *Biochim. Biophys. Acta* **1054**:198–206.

Tholey, G., and Ledig, M., 1990, Plasticité neuronale et astrocytaire: aspects métaboliques, *Ann. Med. Interne* **141**:13–18.

Urenjak, J., Williams, S. R., Gadian, D. G., and Noble, M., 1993, Proton nuclear magnetic resonance spectroscopy unambiguously identifies different neural cell types, *J. Neurosci.* **13**:981–989.

Waelsch, H., Berl, S., Rossi, C. A., Clarke, D. D., and Purpura, D. P., 1964, Quantitative aspects of $CO_2$ fixation in mammalian brain in vivo, *J. Neurochem.* **11**:717–728.

Westergaard, N., Sonnewald, U., Petersen, S. B., and Schousboe, A., 1991a, Characterization of microcarrier cultures of neurons and astrocytes from cerebral cortex and cerebellum, *Neurochem. Res.* **8**:919–923.

Westergaard, N., Fosmark, H., and Schousboe, A., 1991b, Metabolism and release of glutamate in cerebellar granule cells co-cultured with astrocytes from cerebral cortex and cerebellum, *J. Neurochem.* **56**:59–66.

Westergaard, N., Larsson, O. M., Jensen, B., and Schousboe, A., 1992, Synthesis and release of GABA in cerebral cortical neurons co-cultured with astrocytes from cerebral cortex and cerebellum, *Neurochem. Int.* **20**:567–575.

Westergaard, N., Varming, T., Peng, L., Sonnewald, U., Hertz, L., and Schousboe, A., 1993, Uptake, release and metabolism of alanine in neurons and astrocytes in primary cultures, *J. Neurosci. Res.* **35**:540–545.

Westergaard, N., Sonnewald, U., and Schousboe, A., 1994a, Release of α-ketoglutarate, malate and succinate from cultured astrocytes: Possible role in amino acid neurotransmitter homeostasis, *Neurosci. Lett.* **176**:105–109.

Westergaard, N., Sonnewald, U., Unsgård, G., Peng, L., Hertz, L., and Schousboe, A., 1994b, Uptake, release and metabolism of citrate in neurons and astrocytes in primary cultures, *J. Neurochem.* **62**:1727–1733.

Westergaard, N., Sonnewald, U., Petersen, S. B., and Schousboe, A., 1995, Glutamate and glutamine metabolism in cultured GABAergic neurons studied by ¹³C NMR spectroscopy may indicate compartmentation and mitochondrial heterogeneity, *Neurosci. Lett.* **185**:24–28.

Yu, A. C. H., Drejer, J., Hertz, L., and Schousboe, A., 1983, Pyruvate carboxylase activity in primary cultures of astrocytes and neurons, *J. Neurochem.* **41**:1484–1487.

Yu, A. C. H., Hertz, E., and Hertz, L., 1984, Alterations in uptake and release of GABA, glutamate and glutamine during biochemical maturation of highly purified cultures of cerebral cortical neurons, a GABAergic preparation, *J. Neurochem.* **42**:1484–1487.

Yu, A. C. H., George, G. A., and Chan, P. H., 1989, Hypoxia-induced dysfunctions and injury of astrocytes in primary cell cultures, *J. Cereb. Blood Flow Metab.* **9**:20–28.

Yudkoff, M., Nissim, I., Hummler, K., Medow, M., and Pleasure, D., 1986, Utilization of [¹⁵N]glutamate by cultured astrocytes, *Biochem. J.* **234**:185–192.

CHAPTER 2

# MEASUREMENT OF FREE INTRACELLULAR CATIONS

## HERMAN BACHELARD and RONNITTE BADAR-GOFFER

## SUMMARY

Magnetic resonance spectroscopy (MRS) can be used to monitor some intracellular cations noninvasively by exploiting the presence of endogenous chemicals. In $^{31}P$ MRS, $[Mg^{2+}]_i$ is measured from its effects on the chemical shifts of the β- and γ-resonances, relative to the α-resonance, of endogenous ATP, and changes in intracellular pH can be assessed from the chemical shift of the inorganic phosphate ($P_i$) resonance. It is now proving feasible to monitor lithium in the clinic, using the naturally abundant form of lithium ($^{7}Li$). Some progress is also being made in investigations of the possibility of noninvasive measurement of intracellular concentrations of $Na^+$ and $K^+$ using their naturally abundant forms; the difficulties inherent in this approach are discussed. Other currently available methods rely on the use of more invasive techniques. Changes

HERMAN BACHELARD and RONNITTE BADAR-GOFFER • M. R. Centre, Department of Physics, University of Nottingham, Nottingham, NG7 2RD, United Kingdom. Ronnitte Badar-Goffer's present address is Elscint Ltd., Haifa 31004, Israel.

Magnetic Resonance Spectroscopy and Imaging in Neurochemistry, Volume 8 of Advances in Neurochemistry, edited by Bachelard, Plenum Press, New York, 1997.

in $[Mg^{2+}]_i$ can be measured using $^{13}C$-citrate or $^{19}F$-citrate, and changes in the free intracellular concentrations of other divalent cations (e.g., $Ca^{2+}$, $Cd^{2+}$, $Fe^{2+}$, $Pb^{2+}$, and $Zn^{2+}$) can now be routinely followed using $^{19}F$-MRS indicators. These have the advantage over fluorescence techniques that many cations can be measured simultaneously due to their specific chemical shifts. Specific MRS probes have also been developed to enable measurement of intracellular pH and $Na^+$. Whereas the noninvasive techniques can readily be used *in vivo* in the human brain, only a few of these invasive approaches are currently applicable to animal brains *in vivo*.

## 1. INTRODUCTION

Among the intracellular cations of interest to neuroscientists, calcium has received most attention, not only due to its ubiquitous roles in regulation of numerous cellular processes, but particularly because of its pivotal function in information transfer within the brain, to the extent that calcium is often somewhat loosely referred to as a "second messenger." (Yet it was not always perceived as so important: at the final session of a major international neurochemistry meeting in Vancouver in 1983, an eminent neuroscientist was heard to ask plaintively "Why no calcium?" as it had received little if any mention during the conference.)

At about the same time (early 1980s), new avenues for studies on free intracellular calcium were becoming available with the advent of feasible magnetic resonance spectroscopy (MRS), which offered the possibility of studying other aspects of function (the energy state, pH, other cations, and intermediary metabolism) simultaneously or in parallel. An understandable obsession with calcium tended to obscure the desirability of monitoring other cations. Changes in intracellular $K^+$ provide one of the most sensitive indications of impaired cerebral function such as occurs in stroke or epilepsy (McIlwain and Bachelard, 1985; Obrenovitch *et al.*, 1988). Zinc is increasingly invoked as a modulator of many cerebral functions via "zinc-finger" proteins and especially in regulation of the *N*-methyl-D-aspartate (NMDA) subtype of glutamate receptor, and the dangers of lead poisoning in children have become an important environmental issue. The functions of magnesium as an enzyme cofactor and as an endogenous calcium regulator are also of interest to neuroscientists.

All of these cations can now be monitored as a result of new developments in MRS—some noninvasively, such as $H^+$ (pH), $Mg^{2+}$, and, to a limited extent, $Na^+$ and $K^+$, which means that studies on the human brain are possible. Measurements of others ($Ca^{2+}$, $Zn^{2+}$, $Pb^{2+}$) are invasive in that they currently require the use of preloaded (potentially toxic) indicators. However, use of such indicators is providing exciting information on small animals *in vivo* and with model preparations (cultured cells, slices, e.g., of cortex, cerebellum, and hippocampus, and synaptosomes); it may prove possible to modify some of these approaches for human studies in the future.

## 2. DIVALENT CATIONS

### 2.1. Magnesium

The need for accurate measurement of the free intracellular $Mg^{2+}$ concentration ($[Mg^{2+}]_i$) has long been recognized, particularly in progress toward a clearer understanding of cellular mechanisms of regulation of $Mg^{2+}$-requiring enzymes and the role of $Mg^{2+}$ in modulating cellular responses to changes in the free intracellular $Ca^{2+}$ concentration. Many of the regulatory enzymes that use ATP (such as hexokinase, phosphofructokinase, and pyruvate kinase) require MgATP as the substrate; free ATP is usually inhibitory. Changes in the available $Mg^{2+}$ level can be expected to exert considerable influence on flux rates through numerous metabolic pathways (Rose, 1968; Bachelard and Goldfarb, 1969; Watts, 1973). Early attempts to calculate "unbound" $Mg^{2+}$ levels in the brain by conventional biochemical methods resulted in a suggested range of 0.5–1.5 mM, with a total $Mg^{2+}$ content (free plus bound) of 6–7 mM (Bachelard and Goldfarb, 1969; Veloso *et al.*, 1973). We now know that $[Mg^{2+}]_i$ is of the order of 0.3–0.6 mM (see below). More sophisticated and specific techniques were subsequently developed, such as microinjection of appropriate metallochromic dyes (Brinley and Scarpa, 1975) and $Mg^{2+}$-specific microelectrodes (Fry, 1986); these are invasive techniques.

### 2.1.1. Use of Endogenous ATP

An alternative noninvasive MRS method for the measurement of the free intracellular $Mg^{2+}$ concentration has been known for many years and involves measurement of the relative chemical shifts of the $\alpha$- and $\beta$-$^{31}P$ resonances of the nucleoside phosphates. Cohn and Hughes (1962) were the first to recognize that complexing of $Mg^{2+}$ to phosphorus-containing compounds could be observed by $^{31}P$ MRS. They showed that the divalent cations $Mg^{2+}$, $Ca^{2+}$, and $Zn^{2+}$ are able to produce large chemical shifts in the $\beta$- and $\gamma$-phosphate peaks of ATP. There is practically no shift in the position of the $\alpha$-phosphate peak, because the cations chelate essentially the $\beta$- and $\gamma$-phosphates peaks (Fig. 1). Since the levels of free intracellular $Ca^{2+}$ and $Zn^{2+}$ in the brain are much lower than those of free intracellular $Mg^{2+}$, $[Mg^{2+}]_i$ can be calculated from the chemical shifts of the ATP resonances. The chemical shift of the $\beta$-resonance is greater than that of the $\gamma$ resonance, so in practice the chemical shift of the $\beta$- relative to the $\alpha$-resonance

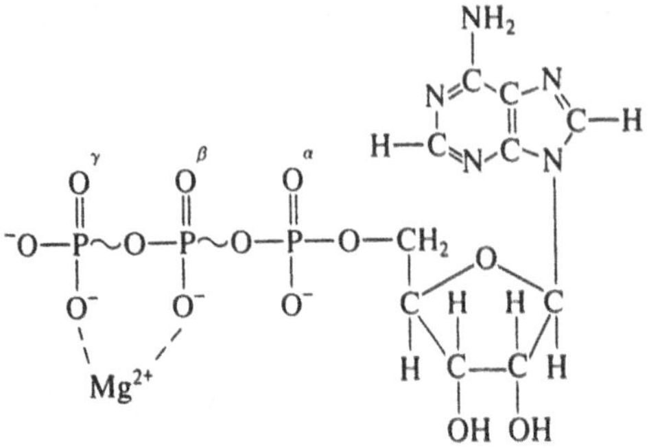

FIGURE 1. Structure of MgATP$^{2-}$

is measured, which titrates the $Mg^{2+}$ concentration (Gupta *et al.*, 1978). Thus, $[Mg^{2+}]_i$ is calculated from the formula

$$[Mg^{2+}]_i = \frac{K_D^{MgATP}(\delta^{ATP} - \delta^{\alpha-\beta})}{(\delta^{\alpha-\beta} - \delta^{MgATP})}$$

where $K_D^{MgATP}$ is the dissociation constant for the Mg–ATP complex, $\delta^{ATP}$ and $\delta^{MgATP}$ are the differences in chemical shift between the $\alpha$ and $\beta$ resonances in the absence of $Mg^{2+}$ and when the ATP is fully complexed with $Mg^{2+}$, respectively, and $\delta^{\alpha - \beta}$ is the measured difference in chemical shift between the $\alpha$ and $\beta$ resonances.

The method also suffers potential limitations. The most important of these is uncertainty of the true $K_D$ for MgATP in the intracellular environment [discussed by Bachelard and Goldfarb (1969)]. Furthermore the generally accepted range for this $K_D$ (10–100 $\mu$M) is low relative to the normally expected range of $[Mg^{2+}]_i$ (0.2–0.6 mM), which means that the chemical shift measurements have to be made at the extreme of the titration range, where errors are maximal (Morris, 1988). Also, it is sometimes difficult to measure the exact chemical shift of the $\beta$-resonance owing to poor signal-to-noise. This can be a problem in *in vivo* studies, and also *in vitro* where experiments on, e.g., ischemia or excitotoxicity cause depletion of ATP. However, with careful experimentation, the method is useful and has the major advantage over any other of being noninvasive: it is the resonances of the naturally occurring intracellular ATP that are monitored.

Using a $K_D$ value of 80 $\mu$M for MgATP gave values of 0.3–0.4 mM for cerebral free intracellular magnesium; the values were found to increase significantly in both hypoglycemia (Fig. 2) and hypoxia. Removal of calcium from the superfusing medium increased the $[Mg^{2+}]_i$ to 0.65 mM (Petroff and Prichard, 1987; Brooks and Bachelard, 1989). Free intracellular magnesium has been observed to decrease in traumatic spinal injury with little change in ATP (Vink *et al.*, 1989).

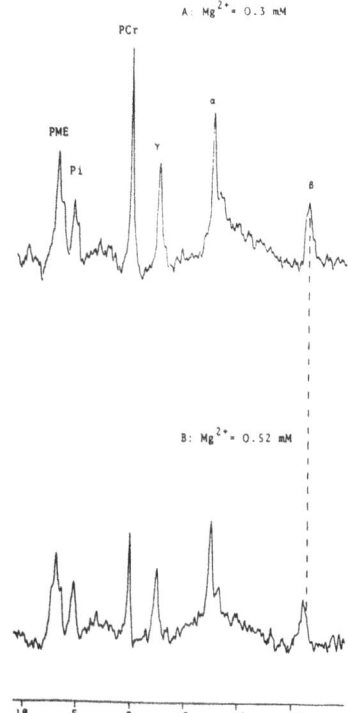

FIGURE 2. $^{31}$P MR spectra showing the small change in the chemical shift of the β resonance of ATP signifying the increase in Mg$^{2+}$ concentration which occurs during hypoglycemia. Resonance assignments: PME, phosphomonesters; Pi, inorganic phosphate; PCr, phosphocreatine; γ-, α-, β-, nucleoside triphosphate groups (mainly ATP). (From Brooks and Bachelard, 1989.)

## 2.1.2. Other MRS Techniques for Measurement of $[Mg^{2+}]_i$

The well-characterized ability of citrate to chelate Mg$^{2+}$ has been exploited in two ways. (Citrate has one advantage over ATP, in its more appropriate $K_D$ of 0.38 mM, which is of the same order as $[Mg^{2+}]_i$.) Changes in the chemical shift of $^{13}$C-labeled citrate which occur when it chelates Mg$^{2+}$ were originally used to titrate free intracellular Mg$^{2+}$ in the liver (Cohen, 1983; Fig. 3A), a feature that we exploited in assigning as citrate a new resonance observed in $^{13}$C MR spectra (Badar-Goffer *et al.*, 1990a). Also, $^{19}$F-labeled citrate has been used to obtain accurate measurements of Mg$^{2+}$ in perfused heart preparations (Kirschenlohr *et al.*, 1988; Fig. 3B). [Fluorocitrate is highly toxic owing to its inhibition of aconitase; however, the enzymically inactive (+) stereoisomer is used.] Interpretation of the results obtained with the citrate chelators may be complicated by uncertainties in the cellular compartmentation of the citrate. Indeed, there is some evidence that glial cells export citrate to the extracellular environment, where it may function as a modulator of extracellular divalent cations (Sonnewald *et al.*, 1991).

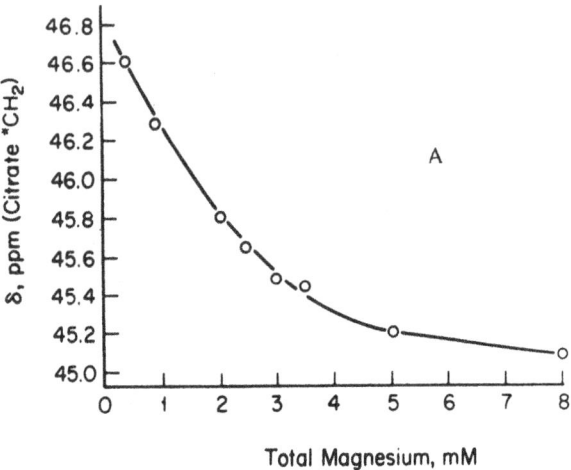

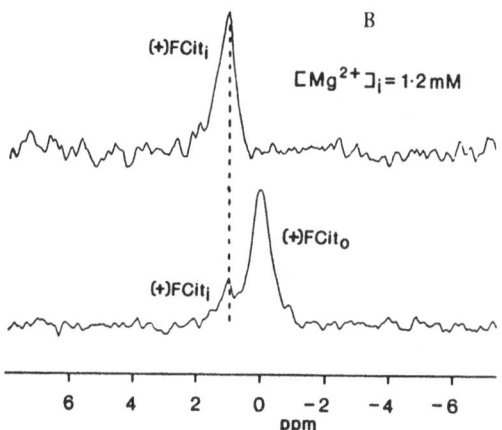

*FIGURE 3.* Use of citrate in MRS measurements of intracellular $Mg^{2+}$ concentration. (A) Titration of the chemical shift ($\delta$) of the $^{13}CH_2$ groups of citrate with $Mg^{2+}$ using $^{13}C$ MRS (Cohen, 1983; with thanks to Sheila Cohen). (B) Change in the chemical shift of (+)-fluorocitrate with $Mg^{2+}$ using $^{19}F$ MRS (Kirschenlohr *et al.*, 1988; with thanks to Heidi Kirschenlohr).

A novel $^{19}F$-labeled chelator (*o*-aminophenol-*N,N,O*-triacetic acid) for MRS measurements of $Mg^{2+}$ has been synthesized (Levy *et al.*, 1987, 1988). The low-naturally-abundance $^{25}Mg$ has been used in MRS studies of $Mg^{2+}$ binding to proteins such as hemoglobin (Bock *et al.*, 1991) but is generally considered to be too insensitive to be applied to intact biological systems.

## 2.2. Calcium

The ability to measure concentrations of free intracellular calcium ($[Ca^{2+}]_i$) accurately in the intact, actively metabolizing tissue is vital to our understanding of brain function. There is increasing awareness of the central role of calcium in altered functional states, such as anesthesia, and especially in the cellular injury known to occur in many pathological conditions. This is of particular importance in our understanding of the mechanisms underlying such clinical disorders as epilepsy, coma, and stroke and in the further damage believed to be associated with partial or complete recovery from these conditions. The importance of the role of calcium in regulation of cellular function has long been recognized (Reznikoff and Chambers, 1927; Heilbrunn, 1930), and approximations of its free intracellular levels (ca. 100 nM) were obtained in resting muscle some 30 years ago using $Ca^{2+}$-EGTA buffering systems (Portzehl *et al.*, 1964; Hagiwara and Nakajima, 1966).

Only relatively recently have methods for the accurate measurement of free intracellular calcium become available. Since $[Ca^{2+}]_i$ (in the range of 100–300 nM) is small in proportion to the total calcium present within the cells (the majority is bound to proteins and organelles) and orders smaller than the extracellular concentration of over 1 mM, techniques need to be highly sensitive and specific. The techniques available (ion-specific electrodes, electron-microscopic X-ray analysis, metallochromic dyes, bioluminescent indicators, fluorescence probes, and MRS indicators) all have their limitations, and none is ideal. The best sensitivity and spatial resolution are provided by electron-microscopic X-ray analysis, but this technique detects total $[Ca^{2+}]_i$ and cannot distinguish between the free and bound species. Also highly sensitive and specific with excellent temporal resolution are the ion-specific electrodes, but these are very invasive (puncturing the cell wall), do not lend themselves readily to studies related to metabolic activity, and are limited in spatial resolution. Metallochromic indicators have less specificity and low signal-to-noise ratio so their use is more limited (Tsien, 1983). Bioluminescent indicators (aequorin, obelin) are photoproteins that emit photons in proportion to $[Ca^{2+}]_i$ and are therefore very sensitive, with rapid response times. This sensitivity, combined with very low $K_D$ values of ca. $10^{-5}$ M, means that low intracellular concentrations of ca. 10 μM can be used (Speksnijder *et al.*, 1989), thus lessening the danger of calcium buffering, which is a potential problem with the fluorescence probes and the MRS indicators. Aequorin suffers one disadvantage in that it is less specific for $[Ca^{2+}]_i$, but transfection of its gene into cells now enables continuous monitoring of $[Ca^{2+}]_i$ with good spatial localization (Button and Brownstein, 1993).

The fluorescent indicators (fura-2, indo-1) are currently the most commonly applied. These chemicals, like the MRS indicators, represent a family derived from EGTA by Tsien (1980). They are highly sensitive and react rapidly to changes in $[Ca^{2+}]_i$. They need to be loaded to intracellular concentrations of up to 0.2 mM, so there is some danger of calcium buffering. As discussed below in the section on MRS indicators, these dangers can be limited by use of appropriate controls. Details of their loading in the form of lipophilic tetraesters are the same as for the MRS indicators described below. Recently, automatic scanners have been developed which enhance routine imaging of these fluorescent probes at light-microscopic level in cells and slices (Mitani *et al.*, 1993). They have limited use in studies related to intermediary metabolism and suffer one major disadvantage in that they cannot distinguish between different divalent cations.

## 2.2.1. Principles of the Use of $^{19}F$-MRS Cation Indicators

Table 1 gives the structures, calcium dissociation constants, and calcium-induced chemical shifts of the more commonly used $^{19}F$-MRS intracellular calcium indicators (Smith and Bachelard, 1996). Like the fluorescent indicators, the $^{19}F$-MRS indicators are prepared as the tetraacetoxymethyl esters, which are sufficiently lipophilic to be transported across cell membranes (Tsien, 1981). They are then hydrolyzed by intracellular esterases to the hydrophilic free tetra-anions, which are membrane-impermeant and are therefore trapped within the cells. The free tetra-anions are intracellular and can only chelate free rather than bound cations with high affinity and specificity according to their structures. On binding the cation, the signal frequency ("chemical shift") of the $^{19}F$-resonance of the indicator changes and is clearly separated in the spectrum from that of the unbound indicator, since the two are in slow exchange. Figure 4 shows the chemical shifts in parts per million (ppm) of 5FBAPTA (see Table 1) bound to the more commonly observed cations. (The ppm scale represents the frequencies of the resonances in parts per million of the applied magnetic field, so the chemical shifts are independent of the field strength of the magnet employed. In practice, field strengths of at least 200 MHz are required.) For example, the $Ca^{2+}$-5FBAPTA resonance is shifted some 5.5 ppm downfield from the free resonance. Two or more $^{19}F$ resonances are observed (depending on the divalent cations present), and the relative concentrations of free and bound indicator are shown quantitatively by the areas of the resonances. The free intracellular concentration of the cation is calculated from its dissociation constant for the indicator, multiplied by the ratio of the bound to the free form of the indicator (Smith *et al.*, 1983; Bachelard *et al.*, 1988). The calculated value of the free intracellular concentration will depend on the value of the dissociation constant ($K_D$) used, which varies from laboratory to laboratory as noted below. The effective range over which the measurement is possible is limited by the signal-to-noise ratio of the resonances, which in turn depends not only on the field strength but also on the amount of intracellular indicator present. If too much is present, problems of calcium buffering are likely to be experienced. These can be checked by the

TABLE 1. Some $^{19}$F MRS Calcium Indicators[a]

| Indicator | Structure | Ca$^{2+}$ affinity (nM at 30°C) | Ca$^{2+}$ induced shift (ppm) |
|---|---|---|---|
| 5-FBAPTA | | 537 | 5.5 |
| DiMe-4-FBAPTA | | 155 | 2.07 |
| DiMe-5-FBAPTA | | 34 | 3.96 |
| DiMe-5-TFMBAPTA | | 724 | −0.8 |

[a]The full range available is given by Smith and Bachelard (1995).

appropriate controls, for example, by parallel monitoring of the energy state ($^{31}$P MRS) and metabolites such as lactate ($^1$H MRS). It becomes therefore a compromise between the amounts of indicator sufficient to give acceptable signal and those which will not affect the tissue's functional or metabolic state.

Loading these indicators can be difficult because the acetoxymethyl esters are insoluble in water and therefore come out of solution when exposed to the

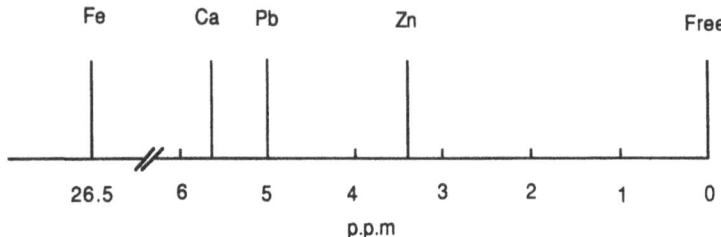

FIGURE 4. Chemical shifts in $^{19}F$ MR spectra of 5FBAPTA bound to divalent cations. The chemical shift of the free 5FBAPTA is set at 0 ppm.

aqueous media required for metabolic studies. For successful loading, the ester is dissolved in dimethyl sulfoxide (DMSO) and added to the medium slowly and intermittently from a Hamilton syringe. If the ester solution is added in this way to a buffer that is gently stirred without vortexing, a colloidal suspension results that is stable for some hours. If the suspension is bubbled with a gas ($O_2/CO_2$) or even stirred too vigorously, the esters are immediately precipitated and are no longer available for uptake into the cells or tissue. We determined optimal loading conditions for our superfusion system using $^3H$-labeled 5FBAPTA (see below). Some 100–150 μl of the indicator (50 mM in DMSO) is slowly added periodically, 10 μl at a time, with gentle stirring and gassing between each addition, to the incubating tissues (about 100 ml of standard incubation medium per gram) and then incubated in fresh gassed medium for a further 30 min to achieve optimal loading. In our hands, we find that loading of 5FBAPTA to intracellular concentrations of 100–150 μM is appropriate, this concentration range being of the same order as that normally employed with the fluorescent indicators. $^{19}F$ spectra can be acquired with reasonable signal-to-noise ratio for many hours; the indicator slowly diffuses out of the tissue with a pseudo-half-life of about 4 hr (Badar-Goffer *et al.*, 1990b).

### 2.2.2. $^{19}F$-MRS Studies Using 5FBAPTA

One of the fluorescent $Ca^{2+}$ chelators originally designed by Tsien (1980, 1981), 1,2-bis(*o*-aminophenoxy)ethane-*N,N,N',N'*-tetraacetic acid (BAPTA), was modified for MR spectroscopy by introduction of $^{19}F$ symmetrically at the 5-position of the two aromatic rings to produce 5FBAPTA (Smith *et al.*, 1983). This proved to provide a most sensitive and highly specific means of accurate measurement of $[Ca^{2+}]_i$ in isolated lymphocytes and in perfused hearts, giving values comparable to the value obtained using fluorescent indicators (Metcalfe *et al.*, 1985). Though the technique at present cannot give intracellular morphological localization, nor good temporal resolution, it has the unique advantage of superb specificity in that $Ca^{2+}$ can be distinguished from other divalent cations

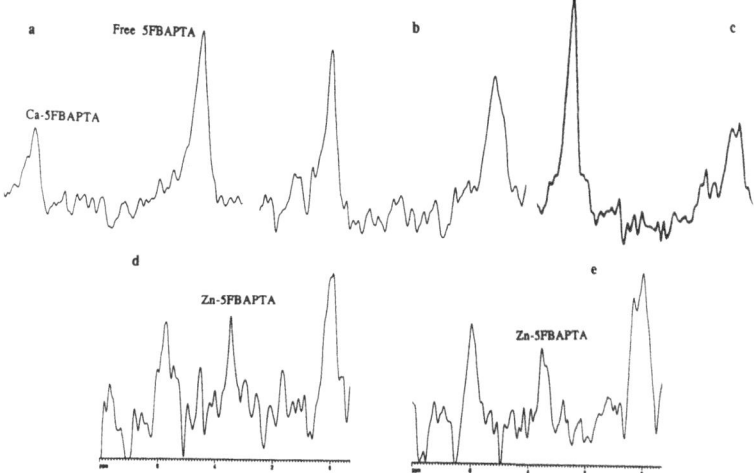

FIGURE 5. Measurement of changes in $[Ca^{2+}]_i$ in superfused brain slices by $^{19}F$ MRS using 5FBAPTA. Changes in $[Ca^{2+}]_i$ from control values (a) were measured on depolarization (b), with low glucose/low $O_2$ (c), and after exposure to glutamate (d) or NMDA (e). (From Bachelard et al., 1994.)

such as iron, lead, and zinc (see below). Its first successful application to cerebral systems was on superfused slices of the guinea pig cerebral cortex (Bachelard et al., 1988). Using a $K_D$ value of 600 nM for $Ca^{2+}$-FBAPTA, this preparation routinely gives values of 150–250 nM (Badar-Goffer et al., 1993, 1994), and similar values of 200 nM in superfused hippocampal slices are now being obtained in our laboratory (N. Thatcher and H. Bachelard, unpublished experiments). Using this approach, changes in $[Ca^{2+}]_i$ have been followed in tissues subjected to depolarization, excitotoxins, and ischemia (Fig. 5; Badar-Goffer et al., 1993, 1994; Espanol et al., 1994). Values of 300–350 nM $[Ca^{2+}]_i$ were also reported from the use of 5FBAPTA in synaptosomes (Denny and Atchison, 1994; these authors used a dissociation constant of 500 nM). These values of $[Ca^{2+}]_i$ (150–350 nM) obtained from $^{19}F$-MRS studies using 5FBAPTA are similar to those previously observed with non-MRS techniques: use of the fluorescent indicators quin-2 and fura-2 yielded values of 100–500 nM for synaptosomes (Ashley et al., 1984; Richards et al., 1984; Ashley, 1986), and similar values (100–500 nM) were reported for bulk isolated neurons (Kudo and Ogura, 1986) or cultured neurons (Perney et al., 1984; Connor, 1986). Use of ion-selective electrodes gave slightly lower values (50–160 nM) for the giant squid axon (DiPolo and Beaugé, 1983).

Free intracellular calcium has only recently been successfully measured in vivo (Song et al., 1995; Fig. 6A). The 5FBAPTA acetoxymethyl ester (100 mg/ml in DMSO) was infused into rat brains through intraventricular cannulae at

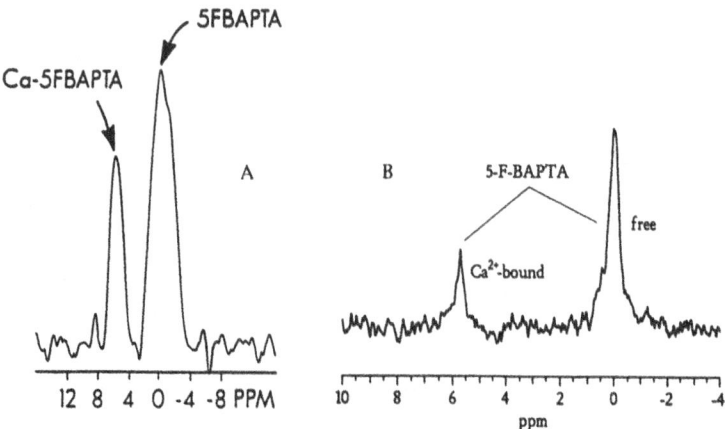

FIGURE 6. Measurement of [Ca$^{2+}$]$_i$ by $^{19}$F-MRS using 5FBAPTA: (A) In rat brain *in vivo* (Song *et al.*, 1995; with thanks to Joseph Ackerman); (B) in superfused immobilized cultured neuroblastoma cells (with thanks to Dieter Leibfritz).

a rate of approximately 0.5 μl/min for 50 min (i.e., a total of 25 μl), and excellent spectra were produced after data were acquired for 30 min. Use of a lower $K_D$ value (337 nM) than those quoted above gave values for [Ca$^{2+}$]$_i$ of 200 nM. Success has also been achieved by Benters, Floegel, and co-workers in loading 5FBAPTA into perfused F98 glioma cells and E367 neuroblastoma cells, immobilized on microcarriers or basement membrane gel threads, which yielded comparable spectra (Fig. 6B). These workers used a $K_D$ value of 500 nM, and values of 100–150 nM [Ca$^{2+}$]$_i$ were calculated. They were also able to detect Cd$^{2+}$ (which has a chemical shift very close to that of Pb$^{2+}$; see Fig. 4) in experiments in which the effects of Cd$^{2+}$ in the media on [Ca$^{2+}$]$_i$ were investigated (Benters *et al.*, 1995; Floegel *et al.*, 1995). These developments should prove most useful since glial cells are also heavily involved in Ca$^{2+}$ homeostasis in the brain (Lazarewicz *et al.*, 1977).

### 2.2.3. Use of Dimethyl 5FBAPTA

The relatively high $K_D$ of 5FBAPTA (500–600 nM at 37°C) used gives some cause for concern about calcium buffering as noted above, and an indicator with a much lower $K_D$ (dimethyl 5FBAPTA) has recently become available (Table 1). This theoretically should provide more sensitive estimations of [Ca$^{2+}$]$_i$ with less concern about buffering. In preliminary unpublished studies with dimethyl 5FBAPTA, after loading under similar conditions to those used for 5FBAPTA, we calculated [Ca$^{2+}$]$_i$ to be of the same order (200 nM) as with 5FBAPTA. However, owing to the much higher $K_D$, the bound peak was much higher than the free peak, and so the measurement was less accurate; we therefore concluded that this indicator would not be useful for measurements of increases in $[Ca^{2+}]_i$ as the free peak would then be almost undetectable, thus rendering accurate calculation impossible. However, the fact that the calculated $[Ca^{2+}]_i$ was so similar using the high- and low-affinity indicators, and that other parameters such as the energy state and lactate levels were also similar, suggests that under these experimental conditions major buffering artifacts are not being introduced by 5FBAPTA.

More detailed descriptions of the principles of the use of these $^{19}$F-MRS calcium indicators and recipes for their chemical synthesis are given in a recent review (Smith and Bachelard, 1995).

### 2.2.4. Other MRS $Ca^{2+}$ Indicators

Concern about the problems of calcium buffering by the fluorescent and MRS indicators described above and limitations in their applications to *in vivo* studies led Robitaille and Jiang (1992) to develop an alternative calcium-sensitive ligand. The $^{13}$C-labeled analog, 1-(2-aminophenoxy)-2-(2-aminoethoxy)ethane-$N,N,N',N'$-tetraacetic acid (AATA), designed for use with the $^1$H-observe carbon-edited technique to improve sensitivity of $^{13}$C measurement, gave a $K_D$ value at 37°C of 350 nM $[Ca^{2+}]_i$. It has not yet been applied to cerebral systems to our knowledge.

$^{43}$Ca MRS has been used to explore Ca-binding sites on proteins (lysozymes and lactalbumin; Aramini *et al.*, 1992), but it is not sufficiently sensitive for biological studies.

### 2.3. Zinc

There is growing evidence from a variety of studies (mainly on cultured brain cells) that, like calcium, zinc may have a multiplicity of roles including enzyme activation, modulation of excitatory and inhibitory neurotransmission, and cytotoxicity, depending on its concentration and the region involved. It seems particularly important as a modulator of glutamate excitotoxicity via the NMDA receptor. Zinc is found enriched in hippocampus, cerebral cortex, cerebellum, and pineal gland at local concentrations of up to 500 mM (Choi *et al.*, 1989; Frederickson, 1989; Westbrook and Mayer, 1987; Weiss *et al.*, 1993). A method for monitoring intracellular zinc in actively metabolizing cerebral preparations is therefore desirable and is afforded by $^{19}$F MRS using 5FBAPTA. As shown in Fig. 4, the $Zn^{2+}$-BAPTA complex has a different chemical shift from that of $Ca^{2+}$-BAPTA, and therefore the two complexes can be clearly distinguished in the same spectra. The zinc observed in this way is within the tissue—

the indicator is intracellular and does not detect divalent cations released to the extracellular environment.

Using this technique, we observed free intracellular $Zn^{2+}$ at calculated concentrations of 10–25 μM, after superfused cerebral cortex slices had been challenged with glutamate or NMDA (Fig. 5), and suggested that it may have been released from protein-bound sites (Badar-Goffer *et al.*, 1994). Pretreatment with MK-801 (an NMDA receptor blocker) prevented the appearance of the zinc-BAPTA resonance, supporting the view of zinc's role in modulation of the functioning of the NMDA subtype of glutamate receptor (Bachelard *et al.*, 1994). Using the 5FBAPTA technique, zinc was also observed in $^{19}F$ MR spectra of synaptosomes after treatment with methylmercury, which strongly indicates that the zinc has been released from protein —SH groups (Denny and Atchison, 1994).

## 2.4. Lead

$Pb^{2+}$ also has a distinct chemical shift in $^{19}F$-5FBAPTA MR spectra (Fig. 4); this resonance has been observed in studies of cultured osteoblasts (Schanne *et al.*, 1990) and has been used to study $Pb^{2+}$ activation of rat brain protein kinase C (Long *et al.*, 1994). Presumably, $Pb^{2+}$ occurs naturally at levels too low to be detected in the normal brain, but this technique offers opportunities of studying aspects of the mechanisms of lead toxicity in animal models and in *in vitro* systems.

## 3. MONOVALENT CATIONS

## 3.1. pH

The $^{31}P$-MRS technique for measurement of intracellular pH is noninvasive, in contrast to other techniques such as microelectrodes and intracellular dyes. The chemical shift of inorganic phosphate ($P_i$) is very sensitive to changes in pH. Under normal physiological conditions, inorganic phosphate ($P_i$) exists primarily as a mixture of $H_2PO_4^-$ and $HPO_4^{2-}$ ($pK = 6.8$), with chemical shifts relative to phosphocreatine (which is pH-insensitive in the physiological range) of 3.29 and 5.81 ppm, respectively. The two ions are in rapid exchange so that a single $P_i$ peak is observed with a chemical shift reflecting the relative proportions of the two ions (Moon and Richards, 1973). The intracellular pH can be readily calculated from this chemical shift, according to the following formula:

$$pH = 6.8 + \log (3.29 - \delta)/(\delta - 5.81)$$

where 6.8 is the p$K$ of $H_2PO_4^-/HPO_4^{2-}$, 3.29 and 5.81 are the chemical shifts from phosphocreatine) of $H_2PO_4^-$ and $HPO_4^{2-}$, respectively, and $\delta$ is the observed chemical shift of the $P_i$ resonance relative to phosphocreatine.

This noninvasive technique has been used to investigate mechanisms of pH regulation in the brain and to monitor changes in pH under various conditions *in vivo* and *in vitro* (Fig. 7A) (Prichard *et al.*, 1983; Behar *et al.*, 1985; Brooks and Bachelard, 1992). Precise quantification of the chemical shift of the $P_i$ resonance *in vivo* and *in vitro* can sometimes be difficult, as this resonance may be low, broad, and poorly resolved. In those situations other components of the [31]P spectra that are pH-sensitive, such as the phosphomonoesters (choline O-phosphate and ethanolamine O-phosphate), can be used. However, the determination of pH would be likely to be less accurate as the two resonances overlap and their p$K$ values are far from the normal physiological pH range (see Brooks and Bachelard, 1989).

An alternative approach for *in vitro* (but not *in vivo*) studies is to use invasive pH probes, such as pH-sensitive phosphorylated compounds: methyl phosphonate, phenyl phosphonate, and fructose 1-phosphate (Shulman *et al.*, 1979; Slonezewski *et al.*, 1981; Thoma *et al.*, 1986). However, we found that methyl phosphonate was not taken up into brain slices sufficiently to be MRS detectable (Brooks and Bachelard, 1989). The chemical shift of 2-deoxyglucose 6-phosphate (DOG6P) has been used in various tissues, including cardiac muscle (Bailey *et al.*, 1981), tumor cells (Gillies *et al.*, 1982), rat brain (Yue *et al.*, 1984), and superfused guinea pig brain slices (Brooks and Bachelard, 1989). 2-De-oxy-D-glucose (DOG) is taken up by the tissues and phosphorylated; the resultant DOG6P is not metabolized further to any appreciable extent, so it is trapped. It can be distinguished from the other phosphomonoesters (choline O-phosphate and ethanolamine O-phosphate) in [31]P MR spectra, and its chemical shift is sufficiently sensitive to changes in pH. Although high concentrations of DOG can cause hypoglycemic effects due to its competition for transport and phosphorylation of glucose (Horton *et al.*, 1973), there is no significant decrease in the levels of high-energy phosphate metabolites if the loading conditions are carefully controlled, and adequate glucose is available to maintain normal rates of intermediary metabolism.

Application of interleaved [1]H and [31]P MRS has proved invaluable in studying possible correlations between pH and lactate under various conditions. Whereas studies on ischemia have confirmed earlier biochemical findings that the fall in pH is associated with increased lactic acid, no clear correlations have emerged from studies on controlled ischemia and reperfusion (Crockard *et al.*, 1987; Hope *et al.*, 1988), on hypoglycemia (Behar *et al.*, 1985; Brooks *et al.*, 1989), or during seizures (Schnall *et al.*, 1988). The dissociation between lactate and pH has been discussed in some detail by Paschen *et al.* (1987).

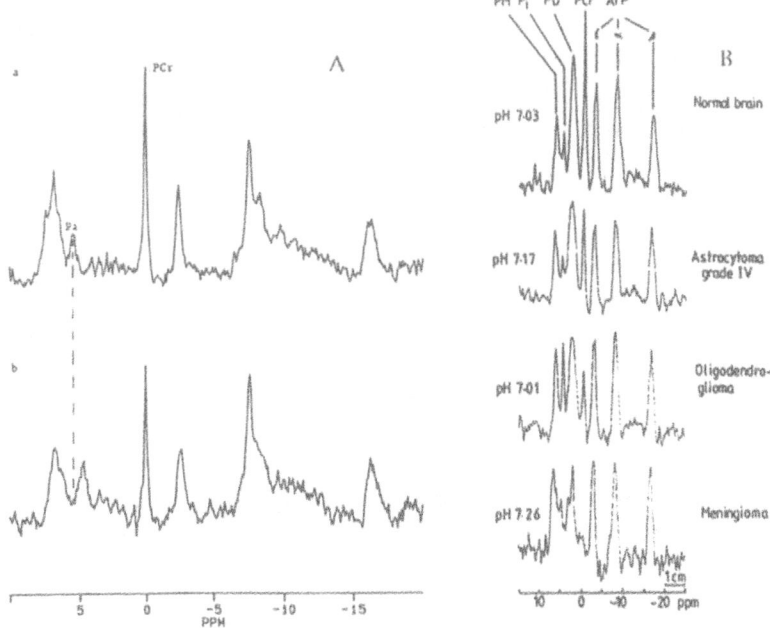

*FIGURE 7.* Measurement of intracellular pH from the $P_i$ resonances in $^{31}P$ MR spectra: (A) Decreased intracellular pH of superfused cortical slices from normal (A) in response to decreased external buffer pH (B) (Brooks and Bachelard, 1992); (B) the alkaline shift observed in some brain tumors (Oberhaensli *et al.*, 1986; with thanks to George Radda).

Brain tumors were earlier thought to be acidic owing to suggestions that their metabolism was essentially anaerobic as the result of an inadequate blood supply, but $^{31}P$ MRS revealed that they may more often be alkaline (Fig. 7B; Oberhaensli *et al.*, 1986, 1987). The reason for this remains unknown, but some workers have suggested that the alkaline pH may stimulate tumor growth. Another possibility is that the elevated intracellular $H^+$ is due to increased $Na^+/H^+$ exchange in the rapidly metabolizing tumor cells.

A specific indicator (F-quene, Fig. 8A) for $^{19}F$-MRS measurement of pH, based on the same principles as the use of 5FBAPTA for the measurement of $[Ca^{2+}]_i$ (see Section 2.2.1), has been developed. It is derived from the fluorescent pH indicator quene-1 and has a $pK$ of 6.8. It has been used in perfused heart and in rat liver (Metcalfe *et al.*, 1985; Beech and Iles, 1991) and has the advantage that it can be co-loaded with 5FBAPTA, giving simultaneous measurement of pH and $[Ca^{2+}]_i$. It does not yet seem to have been applied to brain preparations.

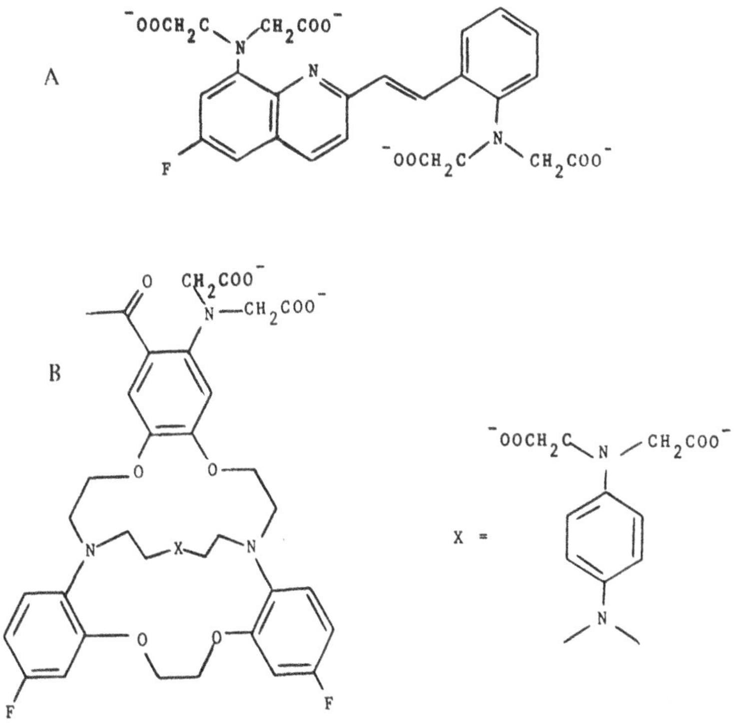

FIGURE 8. $^{19}$F MRS indicators for intracellular monovalent cations: (A) F-quene for pH (Metcalfe *et al.*, 1985); (B) F-cryp-1 for Na$^+$ (Smith *et al.*, 1986). (With thanks to Gerry Smith.)

## 3.2. Sodium and Potassium

The desirability of a noninvasive technique for measurement of intracellular sodium and potassium in clinical assessment and management of such neurological disorders as coma and stroke is self-evident, and it would seem that MRS of $^{23}$Na and $^{39}$K could potentially provide the means. However, studies to date have revealed that not only is it extremely difficult to distinguish the intracellular from the extracellular cations, but particular problems have emerged in that not all of the intracellular cations, if distinguished, for example, with the use of chemical shift reagents (see below), are MRS-detectable. The reasons for this lie in the properties of the $^{23}$Na and $^{39}$K nuclei; these have a spin of $\frac{3}{2}$, in contrast to the more commonly used nuclei ($^1$H, $^{13}$C, $^{19}$F, and $^{31}$P), which have spin $\frac{1}{2}$. The spin $\frac{3}{2}$ results in quadrupolar relaxation of the nuclei with multiexponential components, some of which are "lost" from the spectra. Attempts to resolve these problems have shown improvements in quantitation but have not yet been entirely successful (Burstein *et al.*, 1989; Chu *et al.*, 1990; Rashid *et al.*, 1991).

MRS measurements of intracellular sodium and potassium currently fall into two categories: use of invasive "shift reagents" and use of noninvasive MRS pulse sequences. At present, the shift reagents provide the more precisely defined distinction between intracellular and extracellular ions, but they are too toxic to be applied to human studies. Obviously, the hope is that progress in the noninvasive approach will allow human studies to become routinely performed. The principle of the shift reagents is that they are membrane-impermeable polyanions {e.g., dysprosium tripolyphosphate, $(Dy\text{-}PPP_2)^{7-}$; dysprosium(III) triethylaminetetraamine hexaacetate, $(DyTTHA^{3-})$}, which cause chemical shifts in the resonance of the extracellular cation but not of the intracellular cation. For example, dysprosium tripolyphosphate has been used to distinguish between intracellular and extracellular $^{23}Na$, giving values of 10–30 mM $[Na^+]_i$ in perfused kidney and heart preparations (Gupta *et al.*, 1989; van Echteld *et al.*, 1991). However, as noted above, the difficulties remain in detecting all of the intracellular cation that is present. A careful study of the MRS-detectable $^{39}K$ in different rat tissues (Wellard *et al.*, 1993) revealed that this varied under optimal conditions from 100% in erythrocytes to about 60% in liver; brain and muscle were more encouraging at about 90%. The authors found some correlation between "visibility" and the density of mitochondria and suggested that much of the undetectable cation may be within the mitochondria. Though most studies have been on isolated cells or perfused systems (kidney, heart), some have been performed using shift reagents on animal brains *in vivo* (Eleff *et al.*, 1993).

The alternative noninvasive approach is to employ pulse sequences designed to overcome the quadrupolar interactions, i.e., multiple quantum-filtered techniques [double quantum-filtered (DQ) and triple double quantum-filtered (TQ)] (Pekar *et al.*, 1987). These and analogous techniques have been used in perfused rat heart (Kuki *et al.*, 1990; Payne *et al.*, 1990) and rat brain, where large increases in intracellular $Na^+$ and decreased $K^+$ were observed in postmortem ischemia (Lyon *et al.*, 1991). A similar approach has been taken to imaging $[Na^+]_i$ in rat brain *in vivo* (Boada *et al.*, 1994). The potassium analog rubidium has also been used in $^{87}Rb$-MRS studies to mimic $K^+$ movements in non-neural systems (Syme *et al.*, 1990; Steward *et al.*, 1991).

A $^{19}F$-MRS indicator for $[Na^+]_i$ is also available: F-cryp-1 (Fig. 8B). Like 5FBAPTA and F-quene (see above), it is loaded as the tetraacetoxymethyl ester and has proved successful for accurate measurement of $[Na^+]$ in pig lymphocytes; a value of 13.8 mM was recorded (Smith *et al.*, 1986).

## 3.3. Lithium

Lithium therapy forms a major part of the treatment of bipolar (manic-depressive) psychiatric disorders, and the appropriate dosage is routinely adjusted by monitoring levels in serum or erythrocytes. However, there is not always a good correlation between levels and efficacy of treatment: some patients may suffer neurotoxic symptoms at normal therapeutic levels whereas others do not respond to these levels (Schou, 1968). Therefore, a method of monitoring $Li^+$ in the human brain is highly desirable. This has proved feasible using surface coils to measure $^7Li$ (the naturally abundant form) in the human brain *in vivo* (Renshaw and Wicklund, 1988; Komorosky *et al.*, 1990). In patients with bipolar disorder, the brain (occipital pole-to-serum ratios averaged $0.47 \pm 0.12$ (Gyulai *et al.*, 1991).

## REFERENCES

Aramini, J. M., Drakenberg, T., Hiraoki, T., Ke, Y., Nitta, K., and Vogel, H. J., 1992, Calcium-43 NMR studies of calcium-binding lysozymes and alpha-lactalbumins, *Biochemistry* **31**:6761–6768.

Ashley, R. H., 1986, External calcium, intrasynaptosomal free calcium and neurotransmitter release, *Biochim. Biophys. Acta* **854**:207–212.

Ashley, R. H., Brammer, M. J., and Marchbanks, R. M., 1984, Measurement of intrasynaptosomal free calcium by using the fluorescent indicator quin-2, *Biochem. J.* **219**:149–158.

Bachelard, H. S., and Badar-Goffer, R. S., 1993, NMR spectroscopy in neurochemistry, *J. Neurochem.* **61**: 412–429.

Bachelard, H. S., and Goldfarb, P. S. G., 1969, Adenine nucleotides and magnesium ions in relation to control of mammalian cerebral-cortex hexokinase, *Biochem. J.* **112**:579–586.

Bachelard, H. S., Badar-Goffer, R. S., Brooks, K. J., Dolin, S. J., and Morris, P. G., 1988, Measurement of free intracellular calcium in the brain by $^{19}F$-nuclear magnetic resonance spectroscopy, *J. Neurochem.* **51**:1311–1313.

Bachelard, H., Badar-Goffer, R., Ben-Yoseph, O., Morris, P., and Thatcher, N., 1994, Magnetic resonance spectroscopy studies on $Ca^{2+}$, $Zn^{2+}$, and energy metabolism in superfused brain slices, *Biochem. Soc. Trans.* **22**:988–991.

Badar-Goffer, R. S. Bachelard, H. S., and Morris, P. G., 1990a, Cerebral metabolism of acetate and glucose studied by $^{13}C$-n.m.r. spectroscopy, *Biochem. J.* **266**:133–139.

Badar-Goffer, R. S., Ben-Yoseph, O., Dolin, S. J., Morris, P. G., Smith, G. A., and Bachelard, H. S., 1990b, Use of 1,2-bis(2-amino-5-fluorophenoxy)ethane-*N,N,N',N'*-tetraacetic acid (5FBAPTA) in the measurement of free intracellular calcium in the brain by $^{19}F$-nuclear magnetic resonance spectroscopy, *J. Neurochem.* **55**:878–884.

Badar-Goffer, R. S., Thatcher, N. M., Morris, P. G., and Bachelard, H. S., 1993, Neither moderate hypoxia nor mild hypoglycaemia alone causes any significant increase in cerebral $[Ca^{2+}]_i$; Only a combination of the two insults has this effect. A $^{31}P$ and $^{19}F$ NMR study. *J. Neurochem.* **61**:2207–2214.

Badar-Goffer, R., Morris, P., Thatcher, N., and Bachelard, H., 1994, Excitotoxic amino acids cause appearance of MRS-observable zinc in superfused cortical slices, *J. Neurochem.* **62**:2488–2491.

Bailey, I. A., Williams, S. R., Radda, G. K., and Gadian, D. G., 1981, Activity of phosphorylase in total ischaemia in the rat heart, *Biochem. J.* **196**:171–178.

Beech, J. S., and Iles, R. A., 1991, Hepatic intracellular pH in vivo using FQUENE and $^{19}F$ NMR-spectroscopy, *Magn. Reson. Med.* **19**:386–392.

Behar, K. L., den Hollander, J. A., Petroff, O. A. C., Hetherington, H. P., Prichard, J. W., and Schulman, R. G., 1985, Effect of hypoglycaemic encephalopathy upon amino acids, high-energy phosphates and pH in the rat brain *in vivo:* Detection by sequential $^1H$ and $^{31}P$ NMR spectroscopy, *J. Neurochem.* **44**:1045–1055.

Benters, J., Floegewl, U., Hechtenberg, S., Schaefer, T., Leibfritz, D., and Beyersmann, D., 1995,

Effect of cadmium on intracellular free calcium: A ¹⁹F-NMR spectroscopy study, *Abstract, 3rd Meeting of the Society for Magnetic Resonance, Nice,* p. 1695.

Boada, F. E., Christensen, J. D., Huang-Hellinger, F. R., Reese, T. R., and Thulborn, K. R., 1994, Quantitative in vivo tissue sodium concentration maps: The effects of biexponential relaxation, *Magn. Reson. Med.* **32:**219–223.

Bock, J. L., Crull, G. B., Wishnia, A., and Springer, C. S., 1991, ²⁵Mg NMR studies on magnesium binding to erythrocyte constituents, *J. Inorg. Chem.* **44:**79–87.

Brinley, F. J., and Scarpa, A., 1975, Ionized magnesium concentrations in axoplasm of dialyzed squid axon, *FEBS Lett.* **50:**82–85.

Brooks, K. J., and Bachelard, H. S., 1989, Changes in intracellular free magnesium during hypoglycaemia and hypoxia in cerebral tissue as calculated from ³¹P NMR spectra, *J. Neurochem.* **53:**331-334.

Brooks, K. J., and Bachelard, H. S., 1992, The regulation of intracellular pH studied by ³¹P- and ¹H-n.m.r. spectroscopy in superfused guinea-pig cerebral cortex slices, *Neurochem. Int.* **21:**375–379.

Brooks, K. J., Porteous, R., and Bachelard, H. S., 1989, Effects of hypoglycaemia and hypoxia on the intracellular pH of cerebral tissue as measured by ³¹P nuclear magnetic resonance, *J. Neurochem.* **52:**606–610.

Burstein, D., Litt, H. J., and Fossel, E. T., 1989, NMR characteristics of "visible" intracellular myocardial potassium in perfused rat hearts, *Magn. Reson. Med.* **9:**66–78.

Button, D., and Brownstein, M., 1993, Aequorin-expressing mammalian cell lines used to report Ca²⁺ mobilization, *Cell Calcium* **14:**663–671.

Choi, D. W., Weiss, J. H., Koh, J-Y., Christine, C. W., and Kurth, M. C., 1989, Glutamate neurotoxicity, calcium, and zinc, *Ann. N. Y. Acad. Sci.* **568:**219–224.

Chu, S. C-K., Xu, Y., Balshi, J. A., and Springer, C. S., 1990, Bulk magnetic susceptibility shifts in NMR studies of compartmentalized samples: Use of paramagnetic reagents, *Magn. Reson. Med.* **13:**239–243.

Cohen, S. M., 1983, Simultaneous ¹³C and ³¹P NMR studies of perfused rat liver. Effects of insulin and glucagon and a ¹³C NMR assay of free Mg²⁺, *J. Biol. Chem.* **258:**14294–14308.

Cohn, M., and Hughes, T. R., 1962, Nuclear magnetic resonance spectra of adenosine di- and triphosphate. II. Effect of complexing with divalent metal ions, *J. Biol. Chem.* **237:**176–181.

Connor, J. A., 1986, Digital imaging of free calcium changes and of spatial gradients in growing processes in single, mammalian central nervous system cells, *Proc. Natl. Acad. Sci. USA* **83:**6179–6183.

Crockard, H. A., Gadian, D. G., Frackowiak, R. S. J., Proctor, E., Allen, K., Williams, S. R., and Ross-Russell, R. W., 1987, Acute cerebral ischaemia: Concurrent changes in cerebral blood flow, energy metabolites, pH and lactate measured with hydrogen clearance and ³¹P and ¹H nuclear magnetic resonance spectroscopy. II. Changes during ischaemia, *J. Cereb. Blood Flow Metab.* **7:**394–402.

Denny, M. F., and Atchison, W. D., 1994, Methylmercury-induced elevations in intrasynaptosomal zinc concentrations: An ¹⁹F-NMR study, *J. Neurochem.* **63:**383–386.

DiPolo, R., and Beaugé, L., 1983, The calcium pump and sodium–calcium exchange in squid axons, *Annu. Rev. Physiol.* **45:**313–324.

Eleff, S. M., McLennan, I. J., Hart, G. K., Maruki, Y., Traystman, R. J., and Koehler, R. C., 1993, Shift reagent enhanced concurrence ²³Na and ¹H magnetic resonance spectroscopic studies of transcellular sodium distribution in the dog brain *in vivo, Magn. Reson. Med.* **30:**11–17.

Espanol, M. T., Litt, L., Xu, Y., Chang, L-H., James, T. L., Weinstein, P. R., and Chan, P. H., 1994, ¹⁹F NMR calcium changes, edema and histology in neonatal rat brain slices during glutamate toxicity, *Brain Res.* **647:**172–176.

Floegel, U., Niendorf, T., Benters, J., and Leibfritz, D., 1995, Perfused brain cells embedded in basement membrane gel threads: An experimental model to monitor transient ischemia and subsequent recovery by NMR, *Abstract, 3rd Meeting of the Society for Magnetic Resonance, Nice*, p. 1775.

Frederickson, C. J., 1989, Neurobiology of zinc and zinc-containing neurons, *Int. Rev. Neurobiol.* **31**: 145–238.

Fry, C. H., 1986, Measurement and control of intracellular magnesium ion concentration in guinea pig and ferret ventricular myocardium, *Magnesium* **5**:306–316.

Gillies, R. J., Ogino, T., Shulman, R. G., and Ward, D. C., 1982, $^{31}P$ nuclear magnetic resonance evidence for the regulation of intracellular pH by Ehrlich ascites tumor cells, *J. Cell Biol.* **95**:24–28.

Gupta, R. K., Benovic, J. L., and Rose, Z. B., 1978, The determination of the free magnesium level in the human red blood cell by $^{31}P$ NMR, *J. Biol. Chem.* **253**:6172–6176.

Gupta, R. K., Dowd, T. L., Spitzer, A., and Barac-Nieto, M., 1989, $^{23}Na$, $^{19}F$, $^{35}Cl$ and $^{31}P$ multinuclear magnetic resonance studies of perfused rat kidney, *Renal Physiol. Biochem.* **12**:144–160.

Gyulai, L., Wicklund, S. W., Greenstein, R., Bauer, M. S., Ciccione, P., Whybrow, P. C., Zimmerman, J., Kovachich, G., and Alves, W., 1991, Measurement of tissue lithium concentration by lithium magnetic resonance spectroscopy in patients with bipolar disorder, *Biol. Psychiatry* **29**:1161–1170.

Hagiwara, S., and Nakajima, S., 1966, Effects of the intracellular Ca ion concentration upon the excitability of the muscle fiber membrane of a barnacle, *J. Gen. Physiol.* **49**:807–818.

Heilbrunn, L. F., 1930, The action of various salts on the first stage of the surface precipitation reaction in Arbacia egg protoplasm, *Protoplasma* **11**:558–573.

Hope, P. L., Cady, E. B., Delpy, D. T., Ives, N. K., Gardiner, R. M., and Reynolds, E. O. R., 1988, Brain metabolism and intracellular pH during ischaemia: Effects of systemic glucose and bicarbonate administration studied by $^{31}P$ and $^{1}H$ nuclear magnetic resonance spectroscopy *in vivo* in the lamb, *J. Neurochem.* **50**:1394–1402.

Horton, R. W., Meldrum, B. S., and Bachelard, H. S., 1973, Enzymic and cerebral metabolic effects of 2-deoxy-D-glucose, *J. Neurochem.* **21**:507–520.

Kirschenlohr, H. L., Metcalfe, J. C., Morris, P. G., Rodrigo, G. C., and Smith, G. A., 1988, $Ca^{2+}$ transient, $Mg^{2+}$ and pH measurements in the cardiac cycle by $^{19}F$ NMR, *Proc. Natl. Acad. Sci. USA* **85**:9017–9021.

Komorosky, R. A., Newton, J., Walker, E., Cardwell, D., and Chang, C., 1990, In vivo NMR spectroscopy of lithium-7 in humans, *Magn. Reson. Med.* **15**:347–356.

Kudo, Y., and Ogura, A., 1986, Glutamate-induced increase in intra-cellular $Ca^{2+}$ concentration in isolated hippocampal neurones, *Br. J. Pharmacol.* **89**:191–198.

Kuki, S., Suzuki, E., Watari, H., Takami, H., Matsuda, H., and Kawashima, Y., 1990, Potassium-39 nuclear magnetic resonance observation of intracellular potassium without chemical shift reagents during metabolic inhibition in the isolated perfused rat heart, *Circ. Res.* **67**:401–405.

Lazarewicz, J. W., Kanje, M., Sellström, A., and Hamberger, A., 1977, Calcium fluxes in cultured and bulk isolated neuronal and glial cells, *J. Neurochem.* **29**:495–502.

Levy, L. A., Murphy, E., and London, R. E., 1987, Synthesis and characterisation of $^{19}F$ NMR chelators for measurement of cytosolic free calcium, *Am. J. Physiol.* **252**:C441–C449.

Levy, L. A., Murphy, E., Raju, B., and London, R. E., 1988, Measurement of cytosolic free magnesium ion concentration by $^{19}F$-NMR, *Biochemistry* **27**:4041–4048.

Long, G. L., Rosen, J. F., and Schanne, F. A. X., 1994, Lead activation of protein kinase C from rat brain, *J. Biol. Chem.* **269**:834–837.

Lyon, R. C., Pekar, J., Moonen, C. T., and McLaughlin, A. C., 1991, Double-quantum surface-coil NMR studies of sodium and potassium in the rat brain, *Magn. Reson. Med.* **18**:80–92.

McIlwain, H., and Bachelard, H. S., 1985, "Biochemistry and the Central Nervous System," 5th ed., Churchill Livingstone, Edinburgh.

Metcalfe, J. C., Hesketh, T. R., and Smith, G. A., 1985, Free cytosolic $Ca^{2+}$ measurements with fluorine labelled indicators using $^{19}F$ NMR, *Cell Calcium* **6**:183–195.

Mitani, A., Yanase, H., Sakai, K., Wake, Y., and Kataoka, K., 1993, Origin of intracellular $Ca^{2+}$ elevation induced by in vitro ischemia-like condition in hippocampal slices, *Brain Res.* **601**:103–110.

Moon, R. B., and Richards, J. H., 1973, Determination of intracellular pH by $^{31}P$ magnetic resonance, *J. Biol. Chem.* **248**:7276–7278.

Morris, P. G., 1988, NMR spectroscopy in living systems, *Annu. Rep. NMR Spectrosc.* **20**:1–60.

Oberhaensli, R. D., Hilton-Jones, D., Bore, P. J., Hands, I. J., Rampling, R. P., and Radda, G. K., 1986, Biochemical investigations of human tumors in vivo with phosphorus-31 magnetic resonance spectroscopy, *Lancet* **ii**:8–11.

Oberhaensli, R. D., Galloway, G. F., Hilton-Jones, D., Bore, P. J., Styles, P., Rajagopalan, B., Taylor, D. J., and Radda, G. K., 1987, The study of human organs by phosphorous-31 topical magnetic resonance spectroscopy, *Br. J. Radiol.* **60**:367–373.

Obrenovitch, T. P., Garofalo, O., Harris, R. J., Bordi, L., Ono, M., Momma, F., Bachelard, H. S., and Symon, L., 1988, Brain tissue concentrations of ATP, phosphocreatine, lactate, and tissue pH in relation to reduced cerebral blood flow following experimental acute middle cerebral artery occlusion, *J. Cereb. Blood Flow Metab.* **8**:866–874.

Paschen, W., Djuricic, B., Mies, G., Schmidt-Kasner, R., and Linn, F., 1987, Lactate and pH in the brain: Association and dissociation in different pathophysiological states, *J. Neurochem.* **48**:154–159.

Payne, G. S., Seymour, A.-M. L., Styles, P., and Radda, G. K., 1990, Multiple quantum filtered $^{23}Na$ NMR spectroscopy in the perfused heart, *NMR Biomed.* **3**:139–146.

Pekar, J., Renshaw, F. P., and Leigh, J. S., 1987, Selective detection of intracellular sodium by coherence-transfer NMR, *J. Magn. Reson.* **72**:159–161.

Perney, T. M., Dinerstein, R. J., and Miller, R. J., 1984, Depolarization-induced increases in intracellular free calcium detected in single cultured neuronal cells, *Neurosci. Lett.* **51**:165–170.

Petroff, O. A. C., and Prichard, J. W., 1987, Detection of increased cerebral intracellular free magnesium during hypoglycemia by phosphorus magnetic resonance spectroscopy, in "Abstracts of the Society for Magnetic Resonance in Medicine," p. 527, Society for Magnetic Resonance in Medicine, New York.

Portzehl, H., Caldwell, P. C., and Rüegg, J. C., 1964, The dependence of contraction and relaxation of muscle fibres from the crab *Maia squinado* on the internal concentration of free calcium ions, *Biochim. Biophys. Acta* **79**:581–591.

Prichard, J. W., Alger, J. R., Behar, K. L., Petroff, O. A. C., and Shulman, R. G., 1983, Cerebral metabolic studies in vivo by $^{31}P$ NMR, *Proc. Natl. Acad. Sci. USA* **80**:2748–2751.

Rashid, S. A., Adam, W. R., Craik, D. J., Shehan, B. P., and Wellard, R. M., 1991, Factors affecting $^{39}K$ NMR detectability in rat tissue, *Magn. Reson. Med.* **17**:213–224.

Renshaw, P. F., and Wicklund, S., 1988, In vivo measurement of lithium in man by nuclear magnetic resonance spectroscopy, *Biol. Psychiatry* **23**:465–475.

Reznikoff, P., and Chambers, R., 1927, Microsurgical studies in cell physiology. III. The action of $CO_2$ and some salts of Na, Ca, and K on the protoplasm of *Amoeba dubia*, *J. Gen. Physiol.* **10**:731–738.

Richards, C. D., Metcalfe, J. C., Smith, G. A., and Hesketh, R. T., 1984, Changes in free-calcium levels and pH in synaptosomes during transmitter release, *Biochim. Biophys. Acta* **803**:215–220.

Robitaille, P. M., and Jiang, Z., 1992, New calcium-sensitive ligand for nuclear magnetic resonance spectroscopy, *Biochemistry* **31**:12585–12591.

Rose, J. A., 1968, The state of magnesium in cells as estimated from the adenylate kinase equilibrium, *Proc. Natl. Acad. Sci. USA* **61**:1079–1086.

Schanne, F. A., Dowd, T. L., Gupta, R. K., and Rosen, J. F., 1990, Development of $^{19}$F NMR for measurement of $[Ca^{2+}]_i$ and $[Pb^{2+}]_i$ in cultured osteoblastic bone cells, *Environ. Health Perspect.* **84**:99–106.

Schnall, M. D., Yoshizaki, K., Chance, B., and Leigh, J. S., 1988, Triple nuclear NMR studies of cerebral metabolism during generalized seizure, *Magn. Reson. Med.* **6**:15–23.

Schou, M., 1968, Lithium in psychiatric therapy and prophylaxis, *J. Psychiatry Res.* **6**:67–95.

Shulman, R. G., Brown, T. R., Ugurbil, K., Ogawa, S., Cohen, S. M., and den Hollander, J. A., 1979, Cellular applications of $^{31}$P and $^{13}$C nuclear magnetic resonance, *Science* **205**:160–166.

Slonezewski, J. L., Rosen, B. P., Alger, J. R., and Macnab, R. M., 1981, pH homeostasis in *Escherichia coli:* Measurement by $^{31}$P nuclear magnetic resonance of methylphosphonate and phosphate, *Proc. Natl. Acad. Sci. USA* **78**:6271–6275.

Smith, G. A., and Bachelard, H. S., 1995, The measurement of intracellular calcium by $^{19}$F-NMR, *in* "Methods in Neurosciences" Vol. 27, (J. Kraicer and S. J. Dixon, eds.), Academic Press, London.

Smith, G. A., Hesketh, T. R., Metcalfe, J. C., Feeney, J, and Morris, P. C., 1983, Intracellualr calcium measurements by $^{19}$F NMR of fluorine labeled chelators, *Proc. Natl. Acad. Sci. USA* **80**:7178–7182.

Smith, G. A., Morris, P. G., Hesketh, R. T., and Metcalfe, J. C., 1986, Design of an indicator of intracellular free $Na^+$ concentration using $^{19}$F NMR, *Biochim. Biophys. Acta* **899**:72–80.

Song, S.-K., Hotchkiss, R. S., Neil, J., Morris, P. E., Hsu, C. Y., and Ackerman, J. J. H., 1995, Determination of intracellular calcium *in vivo via* fluorine-19 nuclear magnetic resonance spectroscopy, *Am. J. Physiol.* **269**:C318–322.

Sonnewald, U., Westergaard, N., Krane, J., Unsgard, G., Petersen, S. B., and Schousboe, A., 1991, First direct demonstration of preferential release of citrate from astrocytes using [$^{13}$C]NMR spectroscopy of cultured neurons and astrocytes, *Neurosci. Lett.* **128**:235-239.

Speksnijder, J. E., Miller, A. L., Weisenseel, M. H., Chen, T. H., and Jaffe, L. T., 1989, Calcium buffer injections block fucoid egg development by facilitating calcium diffusion. *Proc. Natl. Acad. Sci. USA* **86**:6607–6611.

Steward, M. C., Seo, Y., Murakami, M., and Watari, H., 1991, NMR relaxation characteristics of rubidium-87 in perfused rat salivary glands, *Proc. R. Soc. London [Biol.]* **243**:115–120.

Syme, P. D., Dixon, R. M., Aronson, J. K., Grahame-Smith, D. G., and Radda, G. K., 1990, Evidence for increased *in vivo* sodium–potassium pump activity and potassium efflux in skeletal muscle of spontaneously hypertensive rats, *J. Hypertens.* **8**:1161–1166.

Thoma, W. J., Steiert, J. G., Crawford, R. L., and Ugurbil, K., 1986, pH measurements by $^{31}$P NMR in bacterial suspensions using phenyl phosphonate as a probe, *Biochem. Biophys. Res. Commun.* **138**:1106–1109.

Tsien, R. Y., 1980, New calcium indicators and buffers with high selectivity against magnesium and protons: Design, synthesis and properties of prototype structures, *Biochemistry* **19**:2396–2404.

Tsien, R. Y., 1981, A non-disruptive technique for loading calcium buffers and indicators into cells, *Nature (London)* **290**:527–528.

Tsien, R. Y., 1983, Intracellular measurements of ion activities, *Annu. Rev. Biophys. Bioeng.* **12**:91–116.

van Echteld, C. J., Kirkels, J. H., Eijelshoven, M. H., van der Meer, P., and Ruigrok, T. J., 1991, Intracellular sodium during ischemia and calcium-free perfusion: a 23Na NMR study, *J. Mol. Cell. Cardiol.* **23**:297–307.

Veloso, D., Guynn, R. W., Oskarsson, M., and Veech, R. L., 1973, The concentration of free and bound magnesium in rat tissues, *J. Biol. Chem.* **248**:4811–4819.

Vink, R., Yum, S. W., Lemke, M., Demediuk, P., and Faden, A. L., 1989, Traumatic spinal cord injury in rabbits decreases intracellular free magnesium concentration as measured by $^{31}$P MRS, *Brain Res.* **490:**144–147.

Watts, D. C., 1973, Creatine kinase (adenosine 5'-triphosphate-creatine phosphotransferase), *in* "The Enzymes," Vol. 8, 3rd ed. (P. D. Boyer, ed.), pp. 428–431, Academic Press, New York.

Weiss, J. H., Hartley, D. M., Koh, J.-Y., and Choi, D. W., 1993, AMPA receptor activation potentiates zinc neurotoxicity, *Neuron* **10:**43–49.

Wellard, R. M., Shehan, B. P., Adam, W. R., and Craik, D. J., 1993, NMR measurement of $^{39}$K detectability and relaxation constants in rat tissues, *Magn. Reson. Med.* **29:**68–76.

Westbrook, G. L., and Mayer, M. L., 1987, Micromolar concentrations of $Zn^{2+}$ antagonize NMDA and GABA responses of hippocampal neurons, *Nature (London)* **328:**640–643.

Yue, G. M., Deuel, R., Schickner, D. M., Sherman, W. R., and Ackerman, J. J. H., 1984, 2-Deoxyglucose metabolism in the awake rat brain *in vivo:* A P-31 n.m.r. investigation, *in* "Abstracts of the Society for Magnetic Resonance in Medicine," pp. 774-775, Society for Magnetic Resonance in Medicine, New York.

# IN VIVO *NITROGEN MRS STUDIES OF RAT BRAIN METABOLISM*

## *KEIKO KANAMORI and BRIAN D. ROSS*

## SUMMARY

Of the important nitrogen-containing neurotransmitters, glutamate and GABA, as well as their precursor glutamine, are at concentrations sufficient for their detection by two new magnetic resonance spectroscopy (MRS) techniques, $^{15}N$ and $^1H$–$^{15}N$ heteronuclear multiple quantum coherence (HMQC) transfer MRS. A brief outline of these MRS procedures and their application *in vivo* is provided. Using these methods, new values for flux through glutamine synthetase and phosphate-dependent glutaminase are provided for the rat brain *in vivo*. Steady-state kinetic studies using a novel $^{15}N$–$^{14}N$ chase experiment lead us to the conclusion that glutamine synthetase is strongly regulated. Special features of $^1H$–$^{15}N$ HMQC studies include the separate identification of upfield ($H_Z$) and

*KEIKO KANAMORI and BRIAN D. ROSS • Magnetic Resonance Spectroscopy Laboratory, Huntington Medical Research Institutes, Pasadena, California 91105.*

*Magnetic Resonance Spectroscopy and Imaging in Neurochemistry,* Volume 8 of *Advances in Neurochemistry,* edited by Bachelard, Plenum Press, New York, 1997.

downfield ($H_E$) protons of glutamine amide, which have different exchange rates with water protons *in vivo*. In addition, the enhanced sensitivity of $^1H-^{15}N$ HMQC permits enzyme activities to be assayed at near-physiological blood ammonia concentration. Finally, inferences concerning the sources of glutamate and GABA in intact brain are made.

## 1. INTRODUCTION

The majority of recognized neurotransmitters, e.g., glutamate, γ-aminobutyrate (GABA), serotonin, and dopamine, are nitrogen-containing compounds. The rates of synthesis and turnover of these neurotransmitters and their precursors in intact brain are largely unknown. Measurement of the rates under specific *in vivo* conditions will contribute to a clearer understanding of how the rates are regulated in response to neuronal activity in intact brain. Nitrogen magnetic resonance spectroscopy (MRS) is emerging as a promising noninvasive method for measuring these rates in living animals. Nitrogen MRS studies of nonmammalian organisms have been reviewed (Blomberg and Rüterjans, 1983; Kanamori and Roberts, 1983), and more recent studies are cited in comprehensive reviews of nitrogen MRS (Witanowski *et al.*, 1992) and of MRS of living systems (Prior, 1995). This chapter will focus on *in vivo* nitrogen MRS studies of mammalian brain. The chapter will cover (1) studies which demonstrate the feasibility of observing some key brain metabolites by nitrogen MRS and, more important, (2) studies which provide new information on metabolite flux in intact brain. Novel techniques that have been used for study of other tissues and are potentially applicable to *in vivo* brain are also discussed.

## 2. METABOLITE FLUX THROUGH THE GLUTAMINE/GLUTAMATE/GABA CYCLE

### 2.1. Significance for Basic Neurochemical Research

The glutamine/glutamate/GABA cycle (Berl *et al.*, 1962; Balázs and Cremer, 1973; Cooper and Plum, 1987; Kvamme *et al.*, 1988) plays an important role in controlling the levels of the major excitatory neurotransmitter, glutamate, and of the inhibitory neurotransmitter, GABA, in the central nervous system. Glutamine is produced in astrocytes, taken up by neurons, and therein converted to glutamate or GABA. Glutamate released into the synaptic space is recycled by astrocytes. Glutamine synthetase (GS), which is localized in astrocytes, phosphate-activated glutaminase (PAG) and glutamate decarboxylase (GAD), which are located in neurons, and glutamate dehydrogenase (GDH), present in both brain compartments, are key enzymes in the cycle. While the spatial distribution of some metabolites (Ottersen and Storm-Mathisen, 1984; Ottersen *et al.*, 1992) and enzymes (Norenberg and Martinez-Hernandez, 1979; Fagg and Foster, 1983; Aoki *et al.*, 1991) involved in the cycle has been mapped on a subcellular level by immunocytochemical studies, MRS can contribute to a clearer understanding of the dynamic aspects of the cycle through measurement of the rates of key reactions as they occur *in vivo*. *In vivo* rates, which depend on *in situ* concentrations of substrates and regulators, are often substantially lower than the *in vitro* rates measured in brain homogenates with enzyme-saturating concentrations of substrates. *In vivo* rates of GS and PAG were not known previously and are amenable to measurement by noninvasive MRS. This has enabled us to explore one part of the cycle discussed above.

## 2.2. Clinical Significance

Disturbance of cerebral ammonia metabolism is clearly implicated in the etiology of hepatic encephalopathy (HE) (Basile *et al.*, 1991). However, it is controversial whether the major cause is (i) the neurotoxicity of ammonia *per se* (Butterworth, 1991), (ii) its conversion to glutamine (Hawkins *et al.*, 1993), or (iii) the resulting depletion of glutamate (Bradford and Ward, 1975). Because GS catalyzes the formation of glutamine from ammonia and glutamate, *in vivo* GS activity is a direct measure of the rate of ammonia and glutamate removal from astrocytes and of glutamine formation. The rate of glutamate replenishment in neurons depends, among other factors, on how ammonia regulates the activity of PAG in intact brain. Thus, measurement of *in vivo* GS and PAG activities at various brain ammonia concentrations in rat as an animal model of HE and examination of correlation with neurobehavioral symptoms of HE will contribute to a better understanding of the mechanism by which hyperammonemia causes HE.

## 2.3. Relevance of Animal Studies

Rat and human brain glutamine synthetases show >90% homology in amino acid sequence (van de Zande *et al.*, 1990; Mill *et al.*, 1991), and glutamate decarboxylases show 96–97% homology (Bu *et al.*, 1992). The kinetic and regulatory properties, as well as cellular compartmentation within the brain, of rat and human GS (Tate *et al.*, 1972; Deuel *et al.*, 1978; Norenberg and Martinez-Hernandez, 1979; Yamamoto *et al.*, 1987), PAG (Benjamin, 1981; Haser *et al.*, 1985; Svenneby *et al.*, 1986), and GAD (Blindermann *et al.*, 1978; Maitre *et al.*, 1978) are remarkably similar. Hence, the *in vivo* rates of key enzymes of the glutamine/glutamate/GABA cycle measured in rat brain will have important implications for the human brain. Definitive rate determination depends upon knowledge of the fractional isotopic enrichment of the reactants and often requires *in vivo* MRS complemented with *in vitro* data. This is often feasible only in animal studies.

## 3. NITROGEN MRS–CHARACTERISTICS FOR *IN VIVO* STUDIES

While the sensitive $^1$H and $^{31}$P MRS techniques provide valuable information on changes in cerebral metabolite concentrations under normal and pathological conditions (Radda, 1986; Ross *et al.*, 1992; Bachelard and Badar-Goffer, 1993; Kreis *et al.*, 1993; Michaelis *et al.*, 1993; Shulman *et al.*, 1993), the low-natural-abundance nuclei [$^{13}$C (Gruetter *et al.*, 1994) and $^{15}$N] are especially useful for measuring the rates of synthesis and turnover of selected metabolites using isotopically enriched precursors. For detailed descriptions of the general characteristics of nitrogen ($^{15}$N and $^{14}$N) MRS, the reader is referred to excellent reviews by Witanowski *et al.* (1992), Levy and Lichter (1979), and Blomberg and Rüterjans (1983). Here, we outline the characteristics that are important for *in vivo* studies.

### 3.1. $^{15}$N MRS

Nitrogen-15 has nuclear spin quantum number $I = \frac{1}{2}$ and a negative gyromagnetic ratio. Its natural abundance is 0.365%.

#### *3.1.1. Direct $^{15}$N Observation*

*3.1.1.1. Advantages for* in Vivo *Studies* The lone-pair electrons of nitrogen render the MRS parameters very sensitive to changes in molecular environment, resulting in a broad chemical shift range (500–900 ppm). This, combined with the narrow linewidth of $^{15}$N, results in high spectral resolution. Low natural abundance permits selective observation of isotopically enriched metabolites *in vivo*. Owing to the negative gyromagnetic ratio of $^{15}$N, the nuclear Overhauser enhancement factor (NOEF) is negative, with a theoretical maximum value of $-4.93$. This NOEF, when added to the original signal intensity ($+1.0$), results in a maximum signal enhancement of $-3.93$. Although signal nulling resulting from an NOEF of $-1$ can be a problem for some azine-type nitrogens, signal enhancement close to the theoretical maximum of $-3.9$ is often attained for primary amine nitrogens present in neurotransmitters such as glutamate and GABA. The spin–lattice relaxation time, $T_1$, of $\alpha$ nitrogens of neurotransmitter amino acids is reasonably short, of the order of <5 sec, allowing acquisition with a relaxation delay of 2–3 sec in combination with a pulse flip angle of about 30°.

*3.1.1.2. Sensitivity* Despite the low sensitivity of $^{15}$N (1/1000th that of $^1$H and 1/15th that of $^{13}$C for equal number of nuclei at constant field), $^{15}$N-enriched brain metabolites in the 1–10 mM range (including glutamine, glutamate, and GABA) can be observed with a time resolution of 5–30 min by proton-decoupled NOE-enhanced $^{15}$N MRS *in vivo* at 20 MHz for $^{15}$N. This permits measurement of reaction rates on the order of 1–10 μmol g$^{-1}$ hr$^{-1}$. The *in vivo* rates of many key reactions in the glutamine/glutamate/GABA cycle in the brain are of this order of magnitude.

Among methods proposed for increasing the sensitivity of $^{15}$N detection is polarization transfer from coupled protons (Morris and Freeman, 1979). For $^{15}$N coupled to nonexchanging protons, this method can provide a signal intensity enhancement of $\gamma_{^1H}/\gamma_{^{15}N} = 10$, compared to the signal obtained by the conventional pulse-acquire sequence without NOE. Hence, the method is particularly useful for signal enhancement of $^{15}$N with an unfavorable NOE. Among brain metabolites of interest, [5-$^{15}$N]glutamine is amenable to this technique at physiological pH. However, because its signal can be enhanced nearly fourfold by NOE, additional gain by polarization transfer was expected to be small. At pH 7.1, the $^{15}$N signal of [5-$^{15}$N]glutamine in water obtained by the polarization transfer method showed only a twofold gain in signal-to-noise (S/N) ratio over the proton-decoupled, NOE-enhanced signal at 20°C (Kanamori *et al.*, 1991). At 37°C, where the rate of exchange of the amide protons with the water is faster, there was no appreciable gain in sensitivity.

*3.1.1.3. Quantitation* Essential to this work is robust and accurate quantitation of MRS data. We have established a direct calibration method for quantitating $^{15}$N-labeled metabolites in rat brain from the observed *in vivo* peak intensity. The method takes into account the differences in coil efficiency and in $T_1$ and NOE values between the *in vivo* and *in vitro* situations, as described below.

The *in vivo* MRS peak intensity, $S_{iv}$, of $^{15}$N-labeled metabolite is proportional to the number of spins (*n*) involved:

$$S_{iv} = k_{iv}n \tag{1}$$

The proportionality constant for an *in vivo* experiment, $k_{iv}$, depends on experimental conditions such as the radio-frequency (RF) coil efficiency *in vivo*, the relaxation delay employed relative to *in vivo* $T_1$, and the contribution of *in vivo* NOE to peak intensity. We determine $k_{iv}$ experimentally for each metabolite, e.g., [5-$^{15}$N]glutamine, by (a) measuring the *in vivo* peak intensity $S_{iv}$ per unit time and (b) determining the number of spins, *n*, expressed as micromoles of [5-$^{15}$N]glutamine, from measurements on the same brain after the *in vivo* measurement (see below for method). These measurements are performed with at least 7 rats under identical experimental conditions in order to determine the mean ± SEM value for $k_{iv} = S_{iv}/n$.

To measure *n*, the rat is sacrificed immediately after the *in vivo* measurement, and the brain is frozen in liquid nitrogen for preparation of perchloric acid extract. The quantity of [5-$^{15}$N]glutamine in the brain extract is determined by taking a separate $^{15}$N spectrum of the extract. In an extract experiment,

$$S_{ext} = k_{ext}n \qquad (2)$$

The proportionality constant for the extracts, $k_{ext}$, is determined with [5-$^{15}$N]glutamine standards. $S_{ext}$ is measured for known quantities ($n$) of [5-$^{15}$N]glutamine dissolved in unlabeled brain extract (to provide the same environment as for the biologically $^{15}$N-enriched metabolite), and $k_{ext}$ is determined from the slope of $S_{ext}$ plotted against $n$. The quantity of [5-$^{15}$N]glutamine in the biologically $^{15}$N-enriched brain extract is then determined from the observed $S_{ext}$ and $k_{ext}$ (Farrow et al., 1990).

Once the quantity of [5-$^{15}$N]glutamine in the brain is determined by the above method for each of 7 rats, the quantity $n$ and the observed in vivo peak intensity $S_{iv}$ per minute of acquisition are used to determine the mean $\pm$ SEM for $k_{iv} = S_{iv}/n$ in Eq. (1). The unit for $k_{iv}$ is in vivo peak intensity per minute of acquisition per micromole of $^{15}$N-labeled metabolite. For [5-$^{15}$N]glutamine in rat brain, the mean $k_{iv}$ was 69.5 with an SEM of less than 7% for 7 rats. This experimental proportionality constant for in vivo experiments was used in subsequent experiments to convert the observed in vivo peak intensity to micromoles of [5-$^{15}$N]glutamine in the brain, using Eq. (1) (Kanamori et al., 1993; Kanamori and Ross, 1993). It is to be noted that the differences in coil efficiency, relaxation, and NOE values between in vivo and extract measurements are reflected in the different observed values of $k_{iv}$ and $k_{ext}$. The values of $k_{iv}$ and $k_{ext}$, checked every 5–6 months, have been stable for 2 years. Accurate quantitation of brain [5-$^{15}$N]glutamine from the observed in vivo peak intensity is ensured by using the experimentally determined and reproducible value of $k_{iv}$.

### 3.1.2. Indirect Detection of $^{15}$N

$^1$H–$^{15}$N heteronuclear multiple quantum coherence (HMQC) transfer MRS (Bax et al., 1983) and the $^1$H–$^{15}$N heteronuclear spin-echo difference method (Brindle et al., 1984) permit selective, indirect detection of $^{15}$N through spin-coupled protons with $^1$H sensitivity, provided that the protons involved are nonlabile at physiological pH and temperature. Under certain conditions, the latter method also allows fractional $^{15}$N enrichment of the metabolite to be determined (Brindle et al., 1984). Among neurotransmitters and their precursors, [5-$^{15}$N]glutamine and [$^{15}$N$_{indole}$]serotonin are detectable by these methods (Kanamori et al., 1991; Street et al., 1993; Kanamori et al., 1995a). In contrast, the amine protons of glutamate, GABA, and dopamine exchange too rapidly with water protons at physiological pH for $^{15}N$ to be detectable through the nitrogen-bound protons.

## 3.2. $^{14}N$ MRS

Nitrogen-14 has nuclear spin quantum number $I = 1$ and an electric quadrupole moment. Its natural abundance is 99.4%. The sensitivity of $^{14}N$ for the same number of nuclei at constant field is comparable to that of $^{15}N$ (1/1000th that of $^{1}H$). Its advantages are high natural abundance and broad chemical shift range. The major limitation is that quadrupolar relaxation normally results in very broad lines, and hence poor spectral resolution, for most biologically interesting metabolites. Exceptions are molecules with more or less symmetrical electronic environments about $^{14}N$, such as $NH_4^+$, cholines, and betaines. *In vivo* observations of these metabolites in mammalian tissue are described in this chapter.

## 4. NITROGEN MRS STUDIES OF RAT BRAIN METABOLISM

### 4.1. $^{15}N$ MRS Studies of *in Vivo* GS Activity

Glutamine synthetase (GS) catalyzes the reaction

$$NH_3 + \text{glutamate} + ATP \xrightarrow{Mg^{2+} \ (Mn^{2+})} \text{glutamine} + ADP + P_i$$

Measurement of the *in vivo* activity of GS is important because of (a) its key role in the glutamine/glutamate/GABA cycle and (b) recent controversy regarding the mechanism of ammonia toxicity in hepatic encephalopathy, as described earlier.

#### 4.1.1. Direct $^{15}N$ MRS

The first *in vivo* $^{15}N$ MR spectra of brain metabolites were obtained in hyperammonemic rat, an animal model of hepatic encephalopathy (Kanamori *et al.,* 1993; Kanamori and Ross, 1993). Hyperammonemia and $^{15}N$ enrichment were induced by intravenous infusion of $^{15}NH_4^+$ (Farrow *et al.,* 1990) in this well-established model. Figure 1A shows a typical *in vivo* $^{15}N$ spectrum obtained from the head of an anesthetized, $^{15}NH_4^+$-infused rat. The proton-decoupled, NOE-enhanced spectrum was acquired in 20 min at 20 MHz for $^{15}N$ on a GE CSI spectrometer. The peaks for [5-$^{15}N$]glutamine ($-271$ ppm) and [2-$^{15}N$]glutamate/glutamine ($-342$ ppm) arise exclusively from the brain (Kanamori *et al.,* 1993). After 3–4 hr of $^{15}NH_4^+$ infusion, cerebral [5-$^{15}N$]glutamine observed *in vivo* reached a steady-state concentration, as shown in Fig. 1B.

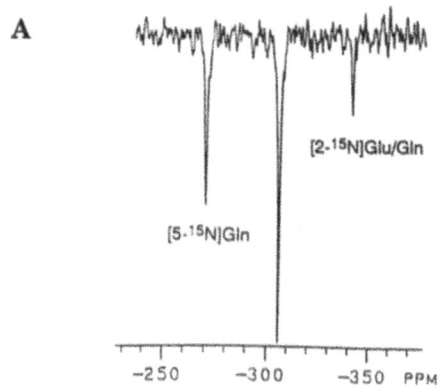

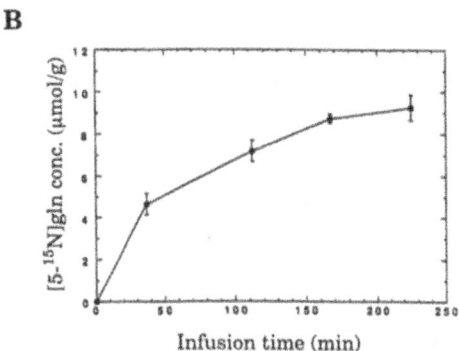

Infusion time (min)

FIGURE 1. (A) An *in vivo* $^{15}N$ spectrum obtained from the head of an anesthetized rat at 20.27 Mz for $^{15}N$ with proton decoupling in 20 min of acquisition. The peaks for $[5-^{15}N]$glutamine ($-271$ ppm) and $[2-^{15}N]$glutamate/glutamine ($-342$ ppm) arise exclusively from the brain (Kanamori *et al.*, 1993). (B) Increase in cerebral $[5-^{15}N]$glutamine pool (mean $\pm$ SEM for 4 rats) as observed *in vivo* during 3.5 hr of $^{15}NH_4^+$ infusion. [Reproduced from Kanamori *et al.* (1993) and Kanamori and Ross (1993).]

*4.1.1.1. Steady-State Rate Measurement* The rate of glutamine synthesis and utilization at steady state was measured *in vivo* in the brains of hyperammonemic rats by $^{15}N$ MRS, using an isotope chase method, in combination with biochemical techniques (Kanamori and Ross, 1993). Rats were given an intravenous $^{15}NH_4^+$ infusion at the rate of 4.8 mmol/hr per kilogram of weight for 3.5 hr, followed by $^{14}NH_4^+$ infusion at the same rate for an additional 5.1 hr (chase period). During the chase period, blood ammonia (0.61 $\mu$mol/g), brain ammonia (2.9 $\mu$mol/g), glutamate (9.4 $\mu$mol/g), and glutamine ($^{15}N + ^{14}N$; 14.4 $\mu$mol/g)

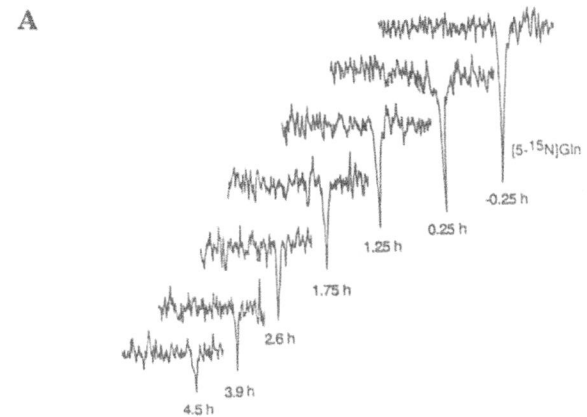

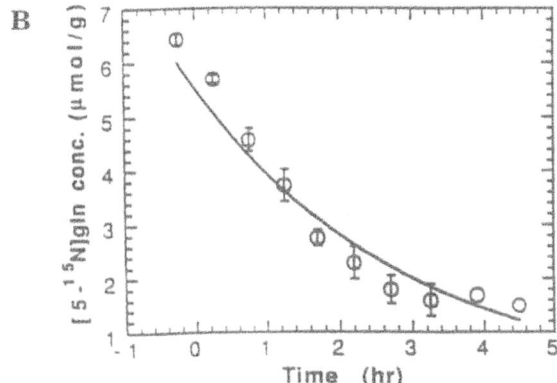

FIGURE 2. (A) $^{15}N$ MR spectra of [5-$^{15}N$]glutamine in rat brain observed *in vivo* during chase by $^{14}N$. Each spectrum was acquired in 30 min. (B) Decrease in cerebral [5-$^{15}N$]glutamine pool during the chase, which was started at $t = 0$. The values are means $\pm$ SEM for 5 rats. [Reproduced from Kanamori and Ross (1993).]

were at steady state. Figure 2A shows changes in cerebral [5-$^{15}N$]glutamine peak intensity observed *in vivo* by MRS during the chase period. Figure 2B shows changes in cerebral [5-$^{15}N$]glutamine concentration ($\mu$mol/g of brain), determined from the *in vivo* peak intensity by the quantitation method described above, and averaged for 5 rats. The *in vivo* rate of glutamine synthesis, $v_1$, which is equal to the rate of glutamine utilization, $v_2$, at steady state, was determined from the relation

$$v_1 = v_2 = (dB^*/dt)/(s_A - s_B) \tag{3}$$

where $B^*$ is brain [5-$^{15}$N]glutamine concentration at time $t$ (Fig. 2B), $s_A$ is the $^{15}$N enrichment of the substrate ammonia and $s_B$ is the $^{15}$N enrichment of brain glutamine determined from $B^*$ and the total steady-state brain glutamine concentration (14.4 μmol/g). To estimate $s_A$, the $^{15}$N enrichments of both blood and brain ammonia were measured by gas chromatography–mass spectrometry (GC-MS). The rate of glutamine synthesis, determined from the observed values using Eq. (3), was 4.8 ± 1.1 μmol/hr per gram of brain if $^{15}$N enrichment of ammonia at the site of GS in astrocytes is equal to that of blood-borne ammonia and 13.0 ± 3.9 μmol/hr per gram if it is equal to that measured for the whole brain. The actual rate is probably closer to the lower estimate because GS is most abundant in astrocytic pericapillary end feet that intervene between capillaries and neurons (Norenberg and Martinez-Hernandez, 1979) and thus act as a metabolic trap for blood-borne ammonia diffusing into the brain (Cooper and Plum, 1987).

*4.1.1.2. Regulation of GS Activity In Vivo* The observed GS activity *in vivo* in the brain of the hyperammonemic rat was 2–5% of the reported optimum activity *in vitro* (Patel *et al.,* 1983; Butterworth *et al.,* 1988) measured at enzyme-saturating concentrations of all substrates. This suggested that the rate of glutamine synthesis is strongly regulated in the intact brain. [For evidence showing that (a) the observed low rate is not due to pentobarbital anesthesia and (b) glutamine efflux from the brain did not occur, see Kanamori and Ross (1993)]. Major factors regulating the rate were inferred from (a) the observed *in vivo* activity, (b) known kinetic and regulatory properties of GS *in vitro,* and (c) the estimated *in situ* concentrations of substrates and cofactors. Rat brain and liver GS have virtually identical amino acid sequences (Mill *et al.,* 1991) and hence kinetic properties. With a $K_m$ of 0.3 mM for ammonia (Deuel *et al.,* 1978), GS was likely to be saturated with ammonia under the hyperammonemic condition. GS has a $K_m$ of 2.3 mM for ATP and requires $Mg^{2+}$ or $Mn^{2+}$ as a cofactor (Tate *et al.,* 1972). Since the average ATP concentration in rat brain is 2.5 mM (Veech *et al.,* 1973), it is likely that *in vivo* GS activity was kinetically limited by *in situ* ATP concentration. A particularly interesting implication of the observed low *in vivo* activity is that *in situ* glutamate concentration also appears to be limiting the rate. This idea is consistent with the recent immunocytochemical evidence that, in rat cerebellum, the glutamate/glutamine ratio is approximately 0.2 in the glia and 2–4.5 in glutamatergic nerve terminals (Ottersen *et al.,* 1992). Thus, glial glutamate concentration may be as low as 3 ± 1 mM. If this is true in the intact cerebral cortex too, *in vivo* activity of GS, with a $K_m$ of 5 mM for glutamate (Deuel *et al.,* 1978), may well be substantially limited by the low *in situ* glutamate concentration. This "feed-forward" regulation of flux through GS could have important consequences for protection of the brain from neurotoxic effects of excess glutamate.

*4.1.1.3. Ammonia Removal Rate and Symptoms of HE* The observed GS activity *in vivo* represents the rate of ammonia removal from the glial compartment under our experimental conditions. At an ammonia infusion rate of 4.8 mmol/hr per kilogram of weight and an ammonia removal rate of 4.8–13.0 $\mu$mol/hr per gram in the brain, the anesthetized rat showed occasional myoclonus and hyperventilation after 2 hr of infusion but was otherwise in good physiological condition for at least 8 hr. This result suggested that at this ammonia removal rate, mild symptoms of HE (stages I–III) occur, but coma (stage IV) can be avoided. To evaluate the hypothesis that ammonia is in itself neurotoxic, it is informative to use these noninvasive techniques to examine correlation of the *in vivo* rate of ammonia removal with the onset of stages I-IV of HE in hyperammonemic rats, (Kanamori *et al.*, 1996).

*4.1.1.4. [15]N versus [13]C or [13]N Labeling* Subsequent to the publication of our [15]N MRS study, Mason *et al.* (1995) reported measurement of the *in vivo* rate of glutamate–glutamine exchange in normal human brain by [13]C MRS after intravenous infusion of [1-[13]C]glucose. The rate was reported to be within the range of 0.139–3.094 $\mu$mol/min per gram, corresponding to 8.3–186 $\mu$mol/hr per gram (95% confidence limits). Their lowest estimate is close to the rate observed in our hyperammonemic rat brain. In the [13]C study, which showed excellent time resolution for observation of brain [13]C-enriched metabolites (Gruetter *et al.*, 1994), the major focus was measurement of the *in vivo* rates of TCA cycle and $\alpha$-ketoglutarate–glutamate exchange, for which more precise values were obtained. The large range for the glutamate–glutamine exchange rate was attributed to the fact that the time course of [4-[13]C]glutamine formation followed closely the time course of [4-[13]C]glutamate formation. With [15]N labeling, the infusate [15]$NH_4^+$ is the substrate for glutamine synthesis and labels glutamine 5-N directly. This permits a more precise measurement of the *in vivo* rate than with [13]C labeling, as numerous metabolic pathways intervene between the infusate [1-[13]C]glucose and [4-[13]C]glutamate.

Another relevant study is one in which the brain ammonia utilization rate in human subjects was measured by [13]N positron-emission tomography (PET) after intravenous (i.v.) injection of [13]$NH_3$. The rate, estimated from the blood [13]$NH_3$ clearance rate and the fraction of [13]N observed in the brain compared to the whole body, was reported to be about 1.4 $\mu$mol/hr per gram of brain in normal subjects (arterial blood ammonia concentration = 0.1 mM), and 2.3 $\mu$mol/hr per gram in HE patients (ammonia concentration = 0.15 mM) (Lockwood *et al.*, 1979). The rate observed in HE patients is slightly lower than, but of the same order of magnitude as, the rate of ammonia incorporation into glutamine observed by [15]N MRS in our hyperammonemic rat. The advantage of [13]N is the high sensitivity, which permits study of ammonia utilization rate at the physiological concentration of blood ammonia. On the other hand, the short half-life of [13]N (10 min) necessitates on-site production of [13]$NH_3$, which poses considerable practical difficulties and may prevent the isotopic steady-state condition from being achieved. While PET measures the total quantity of $^{13}N$ found in the brain, the advantage of $^{15}N$ MRS is its high chemical specificity, which permits *in vivo* monitoring of $^{15}NH_4^+$ incorporation into specific metabolites such as [5-$^{15}N$]glutamine and [2-$^{15}N$]glutamate.

## 4.1.2. Indirect $^{15}N$ Observation

*4.1.2.1. Phase-Cycled $^1H-^{15}N$ HMQC* For measurement of *in vivo* GS activity at closer to physiological concentrations of brain ammonia, $^1H-^{15}N$ HMQC MRS provides a sensitive method of detecting brain glutamine. Figure 3A shows a $^1H$ spectrum of [5-$^{15}N$]glutamine in water (pH 7.1) acquired by $^1H-$ $^{15}N$ HMQC with $^{15}N$ decoupling. The two $^{15}N$-bound amide protons of glutamine, which are chemically nonequivalent and have $^1H$ chemical shifts of 6.89 and 7.61 ppm, are selectively observed, with suppression of carbon-bound proton peaks. The upfield amide proton ($H_Z$) exchanges more slowly with the water proton than $H_E$ and hence has a sharper linewidth and higher S/N ratio. The HMQC method permitted selective observation of [5-$^{15}N$]glutamine amide protons in 2 min of acquisition in isolated brain of $^{15}NH_4^+$-infused rat at 500 MHz (Kanamori *et al.*, 1991). Recently, the amide protons of [5-$^{15}N$]glutamine were selectively observed *in vivo* in the brains of anesthetized, spontaneously breathing rats, after intravenous $^{15}NH_4^+$ infusion and localized shimming on brain water, at 200 MHz for $^1H$ (Kanamori *et al.*, 1995a). Figure 3B–D shows *in vivo* spectra obtained at various brain [5-$^{15}N$]glutamine concentrations. The $H_Z$ peak was observed *in vivo* in 2 min of acquisition, with $H_E$ also detectable by 12.5 min, at a brain [5-$^{15}N$]glutamine concentration of 7.7 μmol/g (Fig. 3B). $^1H$ signals not coupled to $^{15}N$ were suppressed by phase cycling. At brain [5-$^{15}N$]glutamine levels of 4.35 and 2.0 μmol/g (corresponding to 38–82% $^{15}N$ enrichment of the physiological level of glutamine), the $H_Z$ peak was observed in 8.5 min (Fig. 3C) and 25 min (Fig. 3D), respectively. The $H_Z$ peak intensity was proportional to brain [5-$^{15}N$]glutamine concentration, and the demonstrated time resolution of MRS detection is encouraging. Recently, we measured the dependence of *in vivo* GS activity (*v*) on blood ammonia concentration (*s*), using $^1H-$ $^{15}N$ HMQC (Kanamori et al. 1996). The linear increase of $1/v$ with $1/s$ permitted estimation of *in vivo* GS activity at physiological blood ammonia level to be 0.4–2.1 μmol/hr per gram (Kanamori *et al.*, 1995b). The result shows that $^1H-^{15}N$ HMQC is useful for kinetic study of cerebral glutamine synthesis in living animals under near-physiological conditions.

*4.1.2.2. Gradient-Enhanced $^1H-^{15}N$ HMQC* An alternative to phase-cycled HMQC is gradient-enhanced $^1H-^{15}N$ HMQC (Hurd and John, 1991; Ruiz-Cabello *et al.*, 1992), which has not yet been tested *in vivo*. The advantage of phase-cycled HMQC is that both zero and double-quantum transitions are selected so

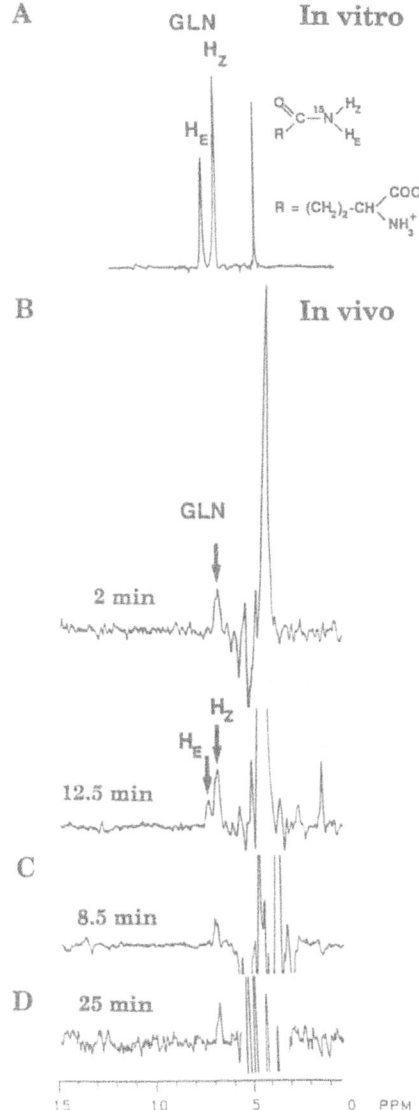

FIGURE 3. $^1$H–$^{15}$N HMQC spectra of [5-$^{15}$N]glutamine at 200 Hz for $^1$H acquired with $^{15}$N decoupling. (A) [5-$^{15}$N]glutamine (430 μmol) in water at pH 7.1. The amide protons of [5-$^{15}$N]glutamine are selectively observed at 6.89 and 7.61 ppm and are assigned to H$_Z$ and H$_E$, respectively (see the structure). (B)–(D) *In vivo* $^1$H–$^{15}$N HMQC spectra of brain [5-$^{15}$N]glutamine in anesthetized, $^{15}$NH$_4$$^+$-infused and spontaneously breathing rats. (B) Spectra obtained from a rat with brain [5-$^{15}$N]glutamine concentration of 7.7 μmol/g, acquired in 2 min and 12.5 min. The amide protons are selectively observed. (C) A spectrum from a rat with brain [5-$^{15}$N]glutamine concentration of 4.35 μmol/g, acquired in 8.5 min. (D) A spectrum from a rat with [5-$^{15}$N]glutamine concentration of 2.0 μmol/g, acquired in 25 min. [Reproduced from Kanamori *et al.* (1995a).]

that the intensity of the final $^1$H signal observed per unit time is theoretically twice that observed by the gradient-enhanced HMQC, which selects one order of transition (Bax *et al.*, 1980). On the other hand, the gradient-enhanced method achieves much better water suppression, as demonstrated for detection of $^{15}$N-bound urea and glutamine protons in tissue biopsies (Freeman *et al.*, 1993).

Whether the excellent water suppression and hence dynamic range attainable by gradient-enhanced HMQC can offset the intrinsic twofold lower sensitivity for observation of $^{15}$N-bound protons in intact animals remains to be demonstrated.

*4.1.2.3. Heteronuclear $^1H$–$^{15}N$ Spin-Echo MRS* This method, described by Brindle *et al* (1984) has been used to monitor label redistribution among extracellular metabolites in cultures of mammalian cells incubated with $^{15}$N-labeled substrates (Street *et al.,* 1993). Detection of [$^{15}$N]glutamate and [$^{15}$N]alanine through β-proton and $^{15}$N enrichment measurements were performed at acidic pH to minimize overlap with adjacent resonances in the spectra of media extracts. Hence, the applicability of this technique to *in vivo* studies is open to question. $^{15}$NH$_4^+$-derived [5-$^{15}$N]glutamine synthesized in the cell and released to the medium was detected through the directly bonded amide proton at neutral pH by the spin-echo method. If fractional $^{15}$N enrichment of glutamine amide nitrogen at physiological pH can also be determined *in vivo,* this would be an advantage of the spin-echo method over HMQC. However, the short spin–spin relaxation time ($T_2$) and broad linewidth of most $^{14}$N-bound protons due to $J[^{14}N$–$^1H]$ coupling and the rapid quadrupolar relaxation of $^{14}$N are likely to make their detection difficult, and the feasibility of $^{15}$N-enrichment determination through nitrogen-bound protons *in vivo* by the spin-echo method remains to be demonstrated.

## 4.2. $^{15}$N MRS Study of *in Vivo* PAG Activity

Phosphate-activated glutaminase (PAG) catalyzes the reaction

$$\text{glutamine} + H_2O \rightarrow \text{glutamate} + NH_4^+$$

PAG (glutaminase I), localized on the inner mitochondrial membrane, is predominantly a neuronal enzyme, particularly enriched in glutamatergic nerve terminals (Bradford *et al.,* 1978). PAG activity in intact brain is of interest because the enzyme is thought to play an important role in replenishing the neuronal pool of glutamate and hence of GABA.

The rate of PAG reaction is expected to be lower than the total rate of glutamine utilization, $v_2$ in Eq. (3), which represents the combined effect of (a) the rates of PAG (glutaminase I) and glutaminase II (glutamine aminotransferase + ω-amidase) reactions, (b) the rates of anabolic amidotransferase reactions in which glutamine 5-N is transferred to purine and pyrimidine nitrogens with concomitant release of glutamate, and (c) the rate of glutamine incorporation into proteins. Among these reactions, only the glutaminase reactions release free ammonia. In rat brain, PAG is the predominant glutaminase pathway, with glutaminase II playing a negligibly small role (Kvamme *et al.,* 1988). $^{15}$N MRS is useful for measuring the rate of PAG reaction in intact brain because changes in

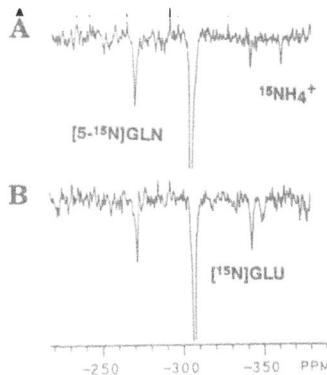

*FIGURE 4. In vivo* $^{15}$N MR spectra of rat brain showing $^{15}$NH$_4^+$ released by PAG-catalyzed hydrolysis of [5-$^{15}$N]glutamine (A) and increase in [$^{15}$N]glutamate peak due to assimilation of $^{15}$NH$_4^+$ by the glutamate dehydrogenase (GDH) reaction (B). [Reproduced from Kanamori and Ross (1995).]

brain [5-$^{15}$N]glutamine concentration as well as PAG-catalyzed production of $^{15}$NH$_4^+$ can be monitored *in vivo*.

The *in vivo* activity of PAG was measured in the brain of hyperammonemic rat by $^{15}$N MRS (Kanamori and Ross, 1995). Brain glutamine was $^{15}$N-enriched by intravenous infusion of $^{15}$NH$_4^+$ until the concentration of [5-$^{15}$N]glutamine reached 6.1 μmol/g. Further glutamine synthesis was inhibited by intraperitoneal injection of L-methionine DL-sulfoximine, an inhibitor of glutamine synthetase, and the infusate was changed to $^{14}$NH$_4^+$ during observation of decrease in brain [5-$^{15}$N]glutamine due to PAG and other glutamine utilization pathways. Progressive decrease in brain [5-$^{15}$N]glutamine, PAG-catalyzed production of $^{15}$NH$_4^+$, and its subsequent assimilation into [2-$^{15}$N]glutamate by glutamate dehydrogenase were monitored *in vivo* by $^{15}$N MRS, as shown in Fig. 4A and B. Brain [5-$^{15}$N]glutamine ($^{15}$N enrichment of 0.35–0.50) decreased at a rate of 1.2 μmol/hr per gram of brain. The *in vivo* PAG activity, determined from the observed rate and the quantity of $^{15}$NH$_4^+$ produced and subsequently assimilated into glutamate and aspartate, was 0.9–1.3 μmol/hr per gram of brain.

### 4.2.1. Regulation of PAG Activity In Vivo

The observed *in vivo* activity is less than 1.1% of the reported *in vitro* activity measured in rat brain homogenate at 10 mM concentrations of the substrate glutamine and of the activator P$_i$ (Ward and Bradford, 1979). This result shows that PAG activity is strongly regulated in the intact brain. Studies with purified or particulate forms of rat brain PAG show that the activity depends on the levels of glutamine, of the activator P$_i$, and of inhibitors ammonia and glutamate (Benjamin, 1981; Haser *et al.*, 1985). With a $K_m$ of 1.6 mM for glutamine (Benjamin, 1981), PAG was probably saturated with the substrate under our experimental conditions. With the estimated *in situ* free P$_i$ concentration of 1.3–2.1 $\mu$mol/g (Veech et al., 1973; Behar et al., 1985), PAG activity is expected to be 10–20% of that measured at 10 mM $P_i$. In brain homogenates, PAG activity decreases to 63–69% of that of the control on addition of 1–2 mM $NH_4Cl$ and to 20% on addition of 1 mM glutamate (Benjamin, 1981). At the observed brain ammonia concentration of 1.4 $\mu$mol/g and glutamate level of 7–10 $\mu$mol/g, similar reduction in activity is expected. These considerations strongly suggest that, in the intact brain of hyperammonemic rat, PAG activity is maintained at a low level by suboptimal in situ concentration of free $P_i$ and strong inhibitory effect of glutamate and ammonia.

### 4.2.2. PAG as Provider of Glutamate

The low in vivo PAG activity observed in hyperammonemic rat brain is consistent with the concept that ammonia-mediated inhibition of PAG decreases the rate of replenishment of the neuronal glutamate pool and thereby contributes to the pathogenesis of HE (Bradford and Ward, 1975). A more definitive conclusion on the effect of ammonia per se on the in vivo PAG activity must await corresponding measurements in normal brain.

Studies with brain slices and intact animals using radioisotope-labeled precursors or inhibitors of GS and of GABA transaminase strongly suggested that glutamine is an important (though by no means the sole) precursor of the glutamate that is subsequently converted to GABA by glutamate decarboxylase (GAD) (Ward et al., 1983; Szerb, 1984; Paulsen et al., 1988; Battaglioli and Martin, 1991). If so, glutamine synthesized in astrocytes and transported to neurons must be converted to glutamate at a rate sufficient to support GABA synthesis. Hence, a new approach to investigating this question is to compare the PAG activity in intact brain with the GAD activity. The rate of GABA synthesis in rat brain, estimated from the rate of accumulation on inhibition of GABA transaminase, is 2.6 ± 1.5 $\mu$mol/hr per gram of brain (Fonnum and Walberg, 1973; Casu and Gale, 1981; Chapman and Evans, 1983; Paulsen et al., 1988). The in vivo rate of PAG in intact brain of hyperammonemic rat was 0.9–1.3 $\mu$mol/hr per gram of brain and is expected to be 30–40% higher in normal brain. Comparison with the GAD activity suggests that PAG activity is not high enough for it to be the sole provider of glutamate for GABA synthesis, especially if the level of PAG in GABAergic neurons is much lower than in glutamatergic neurons, as suggested by immunocytochemical studies (Kaneko et al., 1987; Aoki et al., 1991). Thus, comparison of in vivo activities measured in intact brain suggests, as do studies using different approaches (Ward et al., 1983; Szerb, 1984; Paulsen et al., 1988; Battaglioli and Martin, 1991), that PAG-catalyzed hydrolysis of glutamine cannot be the sole provider of glutamate used for GABA synthesis.

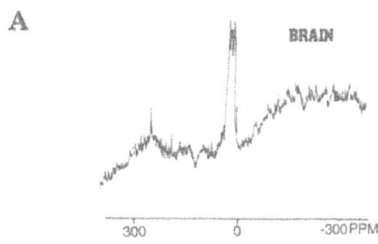

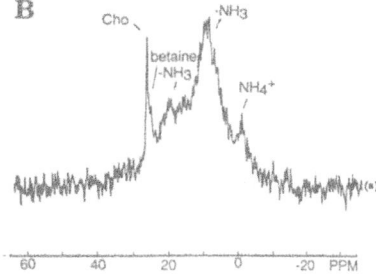

FIGURE 5. (A) *In vivo* $^{14}$N MR spectrum of rat brain obtained with a surface coil at 26 MHz for $^{14}$N (8.4 T), acquired in 44 min. [Reproduced with permission, from Balaban and Knepper (1983).] (B) *In vivo* $^{14}$N MR spectrum of tumor (radiation-induced fibrosarcoma) implanted subcutaneously into the flank of a mouse, obtained at 36.13 MHz for $^{14}$N in 1.1 hr of acquisition. [Reproduced, with permission, from Gamcsik *et al.* (1991).]

## 4.3. $^{14}$N MRS Studies

An *in vivo* $^{14}$N MR spectrum of rat brain, obtained with a surface coil at 26 MHz for $^{14}$N (8.4 T) was reported by Balaban and Knepper (1983). As shown in Fig. 5A, the spectrum, acquired in about 44 min, shows 2-$^{14}$N peaks of amino acids, which are unresolved because of the broad linewidths. Free $NH_4^+$ was described as being relatively low and was not quantitated. More recently, Gamcsik *et al.* (1991) reported *in vivo* $^{14}$N MR spectra of a fibrosarcoma implanted in the flank of a mouse. The $^{14}$N spectrum (Fig. 5B), obtained at 36 MHz for $^{14}$N (11.7 T) with a solenoid coil placed around the tumor, showed relatively sharp resonances for $^{14}NH_4^+$ (estimated quantity 1.3 μmol) and choline (0.56 μmol) in 1.1 hr of acquisition. The short $T_1$ of ammonium $^{14}$N (0.14 sec in 0.1 M phosphate at pH 7.4) permitted rapid signal accumulation. Proton decoupling was not attempted, presumably because (a) rapid exchange of ammonium protons with water protons at physiological pH results in a singlet $^{14}$N resonance and (b) signal intensity enhancement due to the nuclear Overhauser effect (NOE) is not expected owing to the predominantly quadrupolar relaxation of $^{14}$N. Noninvasive and continuous monitoring of brain ammonia by $^{14}$N MRS would be useful for studying the role of ammonia in the glutamine/glutamate cycle in intact brain. Whether the sensitivity and resolution are adequate for quantitative *in vivo* studies at the low to intermediate field strengths currently available for most animal studies remains to be determined.

## 5. FUTURE PROSPECTS

### 5.1. Neurochemical and Clinical Application

The application of *in vivo* nitrogen MRS in the study of mammalian brain is only in its infancy compared with that of $^1$H, $^{31}$P, and $^{13}$C MRS. Nevertheless, $^{15}$N MRS studies of rat brain have provided important novel information regarding *in vivo* rates of GS and PAG under hyperammonemic conditions. The *in vivo* rates were shown to be nearly two orders of magnitude lower than the *in vitro* rates measured in brain homogenates under optimal conditions. The difference could most reasonably be attributed to subsaturating *in situ* concentrations of substrates and cofactors (in the case of GS) and strong *in situ* inhibition of enzyme activity (in the case of PAG). The results highlight the importance of measuring metabolite flux in intact brain.

Studies of metabolite flux through the glutamine/glutamate/GABA cycle in intact brain can contribute to a clearer understanding not only of the etiology of HE but also of diseases in which glutamate excess is regarded as a component of neurotoxicity, such as hypoxia, stroke, epilepsy, Huntington's disease, and amyotrophic lateral sclerosis (Lou Gehrig's disease). Glutamate neurotoxicity in these diseases is mediated by a prolonged depolarizing action of glutamate at its postsynaptic receptor sites, according to the excitotoxic hypothesis (Rothman and Olney, 1987). The accumulation of extracellular glutamate observed in experimental hypoxia, for example, is likely to reflect excessive release from synaptic transmitter pools as well as impaired cellular uptake. Glutamate release occurs by a $Ca^{2+}$-dependent pathway from synaptic vesicles or by $Ca^{2+}$-independent efflux from the cytosolic pool. In the glutamine/glutamate/GABA cycle, PAG is thought to be an important provider of the neurotransmitter pool of glutamate. Hence, its *in vivo* activity can indirectly modulate the level of transmitter release. Glutamate released into the synaptic space is taken up by astrocytes and neurons. Efficient uptake of extracellular glutamate by the $Na^+$-dependent glutamate transporter depends, among other factors, on maintaining low glutamate concentration within astrocytes, where GS is a major pathway of glutamate removal. These considerations suggest that the study of factors that regulate *in vivo* activities of GS and PAG in intact brain can contribute to a better understanding of how intra- and extracellular levels of glutamate are controlled under normal and pathological conditions. Such information, when combined with the wealth of current knowledge on transport kinetics and receptor binding of the transmitter glutamate, can help clarify how glutamate neurotoxicity may be alleviated in these diseases.

## 5.2. MRS Techniques

For broader application of [15]N MRS, the technical challenges include spatial localization and sensitivity enhancement. The feasibility and the potential cost of clinical [15]N MRS depend additionally upon field strength, feasibility of [15]N enrichment by [15]$NH_4Cl$ infusion (in humans), amount, and cost of [15]N-enriched precursors.

### 5.2.1. Spatial Localization

In the [15]N MRS studies described above, *in vivo* [15]N spectra were obtained from the head of an anesthetized, [15]$NH_4^+$-infused rat with a "head probe" consisting of a solenoidal [15]N observe coil and a Helmholtz [1]H coil for shimming and [1]H decoupling. The solenoidal coil was chosen to achieve high sensitivity and $B_1$ homogeneity. Spatial localization was not required for these particular studies because the [5-[15]N]glutamine and [[15]N]glutamate peaks observed *in vivo* were shown to arise exclusively from the brain in preliminary experiments. Separate [15]N spectra taken of the isolated brain, muscle, and blood immediately after the *in vivo* experiment showed [5-[15]N]glutamine and [[15]N]glutamate to be present only in the brain (Kanamori *et al.*, 1993). This result is consistent with the presence of glutamate dehydrogenase (which assimilates ammonia into glutamate) in the brain but not in muscle and the much higher activity of glutamine synthetase in the brain compared to muscle.

For study of other brain [15]N-enriched metabolites that may also occur in adjacent tissues and for exploration of major regional variations within the intact brain, spatial localization is desirable. In the following subsections, potential advantages and limitations of some common spatial localization techniques for observation of [15]N-enriched metabolites are discussed.

*5.2.1.1. Surface Coil* The surface coil, the simplest method of localization, has been used to observe [[15]N]glycine in rat liver, with the surface coil placed between surgically exposed liver lobes (Grunder *et al.*, 1989). After a bolus i.v. injection of [[15]N]glycine, liver [[15]N]glycine at a concentration of 5–10 μmol/g was observed *in vivo* with a time resolution of about 10 min at 9 MHz for [15]N without [1]H decoupling. Sensitivity is expected to improve at higher field and with [1]H decoupling to generate NOE. No attempt was made in that study to quantitate liver [[15]N]glycine from the observed *in vivo* peak intensity. Quantitation methods for [15]N MRS in the perfused rat liver depended upon the great homogeneity provided by a solenoid coil (Geissler *et al.*, 1992). However, quantitation by the direct calibration method described above, should be applicable to signals obtained with a surface coil if the observed *in vivo* peak intensity per micromole of $^{15}$N-labeled metabolite per minute of acquisition is found to be reproducible, despite the $B_1$ inhomogeneity and incomplete $^1$H decoupling expected for surface coils. Use of $B_1$-insensitive adiabatic pulses for excitation (Garwood and Ugurbil, 1992) may be useful in this regard.

*5.2.1.2. Localization by Frequency-Selective Pulse in the Presence of $B_0$ Gradient* If the *in vivo* experiment requires simultaneous observation of $^{15}$N-enriched brain metabolites with large chemical shift dispersions, localization techniques such as image-selected *in vivo* spectroscopy (ISIS) and point-resolved spectroscopy (PRESS) are subject to large spatial displacement error (Bottomley *et al.*, 1989). Frequency-selective $^{15}$N RF pulses in combination with $B_0$ gradients lead to strong chemical shift dispersions in the spatial dimensions so that each chemically shifted species has a different spatial origin. The spatial displacement error, $\Delta X$, is given by

$$\Delta X = \sigma B_0 / G_x$$

where $\sigma$ is the chemical shift value relative to a suitable reference peak, $B_0$ is the main magnetic field strength, and $G_x$ is the gradient strength applied during spatial localization. Thus, for simultaneous observation of the peaks due to brain [5-$^{15}$N]glutamine ($-271$ ppm), [$^{15}$N]glutamate ($-342$ ppm), and $^{15}$NH$_4^+$ ($-362$ ppm), which are 71–91 ppm apart, whole-volume observation was more useful (Kanamori and Ross, 1995). However, for many kinetic experiments involving substrates and products whose peaks are well resolved but relatively close, localization to the brain with negligibly small displacement error should be possible. For example, the $^{15}$N peaks of the substrate (glutamate) and the product (GABA) of the glutamate decarboxylase pathway are only 8 ppm apart. At $B_0 = 4.7$ T, and with $\sigma = 8$ ppm and a moderate $G_x$ of 0.03 mT/mm, the displacement error $\Delta X$ would be only 1.2 mm, which is small relative to the dimensions of rat brain (18 $\times$ 16 $\times$ 10 mm$^3$). A brief discussion follows of three localization techniques that are potentially useful for *in vivo* $^{15}$N observation, viz ISIS (based on subtraction method), and PRESS and stimulated-echo acquisition mode (STEAM) which are based on single-scan excitation of spins in the volume of interest.

*(a) ISIS* ISIS (Ordidge *et al.*, 1986) has been successfully used to obtain localized proton-decoupled NOE-enhanced $^{13}$C spectra from human brain (Gruetter *et al.*, 1994). For a low-sensitivity nucleus like $^{15}$N, its main advantage over the spin-echo methods (PRESS or STEAM) is that it does not entail any $T_2$-induced loss in signal intensity. Possible disadvantages are the requirement for long recycle delay (Ordidge *et al.*, 1986) and incomplete suppression of signals from outside the volume of interest (VOI) due to motion.

*(b) PRESS and STEAM* The signal intensity obtained by PRESS (Bottomley

*et al.,* 1984; Ordidge *et al.,*1985) is twofold higher than that obtained by STEAM (Granot, 1986; Frahm *et al.,* 1987; Kimmich and Hoepfel, 1987) for a fixed echo time. In practice, the gain is less because the minimal echo time achievable with PRESS is generally longer than that achievable with STEAM. Because the spin-echo signal observed by PRESS or STEAM is $T_2$-weighted and *J*-modulated, loss in signal intensity due to these processes for [15]N-enriched brain metabolites of interest must be evaluated. To our knowledge, $T_2$ values of glutamine 5-[15]N and glutamate [15]N at physiological pH and temperature have not yet been reported. For [15]N bound to protons that undergo chemical exchange with water protons, scalar relaxation may result in short $T_2$ (Becker, 1980). Preliminary studies in our laboratory show that for [5-[15]N]glutamine in which the rate of exchange of the amide protons with water protons is slow at pH 7.1 and 20°C (Blomberg *et al.,* 1976; Kanamori *et al.,* 1991), the loss in [15]N signal intensity arising from $T_2$ processes when the spin echo was observed (with proton decoupling and NOE enhancement) amounted to 8% at an echo time of 2 msec and 32% at an echo time of 10 msec. For [[15]N]glutamic acid in which the amine protons exchange rapidly with water protons at pH 7.1 and 20°C, the loss in signal intensity (observed with [1]H decoupling) was 38% at an echo time of 2 msec. The results suggest that PRESS, used at a minimal echo time with fast-switching gradients, is a feasible localization technique for [15]N-enriched brain metabolites with nonex-changing protons, such as [5-[15]N]glutamine and [[15]N$_{indole}$]serotonin. For the amine [15]N nuclei present in glutamate, GABA, and dopamine, ISIS is likely to yield better S/N ratio per unit time.

## 5.2.2. Sensitivity Enhancement

Promising approaches for increasing the sensitivity of [15]N detection include (a) improvement in coil design, (b) indirect detection through coupled protons with effective water suppression for observation of slowly exchanging protons, and (c) higher [15]N enrichment of brain metabolites using other [15]N-labeled precursors. Feasibility of extension to human studies will depend, among other factors, on advances on these fronts. In human studies, we expect substantial gain in sensitivity due to the larger brain volume, which is encouraging for feasibility of studies on the wide-bore 3-T or 4-T spectrometers that are now becoming available. Whether the gain in sensitivity can offset the expected loss due to lower [15]N enrichment remains a question for future investigation.

ACKNOWLEDGMENTS

We are grateful to Dr. Thomas Ernst for critical reading and helpful com-ments on the section on spatial localization. This work was supported by Re-search Grant 1-RO1-NS29048 from the National Institutes of Health and the

Schulte Research Institute of Santa Barbara, California, by the L. K. Whittier Foundation of California, and by the J. W. and Ida M. Jameson Foundation and the Norris Foundation. We are grateful to Dr. John D. Roberts, one of the pioneers of $^{15}$N MRS and Institute Professor of Chemistry, Emeritus, in the Division of Chemistry and Chemical Engineering at the California Institute of Technology, for his continuous encouragement and helpful comments. K. K. and B. D. R. are visiting associates in the Division of Chemistry and Chemical Engineering at the California Institute of Technology.

## REFERENCES

Aoki, C., Kaneko, T., Starr, A., and Pickel, V. M., 1991, Identification of mitochondrial and non-mitochondrial glutaminase within select neurons and glia of rat forebrain by electron microscopic immunocytochemistry, *J. Neurosci. Res.* **28**:531–548.

Bachelard, H., and Badar-Goffer, R., 1993, NMR spectroscopy in neurochemistry, *J. Neurochem.* **61**:412–429.

Balaban, R. S., and Knepper, M. A., 1983, Nitrogen-14 nuclear magnetic resonance spectroscopy of mammalian tissues, *Am. J. Physiol.* **245**:C439–C444.

Balázs, R., and Cremer, J. E., 1973, "Metabolic Compartmentation in the Brain," Macmillan, London.

Basile, A. S., Jones, E. A., and Skolnick, P., 1991, The pathogenesis and treatment of hepatic encephalopathy: Evidence for the involvement of benzodiazepine receptor ligands, *Pharmacol. Rev.* **43**(1):27–71.

Battaglioli, G., and Martin, D. L., 1991, GABA synthesis in brain slices is dependent on glutamine produced in astrocytes, *Neurochem. Res.* **16**(2):151–156.

Bax, A., De Jong, P. G., Mehlkopf, A. F., and Smidt, J., 1980, Separation of the different orders of NMR multiple-quantum transitions by the use of pulsed field gradients, *Chem. Phys. Lett.* **69**:567–570.

Bax, A., Griffey, R. H., and Hawkins, B. L., 1983, Correlation of proton and nitrogen-15 chemical shifts by multiple quantum NMR, *J. Magn. Reson.* **55**:301–315.

Becker, E. D., 1980, "High Resolution NMR—Theory and Chemical Applications," Academic Press, New York.

Behar, K. L., den Hollander, J. A., Petroff, O. A. C., Hetherington, H. P., Prichard, J. W., and Shulman, R. G., 1985, Effect of hypoglycemic encephalopathy upon amino acids, high-energy phosphates, and pH$_i$ in the rat brain *in vivo:* Detection by sequential $^1$H and $^{31}$P NMR spectroscopy, *J. Neurochem.* **44**:1045–1055.

Benjamin, A. M., 1981, Control of glutaminase activity in rat brain cortex *in vitro:* Influence of glutamate, phosphate, ammonium, calcium and hydrogen ions, *Brain Res.* **208**:363–377.

Berl, S., Takagaki, G., Clarke, D. D., and Waelsch, H., 1962, Metabolic compartments *in vivo:* Ammonia and glutamic acid metabolism in brain and liver, *J. Biol. Chem.* **237**:2562–2569.

Blindermann, J. M., Maitre, M., Ossola, L., and Mandel, P., 1978, Purification and some properties of L-glutamate decarboxylase from human brain, *Eur. J. Biochem.* **86**:143–152.

Blomberg, F., and Rüterjans, H., 1983, Nitrogen-15 NMR in biological systems, *Biol. Magn. Reson.* **5**:21–73.

Blomberg, F., Maurer, W., and Rüterjans, H., 1976, $^{15}$N nuclear magnetic resonance investigations on amino acids, *Proc. Natl. Acad. Sci. USA* **73**:1409–1413.

Bottomley, P. A., Foster, T. H., and Leue, W. M., 1984, *In vivo* nuclear magnetic resonance chemical shift imaging by selective irradiation, *Proc. Natl. Acad. Sci. USA* **81**:6856–6860.

Bottomley, P. A., Hardy, C. J., Roemer, P. B., and Mueller, O. M., 1989, Proton-decoupled, Overhauser-enhanced, spatially localized carbon-13 spectroscopy in humans, *Magn. Reson. Med.* **12**:348–363.

Bradford, H. F., and Ward, H. K., 1975, Glutamine as a metabolic substrate for isolated nerve-endings: Inhibition by ammonium ions, *Biochem. Soc. Trans.* **3**:1223–1227.

Bradford, H. F., Ward, H. K., and Thomas, A. J., 1978, Glutamine: A major substrate for nerve endings, *J. Neurochem.* **30**:1453–1459.

Brindle, K. M., Porteous, R., and Campbell, I. D., 1984, $^1$H NMR measurements of enzyme-catalyzed $^{15}$N label exchange, *J. Magn. Reson.* **56**:543–547.

Bu, D.-F., Erlander, M. G., Hitz, B. C., Tillakaratne, N. J. K., Kaufman, D. L., Wagner-McPherson, C. B., Evans, G. A., and Tobin, A. J., 1992, Two human glutamate decarboxylases, 65-kDa GAD and 67-kDa GAD, are each encoded by a single gene, *Proc. Natl. Acad. Sci. USA* **89**:2115–2119.

Butterworth, R. F., 1991, Pathophysiology of hepatic encephalopathy: The ammonia hypothesis revisited, in "Progress in Hepatic Encephalopathy and Metabolic Nitrogen Exchange" (F. Bengtsson, B. Jeppsson, T. Almdal, and H. Vilstrup, eds.), pp. 9–26, CRC Press, Boca Raton, Florida.

Butterworth, R. F., Girard, G., and Giguère, J.-F., 1988, Regional differences in the capacity for ammonia removal by brain following portacaval anastomosis, *J. Neurochem.* **51**:486–490.

Casu, M., and Gale, K., 1981, Intracerebral injection of gamma vinyl GABA: Method of measuring rates of GABA synthesis in specific brain regions *in vivo*, *Life Sci.* **29**:681–688.

Chapman, A. G., and Evans, M. C., 1983, Cortical GABA turnover during bicuculline seizures in rats, *J. Neurochem.* **41**:886–889.

Cooper, A. J. L., and Plum, F., 1987, Biochemistry and physiology of brain ammonia, *Physiol. Rev.* **67**:440–519.

Deuel, T. F., Louie, M., and Lerner, A., 1978, Glutamine synthetase from rat liver, *J. Biol. Chem.* **253**:6111–6118.

Fagg, G., and Foster, A., 1983, Amino acid neurotransmitters and their pathways in the mammalian central nervous system, *Neuroscience* **9**:701–719.

Farrow, N. A., Kanamori, K., Ross, B. D., and Parivar, F., 1990, An N-15 NMR study of cerebral, hepatic and renal nitrogen metabolism in hyperammonemic rats, *Biochem. J.* **270**:473–481.

Fonnum, F., and Walberg, F., 1973, The concentration of GABA within inhibitory nerve terminals, *Brain Res.* **62**:577–579.

Frahm, J., Merboldt, K.-D., and Hänicke, W., 1987, Localized proton spectroscopy using stimulated echoes, *J. Magn. Reson.* **72**:502–508.

Freeman, D., Sailasuta, N., Sukumar, S., and Hurd, R. E., 1993, Proton-detected $^{15}$N NMR spectroscopy and imaging, *J. Magn. Reson. B* **102**:183–192.

Gamcsik, M. P., Constantinidis, I., and Glickson, J. D., 1991, *In vivo* $^{14}$N nuclear magnetic resonance spectroscopy of tumors: Detection of ammonium and trimethylamine metabolites in the murine radiation induced fibrosarcoma 1, *Cancer Res.* **51**:3378–3383.

Garwood, M., and Ugurbil, K., 1992, $B_1$ insensitive adiabatic RF pulses, *in* "NMR: Basic Principles and Progress," (P. Diehl, E. Fluck, H. Günther, R. Kosfeld, and J. Seelig, eds.), pp. 109–148, Springer-Verlag, Berlin.

Geissler, A., Kanamori, K., and Ross, B. D., 1992, Real time study of the urea cycle using $^{15}$N NMR in the isolated perfused rat liver, *Biochem. J.* **287**:813–820.

Granot, J., 1986, Selected volume excitation using stimulated echoes (VEST). Applications to spatially localized spectroscopy and imaging, *J. Magn. Reson.* **70**:488–492.

Gruetter, R., Novotny, E. J., Boulware, S. D., Mason, G. F., Rothman, D. L., Shulman, G. I., Prichard, J. W., and Shulman, R. G., 1994, Localized $^{13}$C NMR spectroscopy in the human brain of amino acid labeling from [1-$^{13}$C]D-glucose, *J. Neurochem.* **63**:1377–1385.

Gründer, W., Krumbiegel, P., Buchali, K., and Blesin, H. J., 1989, Nitrogen-15 NMR studies of rat liver *in vitro* and *in vivo*, *Phys. Med. Biol.* **34**:457–463.

Haser, W. G., Shapiro, R. A., and Curthoys, N. P., 1985, Comparison of the phosphate-dependent glutaminase obtained from rat brain and kidney, *Biochem. J.* **229**:399–408.

Hawkins, R. A., Jessy, J., Mans, A. M., and De Joseph, M. R., 1993, Effect of reducing brain glutamine synthesis on metabolic symptoms of hepatic encephalopathy, *J. Neurochem.* **60**:1000–1006.

Hurd, R. E., and John, B. K., 1991, Gradient enhanced proton-detected heteronuclear multiple quantum coherence: GE HMQC, *J. Magn. Reson.* **91**:648–653.

Kanamori, K., and Roberts, J. D., 1983, $^{15}$N NMR studies of biological systems, *Acc. Chem. Res.* **16**:35–41.

Kanamori, K., and Ross, B. D., 1993, $^{15}$N NMR measurement of the *in vivo* rate of glutamine synthesis and utilization at steady state in the brain of the hyperammonaemic rat, *Biochem. J.* **293**:461–468.

Kanamori, K., and Ross, B. D., 1995, *In vivo* activity of glutaminase in the brain of hyperammonemic rats measured by $^{15}$N NMR, *Biochem. J.* **305**:329–336.

Kanamori, K., Ross, B. D., and Parivar, F., 1991, Selective observation of biologically important $^{15}$N-labelled metabolites in isolated rat brain and liver by $^{1}$H-detected multiple quantum coherence spectroscopy, *J. Magn. Reson.* **93**:319–328.

Kanamori, K., Parivar, F., and Ross, B. D., 1993, A $^{15}$N NMR study of *in vivo* cerebral glutamine synthesis in hyperammonemic rats, *NMR Biomed.* **6**(1):21–26.

Kanamori, K., Ross, B. D., and Tropp, J., 1995a, Selective, *in vivo* observation of [5-$^{15}$N]glutamine amide protons in rat brain by $^{1}$H–$^{15}$N heteronuclear multiple-quantum-coherence transfer NMR, *J. Magn. Reson. B* **107**:107–115.

Kanamori, K., Ross, B. D., and Kuo, E. L., 1995b, Dependence of *in vivo* glutamine synthetase activity on ammonia concentration in rat brain studied by $^{1}$H–$^{15}$N heteronuclear multiple-quantum coherence-transfer NMR, *Biochem. J.* **311**:681–688.

Kanamori, K., Ross, B. D., Chung, J. C., and Kuo, E. L., 1996, Severity of hyperammonemic encephalopathy correlates with brain ammonia level and saturation of glutamine synthetase in vivo. *J. Neurochem.* **67**:1584–1594.

Kaneko, T., Urade, Y., Watanabe, Y., and Mizuno, N., 1987, Production, characterization, and immunohistochemical application of monoclonal antibodies to glutaminase purified from rat brain, *J. Neurosci.* **7**:302–309.

Kimmich, R., and Hoepfel, D., 1987, Volume-selective multipulse spin-echo spectroscopy, *J. Magn. Reson.* **72**:379–384.

Kreis, R., Ernst, T., and Ross, B. D., 1993, Absolute quantitation of water and metabolites in the human brain. Part II: Metabolite concentrations, *J. Magn. Reson. B* **102**:9–19.

Kvamme, E., Svenneby, G., and Torgner, I. A., 1988, Glutaminases, *in* "Glutamine and Glutamate in Mammals" (E. Kvamme, ed.), pp. 53–67, CRC Press, Boca Raton, Florida.

Levy, G. C., and Lichter, R. L., 1979, "Nitrogen-15 Nuclear Magnetic Resonance Spectroscopy," John Wiley & Sons, New York.

Lockwood, A. H., McDonald, J. M., Reiman, R. E., Gelbard, A. S., Laughlin, J. S., Duffy, T. E., and Plum, F., 1979, The dynamics of ammonia metabolism in man: Effects of liver disease and hyperammonemia, *J. Clin. Invest.* **63**:449–460.

Maitre, M., Blindermann, J. M., Ossola, L., and Mandel, P., 1978, Comparison of the structures of L-glutamate decarboxylases from human and rat brains, *Biochem. Biophys. Res. Commun.* **85**:885–890.

Mason, G. F., Gruetter, R., Rothman, D. L., Behar, K. L., Shulman, R. G., and Novotny, E. J., 1995, Simultaneous determination of the rates of the TCA cycle, glucose utilization, alpha-ketoglutarate/glutamate exchange, and glutamine synthesis in human brain by NMR, *J. Cereb. Blood Flow Metab.* **15**:12–25.

Michaelis, T., Merboldt, K.-D., Bruhn, H., Hänicke, W., and Frahm, J., 1993, Absolute concentrations of metabolites in the adult human brain in vivo: Quantification of localized proton MR spectra, *Radiology* **187**:219–227.

Mill, J. F., Mearow, K. M., Purohit, H. J., Haleem-Smith, H., King, R., and Freese, E., 1991, Cloning and functional characterization of the rat glutamine synthetase gene, *Mol. Brain Res.* **9**:197–207.

Morris, G. A., and Freeman, R., 1979, Enhancement of nuclear magnetic resonance signals by polarization transfer, *J. Am. Chem. Soc.* **101**:760–762.

Norenberg, M. D., and Martinez-Hernandez, A., 1979, Fine structural localization of glutamine synthetase in astrocytes of rat brain, *Brain Res.* **161**:303–310.

Ordidge, R. J., Bendall, M. R., Gordon, R. E., and Connelly, A., 1985, Volume selection for *in vivo* biological spectroscopy, *in* "Magnetic Resonance in Biology and Medicine," pp. 387–397, Tata McGraw-Hill, New Delhi.

Ordidge, R. J., Connelly, A., and Lohman, J. A. B., 1986, Image-selected *in vivo* spectroscopy (ISIS). A new technique for spatially selective NMR spectroscopy, *J. Magn. Reson.* **66**:283–294.

Ottersen, O. P., and Storm-Mathisen, J., 1984, Glutamate- and GABA-containing neurons in the mouse and rat brain, as demonstrated with a new immunocytochemical technique. *J. Comp. Neurol.* **229**:374–392.

Ottersen, O. P., Zhang, N., and Walberg, F., 1992, Metabolic compartmentation of glutamate and glutamine: Morphological evidence obtained by quantitative immunocytochemistry in rat cerebellum, *Neuroscience* **46**:519–534.

Patel, A. J., Hunt, A., and Tahourdin, C. S. M., 1983, Regional development of glutamine synthetase activity in the rat brain and its association with the differentiation of astrocytes, *Dev. Brain Res.* **8**:31-37.

Paulsen, R. E., Odden, E., and Fonnum, F., 1988, Importance of glutamine for gamma-aminobutyric acid synthesis in rat neostriatum *in vivo*, *J. Neurochem.* **51**:1294–1299.

Prior, M. J. W., 1995, NMR spectroscopy of living systems, *in* "Nuclear Magnetic Resonance" G. A. Webb, ed.), Vol. 24, pp. 398–443, The Royal Society of Chemistry, London.

Radda, G. K., 1986, The use of NMR spectroscopy for the understanding of disease, *Science* **233**:640–645.

Ross, B. D., Kreis, R., and Ernst, T., 1992, Clinical tools for the 90's: Magnetic resonance spectroscopy and metabolite imaging. *Eur. J. Radiol.* **14**:128–140.

Rothman, S. M., and Olney, J. W., 1987, Excitotoxicity and the NMDA receptor, *Trends Neurosci.* **10**(7):299–302.

Ruiz-Cabello, J., Vuister, G. W., Moonen, C. T. W., van Gelderen, P., Cohen, J. S., and van Zijl, P. C. M., 1992, Gradient-enhanced heteronuclear correlation spectroscopy. Theory and experimental aspects, *J. Magn. Reson.* **100**:282–302.

Shulman, R. G., Blamire, A. M., Rothman, D. L., and McCarthy, G., 1993, Nuclear magnetic resonance imaging and spectroscopy of human brain function, *Proc. Natl. Acad. Sci. USA* **90**:3127–3133.

Street, J. C., Delort, A.-M., Braddock, P. S. H., and Brindle, K. M., 1993, A $^1H/^{15}N$ NMR study of nitrogen metabolism in cultured mammalian cells, *Biochem. J.* **291**:485–492.

Svenneby, G., Roberg, B., Hogstad, S., Torgner, I., and Kvamme, E., 1986, Phosphate-activated glutaminase in the crude mitochondrial fraction (P2 fraction) from human brain cortex, *J. Neurochem.* **47**:1351–1355.

Szerb, J. C., 1984, Storage and release of endogenous and labelled GABA formed from [$^3$H]glutamine and [$^{14}$C]glucose in hippocampal slices: Effect of depolarization, *Brain Res.* **293**:293–303.

Tate, S. S., Leu, F.-Y., and Meister, A., 1972, Rat liver glutamine synthetase: Preparation, properties and mechanism of inhibition by carbamyl phosphate, *J. Biol. Chem.* **247**:5312–5321.

van de Zande, L., Labruyère, W. T., Arnberg, A. C., Wilson, R. H., van den Bogaert, A. J. W., Das, A. T., van Oorschot, D. A. J., Frijters, C., Charles, R., Moorman, A. F. M., and Lamers, W. H.,

1990, Isolation and characterization of the rat glutamine synthetase-encoding gene, *Gene* **87:**225–232.

Veech, R. L., Harris, R. L., Veloso, D., and Veech, E. H., 1973, Freeze-blowing: A new technique for the study of brain *in vivo, J. Neurochem.* **20:**183–188.

Ward, H. K., and Bradford, H. F., 1979, Relative activities of glutamine synthetase and glutaminase in mammalian synaptosomes, *J. Neurochem.* **33:**339–342.

Ward, H. K., Thanki, C. M., and Bradford, H. F., 1983, Glutamine and glucose as precursors of transmitter amino acids: *Ex vivo* studies, *J. Neurochem.* **40:**855–860.

Witanowski, M., Stefaniak, L., and Webb, G. A., 1992, Nitrogen NMR spectroscopy, *Annu. Rep. NMR Spectrosc.* **25:**1–480.

Yamamoto, H., Konno, H., Yamamoto, T., Ito, K., Mizugaki, M., and Iwasaki, Y., 1987, Glutamine synthetase of the human brain: Purification and characterization. *J. Neurochem.* **49:**603–609.

# TRAUMATIC BRAIN INJURY

## ROBERT VINK and TRACY K. McINTOSH

## SUMMARY

Traumatic injury to the central nervous system continues to be the major cause of mortality and morbidity in children and young adults. While much has been accomplished on a preventative level to reduce the incidence of neurotrauma, little progress has been made in terms of developing effective neuroprotective therapies for administration after the traumatic event. Nonetheless, significant advances in understanding the mechanisms associated with the development of irreversible tissue injury after trauma have been made, thereby increasing the likelihood of an effective therapeutic intervention being developed that will prevent, or at least attenuate, the post-traumatic injury process. Much of this progress toward understanding the mechanisms of injury can be ascribed to the development of noninvasive procedures for monitoring physiological and biochemical events after injury. In particular, magnetic resonance spectroscopy has made a significant contribution to the understanding of cell metabolism after traumatic injury and to the elucidation of the effects of experimental pharmacologic interventions. This chapter examines the contributions that magnetic

ROBERT VINK • Department of Physiology and Pharmacology, James Cook University of North Queensland, Townsville, Queensland 4811, Australia.     TRACY K. McINTOSH • Division of Neurosurgery, University of Pennsylvania School of Medicine, Philadelphia, Pennsylvania 19104.

Magnetic Resonance Spectroscopy and Imaging in Neurochemistry, Volume 8 of Advances in Neurochemistry, edited by Bachelard, Plenum Press, New York, 1997.

resonance spectroscopy has made to the characterization of metabolic events after traumatic brain injury and the impact that these findings have had on the development of appropriate interventional strategies.

## 1. INTRODUCTION

Traumatic injury to the brain is the major cause of death in individuals under 44 years of age in developed countries (Ring *et al.*, 1986; Goldstein, 1990). Despite the fact that it is such a significant public health problem, there is no universally accepted pharmacological intervention currently in use to treat victims of brain trauma. This is largely because, until quite recently, little was known about the processes that cause irreversible tissue injury following trauma to the brain. Today it is recognized that neuronal cell death following brain trauma is caused by two processes: primary and secondary injury (Cooper, 1985). Primary events occur at the time of trauma and include mechanical changes such as axonal tearing, shearing, and stretching. Secondary events occur hours to days after the traumatic insult and include a number of biochemical and physiological changes initiated by the original insult. It is now widely accepted that these secondary events are responsible for much of the mortality and morbidity associated with the development of irreversible tissue injury following a traumatic insult. Furthermore, because these secondary events occur over time following trauma, there exists a window of opportunity to counteract such deleterious events by administration of pharmacological agents designed to prevent or at least attenuate such processes, with a resultant improvement in outcome. Before such agents can be developed, it is essential to characterize secondary events following brain trauma. It is in this characterization of secondary events following traumatic brain injury that magnetic resonance spectroscopy (MRS) has been particularly effective.

Although virtually all of the MRS applications to traumatic brain injury have been limited to experimental animal studies, many of the findings have been confirmed by what few clinical studies have been performed to date (Rango *et al.*, 1990; Sutton *et al.*, 1995). This may reflect that the animal models of traumatic brain injury used in the MRS studies have successfully reproduced many of the features typical of clinical brain trauma. Approximately 50% of all cases of traumatic brain injury are caused by motor vehicle accidents, with falls, assaults, and sports injuries (e.g., concussion) accounting in large part for the remaining 50% (Kraus and Arzemanian, 1989). Because these forms of injury primarily involve closed head injury, the animal models have focused on producing shearing, contusion, and axonal damage typical of nonpenetrating injury (Gennarelli, 1994). Since the animal studies have successfully reproduced many of these aspects of clinical trauma, this chapter will focus on the findings reported in these experimental trauma studies.

## 2. PHOSPHATE CONCENTRATION

The first studies of traumatic brain injury appeared in 1987 in reports published separately by Ishige *et al.* (1987) and Vink *et al.* (1987). Both of these groups concentrated on energy metabolism following moderate injury and demonstrated that there is no depletion of either phosphocreatine or ATP following a traumatic insult to the brain (Fig. 1), an observation later supported in another species of animal by a third group of researchers (Unterberg *et al.*, 1988). This was in stark contrast to what had been widely reported in studies of ischemia. Ischemia typically results in depletion of phosphates and severe acidosis (Williams *et al.*, 1989). Trauma had been widely considered to be merely an alternative form of an ischemic insult (Jennet *et al.*, 1973), but these early MRS findings in traumatic injury provided conclusive evidence for the premise that trauma and ischemia are two distinct forms of brain injury. Ishige *et al.* (1987) further demonstrated that ATP changes after trauma could in fact be initiated by exposing the traumatized brain to a second insult such as hypoxia but that trauma on its own did not cause any significant ATP loss. These findings were independently supported by Vink *et al.* (1988c), who later demonstrated in studies of graded injury that significant changes in ATP concentration could only be observed in very severe cases of trauma (associated with 100% mortality) and that this phosphate depletion was in fact caused by a significant ischemic component associated with this level of injury.

Although there was no profound energy depletion after trauma, it was noted that the ratio of phosphocreatine (PCr) to inorganic phosphate ($P_i$) declined after a traumatic insult to the brain (Vink *et al.*, 1987). The decline was biphasic, with a transient decline occurring in the first hour following trauma, followed by a recovery and a second decline after 90 min. The first decline was shown to correspond to a period of transient acidosis (see below), and it was demonstrated that the decline could be accounted for by phosphocreatine buffering of acidosis (Vink *et al.*, 1987). The second decline, however, occurred during a period of no pH changes and was thus assumed to be an indication of bioenergetic dysfunction. Furthermore, while moderate and severely injured animals did not show a recovery in $PCr/P_i$ ratio over an 8-hr MRS monitoring period, mildly injured animals recovered by 6 hr post-trauma (Vink *et al.*, 1988c). It therefore seemed that increasing severity of injury was associated with an increasing period of bioenergetic dysfunction. Moreover, traumatized brain was more sensitive to secondary insults such as post-traumatic hypoxia (Ishige *et al.*, 1987; Andersen *et*

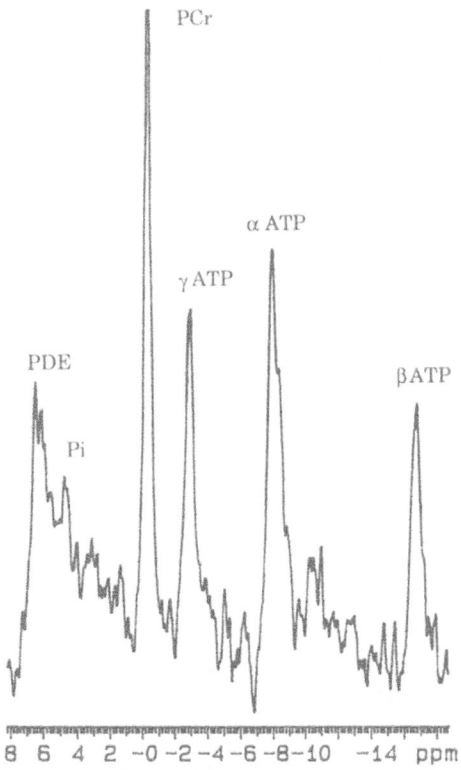

**FIGURE 1.** Phosphorus MR spectrum (7.0 T) of the injured rat brain 60 min after induction of moderate fluid percussion brain injury. Resonance assignments: PDE, phosphodiesters; Pi, inorganic phosphate; PCr, phosphocreatine; ATP, adenosine triphosphate.

*al.*, 1988) and hypovolemic hypotension (Ishige *et al.*, 1988) leading to a more significant loss of bioenergetic control.

## 3. INTRACELLULAR pH

Profound lactic acidosis has been proposed as a deleterious factor in brain ischemia (Myers and Yamaguchi, 1976) and, by extension, in brain trauma. Indeed, in brain ischemia, hyperglycemia has been shown to exacerbate ischemic damage whereas hypoglycemia is known to reduce infarction size following experimental ischemia (Siesjö, 1988). On the basis of these findings, a lactic acid concentration of 15 mM has been proposed as a threshold level at which lactic acid accumulation exacerbates injury (Siesjö, 1988). However, from the discussion of phosphate metabolism above, it is clear that events occurring during ischemia may not necessarily be extrapolated to trauma. This was again demonstrated with respect to the pH profiles of brain following trauma. As opposed to the sustained and profound acidosis reported in ischemia studies (Williams et al., 1989), moderate trauma produced a transient and mild acidosis that existed for approximately the first hour post-trauma before recovering to preinjury levels (Vink et al., 1987). Similar findings were reported in studies of graded traumatic brain injury: intracellular pH never declined by more than 0.3 units at either mild, moderate, or severe injury (Vink et al., 1988c). This transient decline in pH after trauma was linearly related to lactic acid accumulation, and calculations of the lactic acid accumulation (McIntosh et al., 1987) demonstrated that the maximum brain concentration never exceeded 8 mM (Fig. 2). This calculated concentration is well below the threshold level that had been previously proposed as minimum accumulation required for induction of injury. Furthermore, by using a combination of depth pulses and proton lactate editing, the increase in lactic acid was shown to be localized to the injured tissue (McIntosh et al., 1987). These studies were supported by those of Inao (1988), who demonstrated that post-traumatic brain lactate accumulation was transient and correlated with modest alterations in pH. Moreover, these authors demonstrated that cerebrospinal fluid (CSF) lactate, which had been previously shown to be a useful prognostic indicator of outcome in clinical studies (DeSalles et al., 1986), was not predominantly derived from brain tissue but rather that a significant systemic contribution existed after trauma. This may explain the differing results of Cohen et al. (1991), who reported sustained rises in post-traumatic brain lactate in their whole-brain spectroscopic imaging studies of trauma. Since it has been previously shown that CSF lactate increases and remains elevated following trauma (DeSalles et al., 1986), one would expect that in whole-brain images an increase in lactate signal from CSF would be observed. This increase would be sustained for as long as the systemic contribution to lactate production was maintained. Thus, unlike ischemia, mild to severe trauma does not in itself result in significant brain lactate accumulation or acidosis unless there are other complicating factors.

As for the source of the brain lactate production, Hovda et al. (1992) have demonstrated that traumatic brain injury results in hypermetabolism in the form of increased glycolytic rate. This is consistent with what has been previously reported to occur following ischemia (Siesjö, 1978). It was therefore possible that, as in ischemia, blood glucose concentration may contribute to lactic acidosis after trauma. Recent studies by Golding et al. (1994) have addressed this problem in a model of moderate brain injury in rats. Their findings demonstrated that increased blood glucose concentration did not cause increased lactic acid formation after trauma. Similarly, hypoglycemia did not decrease any transient acidosis or confer any degree of neuroprotection after moderate brain trauma. Thus, blood glucose concentration does not seem to influence degree or duration of acidosis following mild trauma. One is left to conclude that traumatic brain injury results in a transient hypermetabolism that rapidly exhausts endogenous supplies of glucose equiva-

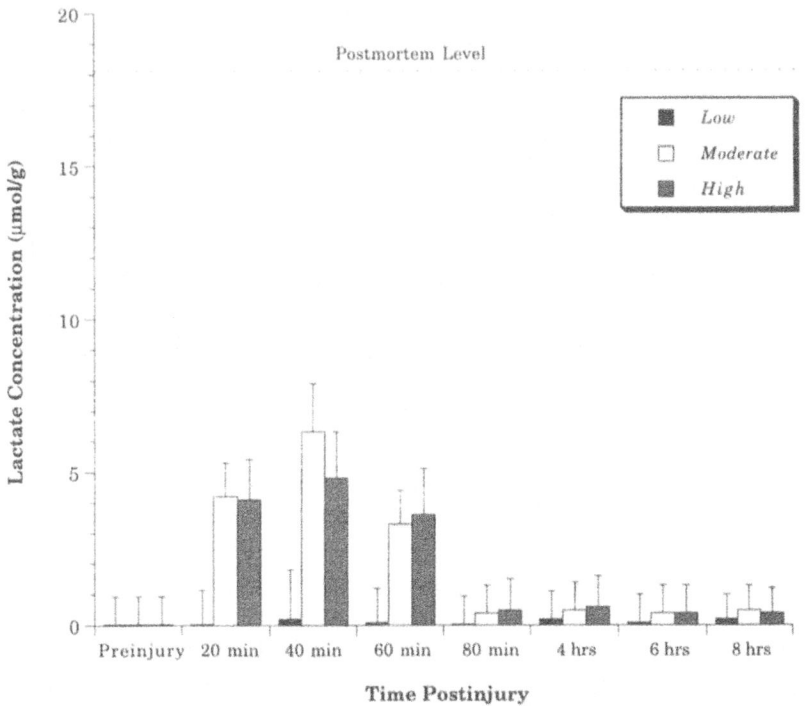

*FIGURE 2.* Brain lactate concentrations as calculated from proton MR spectra obtained following low-, moderate-, and high-severity fluid-percussion-induced lateral brain injury in rats. Typical postmortem brain lactate concentration is indicated by the dotted line. [Adapted from McIntosh *et al.* (1987).]

lents. As proposed by Andersen and Marmarou (1992), this brief period of hypermetabolism may result in the production of pyruvate at levels that exceed the immediate aerobic capacity of the cell and, consequently, lead to lactate production. However, since the traumatized brain is relying on endogenous glucose equivalents (perhaps stored glycogen), the degree and duration of acidosis would be limited by the concentration of the glucose store, and lactate acid production following trauma would therefore be independent of injury severity. This is consistent with previous studies that have shown that lactic acidosis after trauma is transient and similar at all levels of injury (Vink *et al.,* 1988c).

## 4. INTRACELLULAR FREE MAGNESIUM

It has been known for some time that trauma causes an imbalance in ion concentration across the membrane, and since ion homeostasis is critical for membrane energization and therefore normal neuronal function, much effort has been directed toward characterizing these changes. Most studies have concentrated on the sodium and potassium ions and the divalent ion calcium (Hovda *et al.*, 1992; Young and Constantini, 1994). Because of the difficulty in obtaining intracellular concentrations, these studies have largely been total tissue ion measurements with some ion- electrode- and microdialysis-based extracellular determinations. It has generally been found that extracellular sodium and calcium decline after trauma, while extracellular potassium concentration increases (Hovda *et al.*, 1992). From these data, it has been assumed that intracellular calcium increases with injury although this assumption need not be necessarily true because neuronal injury may indeed occur in the absence of any rise in intracellular free calcium concentration (Badar-Goffer *et al.*, 1993). Nonetheless, recent autoradiography studies of calcium transients after trauma do suggest that intracellular calcium does increase after brain injury (Fineman *et al.*, 1993). *In vitro* studies of neuronal cells (Choi, 1988) have demonstrated that intracellular calcium accumulation after induction of some form of stress may occur primarily via a subgroup of glutamate receptors known as the *N*-methyl *D*-aspartate (NMDA) receptors. Furthermore, activity of NMDA receptors has been shown to affect zinc ions, which are known to have a regulatory role in NMDA channel activity (Badar-Goffer *et al.*, 1994). Another divalent cation that has a regulatory role in NMDA channel activity is magnesium, which behaves as a voltage-gated blocker of the NMDA channel (Mayer *et al.*, 1984). One unique application of $^{31}$P MRS is the determination of intracellular free magnesium concentration (see Chapter 2). Because ATP is an endogenous ligand for magnesium and is located almost exclusively within cells, the chemical shift of the metabolite can be used to provide information on the intracellular magnesium status (Cohn and Hughes, 1962; Gupta *et al.*, 1978). Despite the early controversies in the literature regarding the use of this ligand as an intracellular magnesium indicator (Gupta *et al.*, 1983), with careful calibration of the MgATP titration curve under conditions that reflect the *in vivo* situation, the results obtained for a variety of tissues under a number of pathologic conditions are consistent among different researchers and with results obtained by alternative methods (Kushmerick *et al.*, 1986; Jelicks and Gupta, 1990; Headrick and Willis, 1991; London, 1991). Naturally, ATP must be present in the spectrum for a determination to be made. Fortunately, as discussed above, there is no ATP depletion after trauma, which provides a somewhat novel situation in which neuronal damage has occurred (as determined by motor function tests after the event) without significant ATP loss. Furthermore, since pH also affects ATP chemical shifts, the absence of significant acidosis after 1 hr post-trauma eliminates this factor as a complication (Vink *et al.*, 1988c).

Free magnesium concentration in brain prior to injury is reported to be between 0.5 and 1.0 mM, depending on species, age, gender, and diet (Vink, 1993). After all levels of traumatic injury, free magnesium levels decline to approximately 0.3 mM (Fig. 3). This concentration seems to be a tissue free magnesium threshold since tissue depletion studies (Li *et al.*, 1993) and trauma

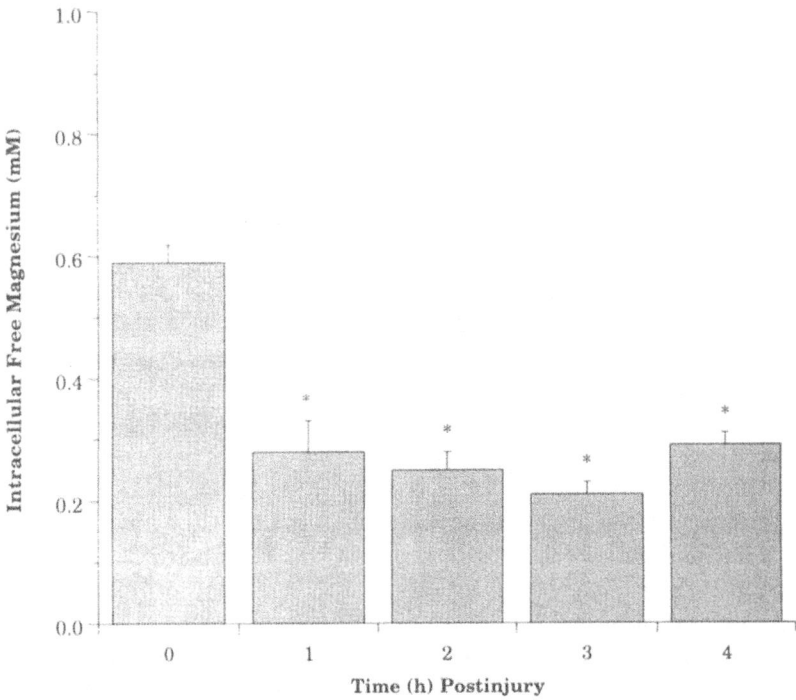

FIGURE 3. Typical decline in brain intracellular free magnesium concentration as determined from phosphorus MR spectra obtained after moderate fluid-percussion-induced brain injury in rats. *P < 0.05 by ANOVA and Student Newman–Keuls tests.

studies (McIntosh *et al.*, 1988a; Vink *et al.*, 1988a) suggest that this free concentration is the minimum that can be achieved. Consistent with the recorded decline in the free concentration, total tissue magnesium concentration also declines after trauma (Vink *et al.*, 1988d) although the decline is only approximately 10% as opposed to the significantly greater decline (40–60%) in the free pool. Corkey *et al.* (1986) have proposed that changes in the total pool, and particularly the membrane binding sites for magnesium, are amplified in the free pool concentration since the total pool is the buffer of the free magnesium concentration. The results from studies of trauma are consistent with that hypothesis. Furthermore, it has been shown that activation of phospholipase C leads to a decline in free magnesium concentration *in vitro* (Vink, 1989). Moreover, this decline was similar to that demonstrated in trauma. It would therefore seem that alteration in membrane binding sites, perhaps in association with phospholipase C activity, is linked to decline in free magnesium concentration following trauma.

Whereas the degree of decline is to threshold levels and seemingly independent of injury severity, the rate of recovery of intracellular magnesium to normal resting levels does seem to reflect severity of injury and resultant neurologic outcome. Induction of mild injury results in the typical 40–50% decline in free magnesium observed at all injury levels (Vink et al., 1991a). However, the magnesium concentration remains depressed for a comparatively short time (3–6 hr) when compared to magnesium depletion at more severe levels of injury (3–6 days). Animals whose free magnesium concentration is pharmacologically raised after trauma also have a shorter period of depressed intracellular free concentration and an associated improvement in neurologic outcome (Vink et al., 1988d). Conversely, animals whose injury is exacerbated by some means (e.g., alcohol) have longer periods of magnesium depletion and a worse neurologic outcome (Halt et al., 1992). Thus, it would appear that, although degree of magnesium decline may be important at the lower levels of traumatic brain injury at which there are no observable motor deficits, it is the duration of magnesium decline after traumatic injury that is primarily associated with neurologic motor outcome after trauma.

## 5. CYTOSOLIC PHOSPHORYLATION RATIO

The early finding that the $PCr/P_i$ ratio declines after trauma suggested that some form of bioenergetic dysfunction occurs and persists for several hours after the traumatic event. However, the use of $PCr/P_i$ ratios as a bioenergetic indicator was first developed in muscle and has been extended to other tissues (Chance et al., 1980). The assumptions used in formulating the $PCr/P_i$ concept are therefore not necessarily true for all tissue and particularly may not hold for those with a relatively low PCr concentration such as brain. A more accurate indicator of bioenergetic state in such tissues would be the cytosolic phosphorylation ratio, defined as $[ATP]/[ADP][P_i]$. The determination of this ratio, however, requires a knowledge of the free phosphate concentrations. Fortunately, it is possible to determine cytosolic phosphorlyation ratios using $^{31}P$ MRS since the phosphorus spectrum provides information on the free ATP, PCr, and $P_i$ concentrations and the ADP concentration can easily be determined from the creatine kinase equation. Such a calculation is possible since the phosphorus spectrum provides the necessary pH and Mg values needed to adjust the creatine kinase equilibrium constant (Lawson and Veech, 1979). Calculations of the phosphorylation ratio after trauma (Vink et al., 1988a) clearly demonstrate a profound decline of almost 50% at all levels of injury (Fig. 4), significantly more than the typical 25% decline observed in the $PCr/P_i$ ratio. Furthermore, the kinetics of decline in the phosphorylation ratio were similar to those observed for free magnesium concentration, and, indeed, the two parameters were linearly correlated (Vink et

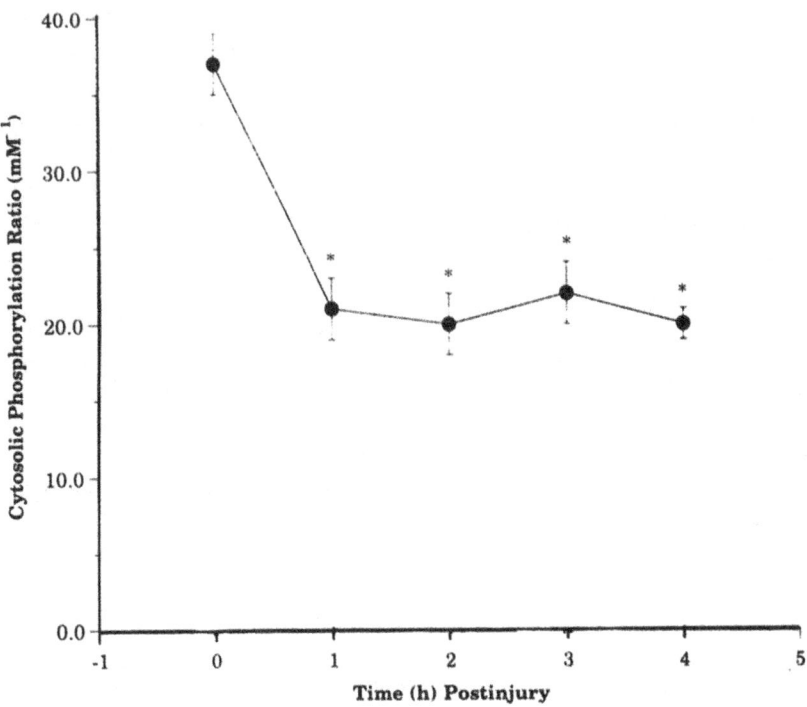

FIGURE 4. Decline in brain cytosolic phosphorylation ratio following moderate fluid-percussion-induced brain injury in rats. *$P < 0.05$ by ANOVA and Student Newman–Keuls tests.

*al.*, 1991a). Decline in cytosolic phosphorylation ratio therefore seems to be associated with the decrease in free magnesium concentration, thus indicating that traumatized brain is under considerable energetic stress after injury. Because trauma did not result in energy failure, mechanisms of energy production must be enhanced after the traumatic event to compensate for the increased energy demands of the cell.

## 6. MITOCHONDRIAL OXIDATIVE CAPACITY

Isolated mitochondria from brain tissue subjected to ischemia characteristically demonstrate a degree of respiratory dysfunction (Sims and Pulsinelli, 1987). Furthermore, it has been suggested that these mitochondria have a decreased magnesium concentration when compared to controls, which may account for their reduced function (Ginsberg *et al.*, 1977). The combination of energy dysfunction together with a possible reduced magnesium concentration in these mitochondria suggests that processes at the mitochondrial level similar to those that exist in ischemia may be responsible for the bioenergetic dysfunction observed after trauma. Indeed, Andersen and Marmarou (1992) have proposed that increased glycolysis after trauma may exceed mitochondrial oxidative capacity and consequently result in lactate formation. One method of examining the mitochondrial capacity is to isolate these organelles from traumatized brain and determine respiratory ratios. Although not being completely representative of the *in vivo* situation, *in vitro* studies of isolated mitochondria after trauma did not demonstrate any uncoupling but rather demonstrated an increased rate of oxygen consumption (Vink *et al.*, 1990a). These results were nonetheless consistent with those published several years earlier by Duckrow *et al.* (1981), who demonstrated increased oxygen consumption after trauma.

An alternative, totally noninvasive method of determining mitochondrial oxidative capacity using $^{31}$P MRS has been formulated by Chance *et al.* (1986b). The synthesis of ATP can be described by the equation:

$$NADH + H^+ + 3ADP + 3P_i + \tfrac{1}{2}O_2 \leftrightarrow 3ATP + NAD^+ + H_2O$$

The mitochondrial oxidative capacity (the relative velocity of ATP synthesis) may then be calculated from the Michaelis–Menten equation (Chance *et al.*, 1986a,b):

$$\left(\frac{V}{V_{max}}\right) = \frac{1}{1 + \dfrac{K_1}{[ADP]} + \dfrac{K_2}{[P_i]} + \dfrac{K_3}{[NADH]} + \dfrac{K_4}{[O_2]}}$$

where $V$ is the velocity of ATP synthesis, $V_{max}$ is the maximum velocity of ATP formation by oxidative phosphorylation, $K_1$ to $K_4$ are the Michaelis–Menten constants $(K_m)$ for ADP, $P_i$, NADH, and $O_2$, respectively, and the square brackets represent concentration. One assumption in the use of the above equation is that the system is at steady state. Although not applicable to all pathologic conditions, this seems to be the case following traumatic brain injury. After trauma, ATP concentration is constant while the small changes in pH are transient and stabilize within 60 minutes. Similarly, the $PCr/P_i$ decline after trauma occurs in the first 90 minutes, with the ratio thereafter stabilizing at the lower value. Presumably, this stabilization reflects the establishment of a new steady state. Furthermore, there can be no steady state based on glycolysis without oxidative metabolism. Should $O_2$ become rate-limiting, ADP and $P_i$ would increase without limit and there would be uncontrolled anaerobic glycolysis and lactic acidosis (Chance *et al.*, 1986b; Nioka *et al.*, 1991). There is very little if any acidosis observed after trauma, ADP does not increase without limit, and the $PCr/P_i$ ratio is always above 1.0 (Vink *et al.,* 1988a). It is therefore clear that after mild to moderate trauma there is no $O_2$ limitation of oxidative phosphorylation and that the system is essentially at steady state.

Taking into account the above limitations, it has been recently demonstrated that traumatic injury actually increases mitochondrial oxidative metabolism (Vink *et al.,* 1994). Indeed, there was a very close correlation between decline in cytosolic phosphorylation ratio and increase in mitochondrial oxidative capacity, which is consistent with classic metabolic control theory. Any increased energy demand induced by trauma, including that caused by membrane depolarization, would result in the hydrolysis of ATP. An increased rate of ATP hydrolysis would increase concentrations of ADP and $P_i$, both of which would stimulate glycolysis. Additionally, the rise in ADP would lead to a significant rise in AMP, another stimulus for glycolysis, owing to the adenylate kinase equilibrium. With increased substrate supply, mitochondrial demand would dictate an increase in oxygen supply or lactic acidosis would result (Chance *et al.,* 1986a). Several studies have shown that there is no profound lactic acidosis after trauma, therefore suggesting that oxygen demand is being met. Thus, the increased energy requirements after trauma result in a stimulation of glycolysis with an associated stimulation of mitochondrial oxidative phosphorylation. These observations are consistent with previous observations in trauma including increased glycolytic rate (Hayes *et al.,* 1988; Kawamata *et al.,* 1992), increased oxygen consumption (Nilsson and Nordstrom, 1977; Duckrow *et al.,* 1981), and maintenance of ATP concentrations and pH homeostasis.

## 7. MAGNESIUM TREATMENT

The evidence accumulated from the MRS data described above strongly suggested that magnesium decline may be a critical factor in the development of irreversible tissue injury after brain trauma. However, irrespective of the amount of supporting data, for magnesium to be regarded as a critical factor in trauma, it remained to be demonstrated that manipulating the decline of brain free magnesium concentration after trauma would affect eventual outcome. In an effort to provide such evidence, a series of studies were performed in which the effect of magnesium supplements on neurologic outcome following trauma was examined (McIntosh *et al.,* 1988a; Vink *et al.,* 1988d; McIntosh *et al.,* 1989c; Smith *et al.,* 1993). Infusion of magnesium sulfate prior to injury attenuated both the degree and duration of decline in free magnesium concentration after trauma. This improvement in post-traumatic magnesium homeostasis was associated with a significant improvement in neurologic outcome. Such an improvement in outcome was also observed if magnesium chloride was administered 30 minutes after the traumatic insult (McIntosh *et al.,* 1989c). Not only did magnesium treatment improve neurologic motor outcome in this case, it also improved cognitive function in moderately injured animals (Smith *et al.*, 1993). Interestingly, treatment with MgATP, which has been extensively shown to be protective in non-neural tissue ischemia, was deleterious in brain trauma, presumably because hypotension reduced blood pressure to levels below that required for autoregulation (McIntosh *et al.*, 1989c). In further support for a role for magnesium in outcome following trauma, another series of studies examined the effects of magnesium deficiency on biochemical and neurologic outcome (McIntosh *et al.*, 1988a). Rats were subject to a magnesium-deficient diet to reduce brain free magnesium concentration. Over a 14-day period, brain free magnesium was reduced by up to 15%. After moderate trauma, magnesium-deficient rats did not demonstrate greater declines in free magnesium concentration than control fed animals, consistent with the hypothesis that there is a threshold for free magnesium concentration below which magnesium does not decline. Nonetheless, these magnesium-deficient rats performed significantly worse on neurologic motor tests at two weeks after trauma. This result suggests that resting brain magnesium levels at the time of insult may be a factor determining outcome.

## 8. MALE VERSUS FEMALE ANIMALS

Most investigations of brain trauma have traditionally used male rats in an effort to avoid the effects of hormonal fluctuations on the data. However, a recent MRS study (Emerson and Vink, 1992) indicates that a further factor to consider is free magnesium concentration. These authors showed that female rats have a lower free magnesium concentration than male siblings raised under identical conditions. Consistent with the hypothesis that resting brain magnesium concentration affects outcome after trauma, female animals did significantly worse than their age- or weight-matched male counterparts. Moreover, a later study demonstrated that estrogen supplementation further reduced free magnesium concentration in females but, paradoxically, increased free magnesium levels in males (Emerson *et al.*, 1993). Indeed, estrogen supplementation in males was protective after trauma whereas it demonstrated no such protection in females. The authors proposed that such differences can be accounted for by the higher levels of brain estrogen receptors in females affecting free magnesium levels, whereas in males estrogen's antioxidant properties dominate. Recently, progesterone has been shown to be protective in brain trauma (Roof *et al.*, 1993). Although no MRS studies have been performed to determine any biochemical mechanism of action, progesterone is a well-known magnesium-stabilizing agent and may confer neuroprotection through magnesium-dependent mechanisms. Taken together, it is possible that hormonal fluctuations during the female estrus cycle affect magnesium homeostasis, in itself previously shown to influence outcome following brain trauma.

## 9. ALCOHOL AND TRAUMA

Alcohol is widely regarded as having deleterious effects on trauma, although the mechanism of action is unknown. Several factors have been implicated, including inhibition of neurotransmitter receptors and membrane-bound enzymes, disruption of ion homeostasis, and generalized effects on membrane fluidity (Flamm *et al.*, 1977; Halt *et al.*, 1992). Recent MRS studies of alcohol effects on outcome following trauma have shed some light on this issue. Initial studies on alcohol-affected brain trauma have demonstrated that in moderate trauma alcohol does not exacerbate the bioenergetic response to trauma (Yamakami *et al.*, 1995). The only significant difference from controls was an alkalotic pH after trauma similar to that observed at more severe levels of injury (Vink *et al.*, 1988c). Consistent with this suggestion of a higher injury level, the animals did significantly worse on neurologic motor tests two weeks after the event. The preinjury levels of magnesium that were recorded immediately prior to the insult but two hours after ethanol infusion were not significantly different from the levels in controls. This may be considered somewhat surprising in view of a report demonstrating that acute alcohol significantly reduces free magnesium concentration and bioenergetic state in rat brain (Altura *et al.*, 1992). However, later studies by Mullins and Vink (1995) provided some explanation for this anomaly. These authors demonstrated that acute alcohol's affect on free magnesium and bioenergetic state in rat brain is in fact transient and that all parameters recover within two hours after initial intoxication. Considering that the earlier studies of ethanol-affected brain trauma examined brain magnesium two hours after ethanol administration, it is therefore not surprising that no affect of alcohol on brain magnesium was noted. Furthermore, traumatic injury in alcohol-affected rats would not be expected to exacerbate the post-traumatic decline in free magnesium level because of the magnesium threshold described earlier. Nonetheless, the rats would be expected to do worse neurologically because the brain may be already energetically stressed from the alcohol. Previous studies done several years earlier and repeated on a number of occasions since clearly support the concept of combined insults exacerbating injury (Ishige *et al.*, 1987; Andersen *et al.*, 1988; Ishige *et al.*, 1988). It would therefore seem that alcohol causes energetic stress, not only through the well-known mechanism of reducing NAD+/NADH ratios, but also by affecting ion gradients. By already being in a stressed state, the intoxicated brain is therefore more susceptible to a second insult. Studies of spinal cord injury seem to support the hypothesis that alcohol adversely affects magnesium homeostasis (Halt *et al.*, 1992). Again, ethanol-intoxicated animals did significantly worse than their nonintoxicated controls.

Moreover, the total tissue magnesium concentration remained depressed for a longer period in the ethanol-treated animals as compared to the controls, implicating magnesium concentration as a critical factor in determining neurologic outcome.

## 10. OTHER TREATMENTS

A number of therapies targeted at various secondary injury factors have demonstrated efficacy following brain trauma (McIntosh, 1993). These include NMDA antagonists, thyrotropin-releasing hormone (TRH) analogs, opiate antagonists, and phospholipase C inhibitors. Many of these have been studied by MRS to determine possible mechanisms of action. The first published MRS report on a therapy in trauma concerned the TRH analog CG3703 (McIntosh et al., 1988b). This compound had previously been shown to improve neurologic outcome following trauma (Faden, 1987). The complementary MRS studies demonstrated that such improvement in outcome was associated with improved bioenergetic state (McIntosh et al., 1988b) and magnesium homeostasis (Vink et al., 1988b). Similarly, studies of the noncompetitive NMDA antagonist MK-801 had demonstrated a degree of neuroprotection following brain trauma (McIntosh et al., 1989b). Again, MRS studies showed that MK-801 improved magnesium homeostasis and bioenergetic state after trauma (McIntosh et al., 1990). Similar observations were found in MRS studies of the NMDA antagonists dextrorphan (Faden et al., 1989) and dextromethorphan (Golding and Vink, 1995a), the opiate antagonists nalmefene (Vink et al., 1990b) and nor-binaltorphimine (Vink et al., 1991b), and the phospholipase C inhibitor neomycin (Golding and Vink, 1995b). Indeed, the neuroprotection offered in each case correlated to the degree in improvement of magnesium homeostasis observed after administration of these compounds (Fig. 5).

Somewhat surprisingly, not all NMDA antagonists have been found to be neuroprotective after trauma. Noncompetitive NMDA antagonists such as MK-801, dextrorphan, dextromethorphan, ketamine, and magnesium all have demonstrated efficacy after traumatic brain injury whereas competitive antagonists such as CG19755 have not shown a beneficial effect on neurologic outcome following trauma (McIntosh et al., 1992). In an attempt to resolve this discrepancy, a recent MRS study compared the effects of the metabolic properties of the compounds following moderate trauma on free magnesium and bioenergetic parameters (Golding and Vink, 1995a). The authors reported that the noncompetitive antagonists all improved magnesium homeostasis after trauma whereas the competitive antagonist had no effect on magnesium homeostasis, bioenergetic state, or neurological motor scores after trauma. The authors suggested a number of possibilities to explain these results. First, they suggested that it is the nature of the block of the NMDA channel (within the channel) that is critical in neuroprotec-

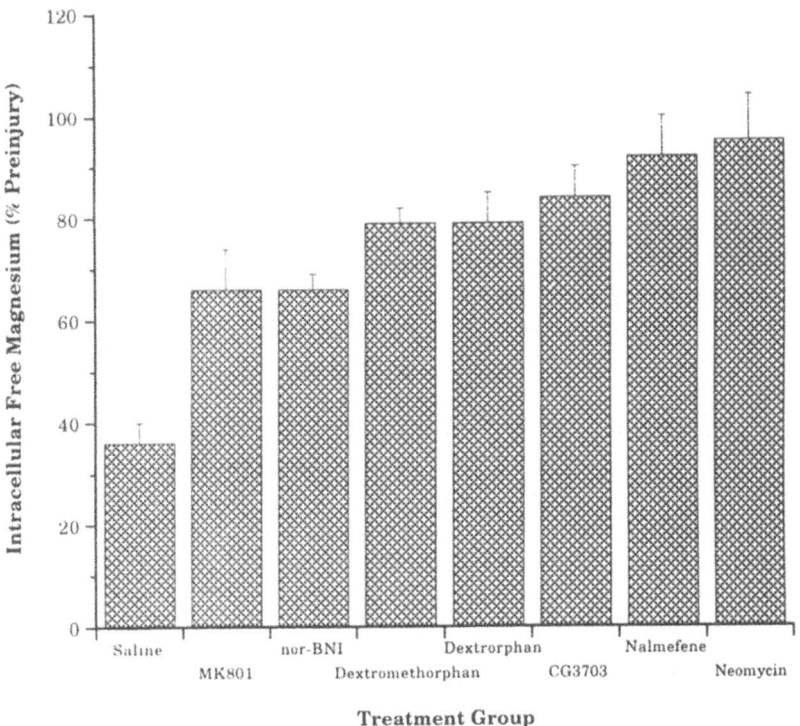

*FIGURE 5.* Mean intracellular free magnesium concentration (expressed as a percentage of the preinjury baseline value) after administration of various treatments following moderate traumatic brain injury.

tion in trauma, and not the block of the receptor in itself. Alternatively, competitive antagonists may not penetrate the blood–brain barrier as readily as the noncompetitive antagonists even though a number of studies have demonstrated blood–brain barrier opening after traumatic injury (Cortez *et al.,* 1989). One final suggestion was that the competitive antagonists are not as effective in blocking the receptor under conditions of high glutamate concentration, although *in vitro* studies would suggest otherwise. Whatever the reason, it seems clear that the noncompetitive antagonists of the NMDA channel seem more readily effective in trauma than the competitive antagonists.

## 11. MODELS OF TRAUMATIC BRAIN INJURY

All of the MRS studies described above have used the lateral fluid percussion injury model in rats. This model involves the rapid injection of a saline bolus into the closed cranial cavity of the rat (Sullivan *et al.*, 1976; McIntosh *et al.*, 1989a). The resultant transient deformation of neural tissue results in a number of pathological alterations that are consistent with those produced in clinical trauma. Nonetheless, this is not the only model that has been used in the study of neurotrauma. A number of groups use a midline model of fluid percussion trauma, which produces much more brain stem damage and less hippocampal damage than the lateral model (Unterberg *et al.*, 1988; Yoshida and Marmarou, 1991). The difficulty with this model, however, is that the region of most injury (brain stem) is rather distant from the surface coil placement used in most MRS studies. Consequently, much of the spectrum is dominated by relatively uninjured tissue, and conclusions drawn may not represent true injury (R. Vink, unpublished results). Indeed, McIntosh *et al.* (1990) have shown in magnesium measurements of total tissue that the decline is restricted to injured tissue and cannot be detected in uninjured cortical tissue adjacent to the site of maximal injury. With the more sophisticated MRS localization techniques that have become available, the model may be useful to more fully characterize the metabolic consequences of midline trauma.

Recently, a model of diffuse axonal injury in rodents has been developed by Marmarou *et al.* (1994). This model successfully reproduces extensive diffuse axonal injury, which is characteristically associated with severe traumatic brain injury and, in particular, appearance of the vegetative state. MRS studies of this model have demonstrated almost identical results to those described in the mild to moderate lateral fluid percussion model, suggesting that the decline in magnesium and bioenergetic state is not model-specific (Heath and Vink, 1994). This view is supported by the results of Smith and colleagues (D. H. Smith, R. Lenkinski, T. A. Gennarelli, and T. K. McIntosh, unpublished results), who have described a linear acceleration model of severe injury in swine and also reported significant declines in both free magnesium concentration and bioenergetic state. Thus, the metabolic effects described in the earlier rat studies are also not species-specific. It therefore seems that, provided the MRS acquisition parameters are carefully set to maximize signal acquisition from injured tissue, the characteristic feature of trauma MRS studies is decline in both free magnesium concentration and bioenergetic state.

## 12. MAGNETIC RESONANCE IMAGING IN TRAUMA

In contrast to MRS studies, the majority of magnetic resonance imaging (MRI) studies in neurotrauma have been of a clinical nature. Zimmerman *et al.* (1986) first illustrated the usefulness of MRI in trauma by demonstrating that the nonhemorrhagic consequences of trauma, including shearing and infarction, are much better displayed by MRI than by X-ray computerized tomography.

These authors further demonstrated that MRI had the capacity to identify causes of unexplained coma and focal neurologic deficits following trauma. Levin *et al.* (1987) later demonstrated the prognostic utility of MRI in clinical management and rehabilitation of head injury by demonstrating that lesion size as determined by MRI paralleled performance in cognition and memory tests. Experimental studies of fluid percussion brain trauma in rats demonstrated that image hyperintensity in $T_2$-weighted MRI images correlated with the area of contusion as shown by conventional histology (Cohen *et al.*, 1991). Moreover, the increase in hyperintensity was also correlated with the injury intensity. Nonetheless, the time to appearance of significant changes was hours after the traumatic event. More recent studies have shown that such changes can be detected within the first hour of trauma when diffusion-weighted sequences are incorporated in the acquisition of MRI images (Hanstock *et al.*, 1994). Images acquired with this technique may also provide information about the type of edema evolving and on metabolic status, although its use for such differentiation is not universally accepted. Nonetheless, diffusion-weighted images do offer rapid identification of regions undergoing pathologic change after trauma. Present-day experimental MRI in trauma concentrates on the use of such functional MRI techniques as prognostic tools. The technique has also been used in a preliminary way to determine cerebral blood flow (perfusion), blood oxygenation and oxygen extraction, and glycolytic rate and for brain mapping (Le Bihan *et al.*, 1993). Indeed, a recent MRI clinical report described the reorganization of the immature brain in response to a focal traumatic brain injury (Cao *et al.*, 1994).

## 13. METHODOLOGICAL CONSIDERATIONS

No attempt will be made here to describe the principles of magnetic resonance since there are a number of excellent books that address these issues (Gadian, 1982; James and Margulis, 1984). However, we will address some issues specific to the application of magnetic resonance to neurotrauma that need consideration: surface coils, quantitation of signals, and chemical shifts.

### 13.1. Surface Coils

Most experimental studies of brain trauma have utilized surface coils, the distinct advantage being the ability to place the coil directly over the region of interest. The signal obtained is representative of tissue directly beneath the coil to a depth of the coil diameter. Therefore, a surface coil with a diameter of 10 mm will have an effective sampling area that would extend to a depth of 10 mm.

While it is readily understood that all tissue (neurons, glia, vasculature) beneath the coil will contribute to the spectrum obtained, it is not so apparent that the signal will also include contributions from tissues above the coil and those within one radius of the coil in the longitudinal plane. Furthermore, the contributions of the tissues are not equal because of the inhomogeneous nature of the field. Tissues very close to the coil (in any direction) will contribute significantly more to the spectrum than tissues farther away from the coil. Conversely, a bulk of tissue at a distance from the coil receiving less excitation may contribute a significant signal to the acquired spectrum simply because of the volume of tissue being excited. It is usual to retract surrounding tissues, like skin and muscle, at least one coil radius away from the coil to minimize contributions from these tissues to a brain spectrum. Nonetheless, studies with large coils may still obtain significant contributions from tissues that are seemingly well clear of the trauma site because of the "wings" of excitation characteristic of surface coils (Bendall, 1984). Alternatively, a variety of localization techniques can be used to concentrate the excitation to the region of interest (Aue, 1986). In either case, it is important for investigators to calibrate the 90° pulse length and state this calibration, together with the size of the coil, in any reports so that it is clear to readers what regions are contributing to the acquired spectrum.

## 13.2. Quantitation

Given the inhomogeneity of surface coils, and of tissue itself, it is extremely difficult to quantitate *in vivo* spectra acquired using surface coils. Field inhomogeneity results in different regions receiving variable degrees of excitation and therefore contributing to each spectrum to a different extent (Bendall, 1984). Furthermore, the relaxation times for each of the metabolites are different, and individual metabolite intensities are therefore differentially affected by acquisition parameters such as repetition rate. Further complicating the issue is that chemical environment and magnet field strength affect relaxation times. Despite these difficulties, several groups have attempted to quantitate spectra, and the reader is referred to these reports for further discussion (Thulborn and Ackerman, 1983; Roth *et al.*, 1989; Cady, 1990; Buchli and Boesiger, 1993).

As a result of these difficulties, most trauma studies have reported metabolic changes in terms of ratios. If one assumes a "normal" preinjury concentration for the metabolites in question using *in vitro* data, the ratios after trauma can be used to calculate approximate relative concentrations. Despite the apparent simplicity of this calculation, it is important that the method used to obtain peak integrals of the individual metabolites in question is made clear. For example, $^{31}$P MR spectra of brain have a broad peak assigned to phospholipids and bone underlying the sharper peaks of the cytosolic metabolites (Gonzalez-Mendez *et al.*, 1984). Failure to remove this peak prior to integration gives completely different ratios from those obtained from spectra analyzed after removal of the broad component. Furthermore, the different methods used to remove the broad component (baseline correction versus convolution difference) gives different ratio values (R. Vink, unpublished results). It is therefore critical that all reports of ratios provide the details of analytical methods used to obtain the data.

### 13.3. Chemical Shifts

As with quantitation, there are important issues involved in the assignment of chemical shift that must be considered. Assuming that phasing of spectra is correct and constant phasing parameters are maintained during a temporal analysis, assignment of chemical shift should be performed on fitted spectra. Methods which assign peak frequencies to raw data often assign the peak shift to the highest intensity point. Because noise may significantly affect such chemical shift assignments, this is not a particularly effective method. Nonetheless, statistically significant changes could still be detected if a sufficient number of experiments were performed because noise is random whereas real changes are not. The preferred method of assigning chemical shifts is to simulate the spectra using computer line-fitting programs that are readily available. This method avoids pitfalls associated with noise affecting chemical shifts. A final issue with interpretation of chemical shifts is that a variety of factors affect peaks, and each of these must be accounted for. For example, the chemical shift of the $\beta$-ATP peak is often used to determine free magnesium concentration (Gupta *et al.*, 1978). However, the chemical shift of this peak is also affected by pH (Cohen and Hughes, 1962), and these effects must be taken into consideration in any interpretation of chemical shift measurements. Furthermore, the standard curve for pH measurements and free magnesium determinations have titration limits within which determination of ion concentration may be considered to be reasonably accurate. Outside of these ranges (6.0–7.0 for pH determinations from the $P_i$ chemical shift and below 1.0 mM for magnesium determinations from the chemical shift of ATP), the values must be treated with caution.

### 14. CONCLUSION

Magnetic resonance investigations of traumatic brain injury have made a significant contribution to the understanding of pathophysiological events associated with brain trauma, particularly with respect to the understanding of energy metabolism. It is now clear that trauma is not necessarily an extension of ischemia but is rather a series of distinct physiological and biochemical changes (secondary injury) that may, or may not, include an ischemic component at the higher injury levels. Moreover, these secondary changes occur over time, with each of the various injury factors being dominant at different time points after the traumatic event. Magnetic resonance spectroscopy and imaging provide a unique opportunity to monitor these changes noninvasively over time and correlate the events with eventual neurologic outcome, be it motor or cognitive. Moreover, the techniques permit the monitoring of the effects of treatment interventions on post-traumatic metabolism and an analysis of each factor's effect on development of irreversible tissue injury. Whereas injury to the brain was not so long ago considered an untreatable and irreversible process, the considerable gain in knowledge of pathophysiological events after trauma over the past 10 to 15 years has resulted in several pharmacologic agents now being trialed in clinical trauma. Magnetic resonance is still the only truly noninvasive technique being successfully used to further characterize the injury process. The continued development of advanced techniques that give functional information on the brain ensure that the technology will continue to have a significant impact in this area.

## REFERENCES

Altura, B. M., Altura, B. T., and Gupta, R. K., 1992, Alcohol intoxication results in rapid loss in free magnesium in brain and disturbances in brain bioenergetics: Relation to cerebrospasm, alcohol-induced strokes, and barbiturate anesthesia-induced deaths, *Magnesium Trace Elem.* **10:**122–135.

Andersen, B. J., and Marmarou, A., 1992, Post-traumatic selective stimulation of glycolysis, *Brain Res.* **585:**184–189.

Andersen, B. J., Unterberg, A. W., Clarke, G. D., and Marmarou, A., 1988, Effect of posttraumatic hypoventilation on cerebral energy metabolism, *J. Neurosurg.* **68:**601–607.

Aue, W. P., 1986, Localization methods for in vivo nuclear magnetic resonance spectroscopy, *Rev. Magn. Reson. Med.* **1:**21–72.

Badar-Goffer, R. S., Thatcher, N. M., Morris, P. G., and Bachelard, H. S., 1993, Neither moderate hypoxia nor mild hypoglycaemia alone causes any significant increase in cerebral $[Ca^{2+}]_i$: Only a combination of the two insults has any effect. A $^{31}P$ and $^{19}F$ NMR study, *J. Neurochem.* **61:**2207–2214.

Badar-Goffer, R. S., Morris, P. G., Thatcher, N. M., and Bachelard, H. S., 1994, Excitotoxic amino acids cause appearance of magnetic resonance spectroscopy observable zinc in superfused cortical slices, *J. Neurochem.* **62:**2488–2491.

Bendall, M. R., 1984, Surface coil and depth resolution using the spatial variation of radiofrequency field, *in* "Biomedical Magnetic Resonance" (T. L. James and A. R. Margulis, eds.), pp. 99–126, Radiology Research and Education Foundation, San Francisco.

Buchli, R., and Boesiger, P., 1993, Comparison of methods for the determination of absolute metabolite concentrations in human muscle, *Magn. Reson. Med.* **30:**552–558.

Cady, E. B., 1990, Absolute quantitation of phosphorus metabolites in the cerebral cortex of the newborn human infant and in the forearm muscles of young adults using a double-tuned surface coil, *J. Magn. Reson.* **87:**433–446.

Cao, Y., Vikingstad, E. M., Huttenlocher, P. R., Towle, V. L., and Levin, D. N., 1994, Functional magnetic resonance studies of the reorganization of the human hand sensorimotor area after unilateral brain injury in the perinatal period, *Proc. Natl. Acad. Sci. USA* **91:**9612–9616.

Chance, B., Eleff, S., and Leigh, J. S., 1980, Noninvasive, nondestructive approaches to cell bio-energetics, *Proc. Natl. Acad. Sci. USA* **77**:7430–7434.

Chance, B., Leigh, J. S., Kent, J., and McCully, K., 1986a, Metabolic control principles and [31]P NMR, *Fed. Proc.* **45**:2915–2920.

Chance, B., Leigh, J. S., Kent, J., McCully, K., Nioka, S., Clark, B. J., Maris, J. M., and Graham, T., 1986b, Multiple controls of oxidative metabolism in living tissues as studied by phosphorus magnetic resonance, *Proc. Natl. Acad. Sci. USA* **83**:9458–9462.

Choi, D. W., 1988, Calcium-mediated neurotoxicity: Relationship to specific channel types and role in ischemic damage, *Trends Neurosci.* **11**:465–469.

Cohen, Y., Sanada, T., Pitts, L. H., Chang, L.-H., Nishimura, M. C., Weinstein, P. R., Litt, L., and James, T. L., 1991, Surface coil spectroscopic imaging: Time and spatial evolution of lactate production following fluid percussion brain injury, *Magn. Reson. Med.* **17**:225–236.

Cohn, M., and Hughes, T. R., 1962, Nuclear magnetic resonance spectra of adenosine di- and triphosphate, *J. Biol. Chem.* **237**:176–181.

Cooper, P. R., 1985, Delayed brain injury: Secondary insults, *in* "Central Nervous System Trauma Status Report—1985" (D. P. Becker and J. T. Povlishock, eds.), pp. 217–228, William Byrd Press/NIH, Bethesda, Maryland.

Corkey, B. E., Duszynski, J., Rich, T. L., Matschinsky, B., and Williamson, J. R., 1986, Regulation of free and bound magnesium in rat hepatocytes and isolated mitochondria, *J. Biol. Chem.* **261**:2567–2574.

Cortez, S., McIntosh, T. K., and Noble, L., 1989, Experimental fluid percussion brain injury: Vascular disruption and neuronal and glial alterations, *Brain Res.* **482**:271–282.

DeSalles, A. A. F., Kontos, H. A., Becker, D. P., Yang, M. S., Ward, J. D., Moulton, R., Gruemer, H. D., Lutz, H., Maset, A. L., Jenkins, L., Marmarou, A., and Muizelaar, P., 1986, Prognostic significance of ventricular CSF lactic acidosis in severe head injury, *J. Neurosurg.* **65**:615–624.

Duckrow, R. B., LaManna, J. C., Rosenthal, M., Levasseur, J. E., and Patterson, J. L., 1981, Oxidative metabolic activity of cerebral cortex after fluid-percussion head injury in the cat, *J. Neurosurg.* **54**:607–614.

Emerson, C. S., and Vink, R., 1992, Increased mortality in female rats after brain trauma is associated with lower free $Mg^{2+}$, *NeuroReport* **4**:957–960.

Emerson, C. S., Headrick, J. P., and Vink, R., 1993, Estrogen improves biochemical and neurologic outcome following traumatic brain injury in male rats but not in females, *Brain Res.* **608**:95–100.

Faden, A. I., 1987, Opiate-receptor antagonists, thyrotropin-releasing hormone (TRH), and TRH analogs in the treatment of spinal cord injury, *Cent. Nerv. Syst. Trauma* **4**:217–226.

Faden, A. I., Demediuk, P., Panter, S. S., and Vink, R., 1989, Excitatory amino acids, *N*-methyl-*D*-asparate receptors and traumatic brain injury, *Science* **244**:798–800.

Fineman, I., Hovda, D. A., Smith, M., Yoshino, A., and Becker, D. P., 1993, Concussive brain injury is associated with a prolonged accumulation of calcium: A [45]Ca autoradiographic study, *Brain Res.* **624**:94–102.

Flamm, E. S., Demopoulos, H. B., Seligman, M. L., Tomasula, J. J., DeCrescito, C., and Ransohoff, J., 1977, Ethanol potentiation of central nervous system trauma, *J. Neurosurg.* **46**:328–335.

Gadian, D. G., 1982, "Nuclear Magnetic Resonance and Its Application to Living Systems," Oxford University Press, New York.

Gennarelli, T. A., 1994, Animate models of human head injury, *J. Neurotrauma* **11**:357–368.

Ginsberg, M. D., Mela, L., Wrobel-Kuhl, K., and Reivich, M., 1977, Mitochondrial metabolism following bilateral cerebral ischemia in the gerbil, *Ann. Neurol.* **1**:519–527.

Golding, E. M., and Vink, R., 1995a, Efficacy of competitive versus noncompetitive blockade of the NMDA channel following traumatic brain injury is associated with restoration of magnesium homeostasis, *Mol. Chem. Neuropathol.* **24**:137–150.

Golding, E. M., and Vink, R., 1995b, Inhibition of phospholipase C with neomycin improves metabolic and neurologic outcome following traumatic brain injury, *Brain Res.* **668**:46–53.

Golding, E. M., McIntosh, T. K., Williams, J. P., and Vink, R., 1994, Blood glucose concentration does not affect outcome following moderate brain trauma in rats, in "Recent Advances in Neurotraumatology—1994" (L. Atkinson and G. S. Merry, eds.), pp. 537–541, World Federation of Neurological Societies, Brisbane.

Goldstein, M., 1990, Traumatic brain injury: A silent epidemic, *Ann. Neurol.* **27**:327.

Gonzalez-Mendez, R., Litt, L., Koretsky, A. P., Von Colditz, J., Weiner, M. W., and James, T. L., 1984, Comparison of ³¹P NMR spectra of in vivo rat brain using convolution difference and saturation with a surface coil. Source of the broad component in the brain spectrum, *J. Magn. Reson.* **57**:526–533.

Gupta, R. K., Benovic, J. L., and Rose, Z. B., 1978, The determination of the free magnesium level in the human red blood cell by ³¹P NMR, *J. Biol. Chem.* **253**:6172–6176.

Gupta, R. K., Gupta, P., Yushok, W. D., and Rose, Z. B., 1983, On the noninvasive measurements of intracellular free magnesium by ³¹P NMR spectroscopy, *Physiol. Chem. Phys. Med. NMR* **15**:265–280.

Halt, P. S., Swanson, R. A., and Faden, A. I., 1992, Alcohol exacerbates behavioral and neurochemical effects of rat spinal cord trauma, *Arch. Neurol.* **49**:1178–1184.

Hanstock, C. C., Faden, A. I., Bendall, M. R., and Vink, R., 1994, Diffusion-weighted imaging differentiates ischemic tissue from traumatized tissue, *Stroke* **25**:843–848.

Hayes, R. L., Katayama, Y., Jenkins, L. W., Lyeth, B. G., Clifton, G. L., Gunter, J., Povlishock, J. T., and Young, H. F., 1988, Regional rates of glucose utilization in the cat following concussive head injury, *J. Neurotrauma* **5**:121–137.

Headrick, J. P., and Willis, R. J., 1991, Cytosolic free magnesium in stimulated, hypoxic, and underperfused rat heart, *J. Mol. Cell. Cardiol.* **23**:991–999.

Heath, D. L., and Vink, R., 1994, NMR characterisation of impact acceleration induced severe traumatic brain injury in rats, *Neurotrauma Soc. Abstr. 1994*:84.

Hovda, D. A., Becker, D. P., and Katayama, Y., 1992, Secondary injury and acidosis, in "Central Nervous System Trauma Status Report—1991" (J. A. Jane, D. K. Anderson, J. C. Torner, and W. Young, eds.), pp. 47–60, Mary Ann Liebert, New York.

Inao, S., Marmarou, A., Clarke, G. D., Andersen, B. J., Fatouros, P. P., and Young, H. F., 1988, Production and clearance of lactate from brain tissue, cerebrospinal fluid and serum following experimental brain injury, *J. Neurosurg.* **69**:736–744.

Ishige, N., Pitts, L. H., Poliani, L., Hashimoto, T., Nishimura, M. C., Bartkowski, H. M., and James, T. L., 1987, The effect of hypoxia on traumatic head injury in rats: Part 2. Changes in high-energy phosphate metabolism, *Neurosurgery* **20**:854–858.

Ishige, N., Pitts, L. H., Berry, I., Nishimura, M., and James, T. L., 1988, The effects of hypovolemic hypotension on high-energy phosphate metabolism of traumatized brain in rats, *J. Neurosurg.* **68**:129–136.

James, T. L., and Margulis, A. R., 1984, "Biomedical Magnetic Resonance," Radiology Research and Education Foundation, San Francisco.

Jelicks, L. A., and Gupta, R. K., 1990, NMR measurement of cytosolic free calcium, free magnesium and intracellular sodium in the aorta of the normal and spontaneously hypertensive rat, *J. Biol. Chem.* **265**:1394–1400.

Jennet, B., Graham, D. I., Adams, H., and Johnston, I. H., 1973, Ischemic brain damage after fatal blunt head injury, in "Cerebral Vascular Diseases: Eighth Conference" (F. H. McDowell and R. W. Brennan, eds.), pp. 163–171, Grune and Stratton, New York.

Kawamata, T., Katayama, Y., Hovda, D. A., Yoshino, A., and Becker, D. P., 1992, Administration of excitatory amino acid antagonists via microdialysis attenuates the increase in glucose utilization seen following concussive brain injury, *J. Cereb. Blood Flow Metab.* **12**:12–24.

Kraus, J. F., and Arzemanian, S., 1989, *in* "Mild to Moderate Head Injury" (J. T. Hoff, T. E. Anderson, and T. M. Cole, eds.), pp. 9–28, Blackwell Scientific Publications, London.

Kushmerick, M. J., Dillon, P. F., Meyer, R. A., Brown, T. R., Krisanda, J. M., and Sweeney, H. L., 1986, $^{31}$P NMR spectroscopy, chemical analysis, and free $Mg^{2+}$ of rabbit bladder and uterine smooth muscle, *J. Biol. Chem.* **261**:14420–14429.

Lawson, J. W., and Veech, R. L., 1979, Effects of pH and free $Mg^{2+}$ on the $K_{eq}$ of the creatine kinase reaction and other phosphate hydrolyses and phosphate transfer reactions, *J. Biol. Chem.* **254**:6528–6537.

Le Bihan, D., Turner, R., Moseley, M. E., and Hyde, J. S., 1993, Functional MRI of the brain, *Magn. Reson. Med.* **30**:405–408.

Levin, H. S., Amparo, E., Eisenberg, H. M., Williams, D. H., High, W. M., McArdle, C. B., and Weiner, R. L., 1987, Magnetic resonance imaging and computerized tomography in relation to the neurobehavioral sequelae of mild and moderate head injury, *J. Neurosurg.* **66**:706–713.

Li, H. Y., Dai, L. J., Krieger, C., and Quamme, G. A., 1993, Intracellular $Mg^{2+}$ concentration following metabolic inhibition in opossum cells, *Biochim. Biophys. Acta 1181*:307–315.

London, R. E., 1991, Methods for measurement of intracellular magnesium: NMR and fluorescence, *Annu. Rev. Physiol.* **53**:241–258.

Marmarou, A., Foda, M. A. A., Van den Brink, W., Campbell, J., Kita, H., and Demetriadou, K., 1994, A new model of diffuse brain injury in rats; Part I: Pathophysiology and biomechanics, *J. Neurosurg.* **80**:291–300.

Mayer, M. L., Westbrook, G. L., and Guthrie, P. B., 1984, Voltage-dependent block by $Mg^{2+}$ of NMDA responses in spinal cord neurons, *Nature (London)* **309**:261–263.

McIntosh, T. K., 1993, Novel pharmacologic therapies in the treatment of experimental traumatic brain injury: A review, *J. Neurotrauma* **10**:215–261.

McIntosh, T. K., Faden, A. I., Bendall, M. R., and Vink, R., 1987, Traumatic brain injury in the rat: Alterations in brain lactate and pH as characterized by $^1$H and $^{31}$P nuclear magnetic resonance, *J. Neurochem.* **49**:1530–1540.

McIntosh, T. K., Faden, A. I., Yamakami, I., and Vink, R., 1988a, Magnesium deficiency exacerbates and pretreatment improves outcome following traumatic brain injury in rats: $^{31}$P magnetic resonance spectroscopy and behavioural studies, *J. Neurotrauma* **5**:17–31.

McIntosh, T. K., Vink, R., and Faden, A. I., 1988b, An analog of thyrotropin-releasing hormone improves outcome after traumatic brain injury: $^{31}$P NMR studies, *Am. J. Physiol.* **254**:R785–R792.

McIntosh, T. K., Vink, R., Noble, L. J., Yamakami, I., Fernyak, S. E., Soares, H., and Faden, A. I., 1989a, Traumatic brain injury in the rat: Characterization of a lateral fluid percussion injury model, *Neuroscience* **28**:233–244.

McIntosh, T. K., Vink, R., Soares, H., Hayes, R., and Simon, R., 1989b, Effects of the *N*-methyl-*D*-aspartate receptor blocker MK-801 on neurologic function after experimental brain injury, *J. Neurotrauma* **6**:247–259.

McIntosh, T. K., Vink, R., Yamakami, I., and Faden, A. I., 1989c, Magnesium protects against neurological deficit after brain injury, *Brain Res.* **482**:252–260.

McIntosh, T. K., Vink, R., Soares, H., Hayes, R. L., and Simon, R. P., 1990, Effect of noncompetitive blockade of *N*-methyl-*D*-aspartate receptors on the neurochemical sequelae of experimental brain injury, *J. Neurochem.* **55**:1170–1179.

McIntosh, T. K., Smith, D. H., Hayes, R. L., Vink, R., and Simon, R. P., 1992, Role of excitatory amino acid neurotransmitters in the pathogenesis of traumatic brain injury, *in* "Excitatory Amino Acids" (R. P. Simon, ed.), pp. 247–253, Thieme Medical Publishers, New York.

Mullins, P. G. M., and Vink, R., 1995, Chronic alcohol exposure decreases brain intracellular free magnesium concentration in rats, *NeuroReport* **6**:1633–1636.

Myers, R. E., and Yamaguchi, M., 1976, Effects of serum glucose concentration on brain response to circulatory arrest, *J. Neuropathol. Exp. Neurol.* **35**:301.

Nilsson, B., and Nordstrom, C.-H., 1977, Rate of cerebral energy consumption in concussive head injury in the rat, *J. Neurosurg.* **47**:274–281.

Nioka, S., Smith, D. S., Mayevsky, A., Dobson, G. P., Veech, R. L., Subramanian, H., and Chance, B., 1991, Age dependence of steady state mitochondrial oxidative metabolism in the *in vivo* hypoxic dog brain, *Neurol. Res.* **13**:25–32.

Rango, M., Lenkinski, R. E., Alves, W. M., and Gennarelli, T. A., 1990, Brain pH in head injury: An image-guided $^{31}$P magnetic resonance spectroscopy study, *Ann. Neurol.* **28**:661–667.

Ring, I. T., Berry, G., Dan, N. G., Kwok, B., Mandryk, J. A., North, J. B., Selecki, B. R., Sewell, M. F., Simpson, D. A., Steining, W. A., and Vanderfield, G. K., 1986, Epidemiology and clinical outcomes of neurotrauma in New South Wales, *Aust. N. Z. J. Surg.* **56**:557–566.

Roof, R. L., Duvdevani, R., and Stein, D. G., 1993, Gender influences outcome of brain injury: Progesterone plays a protective role, *Brain Res.* **607**:333–336.

Roth, K., Hubesch, B., Meyerhoff, D. J., Naruse, S., Gober, J. R., Lawry, T. J., Boska, M. D., Matson, G. B., and Weiner, M. W., 1989, Noninvasive quantitation of phosphorus metabolites in human tissue by NMR spectroscopy, *J. Magn. Reson.* **81**:299–311.

Siesjö, B. K., 1978, "Brain Energy Metabolism," John Wiley & Sons, New York.

Siesjö, B. K., 1988, Acidosis and ischemic brain damage, *Neurochem. Pathol.* **9**:31–88.

Sims, N. R., and Pulsinelli, W. A., 1987, Altered mitochondrial respiration in selectively vulnerable brain subregions following transient forebrain ischemia in the rat, *J. Neurochem.* **49**:1367–1374.

Smith, D. H., Okiyama, K., Gennarelli, T. A., and McIntosh, T. K., 1993, Magnesium and ketamine attenuate cognitive dysfunction following experimental brain injury, *Neurosci. Lett.* **157**:211–214.

Sullivan, H. G., Martinez, J., Becker, D. P., Miller, J. D., Griffith, R., and Wist, A. O., 1976, Fluid-percussion model of mechanical brain injury in the cat, *J. Neurosurg.* **45**:520–534.

Sutton, L. N., Wang, Z., Duhaime, A. C., Costarino, D., Sauter, R., and Zimmerman, R., 1995, Tissue lactate in pediatric head trauma: A clinical study using $^1$H NMR spectroscopy, *Pediatr. Neurosurg.* **22**:81–87.

Thulborn, K. R., and Ackerman, J. J. H., 1983, Absolute molar concentrations by NMR in inhomogenous $B_1$. A scheme for analysis of *in vivo* metabolites, *J. Magn. Reson.* **55**:357–371.

Unterberg, A., Andersen, B. J., Clarke, G. D., and Marmarou, A., 1988, Cerebral energy metabolism following fluid percussion brain injury in cats, *J. Neurosurg.* **68**:594–600.

Vink, R., 1989, Phospholipase C activity reduces free magnesium concentration, *Biochem. Biophys. Res. Commun.* **165**:913–918.

Vink, R., 1993, Nuclear magnetic resonance characterization of secondary mechanisms following traumatic brain injury, *Mol. Chem. Neuropathol.* **18**:279–297.

Vink, R., McIntosh, T. K., Weiner, M. W., and Faden, A. I., 1987, Effects of traumatic brain injury on cerebral high energy phosphates and intracellular pH; A $^{31}$P magnetic resonance spectroscopy study, *J. Cereb. Blood Flow Metab.* **7**:563–571.

Vink, R., Faden, A. I., and McIntosh, T. K., 1988a, Changes in cellular bioenergetic state following graded traumatic brain injury in rats: Determination by phosphorus-31 magnetic resonance spectroscopy, *J. Neurotrauma* **5**:365–380.

Vink, R., McIntosh, T. K., and Faden, A. I., 1988b, Treatment with the thyrotropin-releasing hormone analog CG3703 restores magnesium homeostasis following traumatic brain injury in rats, *Brain Res.* **460**:184–188.

Vink, R., McIntosh, T. K., Demediuk, P., Weiner, M. W., and Faden, A. I., 1988c, $^{31}$P NMR characterization of graded traumatic brain injury in rats, *Magn. Reson. Med.* **6**:37–48.

Vink, R., McIntosh, T. K., Demediuk, P., Weiner, M. W., and Faden, A. I., 1988d, Decline in intracellular free magnesium concentration is associated with irreversible tissue injury following brain trauma, *J. Biol. Chem.* **263**:757–761.

Vink, R., Head, V. A., Rogers, P. J., McIntosh, T. K., and Faden, A. I., 1990a, Mitochondrial metabolism following traumatic brain injury in rats, *J. Neurotrauma* **7**:21–27.

Vink, R., McIntosh, T. K., Rhomhanyi, R., and Faden, A. I., 1990b, Opiate antagonist nalmefene improves intracellular free magnesium, bioenergetic state and neurologic outcome following traumatic brain injury in rats, *J. Neurosci.* **10**:3524–3530.

Vink, R., McIntosh, T. K., and Faden, A. I., 1991a, Magnesium in neurotrauma: Its role and therapeutic implications, *in* "Magnesium and Excitable Membranes" (P. Strata and E. Carbone, eds.), pp. 695–701, Springer-Verlag, Berlin.

Vink, R., Portoghese, P. S., and Faden, A. I., 1991b, Kappa-opioid antagonist improves cellular bioenergetics and recovery after traumatic brain injury, *Am. J. Physiol.* **261**:1527–1532.

Vink, R., Golding, E. M., and Headrick, J. P., 1994, Bioenergetic analysis of oxidative metabolism following traumatic brain injury in rats, *J. Neurotrauma* **11**:265–274.

Williams, S. R., Crockard, A., and Gadian, D. G., 1989, Cerebral ischemia studied by nuclear magnetic resonance spectroscopy, *Cerebrovasc. Brain Metab. Rev.* **1**:91–114.

Yamakami, I., Vink, R., Faden, A. I., Gennarelli, T. A., Lenkinski, R., and McIntosh, T. K., 1995, Effects of acute ethanol intoxication on experimental traumatic brain injury in the rat: Neuro-behavioral and $^{31}$P NMR studies, *J. Neurosurg.* **82**:813–821.

Yoshida, K., and Marmarou, A., 1991, Effects of tromethamine and hyperventilation on brain injury in the cat, *J. Neurosurg.* **74**:87–96.

Young, W., and Constantini, S., 1994, Ionic and water shifts in injured central nervous tissues, *in* "The Neurobiology of Central Nervous System Trauma" (A. I. Faden and S. K. Salzman, eds.), pp. 123–130, Oxford University Press, New York.

Zimmerman, R. A., Bilanuik, L. T., Hackney, D. B., Goldberg, H. I., and Grossman, R. I., 1986, Head injury: Early results of comparing CT and high-field MR, *Am. J. Radiol.* **147**:1215–1222.

# CHAPTER 5

# ANIMAL MODELS
# OF STROKE

## TERRI L. C. LUVISOTTO and GARNETTE R. SUTHERLAND

## SUMMARY

Models of global and focal cerebral ischemia described in the scientific literature are characterized by a broad spectrum of animal species, methodological variations, and experimental designs. In testing a particular research hypothesis, the investigator is presented with a challenging task: the selection of a model that is most suited to the dimension of ischemia pathophysiology under study. The solution to this problem requires some foreknowledge of the paradigms utilized historically, those most widely used today, and those models that have consistently demonstrated the ability to produce predictable and reproducible histopathological outcomes in animal subjects. The goal of the discussion in this chapter is to present a comprehensive and critical overview of the animal models of global and focal cerebral ischemia. Pitfalls associated with inadequate experimental control, poorly constructed study populations, and the applicability of

TERRI L. C. LUVISOTTO and GARNETTE R. SUTHERLAND • Department of Clinical Neurosciences, Division of Neurosurgery, The University of Calgary, Calgary, Alberta, Canada T2N 1N4.

Magnetic Resonance Spectroscopy and Imaging in Neurochemistry, Volume 8 of Advances in Neurochemistry, edited by Rachelard, Plenum Press, New York, 1997.

animal models to the context of human stroke are examined. Finally, future directions in experimental and clinical cerebral ischemia research are presented in light of these considerations.

## 1. INTRODUCTION

Stroke models have been the subject of considerable debate (Wiebers *et al.,* 1970; Millikan, 1992; Zivin and Grotta, 1990). It has been suggested that because the pathophysiology of ischemic cerebrovascular disease is exceedingly complex and its etiology, antecedent risk factors, and clinical manifestations are notable for their considerable diversity, a stroke model designed to address all these associated variables would be virtually impossible to establish. However, this particular view fails to acknowledge that the success of the scientific method depends upon the ability of the researcher to develop a manageable, working representation of real-life events that permits precise control over independent and potentially confounding variables. In short, the puzzle must be assembled piece by piece. To expect to create a meaningful picture without a systematic plan to identify each of its component parts is to ignore a very basic and valid approach to the design and testing of any scientific hypothesis.

Thus, the principal strength of models of cerebral ischemia is their ability to provide a scaled-down version of the human condition while at the same time permitting control over certain physiologic variables. Furthermore, the need for reproducibility of experimental findings requires that attempts be made to produce strokes that are reasonably similar in severity between test animals. The study of therapeutic agents also demands that some degree of uniformity exist among subjects and controls; a possibility which is more plausible in carefully designed and appropriately standardized animal stroke models.

Models of focal and global cerebral ischemia supply the investigator with a number of additional benefits. First, the ability to manipulate the duration and severity of the ischemic insult and alter the timing of therapeutic interventions allows the investigator to study pathophysiological and neurochemical consequences in a stepwise, time-dependent fashion not possible outside of the laboratory. Second, the analysis of these factors often necessitates the use of techniques and surgical procedures which may be invasive and relatively traumatic. Because of the limitations and ethical considerations associated with human experimentation, animal and *in vitro* models become invaluable in supplying the substrate needed for pathophysiologic and neurochemical investigations and for the testing of the efficacy and potential toxicity of various treatment regimens. Third, animal models specifically offer some advantages over their *in vitro* counterparts by virtue of the fact that they, by definition, more closely approximate the human condition. The disturbance in perfusion that constitutes the basis of ischemic injury is most faithfully represented by a functioning vascular network; the

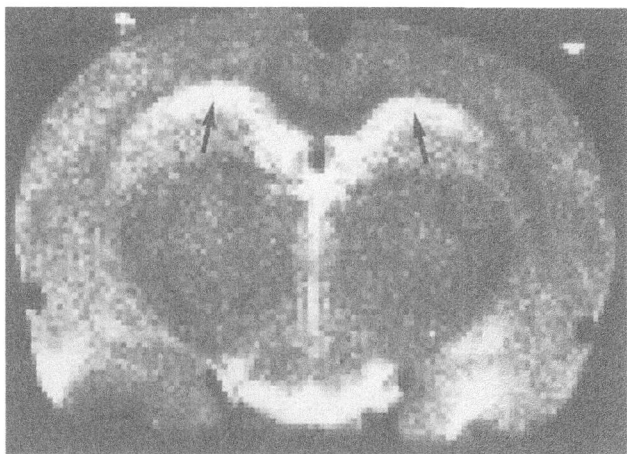

FIGURE 1. $T_2$-weighted MR images of the rat brain taken on day 2 postischemia (global ischemia). The increased intensity in the images is confined to the CA1 region of the hippocampus and corresponds spatially and temporally to the delayed appearance of histopathologically defined neuronal injury. (Reproduced, with permission from the Upjohn Company, Kalamazoo, Michigan, from Current Concepts "Progressing towards Effective Stroke Treatment," 1994.)

biochemical milieu can be approximated in the media of cell cultures, but it is not possible to fully mimic that provided by the dynamically responsive cerebral vasculature. Finally, in addition to providing researchers with the opportunity to safely assess the effects of specific therapeutic agents on the evolution of ischemic brain damage, certain models can be useful in helping to validate technological and investigational advances destined to be applied to humans. For example, animal stroke models can provide the opportunity to correlate postinfarct gross, histopathological, neurochemical, and behavioral changes with characteristic appearances or changes identified by magnetic resonance imaging and spectroscopy (Figs. 1 and 2). Thus, as newer techniques are developed with the ultimate goal of studying the pathophysiology, metabolic alterations, and angioarchitectural interactions which occur following a cerebral ischemic event in living humans, animal subjects will continue to aid in bridging the gap between progress in the laboratory setting and its subsequent clinical application in human stroke.

## 2. GLOBAL VERSUS FOCAL CEREBRAL ISCHEMIA— DEFINITIONS

Global or forebrain ischemia refers to the transient but complete or near complete cessation of cerebral blood flow (CBF). Its clinical correlate is the

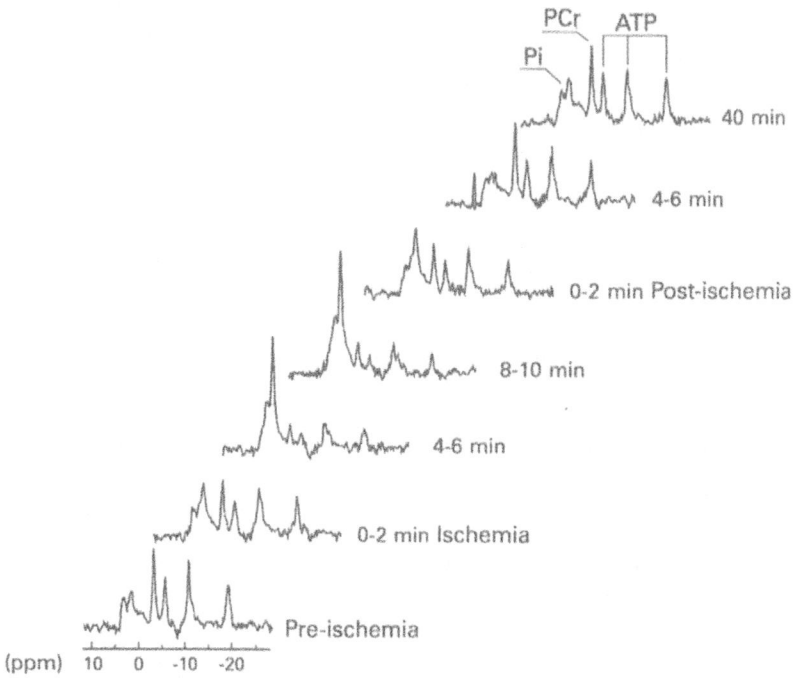

FIGURE 2. Stacked [31]P spectra and plots of ATP and inorganic phosphate (Pi) vs. time. [31]P MR spectra (*left*) of the rat brain show that global ischemia is accompanied by a rapid decrease in the intensities of the peaks due to ATP and phosphocreatine (PCr) and an increase in the Pi peak as the high-energy compounds are hydrolyzed. The position of the Pi peak in the spectrum gives an accurate measure of intracellular pH. Reperfusion following short-duration global ischemia is accompanied by recovery of ATP and PCr. The time course of the changes in the intensities of the ATP and Pi signals gives a measure of the rate of energy failure during ischemia and energy recovery on reperfusion. (Reproduced, with permission from the Upjohn Company, Kalamazoo, Michigan, from Current Concepts "Progressing towards Effective Stroke Treatment," 1994.)

brain's vulnerability to damage within the context of cardiovascular failure, for example, cardiac arrest. The nature of the insult is such that flow must be reinstated within a short interval of time, usually less than 30 minutes, if functional, nonvegetative recovery is to be made (Kirino, 1982; Smith *et al.*, 1984). Thus, animal models designed to study global cerebral ischemia need to be readily and successfully reversible.

The common pathogenic factor underlying models of global or forebrain ischemia is the loss of blood flow and its accompanying energy and cellular membrane failure (Hruska *et al.*, 1992; Sutherland *et al.*, 1992a). Upon reperfusion, CBF often recovers to supranormal levels (*hyperemia*) and, within approx-

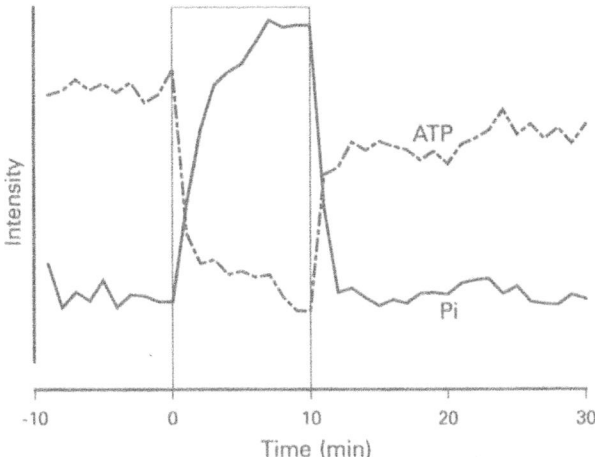

*Figure 2.* Continued

imately 10 minutes, will decrease to half of the level seen in controls (*delayed hypoperfusion*) (Pulsinelli and Jacewicz, 1992).

Histologically, global/forebrain ischemia produces irreversible damage limited to selectively vulnerable neuronal populations; of note is that the extent of injury becomes apparent only after a period of maturation. The time course over which this ischemic maturation occurs may be hours or days, depending on the duration of the ischemia, and varies between different selectively vulnerable regions. Examples of ischemia-sensitive neurons include the hippocampal CA1 pyramidal neurons, the cerebellar Purkinje cells, striatal neurons of medium or small size, and pyramidal neurons in the midneocortical layers.

*Focal cerebral ischemia* can be produced experimentally via the implementation of either extravascular compressive methods or intravascular occlusion. The vessel most commonly affected in stroke syndromes is the middle cerebral artery (MCA). The nature of the injury in focal brain ischemia is characterized by a core of densely and often irreversibly injured tissue surrounded by an "ischemic penumbra": a region of variably ischemic tissue that has the potential to progress to a state of irreversible damage under unfavorable conditions (Fig. 3).

The histopathological sequelae of focal ischemia are distinctive from those seen in global cerebral ischemia. Whereas the damage seen in global models is concentrated to selectively vulnerable neuronal populations, that observed in focal cerebral ischemia involves both neurons and glia—pannecrosis. Variations in histologic outcome occur secondary to differences in the severity and duration of ischemia (Hossmann and Schuier, 1980).

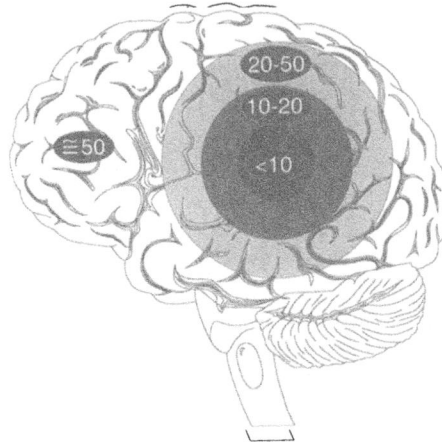

*FIGURE 3.* The ischemic core and penumbra produced by focal cerebral ischemia. At the core of the focal ischemic event, blood flow falls below 10 ml per 100 g of tissue per minute. The ischemic penumbra, containing potentially salvageable tissue, consists of regions of the brain around the core where blood flow is reduced below the normal 50 ml per 100 g of tissue per minute. (Reproduced, with permission from the Upjohn Company, Kalamazoo, Michigan, from Current Concepts "Progressing towards Effective Stroke Treatment," 1994.)

## 3. MODELS OF GLOBAL/FOREBRAIN CEREBRAL ISCHEMIA

An important distinction between global and forebrain ischemia is that in the latter the ischemia is presumed to be incomplete but severe; in contrast, global ischemia results in the transient but complete cessation of blood flow and therefore is characterized by the total absence of substrate delivery to both fore- and hindbrain structures. These differences understandably lead to certain study design and methodological considerations; the additional compromise of brain stem function in global models tends to result in higher mortality as well as the absolute requirement for mechanical ventilation until respiratory centers return to a satisfactory level of coordinated activity following reperfusion. Table 1 presents an overview of different models of global/forebrain ischemia and provides representative examples of each method based on species.

### 3.1. Occlusion of the Extracranial Cerebral Vessels

Bilateral common carotid occlusion with (rat) or without (gerbil) added hypotension (two-vessel occlusion) and bilateral common carotid artery occlusion combined with occlusion of the vertebral arteries (four-vessel occlusion) are the most widely employed models of forebrain ischemia.

#### 3.1.1. The Two-Vessel Occlusion Model

Initially described over two decades ago, temporary surgical ligation of the common carotid arteries, with or without the concomitant use of pharmacologic antihypertensives, results in severe forebrain ischemia (Smith *et al.*, 1984). In the rat model, the addition of systemic hypotension by way of controlled blood

TABLE 1. Global Models of Cerebral Ischemia

| Method | Species references | | | | |
|---|---|---|---|---|---|
| | Primate | Dog | Cat | Rabbit | Rat/gerbil |
| Two-vessel occlusion (carotid arteries +/− hypotension) | Sengupta et al., 1973 | | Ginsberg et al., 1978; Welsh et al., 1977 | | Smith et al., 1984; Kirino, 1982 |
| Four-vessel occlusion (carotid and vertebral arteries) | Wolin et al., 1972 | Fujishima, 1971 | Ginsberg et al., 1979 | Kolata, 1979 | Pulsinelli and Buchan, 1988; Blomqvist et al. 1984 |
| Cardiac arrest | Myers and Yamaguchi, 1977 | Gurvitch et al., 1972 | Hossmann and Hossmann, 1973 | | Blomqvist and Wieloch, 1985 |
| Elevation of cerebrospinal fluid pressure | Heyreh and Edwards, 1971 | Hallenbeck and Bradley, 1977 | van Harreveld, 1947 | Marshall et al., 1975 | Busto and Ginsberg, 1985 |
| Neck tourniquet | Nemoto et al., 1977 | Eleff et al., 1991 | Nemoto et al., 1981 | Kowada et al., 1968 | Siemkowicz and Hansen, 1978 |
| Occlusion of ascending +/− intrathoracic branches | Miller and Myers, 1970 | Jackson et al., 1981 | Jenkins et al., 1979 | Cantu et al., 1969 | de la Torre and Fortin, 1991 |
| Decapitation, isolated head preparations | White et al., 1967 | Hinzen et al., 1972 | Hallenbeck and Bradley, 1977 | | Abe et al., 1983 |

withdrawal into a heparinized syringe produces a consistent injury by decreasing collateral blood flow. In contrast, because the gerbil normally lacks communicating arteries between the carotid and vertebral arterial systems, the induction of hypotension is unnecessary in the gerbil model (Kirino, 1982). The procedure is typically produced in anesthetized, mechanically ventilated rodents and, provided physiologic variables are well controlled, is noted for its success in producing relatively consistent damage to selectively vulnerable neuronal populations.

Advantages of the two-vessel occlusion model are:

- Straightforward surgical preparation
- Well-established models, having undergone rigorous histopathological, neurochemical, and neurobehavioral studies
- Predictable and reproducible neuronal damage
- Correlation of severity of injury with ischemic duration
- Low animal mortality

Disadvantages include:

- Operative stress and requirement for anesthesia
- Potential carotid artery injury
- Necessity of physiological assessment of ischemic severity [e.g., electroencephalogram (EEG)]
- Routine blood gas monitoring difficult in the gerbil due to small vessel size

### 3.1.2. The Four-Vessel Occlusion Model

Occlusion of both common carotid arteries and vertebral arteries is accomplished by a two-stage surgical procedure (Pulsinelli and Buchan, 1988). In the first stage, the rats are anesthetized and have tourniquets placed around both common carotid arteries. The portion of these devices that permits tightening of the tourniquets is then guided externally through a midline neck incision. During this procedure, the vertebral arteries are electrocauterized through the alar foramina of the first cervical vertebrae. The second stage of the procedure results in the generation of global ischemia in the awake rodent by tightening the exteriorized tourniquets for a predetermined length of time, usually 10–20 minutes. The objective is to produce a high-grade forebrain ischemia, which may be assessed by observing the rat's behavioral response (unresponsiveness/loss of righting reflex) and can be verified using an isoelectric EEG as an indicator of the severity of ischemia. By first cauterizing the vertebral arteries and then temporarily occluding the common carotid arteries in two separate surgical procedures with adjunctive mild arterial hypotension, a relatively consistent experimental result can also be obtained in the anesthetized rat (Pulsinelli and Buchan, 1988).

Advantages of the four-vessel occlusion model of forebrain ischemia in the rodent include:

- Option to utilize model in either awake or anesthetized animals
- Well-established model, having undergone rigorous histopathological, metabolic, and neurobehavioral studies
- Correlation of severity of injury with ischemic duration

Disadvantages include:

- Need for a first-stage surgical procedure that is technically demanding
- Inter- and intralaboratory variability in achieving consistent results
- No possibility of blood pressure control during occlusion in the awake animal
- High mortality rate associated with vessel occlusion durations of greater than 20 minutes

## 3.2. Cardiac Arrest

Cardiovascular standstill has been utilized as a means of producing global ischemia in animal models (Gurvitch *et al.*, 1972; Blomqvist and Wieloch, 1985). Unfortunately, cessation of myocardial function using techniques such as electric shock, respiratory arrest, or potassium chloride infusion is much less technically difficult to achieve than successful resuscitation of the animals under study. Other important disadvantages of the model include secondary effects on the brain following ischemic damage to other key organ systems and difficulty in discerning the efficacy of pharmacologic therapy in facilitating neurologic recovery, given this coexistent multiorgan dysfunction. Furthermore, the inability to precisely control the duration of ischemia because of unpredictable lengths of time required for resuscitation and the distinct potential for mortality resulting from associated brain stem injury are undesirable complications of this model. Owing to these significant disadvantages, the cardiac arrest model has been of limited use.

## 3.3. Elevation of Cerebrospinal Fluid Pressure

Investigators have produced cerebral ischemia by infusing artificial cerebrospinal fluid (CSF) into the subarachnoid space to produce a pressure that exceeds arterial blood pressure (Marshall *et al.*, 1975; Busto and Ginsberg, 1985). Depending on the extent of the pressure difference, the procedure results in either complete or incomplete ischemia. Drugs often need to be administered to prevent concomitant reflex arterial hypertension.

A distinct disadvantage of this model is that the infusion of mock CSF to raise intracranial pressure and subsequently cause a reduction in effective cerebral perfusion pressure is nonphysiologic; moreover, the procedure has been shown to increase the risk of meningitis due to the frequently encountered complication of CSF rhinorrhea, felt to be secondary to cribriform plate disruption (Ginsberg and Busto, 1989).

## 3.4. Neck Tourniquet

A cervical pneumatic cuff inflated to high pressures has been used to occlude both cranial arteries and veins, preventing flow to forebrain structures as well as venous return (Nemoto *et al.*, 1977; Kowada *et al.*, 1968). Other surgical manipulations designed to prevent blood flow through the vertebrobasilar system and anterior spinal arteries (Kabat *et al.*, 1941), for example, C2 laminectomy, hypotension via controlled hemorrhage, or surgical occlusion of these vessels, have been added to produce global cerebral ischemia, which does not occur with application of the pressure cuff in isolation. Critics have commented that cerebral venous stasis and significant pressure on other cervical structures are undesirable complications of this particular model, limiting its overall utility.

## 3.5. Surgical Occlusion of the Ascending Aorta

Surgical occlusion of all or selected thoracic vessels produces either global or forebrain ischemia in anesthetized animals ( Jackson *et al.*, 1981). This method also results in total body ischemia with accompaning damage to the heart and lungs. A modification of this experimental procedure has been developed to preserve flow to the lungs and heart (Brockman and Jude, 1960) but does not eliminate variables related to operative stress and the ischemic injury of other involved organ systems.

## 3.6. Decapitation

The model involves the production of total, irreversible cerebral ischemia by way of decapitation. The design is well suited to the generation of *in vitro* biochemical analyses and tissue slice preparation for both physiologic and neurochemical analysis (Abe *et al.*, 1983).

## 4. MODELS OF FOCAL CEREBRAL ISCHEMIA

Methods of producing focal cerebral infarction in animals have evolved considerably over the past half century. Despite numerous variations on this theme, only a few such models have been standardized. Table 2 represents an overview of different models of focal ischemia and provides representative samples of each method based on species. With technological advances and improvements in the physiological monitoring of smaller animals, the rodent has become

TABLE 2. Models of Focal Cerebral Ischemia

| Method | Species references | | | | | Reversible |
|---|---|---|---|---|---|---|
| | Primate | Dog | Cat | Rabbit | Rat | |
| Intraluminal occlusion (wire/thread/retractable emboli) | Bremer et al., 1975 | Molinari, 1970 | | Molnar et al., 1988 | Nagasawa and Kogure, 1989; Kawamura et al., 1991 | Yes |
| Extraluminal occlusion | | | | | | |
| Permanent (ligation/cautery) | Hudgins and Garcia, 1970 | Anthony et al., 1963 | O'Brien and Waltz, 1973 | Ross Russell, 1971 | Chen et al., 1986 | No |
| Microclips/ligatures/hooks | Crowell et al., 1981 | Anthony et al., 1963 | Little, 1977 | Myer et al., 1986 | Buchan et al., 1992; Shigeno et al., 1985 | Yes |
| Vasoactive agents | | Mima et al., 1989 | Robinson and McCulloch, 1990 | | Sharkey et al., 1993 | |
| Thromboembolic techniques | | | | | | |
| Microemboli (artificial/autologous clots) | Brassel et al., 1989 | Hill et al., 1955 | Fritz and Hossmann, 1979 | Hegedus et al., 1985 | Kudo et al., 1982 | Variable |
| Photochemical method/platelet proaggregants | Fritz and Levien, 1989 | | | Fieschi et al., 1975 | Futrell et al., 1988; Furlow and Bass, 1976 | Variable |
| Subarachnoid hemorrhage | Petruk et al., 1972 | Asano and Sano, 1977 | Shigeno et al., 1982 | Diringer et al., 1994 | Solomon et al., 1985; Doczi et al., 1986 | Variable |

the most commonly used model of MCA occlusion (Molinari and Laurent, 1976). Reasons for this popularity include low animal cost and reduced procedural expenses; the demonstration of similarities in physiology, neurochemistry, and cerebrovascular architecture to those seen in higher species; the ability to produce strains that are relatively homogeneous due to inbreeding; and, finally, enhanced acceptance from a bioethical standpoint. The larger-brained animals such as primate, cat, and dog are, however, important in establishing correlation between species and in further evaluating pharmacologic agents prior to their introduction into multicenter clinical trials. It is becoming generally accepted that consistency between animal species will likely translate into similar findings in humans.

## 4.1. Models of Irreversible Focal Cerebral Ischemia

There have been a variety of vascular occlusion techniques developed to interfere with flow through the MCA in the rodent, the most prominent being the subtemporal approach with diathermy occlusion (Tamura *et al.*, 1981). As this particular technique results in permanent MCA occlusion, other investigators have developed procedures that permit reversal of the occlusion and subsequent reperfusion into the ischemic or infarcted region. In larger-brained animals, the transorbital dissection to the carotid bifurcation and proximal MCA is the favored approach. By necessity, these models require enucleation of the eye, with its resulting morbidity and changes in neurochemistry and behavior (Little, 1977; Hudgins and Garcia, 1970).

There are some general advantages of models of permanent MCA occlusion utilizing extravascular compressive techniques with or without the addition of systemic hypotension. These benefits include the ability to visually control the conditions under which ischemia is produced and, using well-standardized models, to produce a relatively predictable lesion. Potential drawbacks associated with the model include production of an ischemic insult that arises by a process which is distinctly unlike clinical stroke; moreover, it necessitates the use of a surgical procedure that is traumatic and can result in the introduction of unwanted variables: blood loss, alterations in intracranial pressure, heat loss, and neurochemical changes associated with anesthesia. By definition, the MCA is irreversibly damaged, precluding the possibility of studying changes associated with reperfusion.

## 4.2. Models of Reversible Focal Cerebral Ischemia

Models incorporating techniques into their design which permit reperfusion are felt to be more representative of the pathophysiological events that occur in human stroke. It is uncommon for flow to a particular territory of the brain to cease permanently; angiographically documented spontaneous recanalization has been shown to occur in up to half of all patients who suffer from an acute occlusion of the MCA (Mhairi Macrae, 1992). Thus, models of reversible cerebral ischemia have, as their principal advantage, the desirable quality of being relatively similar to the clinical condition and may therefore yield clinically relevant information on pathophysiology, neurochemistry, and drug efficacy. Furthermore, extensive surgical manipulations are not usually required to create the vascular occlusion, and the reversibility of the obstruction to blood flow adds an important second dimension to the study. These models address the fact that both the initial ischemia and the secondary reperfusion are critical factors in determining final histopathologic and clinical outcome.

Methods of producing potentially reversible focal cerebral ischemia include:

- intraluminal occlusive techniques by way of thread or wire filaments
- extraluminal occlusive methods using temporary clips, tourniquets, or pharmacologic agents
- thromboembolism using autologous clots, microspheres, and pharmacologic/photochemical techniques

## 4.2.1. Intraluminal Occlusion Techniques

Intraluminal occlusion techniques involve the use of a filament of specific diameter, usually nylon suture material or silicone rubber cylinder, which is introduced into the external or common carotid artery (Longa et al., 1989; Molnar et al., 1988; Nagasawa and Kogure, 1989). The device is advanced until it occludes the origin of the MCA and blocks flow from the internal carotid, anterior cerebral, and posterior cerebral arteries. The filament is left in place for a predetermined length of time and then withdrawn to permit the reestablishment of flow. Variations include efforts to reduce collateral MCA flow by controlled hypotension or by permanently occluding other vessels: for example, the contralateral internal carotid artery and vertebral arteries, the pterygopalatine artery, or the ipsilateral external and common carotid arteries (Mhairi Macrae, 1992). Advantages of these approaches include the absence of extensive and potentially traumatic surgical procedures that introduce additional variables into the experimental design; the ability to accurately control the duration of the occlusive event and the onset of subsequent reperfusion; and, finally, the potential for the procedure to be carried out in the awake animal using exteriorized filaments.

Unfortunately, these models also utilize procedures that may cause vessel injury; this damage can lead to the production of emboli which may disrupt flow more distally and complicate the characterization of the final insult. Furthermore, endothelial injury with platelet aggregation promotes unpredictable changes in reperfusion flow dynamics and may alter the integrity of the blood–brain barrier, resulting in changes that may have a considerable influence on the overall nature of the resulting infarct.

## 4.2.2. Extraluminal Occlusion Techniques

*4.2.2.1. Microvascular Clips* Reversible focal cerebral ischemia has been produced in animals through the use of microvascular clips or snare ligatures around the MCA (Shigeno *et al.,* 1985; Tamura *et al.,* 1981). Results obtained in early experiments utilizing these methods may have produced inconsistent results due to lack of physiological control, occlusion proximal to or beyond lenticulostriate perforators, and local vascular injury. Variability has been largely overcome through recognition of these potential problems and the use of low-closing-pressure serrated microclips which produce minimal or no endothelial damage. Variability has also been reduced through the use of devices which occlude the lenticulostriate perforators, restricting the injury to the basal ganglia. As the models require a craniotomy to gain direct access to the MCA, a number of other potentially confounding variables are encountered: desiccation of exposed brain, temperature variations, blood loss, and changes in normal intracranial pressure and cerebrospinal fluid dynamics.

*4.2.2.2. Pharmacologically Induced Vasoconstriction* Extraluminal application of the vasoconstrictor peptide endothelin-1 has been shown to produce reductions in CBF which are comparable to those demonstrated using more conventional methods of MCA occlusion (Sharkey *et al.,* 1993). Because the duration of action of endothelin-1 is considerably greater than that of other vasoactive peptides, it has been shown to induce severe transient focal ischemia following MCA application (Sharkey *et al.,* 1993).

The resulting ischemia/reperfusion profile can be manipulated by varying the dosage of the peptide, and potential vascular injury is avoided. Although attractive, the model has not been well standardized and requires that the peptide be directly applied to the MCA, with the associated surgical complications. Moreover, the precise duration and degree of occlusion are difficult to control.

*4.2.2.3. Thromboembolic Techniques* Intravascular occlusion of the MCA has been achieved by a variety of methods designed to produce *in situ* thrombosis or emboli. For example, autologous emboli have been introduced into the internal carotid artery, producing distal vascular occlusions (Kudo *et al.,* 1982). Human blood has also been used in these models to facilitate testing of thrombolytic agents that do not appear to react effectively with clots generated from rat blood samples (Papadopoulos *et al.,* 1987). Artificial microspheres of a size and construction mimicking natural emboli have been injected into the internal carotid artery to produce comparable focal ischemic insults (Kogure *et al.,* 1974).

These models are advantageous as they eliminate variables related to complex surgical procedures and anesthesia. Unfortunately, the techniques tend to produce infarcts that are unpredictable in location and size. The tissue damage produced by the ischemia also tends to be compounded by a significant degree of associated brain edema. Thus, because of the limited amount of experimental control inherent in embolic stroke models, their use in research pertaining to ischemia pathogenesis, neurochemistry, and drug therapy should be approached with caution.

*4.2.2.4. Photochemically Initiated Cerebrovascular Thrombosis* Systemic administration of a photoactive dye such as Rose Bengal followed by site-specific irradiation with light has been described (Futrell *et al.,* 1988). The method produces either carotid artery platelet thrombi that subsequently embolize or, if a very localized brain region is illuminated, endothelial injury that initiates platelet aggregation and intravascular thrombosis. The severity of ischemia and resulting infarction has been shown to vary as a function of the dosage of dye, intensity of the irradiating beam, and length of time over which the light beam is applied (Ginsberg and Busto, 1989). The model involves a surgical procedure that is minimally invasive; although the scalp must be incised and retracted, the skull bone need not be violated during the administration of light. Furthermore, it is possible to induce thrombosis in a precisely localized cortical region, consequently providing a degree of experimental control not possible with previously discussed thromboembolic models. A major disadvantage of the model is the propensity of the technique to induce significant microvascular damage, thereby disrupting the blood–brain barrier and producing an excessive degree of vasogenic brain edema.

*In situ* intravascular thrombosis has also been induced by injecting sodium arachidonate, a platelet proaggregant, into the internal carotid artery (Furlow and Bass, 1976). This method has not yet been subjected to appropriate histopathologic validation and consequently has not become widely recognized and accepted.

## 5. INTRACEREBRAL HEMORRHAGE

Ischemic brain injury has also been studied within the context of another very prevalent clinical entity—intracerebral hemorrhage. Intracerebral hemorrhage may be either contained, as in the case of a focal hematoma, or diffuse, as exemplified by subarachnoid hemorrhage (Mendelow, 1993). Models of contained intracerebral hemorrhage have utilized two main techniques to produce experimental hematoma in animals. One method involves the stereotactic insertion of a cannula connected to the femoral artery into a specific region of brain, such as the caudate nucleus (Bullock *et al.,* 1984). Arterial pressure forces blood flow through the cannula and into the brain parenchyma. In the second technique, collegenase, a proteolytic enzyme, is injected directly into brain tissue using stereotactic microinfusion (Rosenberg *et al.,* 1990). Using this design, investigators have examined the effects of calcium antagonists in cerebral ischemia secondary to intracerebral hemorrhage (Elger *et al.,* 1994) and have studied the alterations in brain water content, intracranial pressure, and behavior that occur in animal subjects. Of note, the collegenase infusion model has been shown to produce neuropathologic changes consistent with those described in humans following intracerebral hemorrhage (Rosenberg, *et al.*, 1990).

Subarachnoid hemorrhage with associated vasospasm has been extensively modeled (Solomon *et al.*, 1985; Shigeno *et al.*, 1982). In general, the occurrence of subarachnoid hemorrhage results in two types of cerebral ischemia: global ischemia, by virtue of the concomitant rise in intracranial pressure, and focal ischemia, as a consequence of the development of significant arterial spasm. Vasospasm, in particular, has been produced experimentally using two major techniques: through the introduction of blood or blood products periarterially or, alternatively, following the puncture of an intracranial artery (Boullin, 1980). Preparations in which no craniotomy has been performed, as in the transorbital approach to vessel rupture, have been felt to be less prone to experimental variability due to surgical trauma and more representative of the clinical condition (Shigeno *et al.*, 1982). There are, however, certain advantages and disadvantages unique to each method; such factors must be taken into careful consideration in the selection of a model that best addresses the research hypotheses (Simeone and Vinall, 1980).

## 6. PITFALLS

Although animal models have provided important information concerning the pathophysiology and neurochemistry of ischemic brain injury, considerable inter- and intralaboratory variability has become evident. This variability may also explain why many pharmaceutical agents that have been shown to be effective in animal studies have not demonstrated a similar degree of efficacy in human stroke.

Investigators have repeatedly stressed the importance of accurate physiological monitoring and control in models of cerebral ischemia to decrease experimental variability. Failure of precise physiological control reflects, in part, experimental inconsistencies and the difficulty in bridging the gap between laboratory and clinical medicine.

### 6.1. Brain Temperature

It has been established in stroke paradigms that rigorous control of brain temperature prior to, during, and following the ischemic insult is of paramount importance (Fig. 4). Alterations in brain temperature of as little as a few degrees have a substantial impact on the extent of ischemic damage. Mild to moderate hypothermia with reductions in brain temperature to 33 or 34°C, both during (Welsh *et al.*, 1990) and following cerebral ischemia (Colbourne and Corbett, 1994; Coimbra and Wieloch, 1994), confers a marked neuroprotective effect, whereas modest elevations in temperature above this level greatly accentuate damage (Dietrich *et al.*, 1990; Kuroiwa *et al.*, 1990).

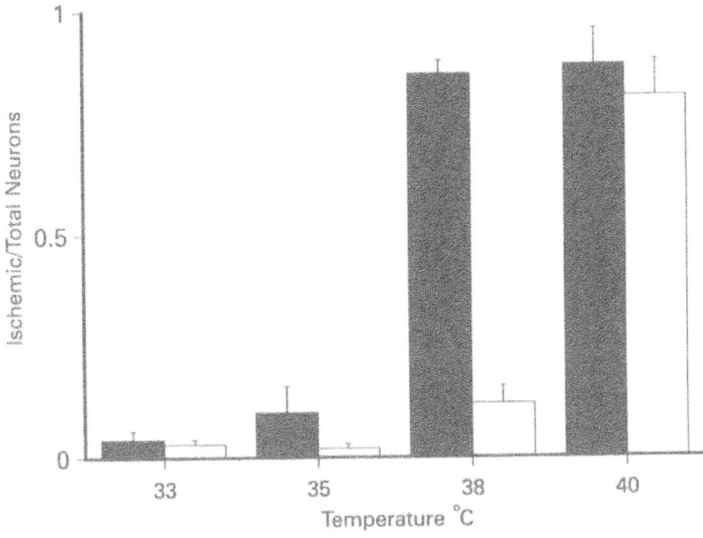

FIGURE 4. Quantitative histopathology following global ischemia at the indicated temperatures. In the CA1 sector of the hippocampus (black), reducing brain temperature by 3°C greatly reduces neuronal injury in a rat model of global ischemia. The more subtle neocortical injury (white) is also reduced by hypothermia; however, even mild hyperthermia greatly exacerbates neuronal damage. Similar temperature effects are seen in models of focal ischemia, with infarct volume being significantly reduced by hypothermia before or immediately following the ischemic event. (Reproduced, with permission from the Upjohn Company, Kalamazoo, Michigan, from Current Concepts "Progressing towards Effective Stroke Treatment," 1994.)

Many previous studies have failed to specifically monitor brain temperature; rectal temperature does not correlate with brain temperature, which, during ischemia, may be up to 5–6°C lower than temperatures recorded rectally. It has been shown that tympanic membrane or middle ear temperature accurately reflects brain temperature (Sutherland et al., 1992a). In a number of experimental models, the brain may be subjected to a considerable degree of heat loss, secondary to (1) the use of anesthesia, which can produce reductions in CBF and metabolic rate, (2) the absence of or reduction in heat supplied to brain tissue by circulating blood (as in global ischemia), and (3) the exposure of the brain to the surrounding environment through the loss of natural insulators such as hair, skin, fat, and skull bone as part of the experimental design. Temperature control should continue for an extended interval of up to six hours following the ischemic insult as anesthetic use and ischemic injury may be associated with prolonged variations in brain temperature. Using radio-frequency temperature probes, several experimental laboratories have extended this practice, precisely following and maintaining brain temperature until the time of perfusion fixation (Colbourne and Corbett, 1994).

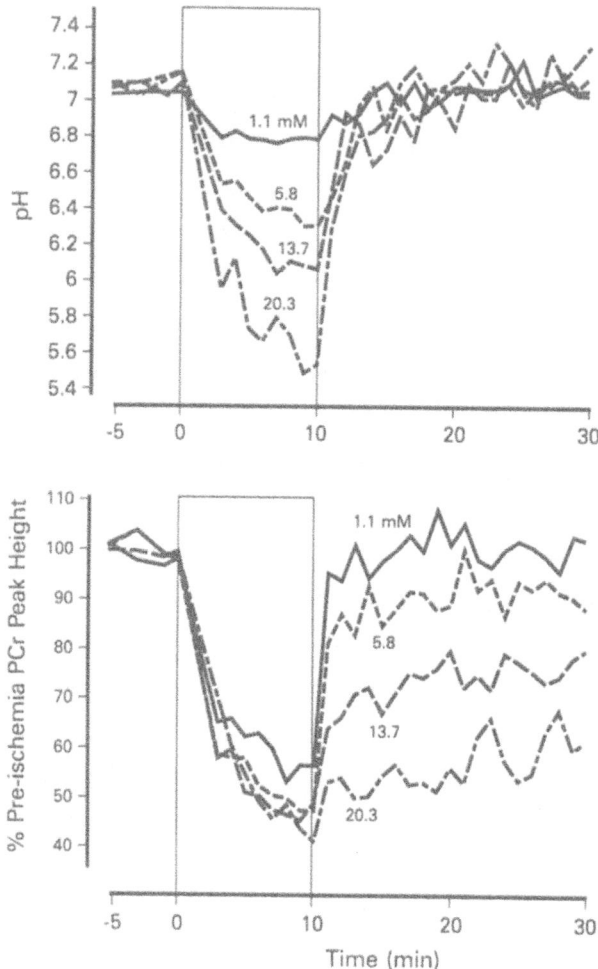

*FIGURE 5.* Changes in pH and phosphocreatine (PCr) at various blood glucose levels. [31]P MR spectra of the rat brain show that the cerebral acidosis produced during global ischemia is greater in animals with higher blood glucose levels. Energy recovery following reperfusion is poorer in these animals, as shown by the progressively poorer recovery of PCr in animals with higher blood glucose levels. Blood glucose concentrations are labeled on the figure. (Reproduced, with permission from the Upjohn Company, Kalamazoo, Michigan, from Current Concepts "Progressing towards Effective Stroke Treatment," 1994.)

## 6.2. Blood Glucose Level

Plasma and brain glucose concentrations greatly modify both experimental (Nedergaard and Diemer, 1987; Sutherland *et al.*, 1992b) and clinical ischemic brain injury (Pulsinelli *et al.*, 1983). In global and focal models of stroke, preischemic hyperglycemia markedly increases the extent of resulting tissue damage, whereas preischemic hypoglycemia protects against ischemic injury. Although glucose-dependent changes in blood viscosity may partially account for such findings, anaerobic lactic acid production, with the associated changes in pH, is felt to be the most significant factor (Fig. 5).

These findings emphasize the importance of making determinations of plasma glucose concentration before, during, and following the ischemic insult. Furthermore, the use of fasted animals is beneficial in reducing variability of blood glucose levels between test animals.

## 6.3. Hematocrit

Cerebral blood flow is inversely related to blood viscosity as determined by the serum hematocrit, while the lower the hematocrit, the less effectively the blood is able to fulfill its role in transporting oxygen to the tissues (Harrison, 1989). Therefore, knowledge of the serum hematocrit becomes important, particularly in experiments where interventions may result in significant hemodilution (Sakas *et al.*, 1993).

## 6.4. Arterial Blood Gases and Blood Pressure

In models that utilize hypotension as a method of impairing collateral flow to ischemic tissue, it has been shown that small variations in blood pressure can produce significant histopathologic variability (Ginsberg and Busto, 1989). Moreover, preischemic alterations in carbon dioxide tension produce substantial changes in cerebral blood flow which may alter ischemic brain injury (Pulsinelli and Jacewicz, 1992). To minimize these effects, such variables need to be both monitored and controlled throughout the experimental procedure.

## 6.5. Anesthesia

Many of the systemic effects of anesthesia can be minimized or controlled if other physiological variables are subjected to close monitoring during the experiment. For example, anesthesia can produce reductions in body temperature and arterial blood pressure and can lead to alterations in arterial blood gases, particularly in nonmechanically ventilated animals. By definition, various anesthetics produce direct central nervous system effects independent of their systemic interactions. Some anesthetics, for example, halothane or barbiturates, are central nervous system depressants and act to decrease cerebral metabolism in a dose-dependent fashion. Others, such as ketamine and nitric oxide, have excitatory effects and may not result in a reduction in cerebral metabolism (Nilsson and Siesjo, 1975). Furthermore, different anesthetics demonstrate varying effects on the cerebral vasculature; some have been shown to produce a degree of vasodilation (e.g., halothane), whereas others vasoconstrict (e.g., barbiturates) (Nilsson and Siesj°, 1975). These alterations ultimately produce changes in CBF which lead to variability in the severity of the final ischemic injury and may confound the observed actions of therapeutic agents with vasoactive properties (Nehls *et al.*, 1990; Park *et al.*, 1989). Because it is neither humane nor technically feasible to eliminate the use of anesthesia in animal experiments, those systemic variables that can be controlled should be identified, while the careful choice and administration of a particular anesthetic may minimize relatively inaccessible variables that also interact to influence neurologic outcome.

## 6.6. Cerebral Blood Flow

Flow reductions from a normal value of 50 ml $(100 \text{ g})^{-1}\text{min}^{-1}$ to about 30 ml $(100 \text{ g})^{-1}\text{min}^{-1}$ result in impaired protein synthesis, progressing to electrical failure when flow falls to 20 ml $(100 \text{ g})^{-1}\text{min}^{-1}$ and reaching energy failure when blood flow is less than about 10 ml $(100 \text{ g})^{-1}\text{min}^{-1}$. The length of time that the brain can tolerate an ischemic event decreases as these thresholds are passed (Siesj°, 1992). While a reduction of blood flow to just below the threshold for impaired protein synthesis may not produce damage for hours or days, reduction to below the energy failure threshold is less well tolerated.

It is essential for investigators to understand these relationships and recognize that variability in results between and within laboratories may be due to the nonuniform production of ischemic insults. The majority of laboratories are not equipped to directly monitor CBF in all experimental protocols. It is important, however, to assess the degree of blood flow decline by some means; this measurement need not involve the use of invasive monitoring techniques but can be achieved indirectly through the use of magnetic resonance spectroscopy, electroencephalography, or transcranial Doppler analysis.

## 6.7. Experimental Design and Methods of Analysis

The failure of drug studies in animal models to translate into effective stroke therapies in humans may be partially explained by flaws which are not unique to ischemia research *per se* but which nonetheless have undermined the credibility of a number of historical experiments. The need to increase the validity of drug research based on animal stroke models by enforcing the use of more stringent criteria developed from guidelines that have come to be mandatory in human therapeutic trials has been emphasized (Hsu, 1992). For example, the need for blinded studies in animal experimentation has been frequently overlooked. The failure of randomization in experimental designs may also lead to the wrong conclusion. Similarly, an appropriate placebo that is physically indistinguishable from the active agent needs to be utilized.

Creation of a biased study population by virtue of differing rates of mortality between groups is another frequently encountered source of error (de Courten-Myers and Wagner, 1992). In studies in which those animals who die are excluded from outcome assessment, results may be skewed to lead to the incorrect assumption that an overall improvement was demonstrated in the surviving members of that group. In the past, inadequate numbers of subjects have also statistically weakened trial results and have lowered the threshold for the potential appearance of type II errors. Through the consideration of both power (1-beta) and significance (alpha), more careful determinations of appropriate sample sizes in animal-based research must be employed in order to address this fundamental problem.

Perhaps most intuitively, one of the simplest yet most powerful ways to enhance the credibility of experimental conclusions is to successfully repeat them at some later time. Moreover, the ability to consistently reproduce results demonstrating therapeutic benefit not only within and between different laboratories but within and between different animal species would dramatically strengthen the confidence with which one would recommend that a particular therapy demonstrating benefit in animals proceeds to clinical testing in humans.

## 6.8. Factors Pertaining to Human Studies of Cerebral Ischemia

Some authors have argued that one of the reasons why human studies have been unable to confirm the results obtained in animal pharmacotherapeutic trials is that different outcome measures are used in clinical versus laboratory settings (de Courten-Myers and Wagner, 1992). While a positive result may be defined experimentally by a reduction in infarct size, in human trials the end point used to establish benefit is clinical improvement. Furthermore, this improvement is frequently determined by employing a scale that is based on neurological examinations and not on actual functional outcome; the criticism here is that the neurologic examination may provide diagnostic or prognostic information but may not be an accurate reflection of the therapeutic potential of a particular drug (Zivin and Grotta, 1990).

Suggestions to improve this situation involve the development of quantifiable end points that are comparable between animal and human research designs; that is, if behavioral scoring is to be used as the desired outcome measure, a system that has been verified and correlated with lesion severity should be developed and then applied to animals and, with appropriate modifications, to human subjects (Hsu, 1992). Furthermore the reliability of outcome determinations may be improved considerably through the evaluation of multiple end points. Thus, functional assessment may be used concurrently with other outcome measures (such as histopathologic analysis and magnetic resonance imaging) to strengthen experimental conclusions.

A criticism of clinical trials of agents that have proven their effectiveness in animal models is the tendency to administer these agents too late to be of benefit in human stroke patients (Biller and Love, 1991). These faulty designs have been utilized despite documented evidence in the preclinical animal studies that the neuroprotective effect of the therapy is lost past a certain point following the onset of the ischemia.

Another important inconsistency between animal and human studies is that, for the most part, the test animals are young, healthy, and without preexisting risk or predisposition to disease. This profile contrasts sharply with that of their human counterparts, who, more often than not, are older individuals who possess underlying risk factors for ischemic cerebrovascular disease and who may also suffer from frequently associated chronic illnesses (Wiebers *et al.,* 1990). As a result, the neuroprotective effect of a particular drug may be diluted or masked by the presence of these coexistent factors which simply do not occur in experiments using young, active, and robust animals as subjects.

Overall, one of the most important suggestions that has been offered in reviews of the inconsistencies between animal-generated data of stroke pharmacotherapy and research in humans is the need to improve communication and interaction between basic scientists and clinical researchers (Zivin and Grotta, 1990). Only by making efforts to merge the creative potentials each side has to offer will we be able to successfully bridge the gap that presently exists between otherwise solidly validated and reproducible therapeutic trials in animal models of stroke and the clinical trials developed to test similar hypotheses in humans.

## 7. CONCLUDING REMARKS

Animal models have contributed tremendously to the understanding of stroke pathogenesis and have fueled the development of novel and focused approaches to the clinical management of ischemic brain injury. The broad spectrum of experimental designs presents the investigator with a dilemma: of all those available, which model would best address the study objectives under consideration? In answering this question, a fundamental guideline is to acknowledge that both global and focal models are important, as they address different dimensions of this exceedingly diverse and inherently complex clinical problem.

Investigators should therefore familiarize themselves with one or two well-standardized models of each type and become skilled in their execution. In general, the global or forebrain model is best suited for the study of pathophysiologic, neurochemical, and molecular biological mechanisms underlying ischemic damage. Focal models, however, tend to produce an injury of greater severity which more accurately reflects clinical strokes. This characteristic makes focal protocols ideally suited to the testing of the therapeutic efficacy of new pharmacologic agents. The graded testing of these agents under conditions of stringent physiological monitoring and control will ultimately result in the development of a clinically effective treatment regime in the management of human stroke.

## REFERENCES

Abe, K., Yoshida, S., Watson, B. P., Busto, R., Kogure, K., and Ginsberg, M. D., 1983, Alpha-tocopherol and ubiquinones in rat brain subjected to decapitation ischemia, *Brain Res.* **273**:166–169.

Anthony, L. U., Goldring, S., O'Leary, J. L., and Schwartz, H. G., 1963, Experimental cerebrovascular occlusion in the dog, *Acta Neurol.* **8**:515–527.

Asano, T., and Sano, K., 1977, Pathogenetic role of no-reflow phenomenon in experimental subarachnoid hemorrhage in dogs, *J. Neurosurg.* **46**:454–466.

Biller, J., and Love, B. B., 1991, Nihilism and stroke therapy, *Stroke* **22**:1105–1107.

Blomqvist, P., and Wieloch, T., 1985, Ischemic brain damage in rats following cardiac arrest using a long-term recover model, *J. Cereb. Blood Flow Metab.* **5**:420–431.

Blomqvist, P., Mabe, H., Ingvar, M., and Seisjø, B. K., 1984, Models for studying long-term recovery following forebrain ischemia in the rat. I. Circulatory and functional effects of a 4-vessel occlusion, *Acta. Neurol. Scand.* **69**:376–384.

Boullin, D. J., 1980, Assessment of animal models, *in* "Cerebral Arterial Spasm" (R. H. Wilkins, ed.), pp. 291–293, Waverly Press, Baltimore.

Brassel, F., Dettmers, C., Nierhaus, A., Hartmann, A., and Solymosi, L., 1989, An intravascular technique to occlude the middle cerebral artery in baboons, *Neuroradiology* **31**:418–424.

Bremer, A. M., Watanabe, O., and Bourke, R. S., 1975, Artificial embolization of the middle cerebral artery in the primates, *Stroke* **6**:387–390.

Brockman, S. K., and Jude, J. P., 1960, The tolerance of the dog brain to total arrest of circulation, *Bull. Johns Hopkins Hosp.* **106**:74–80.

Buchan, A. M., Xue, D., and Slivka, A., 1992, A new model of temporary focal neocortical ischemia in the rat, *Stroke* **23**:273–297.

Bullock, R., Mendelow, A. D., Teasdale, G. M., and Graham, D. I., 1984, Intracranial hemorrhage induced at arterial pressure in the rat. Part 1: Description of technique. ICP changes and neuropathologic findings, *Neurol. Res.* **6**:184–188.

Busto, R., and Ginsberg, M. D., 1985, Graded focal ischemia in the rat by unilateral carotid artery occlusion and elevated intracranial pressure: Hemodynamic and biochemical characterization, *Stroke* **16**:466–476.

Cantu, R. C., Dixon, T., and Ames, A., III, 1969, Reversible ischemia model, *J. Surg. Res.* **9**:121–124.

Chen, S. T., Hsu, C. Y., Hogan, E. L., Maricq, H., and Balentine, J. D., 1986, A model of focal ischemia stroke in the rat: Reproducible extensive cortical infarction, *Stroke* **17**:738–743.

Coimbra, D., and Wieloch, T., 1994, Moderate hypothermia mitigates neuronal damage in the rat brain when initiated several hours following transient cerebral ischemia, *Acta Neuropathol.* **87**:325–331.

Colbourne, F., and Corbett, D., 1994, Prolonged post-ischemic hypothermia prevents CA1 necrosis in the gerbil, *Brain Res.* **654**:265–272.

Crowell, R. M., Marcoux, F. W., and De Girolami, U., 1981, Variability and reversibility of focal cerebral ischemia in unanesthetized monkeys, *Neurology* **31**:1295.

de Courten-Myers, G. M., and Wagner, K. R., 1992, Stroke models: Strengths and pitfalls, *Resuscitation* **23**:91–100.

de la Torre, J. C., and Fortin, T., 1991, Partial or global rat brain ischemia: The SCOT model, *Brain Res. Bull.* **26**:365–372.

Dietrich, W. D., Busto, R., Valdés, F., and Loor, Y., 1990, Effects of normothermic versus mild hyperthermic forebrain ischemia in rats, *Stroke* **21**:1318–1325.

Diringer, M. N., Kirsch, J. R., and Traystman, R. J., 1994, Reduced cerebral blood flow but intact reactivity to hypercarbia and hypoxia following subarachnoid hemorrhage in rabbits, *J. Cereb. Blood Flow Metab.* **14**:59–63.

Doczi, T., Joo, F., Adam, G., Bozoky, B., and Szerdahelyi, P., 1986, Blood–brain barrier damage during the acute stage of subarachnoid hemorrhage, as exemplified by a new animal model, *Neurosurgery* **18**:733–739.

Eleff, S. M., Maruki, Y., Monsein, L. H., Traystmann, R. K., Bruan, R. N., and Koehler, R. C., 1991, Sodium, ATP, and intracellular pH transients during reversible complete ischemia of dog cerebrum, *Stroke* **22**:233–241.

Elger, B., Seega, J., and Brendel, R., 1994, Magnetic resonance imaging study on the effect of levemopamil on the size of intracerebral hemorrhage in rats, *Stroke* **25**:1836–1841.

Fieschi, I. C., Battistini, N., Volante, F., Zanette, F., Weber, G., and Passero, S., 1975, Animal models of TIA: An experimental study with intracarotid ADP infusion in rabbits, *Stroke* **6**:617–621.

Fritz, H., and Hossmann, K.-A., 1979, Arterial air embolism in the cat brain, *Stroke* **10**:581–589.

Fritz, V. U., and Levien, L. J., 1989, Pathogenesis of transient ischemic attacks and stroke in baboons, *Stroke* **20**:386–389.

Fujishima, M., 1971, Effect of constricting carotid arteries on cerebral blood flow and on cerebrospinal fluid pH, lactate, and pyruvate in dogs, *Jpn. Heart J.* **12**:467–473.

Furlow, T. W., and Bass, N. H., 1976, Arachidonate-induced cerebrovascular occlusion in the rat, *Neurology* **26**:297–304.

Futrell, N., Watson, B. D., Dietrich, W. D., Prudo, R., Millikan, C., and Ginsberg, M. D., 1988, A new model of embolic stroke produced by photochemical injury to the carotid artery in the rat, *Ann. Neurol.* **23**:251–257.

Ginsberg, M. D., Budd, M. W., and Welsh, F. A., 1978, Diffuse cerebral ischemia in the cat: I. Local blood flow during severe ischemia and recirculation, *Ann. Neurol.* **3**:482–492.

Ginsberg, M. D., and Busto, R., 1989, Rodent models of cerebral ischemia, *Stroke* **20**:1627–1642.

Ginsberg, M. D., Graham, D. I., and Welsh, F. A., 1979, Diffuse cerebral ischemia in the cat: III. Neuropathological sequelae of severe ischemia, *Ann. Neurol.* **5**:350–358.

Gurvitch, A. M., Romanova, N. P., and Mutuskina, E. A., 1972, Quantitative evaluation of brain damage resulting from circulatory arrest to the central nervous system of the entire body: I. Electroencephalographic and histological evaluation of the severity of permanent postischemic damage, *Resuscitation* **1**:205–218.

Hallenbeck, J. M., and Bradley, M. E., 1977, Experimental model for systemic study of impaired microvascular reperfusion, *Stroke* **8**:238–243.

Harrison, M. J., 1989, Influence of hematocrit in the cerebral circulation, *Cerebrovasc. Brain Metab. Rev.* **1**:55–67.

Hegedus, K., Fekete, I., Tury, F., and Molnar, L., 1985, Experimental focal cerebral ischemia in rabbits, *J. Neurol.* **232**:223–230.

Heyreh, S. S., and Edwards, J., 1971, Vascular responses to acute intracranial hypertension, *J. Neurol. Neurosurg. Psychiatry* **34**:587–601.

Hill, N. C., Millikan, C. H., Wakim, K. G., and Sayre, G. P., 1955, Studies in cerebrovascular disease. VII. Experimental production of cerebral infarction by intracarotid injection of homologous blood clot. Preliminary report, *Staff Meet. Mayo Clin.* **30**:625–633.

Hinzen, D. H., Müller, U., Sobotka, P., Gebert, E., Lang, R., and Hirsch, H., 1972, Metabolism and function of dog's brain recovering from long-time ischemia, *Am. J. Physiol.* **223**:1158–1164.

Hossmann, V., and Hossmann, K.-A., 1973, Return of neuronal functions after prolonged cardiac arrest, *Brain Res.* **60**:423–438.

Hossmann, K.-A., and Schuier, F. J., 1980, Experimental brain infarcts in cats. I. Pathophysiologic observation, *Stroke* **11**:583–592.

Hruska, F. E., Buist, R. J., Sutherland, G. R., Yang, F.-W., and Peeling, J., 1992, Mannitol does not affect energy metabolism in forebrain ischemia, *Neuroreport* **3**:897–900.

Hsu, C. Y., 1992, Criteria for valid preclinical trials using animal stroke models, *Stroke* **24**:633–636.

Hudgins, W. R., and Garcia, J. H., 1970, Transorbital approach to the middle cerebral artery of the squirrel monkey: A technique for experimental cerebral infarction applicable to ultrastructural studies, *Stroke* **1**:107–111.

Jackson, D. L., Dole, W. P., McGloin, J., and Rosenblatt, J. I., 1981, Total cerebral ischemia: Application of a new model system to studies of cerebral microcirculation, *Stroke* **12**:66–72.

Jenkins, L. W., Povlishock, J. T., Becker, D., Miller, J. D., and Sullivan, H. G., 1979, Complete cerebral ischemia: An ultrastructural study, *Acta. Neuropathol.* **48**:113–125.

Kabat, H., Dennis, C., and Baker, A. B., 1941, Recovery of function following arrest of the brain circulation, *Am. J. Physiol.* **132**:737–747.

Kawamura, S., Yashi, N., Shirasawa, M., and Fukasawa, H., 1991, Rat middle cerebral artery occlusion using an intraluminal thread technique, *Acta. Neurochir.* **109**:126–132.

Kirino, T., 1982, Delayed neuronal death in the gerbil hippocampus following ischemia, *Brain Res.* **239**:57–59.

Kogure, K., Busto, R., Scheinberg, P., and Reinmuth, O. M., 1974, Energy metabolites and water content in rat brain during the early stage of development of cerebral infarction. Homologous blood clot emboli in rats, *Stroke* **13**:505–508.

Kolata, R. J., 1979, Survival of rabbits after prolonged cerebral ischemia, *Stroke* **10**:272–277.

Kowada, M., Ames, A., III, Majno, G., and Wright, R. L., 1968, Cerebral ischemia. I. An improved experimental method for study: Cardiovascular effects and demonstration of an early vascular lesion in the rabbit, *J. Neurosurg.* **28**:150–157.

Kudo, M., Ayoyama, A., Ichimori, S., and Fukunaga, N., 1982, An animal model of cerebral infarction. Homologous blood clot emboli in rats, *Stroke* **13**:505–508.

Kuroiwa, T., Bonnekoh, P., and Hossmann, K.-A., 1990, Prevention of postischemic hyperthermia prevents ischemia injury of CA1 neurons in gerbils, *J. Cereb. Blood Flow Metab.* **10**:550–556.

Little, J. R., 1977, Implanted device for middle cerebral artery occlusion in conscious cats, *Stroke* **8**:258–260.

Longa, E. Z., Weinstein, P. R., Carlson, S., and Cummins, R., 1989, Reversible middle cerebral artery occlusion without craniectomy in rats, *Stroke* **20**:84–91.

Marshall, L. F., Durity, F., Lounsbury, R., Graham, D. I., Welsh, F., and Langfitt, T. W., 1975, Experimental cerebral oligemia and ischemia produced by intracranial hypertension. I. Pathophysiology, electroencephalography, cerebral blood flow, blood brain barrier, and neurological function, *J. Neurosurg.* **43**:308–317.

Mendelow, A. D., 1993, Mechanisms of ischemic brain damage with intracerebral hematoma, *Stroke* **24**(12 Suppl.):I115–I119.

Mhairi Macrae, I., 1992, New models of focal cerebral ischemia, *Br. J. Clin. Pharmacol.* **34**:302–308.

Miller, J. R., and Myers, R. E., 1970, Neurological effects of systemic circulatory arrest in the monkey, *Neurology* **20**:715–724.

Millikan, C., 1992, Animal stroke models, *Stroke* **23**:795–797.

Mima, T., Yanagisawa, M., Shigeno, T., Saito, A., Goto, K., Takakura, K., and Masaki, T., 1989,

Endothelin acts on feline and canine cerebral arteries from the adventitial side, *Stroke* **20:**1553–1556.

Molinari, G. F., 1970, Experimental cerebral infarction. I. Selective segmental occlusion of intracranial arteries in the dog, *Stroke* **1:**224–231.

Molinari, G. R., and Laurent, J. P., 1976, A classification of experimental models of brain ischemia, *Stroke* **7:**14–17.

Molnar, L., Hegedus, K., and Fekete, I., 1988, A new model for introducing transient cerebral ischemia and subsequent reperfusion in rabbits without craniectomy, *Stroke* **19:**1262–1266.

Myer, F. B., Anderson, R. E., Sundt, T. M., Jr., and Yaksh, T. L., 1986, Intracellular brain pH, indicator tissue perfusion, electroencephalography, and histology in severe and moderate focal cortical ischemia in the rabbit, *J. Cereb. Blood Flow Metab.* **6:**71–78.

Myers, R. E., and Yamaguchi, S., 1977, Nervous system effects of cardiac arrest in monkeys, *Arch. Neurol.* **34:**65–74.

Nagasawa, H., and Kogure, K., 1989, Correlation between cerebral blood flow and histologic changes in a new rat model of MCA occlusion, *Stroke* **20:**1037–1043.

Nedergaard, M., and Diemer, N. H., 1987, Focal ischemia of the rat brain, with special reference to the influence of plasma glucose concentration, *Acta. Neuropathol.* **73:**131–137.

Nehls, D. G., Park, C. K., MacCormack, A. G., and McCulloch, J., 1990, The effects of *N*-methyl-*D*-aspartate receptor blockage with MK-801 upon the relationship between cerebral blood flow and glucose utilization, *Brain Res.* **511:**271–279.

Nemoto, E. M., Bleyaert, A. L., Stezoski, S. W., Moossy, J., Rao, G. R., and Safar, P., 1977, Global brain ischemia. A reproducible monkey model, *Stroke* **8:**558–564.

Nemoto, E. M., Hossmann, K.-A., and Cooper, H. K., 1981, Post-ischemic hypermetabolism in cat brain, *Stroke* **12:**666–676.

Nilsson, L., and Siesjö, B. K., 1975, The effect of phenobarbitone anesthesia on blood flow and oxygen consumption in the rat brain, *Acta Anesthesiol. Scand. (Suppl.)* **57:**18–24.

O'Brien, M. D., and Waltz, A. G., 1973, Transorbital approach for occluding the middle cerebral artery without craniectomy, *Stroke* **4:**201–206.

Papadopoulos, S. M., Chandler, W. F., Salamat, M. S., Topol, E. J., and Sackellares, J. C., 1987, Recombinant human tissue-type plasminogen activator therapy in acute thromboembolic stroke, *J. Neurosurg.* **67:**394–398.

Park, C. K., Nehls, D. G., Teasdale, G. M., and McCulloch, J., 1989, Effect of the NMDA antagonist MK-801 on local cerebral blood flow in focal cerebral ischemia in the rat, *J. Cereb. Blood Flow Metab.* **9:**617–622.

Petruk, K. C., West, G. R., Marriot, M. R., McIntyre, J. W., Overton, T. R., and Weir, B. K., 1972, Cerebral blood flow following induced subarachnoid hemorrhage in the monkey, *J. Neurosurg.* **37:**316–324.

Pulsinelli, W. A., and Buchan, A. M., 1988, The four-vessel occlusion rat model: Method for complete occlusion of vertebral arteries and control of collateral circulation, *Stroke* **19:**913–914.

Pulsinelli, W. A., and Jacewicz, M., 1992, Animal models of brain ischemia, in "Stroke: Pathophysiology, Diagnosis, and Management," 2nd ed. (H. J. M. Barnett, J. P. Mohr, B. M. Stein, and F. M. Xatsu, eds.), pp. 49–67, Churchill Livingstone, New York.

Pulsinelli, W. A., Levy, D. E., Sigsbee, B., Scherer, P., and Plum, F., 1983, Increased damage after ischemic stroke in patients with hyperglycemia with or without established diabetes mellitus, *Am. J. Med.* **74:**540–544.

Robinson, M. J., and McCulloch, J., 1990, Contractile responses to endothelin in feline cortical vessels in situ, *J. Cereb. Blood Flow Metab.* **10:**285–289.

Rosenberg, G. A., Mun-Bryce, S., Wesley, M., and Kornfield, M., 1990, Collagenase-induced intracerebral hemorrhage in rats, *Stroke* **21:**801–807.

Ross Russell, R. W., 1971, The reactivity of the pial circulation of the rabbit to hypercapnia and the effect of vascular occlusion, *Brain* **94:**622–634.

Sakas, D. E., Crowell, R. M., Kim, K., Korosue, K., and Zervas, N. T., 1993, The perfluorocarbon fluoromethyloadamantane offers cerebral protection in a model of ischemic hemodilution in rabbits, *Stroke* **25:**197–201.

Sengupta, D., Harper, M., and Jennett, B., 1973, Effect of carotid ligation on cerebral blood flow in baboons. I. Response to altered arterial $PCO_2$, *J. Neurol. Neurosurg. Psychiatry* **36:**736–741.

Sharkey, J., Ritchie, I. M., and Kelly, P. A., 1993, Perivascular microapplication of endothelin-1: A new model of focal cerebral ischemia in the rat, *J. Cereb. Blood Flow Metab.* **13:**865–871.

Shigeno, T., Fritschka, E., Brock, M., Schramm, J., Shigeno, S., and Cervos-Navarro, J., 1982, Cerebral edema following experimental subarachnoid hemorrhage, *Stroke* **13:**368–379.

Shigeno, T., Teasdale, G. M., McCulloch, J., and Graham, D. I., 1985, Recirculation model following MCA occlusion in rats. I. Cerebral blood flow, cerebrovascular permeability, and brain edema, *J. Neurosurg.* **63:**272–277.

Siemkowicz, E., and Hansen, A. J., 1978, Clinical restitution following cerebral ischemia in hypo-, normo-, and hyperglycemic rats, *Acta. Neurol. Scand.* **58:**1–8.

Siesjö, B. K., 1992, Pathophysiology and treatment of focal cerebral ischemia. I. Pathophysiology, *J. Neurosurg.* **77:**169–184.

Simeone, F. A., and Vinall, P. E., 1980, Evaluation of animal models of cerebral vasospasm, *in* "Cerebral Arterial Spasm" (R. H. Wilkins, ed.), pp. 284–286, Waverly Press, Baltimore.

Smith, M.-L., Auer, R. N., and Siesjö. B. K., 1984, The density and distribution of ischemic brain injury in the rat following 2–10 minutes of forebrain ischemia, *Acta. Neuropathol.* **64:**319–332.

Solomon, R. A., Antunes, J. L., Chen, R. Y., Bland, L., and Chien, S., 1985, Decrease in cerebral blood flow in rats after experimental subarachnoid hemorrhage: A new animal model, *Stroke* **16:**58–64.

Sutherland, G. R., Lesiuk, H., Hazendonk, D., Peeling, J., Buist, R., Kozlowski, P., Juzinski, A., and Saunders, J. K., 1992a, MRI and $^{31}P$ magnetic resonance spectroscopy study of the effect of temperature on ischemic brain injury, *Can. J. Neurol. Sci.* **19:**317–325.

Sutherland, G. R., Peeling, J., Sutherland, E., Tyson, R., Dai, F., Kozlowski, P., and Saunders, J. K., 1992b, Forebrain ischemia in diabetic and nondiabetic BB rats studied with $^{31}P$ magnetic resonance spectroscopy, *Diabetes* **41:**1328–1334.

Tamura, A., Graham, D. I., McCulloch, J., and Teasdale, G. M., 1981, Focal cerebral ischemia in the rat. Description of technique and early neuropathological consequences following middle cerebral artery occlusion, *J. Cereb. Blood Flow Metab.* **1:**53–60.

van Harreveld, A., 1947, The electroencephalogram after prolonged brain asphyxiation, *J. Neurophysiol.* **10:**361–370.

Welsh, F. A., O'Connor, M. J., and Marcy, R., 1977, Effect of oligemia on regional metabolite levels in cat brain, *J. Neurochem.* **31:**311–319.

Welsh, F. A., Sims, R. E., and Harris, V. A., 1990, Mild hypothermia prevents ischemic injury in gerbil hippocampus, *J. Cereb. Blood Flow Metab.* **10:**557–563.

While, R. J., Albin, M. S., Verdura, J., and Locke, G. E., 1967, The isolated monkey brain: Operative preparation and design of support systems, *J. Neurosurg.* **27:**216–255.

Wiebers, D. O., Adams, H. P., Jr., and Whisnant, J. P., 1990, Animal models of stroke: Are they relevant to human disease? *Stroke* **21:**1–3.

Wolin, L. R., Massopust, L. C., Jr., White, R. J., and Taslitz, N., 1972, Arrest of cerebral blood flow and reperfusion of the brain in the Rhesus monkey, *Resuscitation* **1:**39–44.

Zivin, J. A., and Grotta, J. C., 1990, Animal stroke models. They are relevant to human disease, *Stroke* **31:**981–983.

CHAPTER 6

# IN VIVO *MAGNETIC RESONANCE IMAGING AND SPECTROSCOPY:*
## Application to Brain Tumors

BRIAN D. ROSS, ODED BEN-YOSEPH, and
THOMAS L. CHENEVERT

## SUMMARY

This chapter contains an overview of the relatively recent applications of magnetic resonance imaging and spectroscopy for studies of experimental and human brain tumors. Specific examples have been chosen to demonstrate the unique possibilities offered by magnetic resonance techniques to advance our understanding of the physiology, biochemistry, and therapeutic response of brain tumors.

BRIAN D. ROSS, ODED BEN-YOSEPH, and THOMAS L. CHENEVERT • Department of Radiology, University of Michigan Medical Center, Ann Arbor, Michigan 48109-0648.

Magnetic Resonance Spectroscopy and Imaging in Neurochemistry, Volume 8 of Advances in Neurochemistry, edited by Bachelard, Plenum Press, New York, 1997.

## 1. INTRODUCTION

The management of high-grade malignant tumors of the central nervous system is currently one of the more frustrating areas of cancer therapy, and, despite the use of multimodality therapy, malignant gliomas remain uniformly fatal (Walker *et al.,* 1986). Furthermore, quantitation of therapeutic response is more difficult in brain tumors than in systemic cancers because an improvement in patient function is multifactorial and changes in neurological deficits may be unrelated to changes in tumor volume (Grossman and Burch, 1988; Levin *et al.,* 1977; Wilson *et al.,* 1977; Grossman, 1991). Magnetic resonance imaging (MRI) and X-ray computed tomographic (CT) scans of malignant human brain tumors do not allow quantitation of the actual tumor volume. Following administration of a contrast agent, the largest cross-sectional area of contrast enhancement is usually used to "estimate" the area of the tumor. However, the blood–brain barrier (BBB) may be altered by chemotherapeutic agents (Grossman, 1991) or corticosteroids, and thus measurements of tumor dimensions by the boundaries of contrast enhancement are an indirect estimate of tumor size. Because accurate measurements of intracranial tumor volumes are difficult, brain tumor therapeutic trials usually report median survival time and median time to progression as a quantitative measure of response. Clearly, a need exists for improved methods which would allow for earlier and accurate quantitation of the therapeutic response in individual patients. In this regard, recent technical developments in MRI and magnetic resonance spectroscopy (MRS) have added to the number of available techniques for studying brain tumors. In this chapter, we will provide examples of the application of MRI and MRS methodologies to *in vivo* studies of both experimental and human brain tumors. The examples presented were chosen to convey the unique possibilities that exist for using MR techniques to advance our understanding of brain tumor biology and biochemistry as well as to provide new insights into the effectiveness of standard and novel therapeutic approaches. The focus of this chapter will be on the relatively recent applications of MRI and MRS, some of which are still in the process of clinical evaluation.

## 2. USE OF MRI FOR STUDYING BRAIN TUMORS

### 2.1. Quantitation of Growth Kinetics and Cell Kill in Experimental Brain Tumors Following Therapeutic Intervention

The development and use of animal models of brain tumors over the past 30 years has helped in treatment evaluation and has furthered our understanding of the biology of the disease (Peterson *et al.,* 1994). The therapeutic response of

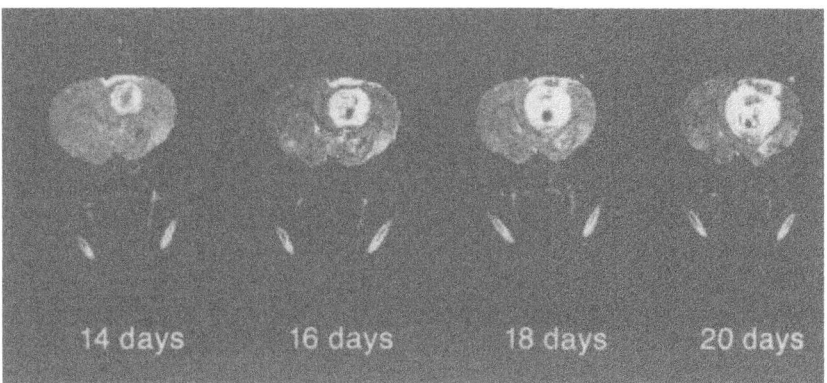

FIGURE 1. Coronal $T_2$-weighted images (0.5 mm thick) of a rat brain at various times postimplantation of 9L tumor cells. Images are from an untreated 9L tumor rat at 14, 16, 18, and 20 days
postimplantation. The tumor is clearly delineated from normal brain parenchyma as a hyperintense
region in the right hemisphere. Images were obtained on a 7-T MR system.

experimental orthotopic brain tumors has traditionally been quantitated using
tumor weights following excision, animal survival, or colony-forming assays of
cells cultured from the *in vivo* tumor as biological end points (Barker *et al.*, 1973;
Rosenblum *et al.*, 1975, 1976, 1977, 1980; Weizsaecker *et al.*, 1981). These
methods have proved valuable for *in vivo* testing of new therapeutic approaches
but require large numbers of animals because the animal-to-animal tumor growth
rate can vary significantly. Recently, MRI has been shown (Kim *et al.*, 1995) to
be useful for noninvasively monitoring the growth and therapeutic response of
the intracranial rat 9L brain tumor model (Barker *et al.*, 1973; Rosenblum *et al.*,
1977). Shown in Fig. 1 is a representative series of 0.5-mm-thick coronal $T_2$-
weighted MR images of a rat with an untreated intracranial 9L tumor from 14 to
20 days post implantation of $10^5$ cells. The tumor is evident in the right hemisphere as a hyperintense mass which expands over time. The area of tumor in
each cross-sectional image was quantitated, multiplied by the slice separation,
and summed to yield the total tumor volume. Acquiring multislice image data
sets every 2–3 days during the growth of the tumor allows accurate quantitation
of the intracranial tumor doubling time ($T_d$) (Kim *et al.*, 1995). Shown in Fig. 2 is
a plot of the volume of an untreated intracerebral 9L tumor versus time post cell
implantation. Individual points were obtained from MRI tumor volume measurements, and the line represents the least-squares fit to an exponential growth
function. Figure 2 reveals that 9L tumors grow exponentially over the entire life
span of the animal so that treated animals could serve as their own control
following determination of $T_d$ from several MRI scans before therapeutic inter-

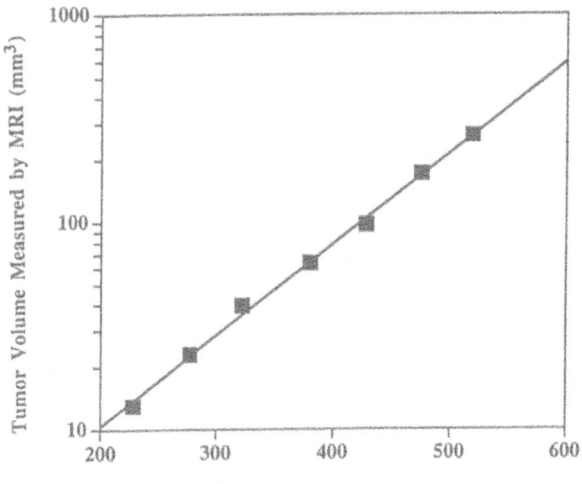

**Hours Post Implantation of 9L Cells**

*FIGURE 2.* A plot of the MRI-determined untreated intracerebral 9L tumor volume vs. time post-implantation. The line corresponds to the least-squares fit to the experimental data. This plot reveals the exponential growth of the intracerebral 9L tumor, which is characteristic of all 9L tumors examined to date. [Reprinted, with permission, from Kim *et al.*, (1995).]

vention. An example of this approach is presented in Fig. 3, which shows the plot obtained when a rat with an intracerebral 9L tumor was treated with a single dose of 1,3-bis(2-chloroethyl)-1-nitrosourea (BCNU) after four MRI volumetric tumor measurements were made. MRI scans were repeated at approximately 2-day intervals. The tumor $T_d$ decreased dramatically following treatment, and a regression of the tumor occurred, followed by regrowth at about 760 hr. Extrapolation of the tumor regrowth curve (b) to the time of treatment reveals the log kill of 0.9, which is approximately equivalent to killing 90% of the cells. MRI offers important advantages for determination of cell kill following therapy since relatively few animals are needed to evaluate a given treatment, whereas such an evaluation was previously obtainable only through the use of large numbers of animals (see Fig. 4 in Rosenblum *et al.*, 1977). Since each animal can serve as its own pretreatment control, interanimal variations in intracerebral tumor $T_d$ values can also be assessed. Finally, MRI provides a more sensitive approach for detecting changes in tumor growth rates as treatments producing >0.1 log kill can be quantitated (Kim *et al.*, 1995) whereas animal survival studies require >1.0 log kill to produce significant increases in animal life span (Rosenblum *et al.*, 1975). The MRI approach should be applicable to a wide variety of therapeutic interventions (Kimler, 1994) and should greatly facilitate the evaluation of novel treatment approaches.

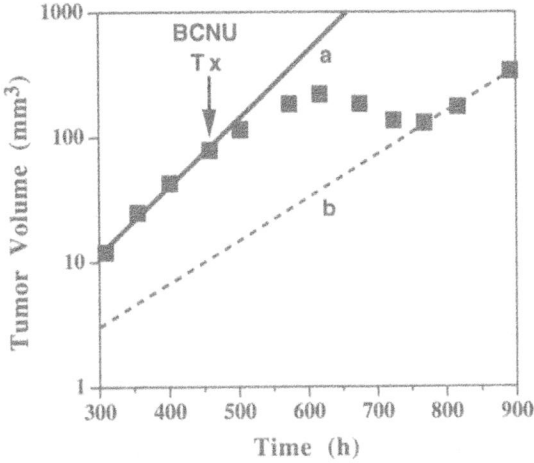

FIGURE 3. A plot of the MRI-determined intracerebral tumor volume vs. time postimplantation for an intracranial 9L tumor treated with an $LD_{10}$ dose (13.3 mg/kg, i.p.) of BCNU at about 450 hr postimplantation. From the initial four tumor volume measurements (a), the pretreatment $T_d$ was determined to be 55 hr. Following BCNU administration (denoted by downward-pointing arrow), the tumor volume continued to increase, but at a greatly reduced rate, followed by a slight tumor regression. Regrowth of the tumor occurred at approximately 760 hr postimplantation at a calculated $T_d$ of 86 hr as shown by the regression line (b). A growth delay of 356 hr (approximately 15 days) was observed for this tumor.

## 2.2. Diffusion MRI: Insights into Tumor Structure at the Cellular Level

As described above, immediate clinical value of conventional MR imaging stems from the ability to noninvasively determine gross tumor morphology and volume and monitor their change with time and/or treatment. Largely untapped potential also resides in exploratory methods known to be sensitive to tissue structure at the cellular level. Such information may be derived via quantitation of tissue properties that are reflective of dynamics in the microscopic environment. Magnetic resonance image contrast has this dependence embodied in MR relaxation times $T_1$ and $T_2$. However, physical processes that determine $T_1$ and $T_2$ involve complex interactions of chemical, physical (i.e., molecular translation and rotation), and magnetic environments experienced by water. Water diffusion in tissue is also a complex process. However, determination of water diffusion properties may provide a more direct measure of tissue/tumor cellularity and integrity of cellular membranes that impede water translational mobility. Water diffusion measurements have already been shown to be sensitive to tissue cellular size, water permeability, and structural anisotropy that is manifest as diffusion anisotropy (Stejskal, 1965; Tanner, 1978; Moseley et al., 1990a; Cooper et al.,

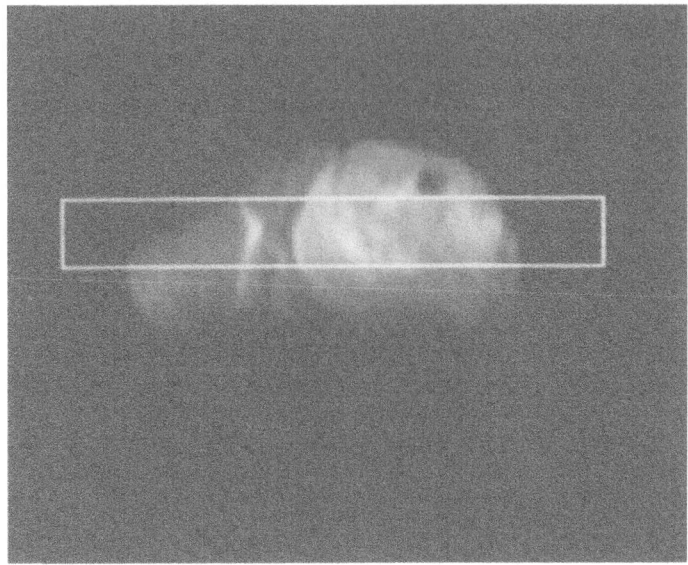

FIGURE 4. (a) Surface coil coronal $T_2$-weighted MR image of a rat 9L tumor with the column used for obtaining the diffusion data scribed on the image. (b) Plot of water $ADC_{mean}$ (mean of $ACD_x$, $ADC_y$, and $ADCz$) as a function of position along the column shown in (a). Note, the $ADC_{mean}$ in the tumor is only moderately elevated relative to that in the contralateral brain tissue. (c) A BCNU-treated intracerebral 9L tumor at 10 days after BCNU administration and 25 days after tumor implant. Plot of water $ADC_{mean}$ as a function of position through the 9L tumor and contralateral brain. Note, the ADC in the treated tumor is significantly elevated relative to the contralateral brain tissue and nontreated 9L tumor values shown in (b).

1974; Chenevert *et al.*, 1990). Diffusion MRI has already been shown to be extremely valuable in cerebral stroke (Mintorovitch *et al.*, 1990; Moseley *et al.*, 1990b). It is hoped that quantitative apparent diffusion coefficient (ADC) measurements may also be a valuable probe in the study of central nervous system (CNS) tumors (Brunberg *et al.*, 1995) and, more importantly, may serve to guide more effective treatment protocols.

Recently, studies of rats bearing intracranial 9L and C6 gliomas have shown that diffusion is sensitive to heterogeneity within a lesion (e.g., solid tumor vs. necrosis), to differences between tumor type, and to changes over time with treatment (Chenevert *et al.*, 1991; Ross *et al.*, 1994). The known histologic properties of the 9L glioma include homogeneous cellularity with minimal necrosis, whereas the C6 glioma is more heterogeneous, with large regions of spontaneous necrosis. These glioma models were used to study water properties in cellular and necrotic tumor environments, as well as the evolution toward necrosis in 9L tumors following administration of a chemotherapeutic agent known to induce significant cell death. Shown in Fig. 4a is a $T_2$-weighted image of an

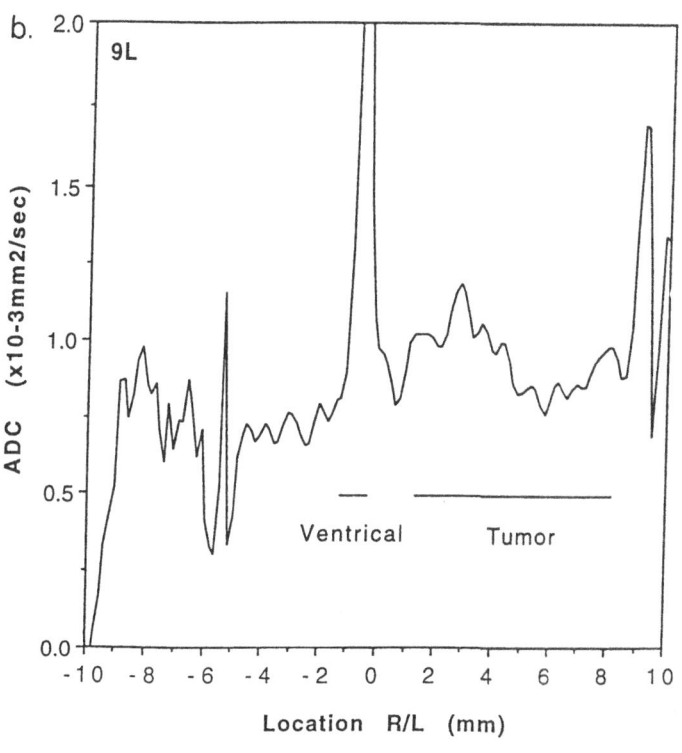

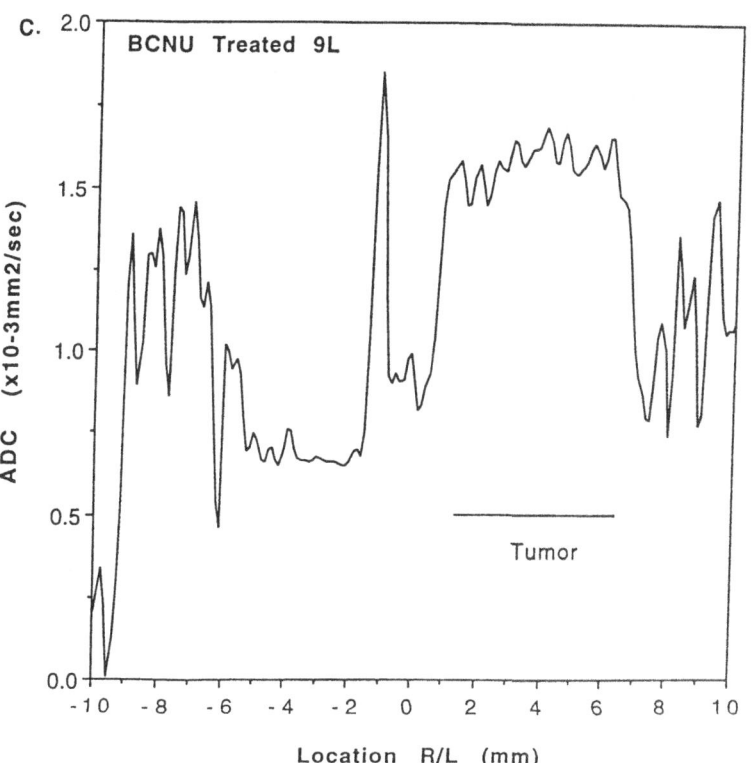

untreated 9L rat brain tumor indicating the column used for localized ADC measurements. It is well known that diffusion in the brain is anisotropic diffusion in ordered tissue structures such as white matter tracts (Moseley *et al.*, 1990a; Chenevert *et al.*, 1990). A robust quantity that is independent of the relative structure orientation is $ADC_{mean} = (ADC_x + ADC_y + ADC_z)\beta$, where *x*, *y*, and *z* refer to ADC measurements along orthogonal directions (Basser *et al.*, 1994). Consistent with the observed 9L histologic properties, the $ADC_{mean}$ values within the tumor are only moderately elevated ($0.8 \times 10^{-3}$–$1.3 \times 10^{-3}$ mm²/sec) relative to the values within contralateral brain tissue ($0.7 \times 10^{-3}$–$0.8 \times 10^{-3}$ mm²/sec) (Fig. 4b). In contrast, the histologic characteristics of the C6 tumor are a reduced cellularity and increased interstitial water space in the necrotic regions, producing $ADC_{mean}$ values of approximately $1.8 \times 10^{-3}$ mm²/sec (data not shown), which imply greater water translational mobility. (For reference, completely unrestricted water diffusion would have a value of $2.5 \times 10^{-3}$–$3 \times 10^{-3}$ mm²/sec.) It is anticipated that the diffusion values would increase as a cellular tumor evolves toward necrosis in response to treatment. This is supported by observations of a BCNU-treated 9L glioma, where there is a distinct increase in $ADC_{mean}$ (Fig. 4c) relative to pretreatment levels (Fig. 4b). The observed increase in water mobility is expected to be a direct result of increased cellular permeability and interstitial water volume as the tumor cells necrose.

Similar observations have been made in humans by the same technique (Chenevert *et al.*, 1992; Brunberg *et al.*, 1995). Figures 5a and 5b illustrate representative $T_2$-weighted MR images of a patient with a grade III astrocytoma and show the locations of two columns (from two different regions of the same tumor patient's brain) superimposed upon the $T_2$-weighted images which correspond to the locations from which diffusion data were collected. The corresponding diffusion properties through the tumor, peritumor edema, and contralateral brain are shown in Figs. 5c and 5d. Superimposed directional ADC plots in Figs. 5c and 5d reveal isotropic ADC within the tumor and cyst, as opposed to anisotropic diffusion in normal and edematous white matter.

The strong contrast in MRI diffusion ADC values between normal white matter, necrosis or cyst formation, edematous tissue, and solid enhancing tumors provides information not obtainable from standard MR imaging. This information is invaluable for the design of both surgical and radiation treatment of glioma patients. Furthermore, the sensitivity of diffusion MRI to relative intra- and extracellular water volumes and membrane integrity between these volumes offers the potential to apply diffusion measurements for the monitoring of treatment responsiveness in human gliomas. In this regard, the preliminary data presented in Fig. 4c reveal that diffusion MRI is indeed sensitive to therapeutic-induced changes in 9L rat glioma tissue integrity. Future experiments correlating diffusion changes in human gliomas with clinical outcome following therapeutic intervention would be a very promising avenue to pursue.

a. 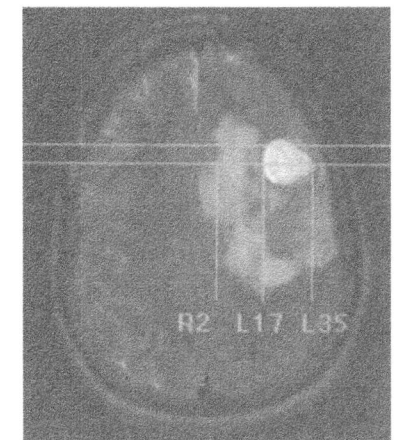   b.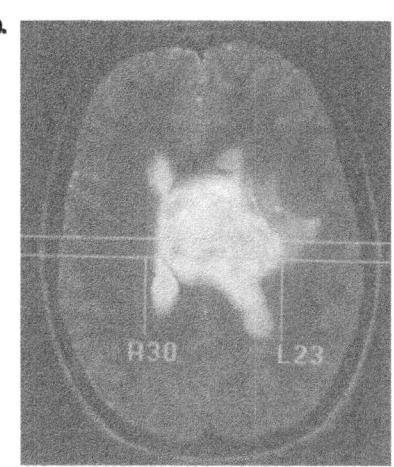

FIGURE 5. Diffusion study of a patient with a grade III astrocytoma: (a) and (b) $T_2$-weighted MR images showing column selected for diffusion study through a cyst (a) and solid tumor (b); (c) and (d) corresponding diffusion properties along tissue columns intersecting cyst (c) and solid tumor (d). Diffusion sensitivity along R/L (right/left), A/P (anterior/posterior), and S/I (superior/inferior) directions are superimposed to show relative isotropic ADC within tumor and cyst, as opposed to anisotropic diffusion in normal and edematous white matter.

## 2.3. MRI Measurements of Tumor Blood Flow (TBF) and Blood–Brain Barrier Permeability

Characterization of tumor vascularity and BBB integrity has clear clinical importance in diagnosis of tumor type, grade, and determination of treatment strategy. As indicated by stroke studies, diffusion MRI is sensitive to brain perfusion, although this is considered by most to be a direct result of cytotoxic edema in metabolically challenged cells that is secondary to perfusion changes (Mintorovitch *et al.*, 1990; Moseley *et al.*, 1990b; Benveniste *et al.*, 1992). Several methods have been developed to measure hemodynamic properties in the brain more directly. Methods designed for sensitivity to pseudo-random motions of endogenous intracapillary water have been proposed (Le Bihan *et al.*, 1986). Unfortunately, these are prone to artifact and have inherently low sensitivity owing to the small signal contribution by water in the capillary space (typically, a few percent of the MR signal). Alternatively, arterial water can be magnetically labeled by regionally selective radio-frequency (RF) pulses applied upstream before it mixes with water in tissues of interest (Williams *et al.*, 1992; Detre *et al.*, 1992). Regional perfusion rates are estimated using theory developed in close analogy to that for other diffusible tracer techniques from $T_1$ maps and MR images acquired with and without arterial water labeling. This approach shows great promise and has been successfully applied to measure cerebral perfusion in rats, where transit times of labeled arterial water to the brain are short, in satisfac-

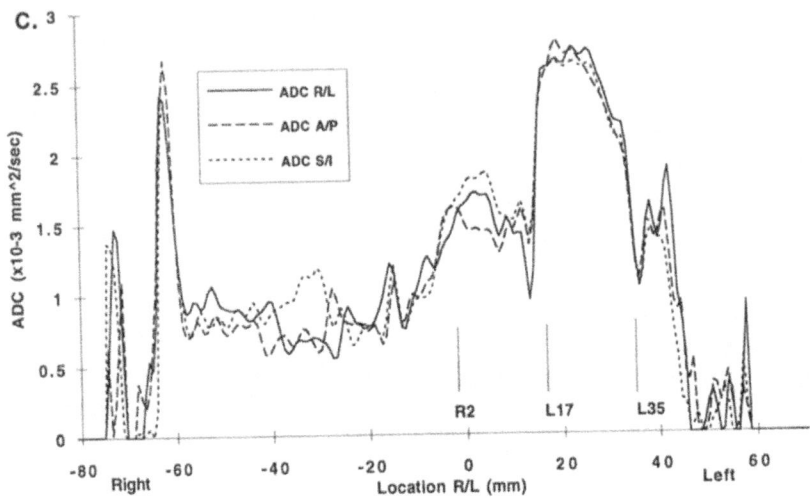

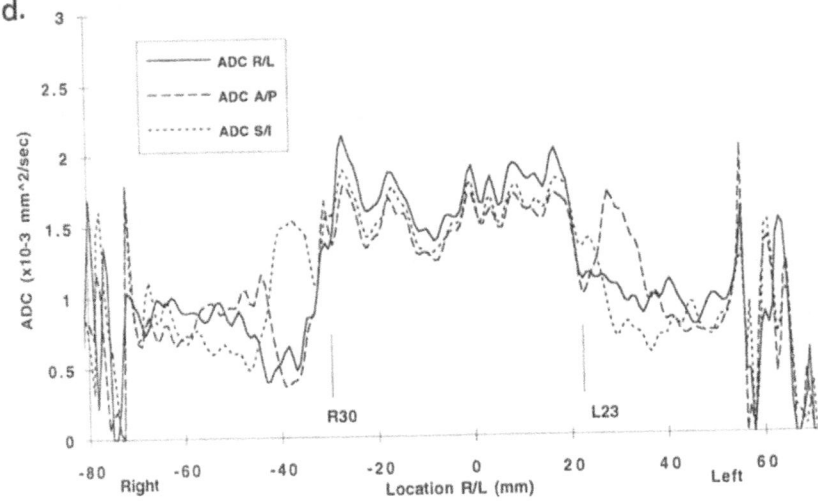

*Figure 5.* Continued

tion of critical assumptions. This method may be extended to human studies, although the long transit time of blood relative to $T_1$ presents a technical challenge (Roberts *et al.,* 1994).

The perfusion MRI methods described above utilize endogeneous contrast sources. These methods suffer from low sensitivity to the limited blood space signal. Perfusion sensitivity can be greatly enhanced by use of exogenous contrast agents (Rosen *et al.*, 1989, 1991a, b). Several gadolinium-based paramagnetic agents in use for the majority of clinical MRI examinations offer the possibility of probing brain function at substantially higher sensitivity than offered by endogenous sources. The primary contrast mechanism in conventional contrast-enhanced MRI is dipolar interaction of the unpaired electrons of the paramagnetic compound with tissue water protons. The dominant effect is the shortening of $T_1$ of tissue water, and thus a signal enhancement on $T_1$-weighted MR images. A rather distinct contrast mechanism is employed when these paramagnetic agents are used in perfusion MRI studies. The large magnetic moment of the agent when concentrated in the vasculature alters magnetic susceptibility patterns around capillaries. Tissue water that diffuses through these susceptibility gradients during the echo period, TE, will dephase relative to that following other random diffusion paths (Fisel *et al.*, 1991). Although the gadolinium is contained within the small capillary space (a few percent of tissue volume), it can have a profound impact on the net water signal received. For example, brain signal may drop approximately 50% following intravenous (i.v.) bolus administration of 0.1–0.2 mmol of gadolinium-based contrast agent per kilogram (Rosen *et al.*, 1989, 1991a,b). This is a transient process (10 sec in humans) because the concentration of agent is sufficiently high only during the first pass of the bolus.

For tumors with a compromised BBB, there is competition between $T_1$ shortening, which enhances MR signal, and $T_2$ (and $T_2^*$) shortening, which reduces signal. The dominant effect is determined by the concentration of agent, its compartmentalization, and MRI acquisition parameters. During bolus passage through an intact BBB, the susceptibility contrast mechanism ($T_2$ shortening) dominates. If one is allowed to ignore $T_1$ effects the reduction of MR signal by gadolinium ($S_{Gad}$) is directly related to $T_2$ change as $\triangle R_2 = \triangle(1/T_2) = [\ln(S_{Gad}/S_{preGad})] / TE$. It is fortuitous for MR perfusion measurements that the change in $T_2$ relaxivity, $\triangle R_2$, is proportional to gadolinium concentration. Using classic tracer kinetic models, investigators have provided strong theoretical and empirical evidence that temporal integration of the gadolinium concentration curves reflects cerebral blood volume (CBV). Since these signal changes are transient, rapid MR imaging is essential. Echo-planar imaging (EPI; see Chapter 8) is one specialized MRI technique that meets the technical demands to measure transient signal changes at sufficiently high temporal and spatial resolution to generate human CBV images (Rosen *et al.*, 1989, 1991a,b). Examples of the application of this methodology to distinguish recurrent tumor from radiation necrosis are illustrated below.

EPI was used for interleaved acquisition of 10 axial sections through the human brain before and during an intravenous bolus administration of gadolinium (Gd) contrast material at 1.5-sec temporal resolution. Temporal integration of $\triangle R_2$ time-course curves for each pixel was used to generate regional CBV maps (Aronen *et al.*, 1994). Figure 6a illustrates the CBV map of a patient post

**a.**

**b.**

**c.**

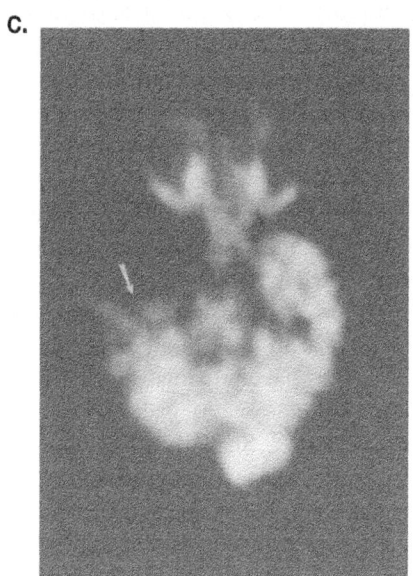

*FIGURE 6.* CBV map (a), $T_1$-weighted image (b), and FDG-PET image (c) of a patient post resection of a grade II/IV astrocytoma and radiation. The CBV map shows no increased CBV near the focus of enhancement observed along the posterior aspect of the resection site on the conventional Gd-enhanced $T_1$-weighted MR image, which is also cold on FDG-PET (see arrows on the respective panels of the figure). These observations are consistent with radiation necrosis. (Figure courtesy of Drs. J. D. Rabinov and B. R. Rosen.)

surgical resection of a grade II/IV astrocytoma of the right temporal lobe. While there is a focus in enhancement in the conventional post-Gd $T_1$-weighted image (Fig. 6b), there is no increased CBV. This region was also "cold" on the [18]F-fluorodeoxyglucose (FDG) positron-emission tomography (PET) study (Fig. 6c). These results are consistent with radiation necrosis (Rosen et al., 1991a; Aronen et al., 1994; Wenz et al., 1994).

The same methodology was applied to another patient who had a grade II/IV astrocytoma. Although there is minimal Gd enhancement in the anterior right temporal lobe (Fig. 7b), there is regional increased CBV (Fig. 7a) and uptake on the FDG-PET image (Fig. 7c) in the anterior right temporal lobe. These results are suggestive of recurrent tumor in this area, which was subsequently proven by surgical biopsy. Investigators have shown that CBV maps correlate with tumor grade (Rosen et al., 1991a; Aronen et al., 1994). High-grade gliomas tend to have comparable or increased CBV relative to normal brain. In addition, high-grade gliomas show considerable CBV heterogeneity across individual lesions (Aronen et al., 1994). Although excellent agreement between FDG-PET and CBV maps is often observed (Rosen et al., 1991a), complete congruence is not guaranteed (c.f. Fig. 7). Presumably, this is a result of lesion heterogeneity and the fact that CBV and FDG-PET modalities probe different physiologic processes.

Extension of relative CBV to absolute measures of tissue perfusion requires determination of the arterial input function and correct assumptions regarding agent compartmentalization. Several investigators have extended dynamic susceptibility MR imaging to simultaneously monitor signal changes in the carotid artery and the brain slice of interest (Perman et al., 1992; Rempp et al., 1993, 1994). This general approach has the potential to improve initial diagnosis and may be valuable in serial studies of treated patients where absolute quantification of tissue perfusion or CBV is desired. As noted above, a compromised BBB leads to leakage of contrast material into the interstitial space, in violation of key assumptions used in the MR-derived CBV model. A leaky BBB is clearly a common occurrence in brain tumors, particularly high-grade gliomas. Since current gadolinium-based contrast is an effective $T_1$ relaxer as well as a susceptibility agent, signal enhancement may mask signal loss, leading to an underestimation of CBV. Signal time-course curves (Fig. 8a) derived from an intracerebral C6 tumor in the rat following Gd i.v. bolus injection illustrate a transient susceptibility signal dip followed by enhancement as the agent diffuses into the interstitium. Corresponding curves of $\Delta R_2$ (Fig. 8b) and its partial integral (Fig. 8c) illustrate that these competitive effects can lead to erroneous relative CBV values depending on integration limits. Several strategies have been developed to reduce CBV errors owing to $T_1$ enhancement. Patients have been "preloaded" with Gd prior to the dynamic study to saturate the interstitial space with Gd (Aronen et al., 1994). Gamma-variate function fits to $\Delta R_2(t)$ also can reduce CBV effects of Gd

a.

b.

c.

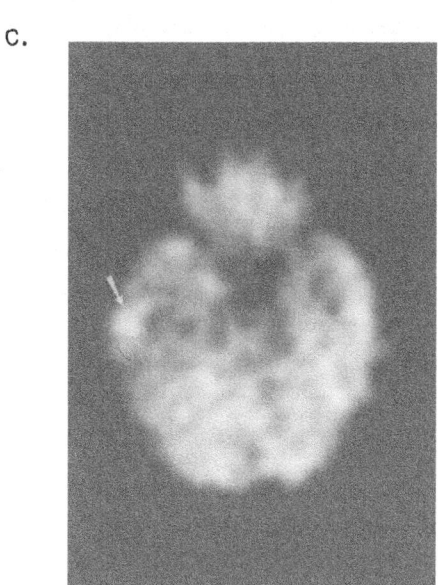

FIGURE 7. CBV map (a), $T_1$-weighted image (b), and FDG-PET image (c) of a patient who had a grade II/IV astrocytoma and radiation. The CBV map shows increased CBV in the right temporal lobe (see arrow). There is little enhancement on the Gd-enhanced $T_1$-weighted MR image, although there is increased uptake on the FDG-PET image (see arrows on the respective images). These observations are consistent with recurrent tumor. (Figure courtesy of Drs. J. D. Rabinov and B. R. Rosen.)

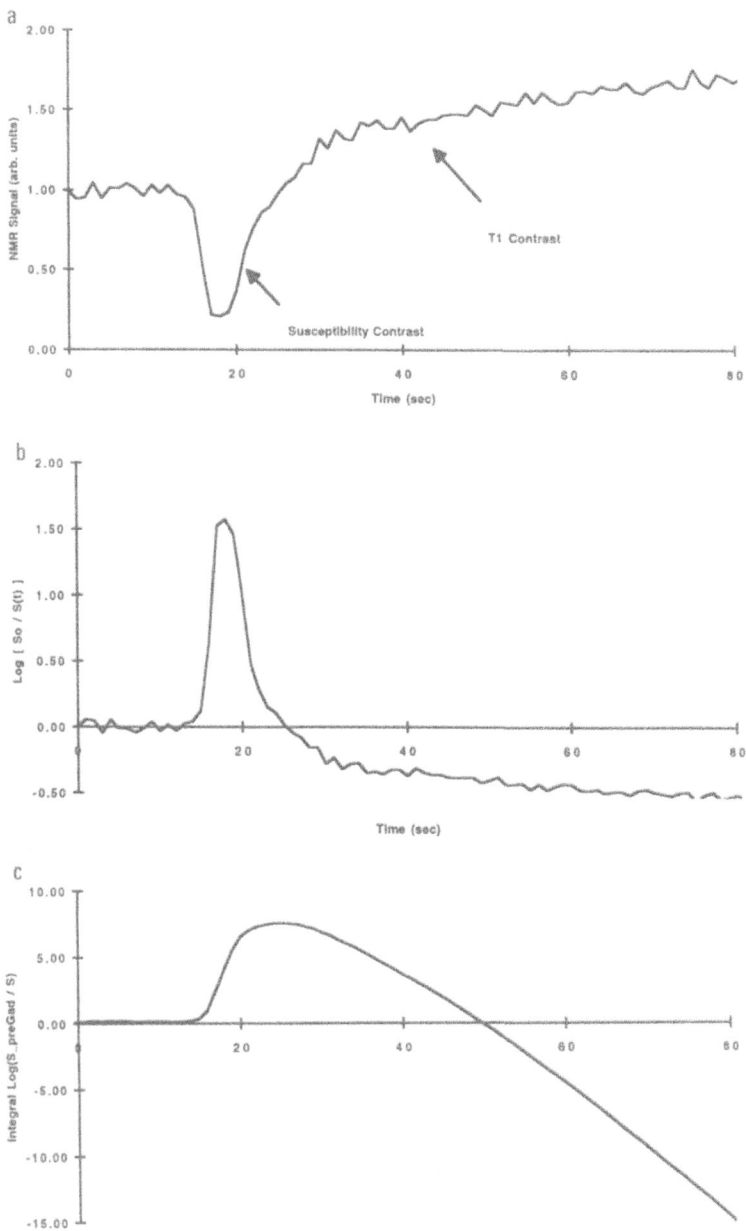

FIGURE 8. Dynamic gadolinium time course in a rat intracranial C6 glioma, (a) illustrating CBV underestimation when the blood–brain barrier is permeable to gadolinium. Time course of NMR signal amplitude, $S(t)$, exhibiting transient signal drop due to susceptibility contrast effect and signal enhancement as gadolinium diffuses into interstitium. (b) Time course of $\log[S_{preGad} / S(t)]$, which would be proportional to gadolinium concentration if blood–brain barrier were intact. (c) Temporal integral of $\log[S_{preGad} / S(t)]$. Underestimate of CBV is dependent on amount of integrated $T_1$ enhancement.

recirculation and $T_1$ enhancement. Lastly, more realistic models of both susceptibility and $T_1$ relaxivity effects can be used to provide corrected CBV images
and maps of relative permeability (Weisskoff *et al*, 1994). These methods may
offer a more accurate assessment of tumor grade and response to treatment.

## 2.4. *In Vivo* MRI of Particle Delivery and Phagocytosis in Experimental Gliomas

MRI has also been used to noninvasively map and follow the time course
and distribution of intra-arterially delivered monocrystalline iron oxide nanopar-
ticles (MION) in rat brain (Zimmer *et al.*, 1995) and experimental rat brain
tumors (Rainov *et al.*, 1995). The MION used in these studies consisted of a
superparamagnetic iron oxide crystal core with a stable dextran coat, yielding an
overall hydrodynamic diameter in the nanometer range. Because the transfer of
therapeutic genes into brain tumors using viral vectors (also having diameters in
the nanometer range) is of tremendous therapeutic interest, a comparison of the
distribution of MION and viral particles, in order to ascertain whether MION can
serve as noninvasive MRI visible marker that reflects the distribution of gene
transfer following viral delivery, was recently reported (Rainov *et al.*, 1995).
Results from this study revealed that the intra-arterial delivery of viral and
nonviral particles (MION) to experimental brain tumors had essentially the same
distribution (Rainov *et al.*, 1995). This approach represents a useful and novel
technique to study the delivery of particles to both the CNS and brain tumors.

MION that have been modified to increase their circulatory half-life have
been used to observe *in vivo* phagocytosis of glioma cells (Zimmer *et al.*, 1995).
Shown in Fig. 9 is an example of the type of images produced by this approach.
Figure 9 displays $T_2$-weighted spin-echo images (TR/TE = 1500/40) of an
intracerebral C6 glioma before (left image) and 24 hr after i.v. administration of
MION (10 mg Fe/kg body weight) (right image). Prior to MION administration,
the tumor appeared uniformly hyperintense on the $T_2$-weighted images (Fig. 9,
left). Following i.v. administration of MION, a hypointense middle ring was
evident, which was histologically shown to correspond to the highest accumulation of iron and to be the most vascular and most cellular region of the tumor. A
central area of necrosis and infiltrating tumor cells in the surrounding normal
brain tissue corresponded to the hyperintense inner and outer circles, respectively, and was histologically shown to have relatively lower amounts of MION
accumulation. Thus, the authors pointed out that this approach images both
functional (phagocytosis, neovascularity) and anatomic (tumor margin) information (Zimmer *et al.*, 1995b). This can provide vital information not obtainable by
Gd contrast studies which only demonstrate disrupted BBB; including the potential for differential phagocytosis of MION by tumor cells versus normal brain,
which would allow delineation of tumor margins. Furthermore, as tumor pha-

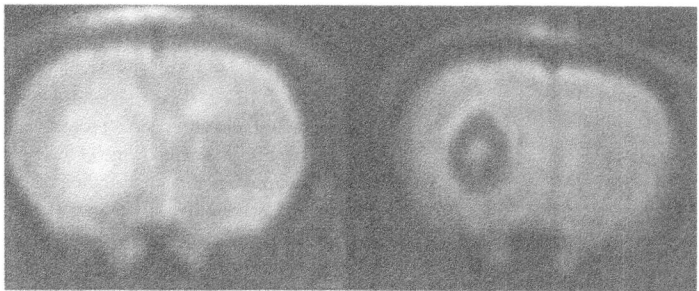

FIGURE 9. $T_2$-weighted MR images (SE 1500/40) of MION uptake by intracerebral C6 gliomas in rats before (*left*) and 24 hr after intravenous administration of 10 mg Fe/kg body weight (*right*). The periphery of the tumor appears hypointense while the center appears hyperintense. (Figure courtesy of Dr. R. Weissleder.) SE = spin echo.

gocytosis may vary with tumor grade, this method may potentially provide a means to monitor tumor cell differentiation. Finally, MION may also have therapeutic potential as carriers of antineoplastic agents which could be released within the tumor cells upon degradation following phagocytosis.

## 3. APPLICATION OF MULTINUCLEAR MRS TO STUDY BRAIN TUMORS

### 3.1. $^{31}$P MRS Studies of Brain Tumors

Tumor metabolites detected by $^{31}$P MRS consist primarily of the phosphomonoesters (PME) phosphocholine (PC) and phosphoethanolamine (PE), inorganic phosphate ($P_i$), the phosphodiesters (PDE) glycerophosphoethanolamine (GPE) and glycerophosphocholine (GPC), nucleoside triphosphates (predominately the three phosphorus nuclei of ATP), and diphosphodiesters (DPDE). In view of the unique capability of $^{31}$P MRS for allowing *in vivo* measurements of tissue energy metabolites, it was hoped that this method would become an important clinical tool for diagnosis, prognosis, and individualization of tumor treatment, and its potential clinical use has been the subject of review (Negendank, 1992). The early expectation of using the $^{31}$P spectral profile as a diagnostic aide to discriminate brain tumor type or degree of malignancy (histologic grade) has not been realized. Furthermore, reports on the use of $^{31}$P for evaluating brain tumor treatment have been relatively few and anecdotal (Segebarth *et al.*, 1987; Arnold *et al.*, 1987; Ross *et al.*, 1989a). More recently, it has been reported that tumor cells dying by apoptosis and during mitosis probably do not directly affect the levels of metabolites observed by $^{31}$P MRS and that necrotic

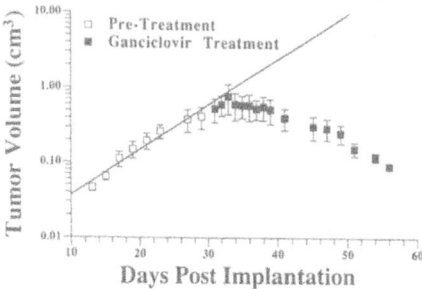

*FIGURE 10.* Subcutaneous C6BSTK glioma tumor volume vs. time post implantation of $10^5$ cells in the hind limb of a rat: before (□) and during (■) administration of ganciclovir (i.p. twice daily at 15 mg/kg per dose). Note the significant reduction in tumor volume obtained by this gene therapy approach. (Figure courtesy of L. D. Stegman.)

cells do not contribute to the $^{31}P$ spectrum (Tozer and Griffiths, 1992). This is in agreement with a $^{31}P$ MRS study of the rat C6 glioma model in which no correlation was found between $^{31}P$ metabolite ratios or pH with the amount of necrosis in either subcutaneous or intracerebral gliomas (Ross *et al.,* 1988a). Furthermore, a recent study investigating a gene therapy approach for treating brain tumors using a suicide gene to encode for a protein that transforms a relatively nontoxic pro-drug into a cytotoxic compound illustrated the lack of changes in tumor energy metabolites following significant cell killing (Stegman *et al.,* 1997). This approach involved the *in vitro* transfer of the herpes simplex virus thymidine kinase gene (HSV*tk*) into rat C6 glioma cells (C6BSTK; Ezzeddine *et al.,* 1991) that were then implanted subcutaneously in the hind limb of a rat. The tumor cells expressing HSV*tk* are able to phosphorylate the antiherpetic drug ganciclovir (GCV), an analog of guanosine, and upon further phosphorylation, this analog can be incorporated into DNA, which causes inhibition of DNA polymerase, resulting in cytotoxicity (Moolten, 1994). Shown in Fig. 10 is a plot of the subcutaneous C6BSTK glioma tumor volume versus time before and during administration of GCV (Stegman *et al.,* 1997). Tumor growth was arrested approximately six days after initiation of GCV administration followed by significant regression of the tumor mass. Representative *in vivo* $^{31}P$ MR spectra obtained before and 12 days after GCV treatment are shown in Fig. 11 (Stegman *et al.,* 1977). No significant changes in ATP, phosphocreatine (PGr), or tumor pH were observed during GCV treatment despite significant tumor regression, which is consistent with the notion that apoptotic or necrotic (Ross *et al.,* 1988a; Tozer and Griffiths, 1992) mechanisms of cell death probably do not affect the $^{31}P$ spectrum. Preliminary results were also reported indicating that changes in the PE/PC ratio occurred during GCV treatment and preceded the reduction in tumor volume (Stegman *et al.,* 1997). Although further work is needed to confirm this observation, it is consistent with the finding in a study using 9L glioma spheroid extracts that the PE/PC ratio declined with increased fraction of proliferating cells (Freyer *et al.,* 1994).

$^{31}P$ MRS can provide useful information for tumors in which a large fraction of cells are killed (e.g., following hyperthermia) or rendered ischemic (e.g.,

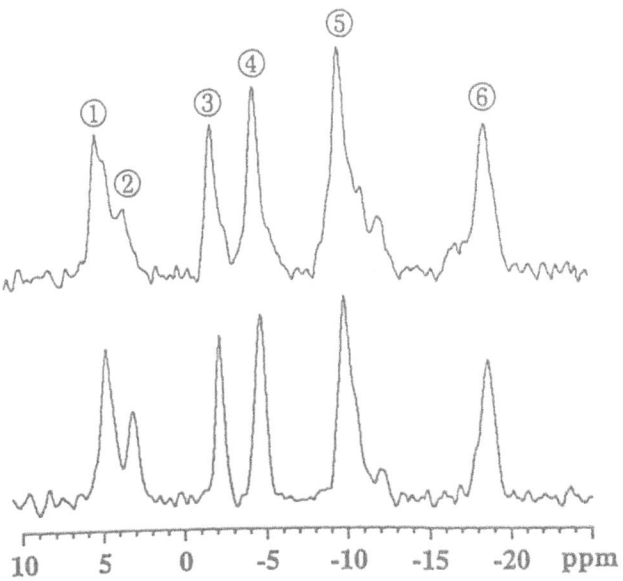

FIGURE 11. *In vivo* [31]P MR spectra of a subcutaneous C6BVIK glioma obtained during exponential growth (*top*) and during regression due to 12 days of ganciclovir treatment (*bottom*) as shown in Fig. 12. No significant differences in glioma pH or ATP levels could be detected although complete retardation in tumor growth had been achieved. Peak assignments: 1, Phosphomonoesters (PME); 2, inorganic phosphate ($P_i$); 3, phosphocreatine (PCr); 4, γ-ATP and β-ADP; 5, α-ATP, α-ADP, and $NAD^+$/NADPH; 6, β-ATP. (Figure courtesy of L. D. Stegman.)

during hyperglycemia) or following therapies that inhibit the enzymes (e.g., hexokinase) involved in maintaining the energy state or lactate transport. For example, lonidamine (LND) has been reported to be an inhibitor of mito-chondrially bound hexokinase (Paggi *et al.,* 1987) and more recently an inhibitor of lactate efflux (Ben-Horin et al., 1995). Since the amount of mitochondrially bound hexokinase increases with tumor malignancy and since gliomas metabol-ize a significant portion of glucose utilized to lactate, LND was thought to be an attractive agent for achieving selective inhibition of brain tumor energy metabo-lism (Pedersen, 1978). [31]P MRS studies of the effect of LND on the energy metabolism of subcutaneous rat 9L gliomas revealed that significant declines in tumor ATP/$P_i$ ratio (65%) and pH ($-0.45$ pH units) occurred following LND administration, as shown in Fig. 12 (Ross *et al.,* 1994). This effect was shown to be associated with inhibition of lactate efflux (Ben-Yoseph, *et al.,* 1997).

Treatment of tumors with vasoactive agents resulting in a reduction in tumor blood flow (TBF) has also been shown to affect the tumor energy state. One such treatment that has been widely investigated in experimental tumors is the bolus administration of glucose in order to induce acute hyperglycemia (Ward and Jain,

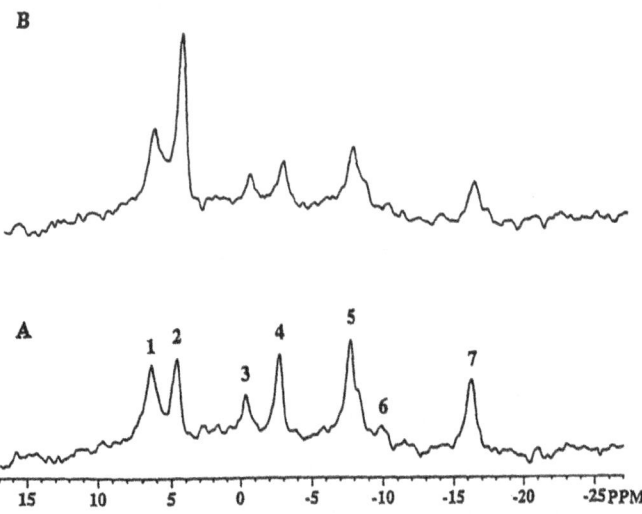

FIGURE 12. *In vivo* [31]P MR spectra of a subcutaneous rat 9L glioma before (A) and 3 hr following intraperitoneal administration of a 100-mg/kg dose of lonidamine. Note the significant reduction in tumor ATP and the concomitant increase in $P_i$ levels. Peak assignments are as in caption to Fig. 11.

1988). Hyperglycemia has been shown to increase the sensitivity of experimental tumors to hyperthermia through reductions in TBF and pH. The belief was that all tumors, regardless of cell type and host tissue, were selectively sensitive to hyperglycemia-induced reductions in TBF and tumor pH. [31]P MRS studies of rats bearing subcutaneous C6 gliomas revealed the classic reduction in tumor pH following glucose administration (Ross *et al.*, 1989b). However, the same treatment for rats with intracerebral C6 gliomas revealed that when the tumor was within the brain parenchyma, hyperglycemia had no significant effect on tumor pH (Ross *et al.*, 1989b). These [31]P MRS results provided new insights into the role of the host tissue in hyperglycemia-induced reduction in TBF and the utility of [31]P MRS for such studies.

To summarize, it appears that treatments involving DNA damage are not likely to be detected as changes in energy metabolites or pH by [31]P MRS. However, changes in PME resonances of tumors have been reported to be linked to therapeutic response, possibly due to alteration of the fraction of proliferating cells following treatment (Negendank, 1992). This concept has recently gained further momentum from high-resolution [31]P MRS studies of extracts obtained from rat 9L glioma spheroid cultures (Freyer *et al.*, 1994). Cell killing following therapy may, however, indirectly produce significant changes in the [31]P MRS-observed tumor energy metabolite levels by inducing changes in TBF and alterations in oxygen consumption rates and diffusion distances (Tozer and Griffiths,

1992). Finally, $^{31}$P MRS can provide important information for tumors in which a large fraction of the tumor cells are killed in a short period of time (Ben-Yoseph and Ross, 1994) or rendered ischemic or following therapies that inhibit the enzymes involved in maintaining the energy state or lactate transport.

## 3.2. Localized $^1$H MRS of Intracranial Tumors

The predominant tumor metabolites detected by $^1$H MRS are mobile lipids (0.0–2.0 ppm), lactate (1.3–1.5 ppm), *N*-acetylaspartate (NAA, 2.0 ppm), glutamate and glutamine (2.1–2.5 ppm), total creatine (3.0 ppm), and choline-containing compounds (3.2 ppm) (Miller, 1991; Howe *et al.*, 1993). The choline signal can include contributions from choline, glycerophosphocholine, phosphocholine, possibly phosphatidylcholine, and metabolites involved in phospholipid metabolism (Miller, 1991). The relatively high sensitivity of $^1$H MRS allows for spectra to be obtained from smaller volumes of tumor tissue. This feature provides the capability for potential investigations of tumor heterogeneity, which can arise from both histological and metabolic variations within the tumor mass. Furthermore, as the therapeutic response of a tumor may vary regionally, spatially localized $^1$H spectra may afford the opportunity for detecting those tumor regions which are sensitive or resistant to treatment.

The majority of orthotopic *in vivo* brain tumor models utilize rodents. In these models, an intracerebral tumor can reach approximately 5–7 mm in diameter before the animal succumbs to the tumor burden. Therefore, it is advantageous to utilize a surface coil because the "filling factor" is optimal for maximal MR sensitivity. Because the $B_1$ field of the surface coil is highly inhomogeneous and decreases in magnitude with distance from the coil, the use of RF pulses that are insensitive to $B_1$ amplitude is vital. Amplitude- and frequency/phase-modulated pulses based on adiabatic passage principles which are insensitive to variation in $B_1$ amplitude have been used for both imaging (Garwood *et al.*, 1989) and localized $^1$H MRS studies (Ross *et al.*, 1992; Schupp *et al.*, 1993). Thus, the brain can be imaged using a single surface coil for both RF transmission and reception in order to determine the spatial coordinates necessary for obtaining water-suppressed, spatially localized $^1$H spectra. This approach was recently utilized using the rat intracerebral 9L glioma model to investigate if therapeutic intervention could produce changes in the tumor $^1$H spectra and if any such changes could be correlated with therapeutic efficacy as quantitated by an increase in tumor doubling time $(T_d)$ (Ross *et al.*, 1995). Figure 13 shows the spatially localized *in vivo* $^1$H MR spectra of a rat brain with an untreated 9L tumor (Fig. 13A) and a 9L tumor treated with a recombinant adenoviral vector (Ad.RSV*tk*) (used to transfer the HSV*tk* gene) followed by 4 days of systemic GCV administration (Fig. 13B). Each spectrum arises from 25-$\mu$l voxels along a column positioned through the cerebral hemispheres of the

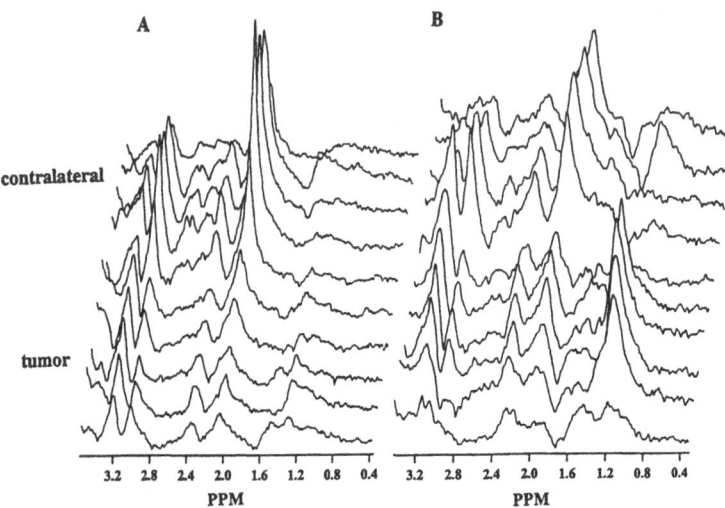

FIGURE 13. *In vivo* localized ¹H spectra arising from 25-μl voxels along a column through a rat brain with an untreated 9L glioma (A) and a 9L glioma injected with Ad.RSV*tk* and systemically treated with ganciclovir for 4 days (B). A large therapeutic-induced increase in the lipid/lactate resonance can be seen in the spectra obtained from the treated 9L tumor. Resonance assignments are as follows: Choline-containing compounds, 3.2 ppm; total creatine, 3.0 ppm; glutamate and glutamine, 2.1–2.5 ppm; NAA, 2.0 ppm; lipid/lactate, 1.0–1.4 ppm. [Spectra were reprinted, with permission, from Ross *et al.*, (1995).]

rat. Spectra of untreated 9L tumors (Fig. 13A) revealed reduced levels of creatine and NAA as compared to the levels in the contralateral hemisphere. The lipid/lactate resonance at 0.9–1.3 ppm was barely visible in the untreated tumor and contralateral brain. However, following gene therapy, this peak dramatically increased, as seen in Fig. 13B. Concomitantly, an increase in the intracranial 9L tumor $T_d$ from 43 to 213 hr was observed, indicating that the treatment had a significant effect on the tumor growth which correlated with an increased lipid/lactate peak. Histological analysis of the treated 9L tumor revealed a large area of necrosis (Ross *et al.*, 1995). These findings are consistent with a recent high-field MRS study of biopsies obtained from high-grade astrocytomas in which it was found that increased amounts of necrosis resulted in an increased intensity of the lipid signal (Kuesel *et al.*, 1994). It is also possible that the increased resonance intensity at 0.9–1.3 ppm following treatment could have a significant contribution from lactic acid. While further studies using spectral editing techniques are needed to fully characterize the relative contributions of lipid and lactate following this gene therapy approach, it is clear that ¹H MRS can detect gene therapy-induced changes in a rodent brain tumor that correlate with a reduction in tumor growth rate.

The lactate component in the lipid/lactate resonance of brain tumors can be conclusively identified and quantitated using spectral editing techniques. In this regard, an adiabatic homonuclear spectral editing method for quantitating the lactate resonance (1.3 ppm), combined with two-dimensional (2-D) chemical shift imaging, has been recently used to investigate the spatial distribution of lactate within an intracerebral C6 glioma in the rat (deGraaf *et al.*, 1995). Shown in Fig. 14a is a $T_2$-weighted image of a C6 glioma obtained using a surface coil. In Fig. 14b, a contour plot is shown overlaid with an expanded area of the MR image, revealing the distribution of lactate within the tumor region. The contour levels are described in the caption to Fig. 14b. Shown in Fig. 14c are the lactate spectra arising from 1.5-$\mu$l voxels of tissue from which the contour levels were derived. Histological analysis of the rat brain revealed that the abnormal lactate levels were confined to the tumor and that the region of adjacent peritumoral edema did not have increased lactate levels over that of normal brain (Luo *et al.*, 1995). Furthermore, the lactate measured in individual voxels revealed a positive correlation with the density of neoplastic cells contained in those voxels. Conversely, regions histologically determined to have low numbers of neoplastic cells and/or large amounts of necrotic tissue were found to have low lactate signal intensities. These results reveal the exquisitely high spatial resolution offered by $^1$H chemical shift imaging (CSI) when combined with adiabatic pulses and surface coils for RF transmission and reception. This methodology now provides the capability for minimizing partial-volume effects arising from the different cell types within heterogeneous tissues such as tumors. This spatial resolution, coupled with histological confirmation, provides an important opportunity for improving our understanding of the metabolic profiles associated with each cell type within brain tumors and the contribution of these cell types to the overall $^1$H spectra.

$^1$H MRS has also been evaluated as a discriminator of brain tumor grade and, more recently, as a monitor of therapy in humans (Fulham *et al.*, 1992; Heesters *et al.*, 1993; Bizzi *et al.*, 1995; Nelson *et al.*, 1995). Although the use of *in vivo* $^1$H MRS for identification of tumor type or grade has not been tremendously successful a recent study in which linear discriminant analysis of brain tumor proton spectra showed promise for tumor classification (Preul *et al.*, 1996). Moreover, recent independent studies have reported that a response to chemotherapeutic or radiation treatment may be gauged by a decline in the choline resonance intensity (Fulham *et al.*, 1992; Heesters *et al.*, 1993; Bizzi *et al.*, 1995; Nelson *et al.*, 1995). An illustrative example is presented in Fig. 15, in which MR images and water-suppressed $^1$H MR spectra from three serial MR examinations of a single patient with a recurrent anaplastic astrocytoma are shown. All data in Fig. 15 were acquired on a standard 1.5-T GE Signa clinical MR system. The time points shown are immediately prior to radiosurgery to region R2 (left) and one month (middle) and four months following radiosurgery (right). This patient also had been treated a year previously with radiosurgery to region R1. On the volume spoiled gradient-echo post-Gd MR images (TR/TE = 32/8, flip angle = 45°, 124 1.5-mm slices, 180 × 240 mm FOV (field of view),

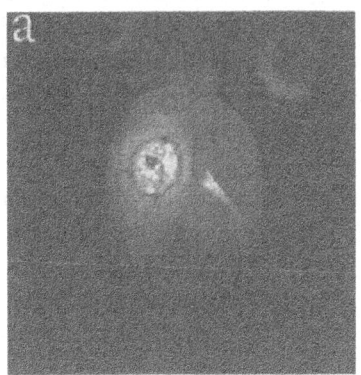
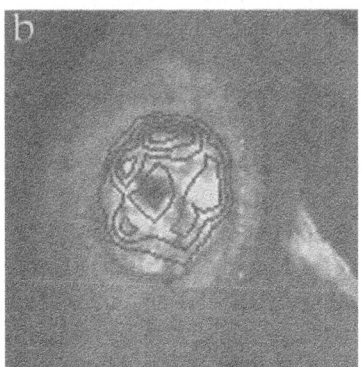

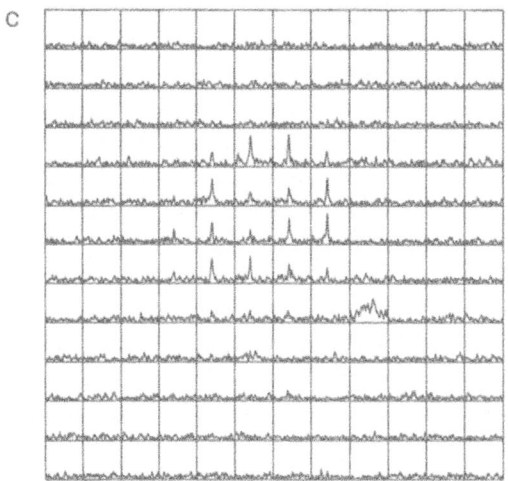

FIGURE 14. (a) $T_2$-weighted MR transverse image of a rat with an intracerebral C6 glioma acquired with a surface coil. (b) An expanded view of the tumor overlaid with a contour plot with five different levels indicating the amount of lactate. The lactate levels of 33, 46, 60, 74, and 86% are the percentages of the difference between the highest and lowest lactate levels. The outer contour and the small central contour correspond to 33%. The second contour and the larger contour in the middle have a level of 46%. For the other contours, the amount of lactate increases level by level, with the highest level (86%) corresponding to the most anterior (topmost) portion of the tumor. (c) The spatial distribution and intensities of the lactate resonances arising from 1.5-μl voxels from which the contour levels were derived in (b). Spectra were acquired on a 4.7-T MR system using an adiabatic multiple-quantum technique as previously described (deGraaf *et al.*, 1995). Acquisition parameters: $32 \times 32$ phase-encode steps, slice thickness of 1.5 mm, 2 transients per phase-encode step, TR/TE = 2000/144, FOV = $32 \times 32$, total accumulation time of 52 min (nominal voxel size = $1.0 \times 1.0 \times 1.5$ mm$^3$ = 1.5 μl). (Figure courtesy of Drs. M. Garwood and Y. Luo.)

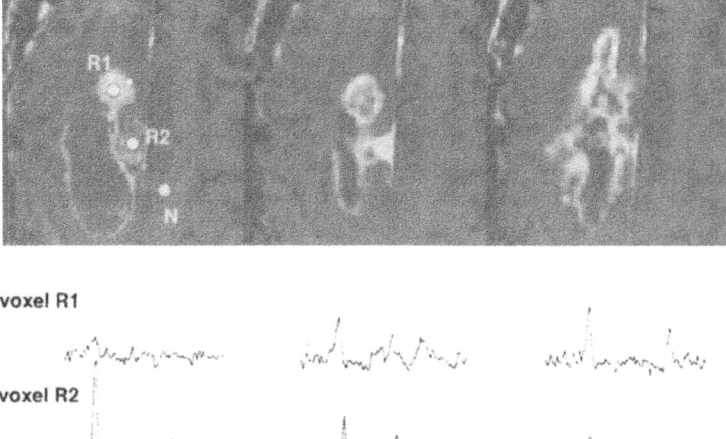

**voxel R1**

**voxel R2**

**voxel N**

*FIGURE 15.* MR images and water-suppressed ¹H MR spectra from three serial MR examinations of a single patient with a recurrent anaplastic astrocytoma. The time points shown are immediately prior to radiosurgery to region R2 (*left*), one month following radiosurgery (*middle*), and four months following radiosurgery (*right*). This patient had also been treated a year previously with radiosurgery to region R1. On the volume spoiled gradient-echo post-Gd MR images, there are significant increases in the size and changes in the shape of the contrast-enhancing lesion with time, as well as alterations in the cystic component. Extracted spectra centered on three selected voxels (radiosurgery targets R1 and R2 and normal appearing brain N) are shown below their corresponding images. Clear differences in metabolite patterns are seen in the spectra. Voxel N shows approximately constant levels of choline, creatine, and NAA, demonstrating the reproducibility of the technique. In the region R2, there is initially a very high level of choline, negligible creatine and NAA, and a small peak which could be either lactate or lipid. After treatment, the choline decreases, with minor changes in the other metabolites, suggesting successful local therapy. In region R1, there is initially a low-level, broad peak at the position of choline. After one month this has increased significantly and increases again at the third time point, suggesting tumor recurrence. Resonance assignments are the same as in Fig. 13. (Figure courtesy of Dr. S. J. Nelson.)

192 × 256 matrix, 0.75 NEX (number of averages)), there are significant increases in the size and changes in shape of the contrast-enhancing lesion with time, as well as alterations in the cystic component. The spectroscopy data were acquired with the standard head coil using PRESS volume selection and 3-D phase encoding (TR = 1 sec, TE = 272 msec, 8 × 8 × 8 matrix, 2-cc spatial resolution, 2 NEX, $T_{acq}$ = 17 min). By extracting and examining spectra that

were centered on three selected voxels (radiosurgery targets R1 and R2 and normal appearing brain N), clear differences in the metabolite patterns were seen. Voxel N shows approximately constant levels of choline, creatine, and NAA, demonstrating the reproducibility of the technique. In the region R2, there is initially a very high level of choline, negligible creatine and NAA, and a small peak which could be either lactate or lipid. After treatment, the choline decreases, with minor changes in the other metabolite levels, suggesting successful local therapy. In region R1, there is initially a low-level, broad peak at the position of choline. After one month this peak has increased significantly, and there is a further increase at the third time point. These increases suggest the presence of recurring, proliferating tumor cells since it is generally believed that an increased choline resonance intensity reflects an increased rate of membrane synthesis and/or tissue cellularity. Thus, in the same patient, different regions of tissue within and close to the lesion are showing a heterogeneous pattern of response. The data presented in Fig. 15 also confirm that different regions of the contrast-enhancing lesion may show quite distinct patterns of metabolites due to spatial heterogeneity of the tumor tissue. This fact underscores the importance of using sequential, combined MRI and multivoxel localized MRS examinations for non-invasively monitoring treatment response in patients with brain tumors.

It is interesting to note that the dominant change in the rat glioma $^1$H MR spectra following treatment was an increase in the lipid/lactate resonance intensity (Fig. 13B). In contrast, a decrease in the observed choline signal appeared to consistently correlate with the successful treatment of human brain tumors. This discrepancy may be due to differences in time between the acquisition of the pre- and post-treatment $^1$H spectra. In the rodent brain tumor studies, $^1$H MRS was usually accomplished a relatively short time after therapy. As the clearance of macromolecular debris (e.g., necrotic cells) occurs rather slowly within the brain (Kumar *et al.*, 1974), it is reasonable to assume that the early increase in the lipid/lactate signal following treatment is due to treatment-induced cell death resulting in necrosis. Furthermore, as the gene therapy approach involved a localized injection of viral vector into the tumor mass, a significant partial-volume effect in the localized $^1$H spectra from both treated and untreated tumor cells was likely a confounding factor in determining the specific change(s) that best predicts eventual therapeutic outcome.

From the above discussion, $^1$H MRS appears to be useful for monitoring the metabolism and therapeutic response of experimental and human brain tumors. Further evaluation of $^1$H MRS with different tumor types and treatment interventions should be accomplished; however, it would be reasonable to assume that this spectroscopic approach will become an important and more routine contributor to the array of accepted radiological techniques used in the management of brain tumor patients in the near future.

## 3.3. MRS Metabolic Studies of Experimental Brain Tumors Using [13]C-Labeled Precursors

Positron-emission tomography (PET) studies have revealed that brain tumors have abnormally high rates of glucose utilization and an increased capacity for aerobic glycolysis compared to the brain (Rhodes *et al.*, 1983; Di Chiro *et al.*, 1984). Insights into the biochemical pathways by which malignant gliomas maintain essential energy metabolite levels through abnormally high rates of glucose utilization and aerobic glycolysis could aid in our understanding of how these tumors rapidly proliferate and resist therapeutic intervention. In this regard, [13]C MRS offers intriguing possibilities since it allows intermediary metabolism to be studied; however, unlike [31]P or [1]H, [13]C is only 1.1% naturally abundant so that [13]C-labeled precursors must be used. For example, [13]C MRS was used to follow the metabolism of [1-[13]C]glucose in the intracerebral C6 glioma model (Ross *et al.*, 1988b). In this study, [1]H-decoupled spectra were acquired during an infusion of [1-[13]C]glucose through the femoral vein during acquisition of [13]C spectra from the C6 glioma. *In vivo* [13]C MR spectra of the C6 glioma before and during [1-[13]C]glucose infusion are shown in Fig. 16A and B, respectively. The [13]C-enriched metabolites can be more clearly seen in the background-subtracted spectrum shown in Fig. 16C. Metabolites present included the α and β C-1 anomers of glucose at 92.9 and 96.8 ppm, respectively, C-3-labeled lactate at 20.8 ppm, C-2-, C-3-, and C-4-labeled glutamate/glutamine at 55.3, 27.5, and 34.2 ppm, respectively, and [1-[13]C]glycogen at 100.5 ppm. These results indicated that in this glioma, glucose is utilized by glycolysis, by the Krebs cycle, and for glycogen production. Since [13]C-labeled glycogen is not observed in the brain by MRS following infusion with [1-[13]C]glucose, the ability of [13]C MRS to detect labeled glycogen in gliomas may reflect a lack of metabolic regulation due to dedifferentiation of the tumor cells. Thus, delineation of the relative utilization of glucose via glycolysis, the Krebs cycle, etc., could prove diagnostically important for grading tumors and for monitoring therapeutic efficacy. In fact, PET studies using 2-[[18]F]fluorodeoxyglucose ([[18]F]-FDG) have found that tumors which had an acute increase in glucose utilization within 24 hr of radiation or chemotherapy was predictive of therapeutically non-responsive tumors (Rozental *et al.*, 1993). The authors believe that the post-treatment increase in FDG uptake results from an augmented cellular metabolism including increased glucose flux through glycolysis, the Krebs cycle, and the pentose phosphate pathway as the tumor repairs the damage induced by the treatment. Whereas PET is limited to monitoring only changes in glucose uptake, [13]C MRS can provide complementary information related to how the glucose is utilized within the tumor; specifically, it can be used to examine whether glycolytic and/or Krebs cycle rates are affected and to ascertain whether any such changes are predictive of therapeutic

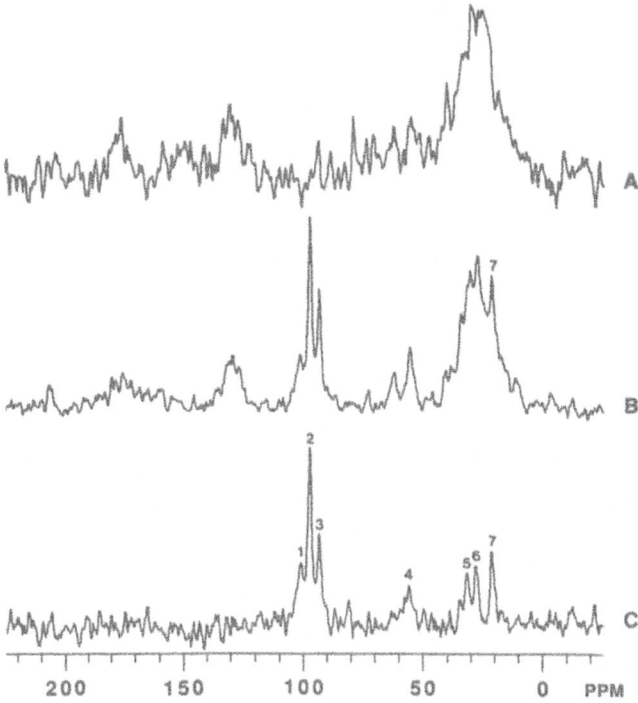

FIGURE 16. *In vivo* [13]C MR spectra of an intracerebral C6 glioma. (A) Natural abundance spectrum obtained with a total acquisition time of 46.8 min. (B) Spectrum from the same C6 glioma acquired over a 93.6-min time interval (twice the time of the natural abundance spectrum) during a constant infusion with [1−13C]glucose. (C) Difference spectrum obtained following subtraction of 2 × spectrum (A) from spectrum (B). Resonance assignments: 1, C-1 of glycogen; 2, β-glucose; 3, α-glucose; 4, C-2 of Glu/Gln; 5, C-4 of Glu/Gln; 6, C-3 of Glu/Gln; 7, C-3 of lactate. [Reprinted, with permission, from Ross *et al.*, (1988b).]

efficacy. Furthermore, the use of a recently reported method for quantitating pentose phosphate pathway (PPP) activity (which has traditionally been difficult to monitor *in vivo* by MRS) in intracerebral rat gliomas would allow for the role of this pathway to also be evaluated following therapy (Ben-Yoseph *et al.*, 1995).

For increased temporal and spatial sensitivity, the glycolytic activity of intracerebral gliomas can also be monitored using spectral editing [1]H MRS techniques to detect [3−13C]lactate produced by the metabolic degradation of [1−13C]glucose (Schupp *et al.*, 1993). Rats with intracerebral C6 gliomas were infused intravenously with [1−13C]glucose over time during the acquisition of spatially localized heteronuclear edited [1]H MR spectra. Shown in Fig. 17 is the time course of [3−13C]lactate production in an intracerebral C6 glioma as mon-

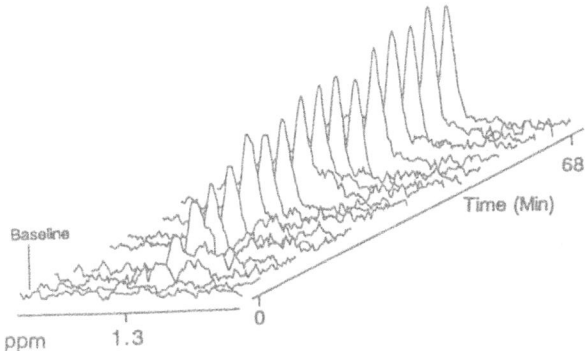

FIGURE 17. Localized heteronuclear edited ¹H MR spectra of an intracerebral C6 glioma in the rat following intravenous infusion with [1 − ¹³C]glucose. Spectra displayed were acquired over a 68-min time period. [Reprinted, with permission, from Schupp *et al.*, (1993).]

itored by localized 3-D image-selected *in vivo* spectroscopy (ISIS) and heteronuclear spectral editing using an adiabatic pulse (Schupp *et al.*, 1993). The [3 − ¹³C]lactate peak at 1.3 ppm could be observed within 4–10 min following initiation of [1 − ¹³C]glucose infusion. This study reveals that editing of ¹³C metabolites by ¹H MRS can be accomplished in a localized region of the tumor with excellent time resolution. These types of experimental approaches should greatly facilitate our understanding of the unique biochemistry associated with malignant neoplasms of the CNS.

## 4. CONCLUSIONS

MRI and MRS are important and complementary techniques for studying many aspects of brain tumors and can provide information not easily obtained by other methods. These techniques bridge the gap between cultured tumor cells, animal models, and human studies and are invaluable for aiding in the diagnosis and therapeutic management of patients with brain tumors and for endeavors involving the quest for a cure for this disease. The results obtained thus far from MR studies of experimental brain tumors in rodents have already provided novel and fascinating insights into their biology and biochemistry and the effects of treatment on these model systems. In particular, as gene transfer techniques are increasingly being developed and explored for the treatment of malignant primary tumors of the brain such as glioblastoma, the potential of using MR for monitoring *in vivo* gene expression is of particular interest. Because the efficiency of gene transfer can vary for many reasons (including type of vector used, infection technique, stability of the particular gene and expressed protein, and viral titer of injected solution, it would be especially valuable to have a method for noninvasive assessment of gene transfer effectiveness. The capability of *in vivo* MRS for monitoring cellular metabolites in intact biological tissues is especially attractive for such purposes. A specific example could be the use of localized MRS for the detection of the conversion of $^{13}$C-labeled substrates into $^{13}$C-labeled products by an enzyme expressed by gene transfer into a solid tumor. This approach would offer an exciting and powerful method for regional assessment of the effectiveness of gene transfer into tumors and could merge the two seemingly disparate worlds of molecular biology and radiology. Furthermore, the application of recent MRI and MRS techniques for the serial assessment of morphological and functional characteristics and metabolism of human brain tumors and their response to therapy offers new and exciting dimensions for clinical research and patient care.

ACKNOWLEDGMENTS

The authors are grateful for the willing contributions of figures from Dr. Michael Garwood, Dr. Sara J. Nelson, Dr. James D. Rabinov, Lauren D. Stegman, and Dr. Ralph Weissleder and thank Dr. Jeffry R. Alger for helpful discussions. The work from the authors' laboratory was supported by Research NIH Grants R29 CA59009 and P20 NS31114, BE-149A from the American Cancer Society, and support from GE Medical Systems. O. B.-Y. is a fellowship recipient from the American Brain Tumor Association.

## REFERENCES

Arnold, D. L., Shoubridge, E. A., Feindel, W., and Villemure, J.-G, 1987, Metabolic changes in cerebral gliomas within hours of treatment with intra-arterial BCNU demonstrated by phosphorus magnetic resonance spectroscopy, *Can. J. Neurol. Sci.* **14:**570–575.

Aronen, H. J., Gazit, I. E., Louis, D. N., Buchbinder, B. R., Pardo, F. S., Weisskoff, R. M., Harsh, G. R., Cosgrove, G. R., Halpern, E. F., Hochberg, F. H., and Rosen, B. R., 1994, Cerebral blood volume maps of gliomas: Comparison with tumor grade and histologic findings, *Radiology* **191:**41–51.

Barker, M., Hoshino, T., Gurcay, O., Wilson, C. B., Nielsen, S. L., Downie, R., and Eliason, J., 1973, Development of an animal brain tumor model and its response to therapy with 1,3-bis(2-chloroethyl)-1-nitrosourea, *Cancer Res.* **33:**976–986.

Basser, P. J., Mattiello, J., and LeBihan, D., 1994, MR diffusion tensor spectroscopy and imaging, *Biophys. J.* **66:**259–267.

Ben-Horin, H., Tassini, M., Vivi, A., Navon, G., and Kaplan, O., 1995, Mechanism of action of the antineoplastic drug lonidamine: $^{31}$P and $^{13}$C nuclear magnetic resonance studies, *Cancer Res.* **55:**2814–2821.

Benveniste, H., Hedlund, L. W., and Johnson, G. A., 1992, Mechanism of detection of acute cerebral ischemia in rats by diffusion-weighted magnetic resonance microscopy, *Stroke* **23:**746–754.

Ben-Yoseph, O., and Ross, B. D., 1994, Oxidation therapy: The use of a reactive oxygen species-generating system for tumor treatment, *Br. J. Cancer* **70:**1131–1135.

Ben-Yoseph, O., Camp, D. M., Robinson, T. E., and Ross, B. D., 1995, Dynamic measurements of cerebral pentose phosphate pathway activity *in vivo* using $(1,6-^{13}C_2,6,6-^2H_2)$glucose and microdialysis, *J. Neurochem* **64**:1336–1342.

Ben-Yoseph, O., Lyons, J. C., Song, C. W., and Ross, B. D., 1997, Mechanism of action of lonidamine in the 9L brain tumor model involves inhibition of lactate efflux and intracellular acidification, *J. Neuro-Oncol.* in press.

Bizzi, A., Movsas, B., Tedeschi, G., Phillips, C. L., Okunieff, P., Alger, J. R., and Chiro. G. D., 1995, Response of non-Hodgkin lymphoma to radiation therapy: Early and long-term assessment with H-1 MR spectroscopic imaging, *Radiology* **194**:271–276.

Brunberg, J. A., Chenevert, T. L., McKeever, P. E., Ross, D. A., Junck, L. R., Muraszko, K. M., Dauser, R., Pipe J. G., and Betley, A. T., 1995, *In vivo* MR determination of water diffusion coefficients and diffusion anisotropy: Correlation with structural alteration in astrocytomas of the cerebral hemispheres, *Am. J. Neuroradiol.* **16**:361–371.

Chenevert, T. L., Brunberg, J. A., and Pipe, J. G., 1990, Anisotropic diffusion in human white matter: Demonstration with MR techniques *in vivo*, *Radiology* **177**:401–405.

Chenevert, T. L., Ross, B. D., Pipe, J. G., and Simerville, S. J., 1991, Quantitative diffusion anisotropy in rat gliomas, 10th Annual Meeting of the Society of Magnetic Resonance in Medicine, San Francisco, Book of Abstracts, Vol. 2, p. 787.

Chenevert, T. L., Brunberg, J. A., and Pipe, J. G., 1992, Quantitative diffusion and anisotropy of human CNS lesions, Society of Magnetic Resonance in Medicine, Book of Abstracts, Vol. 1, p. 1008.

Cooper, R. L., Chang, D. B., Young, A. C., Martin, C. J., and Ancker-Johnson, D., 1974, Restricted diffusion in biophysical systems, *Exp. Biophys. J.* **14**:161–177.

deGraaf, R. A., Luo, Y., Terpstra, M., and Garwood, M., 1995, Spectral editing with adiabatic pulses, *J. Mag. Reson.*, series B. **109**:184–193.

Detre, J. A., Leigh, J. S., Williams, D. S., and Koretsky, A. P., 1992, Perfusion imaging, *Magn. Reson. Med.* **23**:37–45.

Di Chiro, G., Brooks, R. A., Patronas, H. J., Bairamian, D., Kornblith, P. L., Smith, B. H., Mansi, L., and Barker, J., 1984, Issues in the *in vivo* measurement of glucose metabolism of central nervous system tumors, *Ann. Neurol.* **15**(Suppl.):S138–S146.

Ezzeddine, Z. D., Martuza, R. L., Platika, D., Short, M. P., Malick, A., Choi, B., and Breakefield, X. O., 1991, Selective killing of glioma cells in culture and *in vivo* by retrovirus transfer of the herpes simplex virus thymidine kinase gene, *New Biol.* **3**:608–614.

Fisel, C. R., Ackerman, J. L., Buxton, R. B., Garrido, L., Belliveau, J. W., Rosen, B. R., and Brady, T. J., 1991, MR contrast due to microscopically heterogeneous magnetic susceptibility: Numerical simulations and applications to cerebral physiology, *Magn. Reson. Med.* **17**:336–347.

Freyer, J. P., Linford, H., Lovejoy, V., and Moore, G. J., 1994, Correlation between alterations in phosphomonoesters and cellular proliferation during recovery from quiescence in a tumor model, Society of Magnetic Resonance 2nd Annual Meeting, Abstracts, p. 451.

Fulham, M. J., Bizzi, A., Dietz, M. J., Shih, H. H.-L, Raman, R., Sobering, G. S., Frank, J. A., Dwyer, A. J., Alger, J. R., and Chiro, G. D., 1992, Mapping of brain tumor metabolites with proton MR spectroscopic imaging: Clinical relevance, *Radiology* **185**:675–686.

Garwood, M., Ugurbil, K., Rath, A. R., Bendall, M. R., Ross, B. D., Mitchell, S. L., and Merkle, H., 1989, Magnetic resonance imaging with adiabatic pulses using a single surface coil for RF transmission and signal detection, *Magn. Reson. Med.* **9**:25–34.

Grossman, S. A., 1991, Chemotherapy of brain tumors, *In* "Concepts in Neurosurgery, Vol. 4, Neurobiology of Brain Tumors" (M. Salcman, ed.), pp. 321–340, Williams and Wilkins, Baltimore.

Grossman, S. A., and Burch, P. A., 1988, Quantitation of tumor response to antineoplastic therapy, *Semin. Oncol.* **15**:441–454.

Heesters, M. A., Kamman, R. L., Mooyaart, E. L., and Go, K. G., 1993, Localized proton spectroscopy of inoperable brain gliomas. Response to radiation therapy, *J. Neuro-Oncol.* **17:** 27–35.

Howe, F. A., Maxwell, R. J., Saunders, D. E., Brown, M. M., and Griffiths, J. R., 1993, Proton spectroscopy *in vivo, Magn. Reson. Q.* **9:**31–59.

Kim, B., Chenevert, T. L., and Ross, B. D., 1995, Growth kinetics and treatment response of the intracerebral rat 9L brain tumor model: A quantitative *in vivo* study using magnetic resonance imaging, *Clin. Cancer Res.* **1:**643–650.

Kimler, B. F., 1994, The 9L rat brain tumor model for pre-clinical investigation of radiation–chemotherapy interactions, *J. Neuro-Oncol.* **20:**103–109.

Kuesel, A. C., Donnelly, S. M., Halliday, W., Sutherland, G. R., and Smith, I. C. P., 1994, Mobile lipids and metabolic heterogeneity of brain tumours as detectable by *ex vivo* [1]H MR spectroscopy, *NMR Biomed.* **7:**172–180.

Kumar, A. R. V., Hoshino, T., Wheeler, K. T., Barker, M., and Wilson, C. B., 1974, Comparative rates of dead tumor cell removal from brain, muscle, subcutaneous tissue, and peritoneal cavity, *J. Natl. Cancer Inst.* **52:**1751–1755.

Le Bihan, D., 1991, Molecular diffusion nuclear magnetic resonance imaging, *Magn. Reson. Q.* **7:**1–30.

Le Bihan, D., Breton, E., Lallemond, D., Grenier, P., Cabanis, E., and Laval-Jeantet, M., 1986, MR imaging of intravoxel incoherent motions: Application to diffusion and perfusion in neurologic disorders, *Radiology* **161:**401–407.

Levin, V. A., Crafts, D. C., Norman, D. M., Hoffer, P. B., Spire, J. P., and Wilson, C. B., 1977, Criteria for evaluating patients undergoing chemotherapy for malignant brain tumors, *J. Neurosurg.* **47:**329–335.

Luo, Y., High, W. B., deGraaf, R. A., Terpstra, M., and Garwood, M., 1995, Lactate distribution correlates with the density of neoplastic cells in rat glioma, Third Annual Meeting of the Society of Magnetic Resonance, Nice, France.

Miller, B. L., 1991, A review of chemical issues in [1]H NMR spectroscopy: *N*-acetyl-l-aspartate, creatine and choline, *NMR Biomed.* **4:**47–52.

Mintorovitch, J., Wendland, M. F., Moseley, M. E., Asgari, H., Cohen, Y., and Kucharczyk, J., 1990, Spin-echo and echo-planer diffusion-weighted MRI of experimental ischemia, Abstract, Society of Magnetic Resonance in Medicine, p. 1120.

Moolten, F. L., 1994, Drug sensitivity ("suicide") genes for selective cancer chemotherapy, *Cancer Gene Ther.* **1:**279–287.

Moseley, M. E., Cohen, Y., Kucharczyk, J., Mintorovitch, J., Asgari, H. S., Wendland, M. F., Tsuruda, J., and Norman, D., 1990a, Diffusion-weighted MR imaging of anisotropic water diffusion in cat central nervous system, *Radiology* **176:**439–445.

Moseley, M. E., Kucharczyk, J., Mintorovitch, J., Cohen, Y., Kurhanewicz, J., Derugin, N., Asgari, H., and Norman, D., 1990b, Diffusion-weighted MR imaging of acute stroke: Correlation with T2-weighted and magnetic susceptibility-enhanced MR imaging in cats, *Am. J. Neuroradiol.* **11:**423–429.

Negendank, W., 1992, Studies of human tumors by MRS: a review, *NMR Biomed.* **5:**303–324.

Nelson, S. J., Vigneron, D. B., Wald, L. L., Day, M. R., Moyher, S. E., and Dillon, W. P., 1995, Serial assessment of brain tumor response to therapy using volume MRI and MRS, *in* "MR '95 Internationales Kernspintomographie Symposium," Garmisch-patenkirchen Schnetztor-Verlag, GmbH Konstanz.

Paggi, M. G., Zupi, G., Fanciulli, M., Carlo, C. D., Giorno, S., Laudonio, N. Silvestrini, B., Caputo, A., and Floridi, A., 1987, Effect of lonidamine on the utilization of [14]C-labeled glucose by human astrocytoma cells, *Exp. Mol. Pathol.* **47:**154–165.

Pedersen, P. L., 1978, Tumor mitochondria and the bioenergetics of cancer cells, *Prog. Exp. Tumor Res.* **22:**190–274.

Perman, W. H., Gado, M. H., Larson, K. B., and Perlmutter, J. S., 1992, Simultaneous MR acquisition of arterial and brain signal–time curves, *Magn. Reson. Med.* **28:**74–83.

Peterson, D. L., Sheridan, P. J., and Brown, W. E., Jr., 1994, Animal models for brain tumors: Historical perspectives and future directions, *J. Neurosurg.* **80:**865–876.

Preul, M. C., Caramanos, Z., Collins, D. L., Villemure, J.-G., Leblanc, R., Oliver, A., Pokrupa, R., and Arnold, D. L., 1996, Accurate, noninvasive diagnosis of human brain tumors by using proton magnetic resonance spectroscopy, *Nature Medicine* **2:**323–325.

Rainov, N. G., Zimmer, C., Chase, M., Kramm, C. M., Chiocca, E. A., Weissleder, R., and Breakefield, X. O., 1995, Selective uptake of viral and monocrystalline particles delivered intraarterially to experimental brain neoplasms, *Human Gene Therapy* **6:**1543–1552.

Rempp, K., Brix, G., Becker, C., Wenz, F., Guckel, F., Bader, R., and Lorenz, W. J., 1993, Quantitative evaluation of the regional cerebral blood volume with dynamic MRI, *Proc. Soc. Magn. Reson. Med. 1993:*631.

Rempp, K., Wenz, F., Brix, G., Becker, C., Guckel, F., and Lorenz, W. J., 1994, Quantification of regional cerebral blood flow and volume by dynamic enhanced MR imaging, *Proc. Soc. Magn. Reson. 1994:*278.

Rhodes, C. G., Wise, R. J. S., Gibbs, J. M., Frackowiak, R. S., Hatazawa, J., Palmer, A. J., Thomas, D. G. T., and Jones, T., 1983, In vivo disturbance of the oxidative metabolism of glucose in human cerebral gliomas, *Ann. Neurol.* **14:**614–626.

Roberts, D. A., Detre, J. A., Bolinger, L., Insko, E. K., and Leigh, J. S., Jr., 1994, Quantitative magnetic resonance imaging of human brain perfusion at 1.5 T using steady-state inversion of arterial water, *Proc. Natl. Acad. Sci. USA* **91:**33–37.

Rosen, B. R., Belliveau, J. W., and Chien, D., 1989, Perfusion imaging by nuclear magnetic resonance, *Magn. Reson. Q.* **5:**263–281.

Rosen, B. R., Belliveau, J. W., Aronen, H. J., Kennedy, D., Buchbinder, B. R., Fischman, A., Gruber, M., Glas, J., Weisskoff, R. M., Cohn, M. S., Hochberg, F. H., and Brady, T. J., 1991a, Susceptibility contrast imaging of cerebral blood volume: Human experience, *Magn. Reson. Med.* **22:**293–299.

Rosen, B. R., Belliveau, J. W., Buchbinder, B. R., McKinstry, R. C., Porkka, L. M., Kennedy, D. N., Neuder, M. S., Fisel, C. R., Aronen, H. J., Kwong, K. K., Weisskoff, R. M., Cohn, M. S., and Brady, T. J., 1991b, Contrast agents and cerebral hemodynamics, *Magn. Reson. Med.* **19:**285–292.

Rosenblum, M. L., Wheeler, K. T., Wilson, C. B., Barker, M., and Knebel, K. D., 1975, In vitro evaluation of in vivo brain tumor chemotherapy with 1,3-bis(2-chloroethyl)-1-nitrosourea, *Cancer Res.* **35:**1387–1391.

Rosenblum, M. L., Knebel, K. D., Vasquez, D. A., and Wilson, C. B., 1976, In vivo clonogenic tumor cell kinetics following 1,3-bis(2-chloroethyl)-1-nitrosourea brain tumor therapy, *Cancer Res.* **36:**3718–3725.

Rosenblum, M. L., Knebel, K. D., Vasquez, D. A., and Wilson, C. B., 1977, Brain-tumor therapy. Quantitative analysis using a model system, *J. Neurosurg.* **46:**145–154.

Rosenblum, M. L., Dougherty, D. A., Brown, J. M., Barker, M., Deen, D. F., and Hoshino, T., 1980, Improved methods for disaggregating single cells from solid tumors, *Cell Tissue Kinet.* **13:**667.

Ross, B. D., Higgins, R. J., Boggan, J. E., Knittel, B., and Garwood, M., 1988a, $^{31}$P NMR spectroscopy of the in vivo metabolism of an intracerebral glioma in the rat, *Magn. Reson. Med.* **6:**403–417.

Ross, B. D., Higgins, R. J., Boggan, J. E., Willis, J. A., Knittel, B., and Unger, S. W., 1988b, Investigation of the carbohydrate metabolism of the C6 glioma: An in vivo $^{13}$C and in vitro $^1$H magnetic resonance spectroscopy study, *NMR Biomed.* **1:**20–26.

Ross, B. D., Tropp, J., Derbey, K. A., Sugiura, S., Hawryszko, C., Jacques, D. B., and Ingram, M., 1989a, Metabolic response of glioblastoma to adoptive immunotherapy: Detection by phosphorus MR spectroscopy. *J. Comput. Assist. Tomogr.* **13:**189–193.

Ross, B. D., Mitchell, S. L., Merkle, H., and Garwood, M., 1989b, in vivo $^{31}$P and $^2$H NMR studies of rat brain tumor pH and blood flow during acute hyperglycemia: Differential effects between subcutaneous and intracerebral locations, *Magn. Reson. Med.* **12:**219–234.

Ross, B. D., Merkle, H., Hendrich, K., and Garwood, M., 1992, Spatially localized *in vivo* ¹H magnetic resonance spectroscopy of an intracerebral rat glioma, *Magn. Reson. Med.* **23**:96–108.

Ross, B. D., Chenevert, T. L., Kim, B., and Ben-Yoseph, O., 1994, Magnetic resonance imaging and spectroscopy: Application to experimental neuro-oncology, *Q. Magn. Reson. Biol. Med.* **1**:89–106.

Ross, B. D., Kim, B., and Davidson, B. L., 1995, Assessment of ganciclovir toxicity to experimental intracranial gliomas following recombinant adenoviral mediated transfer of the herpes simplex virus thymidine kinase gene by magnetic resonance imaging and proton magnetic resonance spectroscopy, *Clin. Cancer Res.* **1**:651–657.

Rozental, J. M., Cohen, J. D., Mehta, M. P., Levine, R. L., Hanson, J. M., and Nickles, R. J., 1993, Acute changes in glucose uptake after treatment: The effects of carmustine (BCNU) on human glioblastoma multiforme, *J. Neuro-Oncol.* **15**:57–66.

Schupp, D. G., Merkle, H., Ellermann, J. M., Ke, Y., and Garwood, M., 1993, Localized detection of glioma glycolysis using edited ¹H MRS, *Magn. Reson. Med.* **30**:18–27.

Segebarth, C. M., Baleriaux, D. F., Arnold, D. L., Luyten, P. R., and den Hollander, J. A., 1987, MR image-guided P-31 MR spectroscopy in the evaluation of brain tumor treatment, *Radiology* **165**:215–219.

Stegman, L. D., Ben-Yoseph, O., Freyer, J. P., and Ross, B. D., 1997, ³¹Phosphorus MRS assessment of ganciclovir toxicity in gliomas stably expressing the herpes simplex thymidine kinase gene, *NMR Biomed.*, in press.

Stejskal, E. O., 1965, Use of spin echoes in a pulsed magnetic-field gradient to study anisotropic, restricted diffusion and flow, *J. Chem. Phys.* **43**:3597–3603.

Tanner, J. E., 1978, Transient diffusion in a system partitioned by permeable barriers. Application to NMR measurements with a pulsed field gradient, *J. Chem. Phys.* **69**:1748–1754.

Tozer, G. M., and Griffiths, J. R., 1992, The contribution made by cell death and oxygenation to ³¹P MRS observations of tumour energy metabolism, *NMR Biomed.* **5**:279–289.

Walker, M. D., Green, S. B., Byar, D. P., Alexander, E., Jr., Batzdorf, U., Brooks, W. H., Hunt, W. E., MacCarty, C. S., Mahaley, M. S., Jr., Mealey, J., Jr., Owens, G., Ransohoff, J., Robertson, J. T., Shapiro, W. R., Smith, K. R., Jr., Wilson, C. B., and Strike, T. A., 1986, Randomized comparisons of radiotherapy and nitrosoureas for the treatment of malignant glioma after surgery, *N. Engl. J. Med.* **303**:1323–1329.

Ward, K. A., and Jain, R. K., 1988, Response of tumors to hyperglycemia: Characterization, significance and role in hyperthermia, *Int. J. Hyperthermia* **4**:223–250.

Weisskoff, R. M., Boxerman, J. L., Sorensen, A. G., Kulke, S. M., Campbell, T. A., and Rosen, B. R., 1994, Simultaneous blood volume and permeability mapping using a single Gd-based contrast injection, *Proc. Soc. Magn. Reson. 1994:*279.

Weizsaecker, M., Deen, D. F., Rosenblum, M. L., Hoshino, T., Gutin, P. H., and Barker, M., 1981, The 9L rat brain tumor: Description and application of an animal model, *J. Neurol.* **224**:183–192.

Wenz, F., Brix, G., Hess, T., Weisser, G., Debus, J., Knopp, M. V., Engenhart, R., and van Kaick, G., 1994, Radiation induced rCBV changes of low grade astrocytomas and normal brain tissue, *Proc. Soc. Magn. Reson. 1994:*523.

Williams, D. S., Detre, J. A., Leigh, J. S., and Koretsky, A. P., 1992, Magnetic resonance imaging of perfusion using spin inversion of arterial water, *Proc. Natl. Acad. Sci. USA* **89**:212–216.

Wilson, C. B., Crafts, D., and Levin, V. A., 1977, Brain tumors: Criteria of response and definition of recurrence, *Natl. Cancer Inst. Monogr.* **46**:197–203.

Zimmer, C., Weissleder, R., Poss, K., Bogdanova, A., Wright, S. C., Jr., and Enochs, W. S., 1995, MR imaging of phagocytosis in experimental gliomas, *Radiology* **197**:533–538.

# DIFFUSION-WEIGHTED MAGNETIC RESONANCE IMAGING

## MARTIN KING, NICK VAN BRUGGEN, ALBERT BUSZA, and ROBERT TURNER

## SUMMARY

Contrast in magnetic resonance imaging (MRI) is generated by exploiting a variety of physicochemical properties. Conventional clinical MRI techniques are largely based upon disease-induced changes in water relaxation, but these have been complemented by a number of other approaches, including a sensitization to the diffusion of water. It has been shown that diffusion-weighted (DW) imaging can be used to advantage in the diagnosis of a number of pathologies that are undetectable using standard imaging protocols. Moreover, DW imaging can provide information concerning the nature of the pathology, in addition to the

MARTIN KING, ALBERT BUSZA, and ROBERT TURNER • RCS Unit of Biophysics, Institute of Child Health, London, WC1N 1EH, United Kingdom.
NICK VAN BRUGGEN • Department of Neuroscience, Genentech Inc., South San Francisco, California 94080.

Magnetic Resonance Spectroscopy and Imaging in Neurochemistry, Volume 8 of Advances in Neurochemistry, edited by Bachelard, Plenum Press, New York, 1997.

spatial information that is obtained from the image *per se*. This chapter starts with a brief description of the diffusion phenomenon and nuclear magnetic resonance (NMR) methods for measuring diffusion coefficients. This is followed by sections devoted to some theoretical and instrumental aspects of DW imaging. The remainder of the chapter is concerned with some biomedical applications of DW imaging, with an emphasis on cerebral pathophysiology. No attempt has been made to provide an exhaustive review of the literature on DW imaging. On the contrary, we have identified a few studies that specifically serve to illustrate its potential application to neuroscience.

## 1. INTRODUCTION

Diffusion is the process by which any gradient in the concentration of a solute spontaneously decreases until a uniform distribution is obtained. Fick's first law is fundamental to diffusion. It states that the diffusive flux ($J$), i.e., the net amount of a solute that diffuses through a unit area in a unit time, is proportional to the concentration gradient, Thus,

$$J \propto -(\partial C / \partial x)_t \tag{1}$$

$$J = -D (\partial C / \partial x)_t \tag{2}$$

where $C$ is the solute concentration, $\partial C / \partial x$ is the concentration gradient, and $D$ is the diffusion coefficient.

The role of Brownian motion in diffusion is well known. An important parameter is the distance moved by any solute molecule in a given time $t$. Free diffusion is a random process, and because the displacements occur in all directions, the mean distance moved is zero. For this reason it is the mean-square distance, $\overline{x^2}$, that is used as a measure of diffusive displacement. The relationship between $\overline{x^2}$ and $D$ is $\overline{x^2} = (2Dt)^{1/2}$.

Among the reasons why diffusion has occupied the attention of chemical physicists for many decades is that an important class of chemical reactions is diffusion-controlled. These reactions have a negligible activation energy, with the result that the rate at which they occur is determined by the rate at which the reacting species diffuse through the medium. It should be noted that diffusion-controlled processes are particularly important in biological systems. In addition, molecular diffusion is rather sensitive to binding and to the presence of structure. Consequently, diffusion measurements have been central to the study of colloidal and macromolecular systems.

Usually, the solute and solvent are chemically distinct. In contrast, the term *self-diffusion* has been used in the special context in which the solute and solvent are identical; this chapter is mainly concerned with the diffusion of water in water.

## 2. DIFFUSION AND THE NUCLEAR MAGNETIC RESONANCE EXPERIMENT

In order to measure the self-diffusion coefficient of a liquid, it is necessary to tag or label specific molecules in order that their displacements can be measured over some time period. Radioisotopes are traditionally used for this purpose. Pulsed-gradient nuclear magnetic resonance (NMR) provides an alternative and particularly elegant method for labeling water and other molecules. Each water molecule in the system is labeled in terms of its position with respect to an applied magnetic field gradient. The following section gives an outline of how this works when the pulsed-field gradient spin echo is used to measure a diffusion coefficient.

### 2.1 The Spin-Echo Method for Measuring Self-diffusion

The behavior of a population of spins in a spin-echo experiment is illustrated in the cartoon shown in Fig. 1. This figure is taken from the front cover of the issue of *Physics Today* (November 1953) in which Hahn published his famous paper on the spin echo (Hahn, 1953). The position of each runner around the circular track is analogous to the phase of a spin packet in an NMR experiment. Phase is expressed in angular units (degrees or radians) and is a measure of the angular separation between two vectors. In the context of NMR, we are concerned with (1) the angular separation in the phase of two spins, i.e., the phase difference between the spins, and (2) the phase of a spin at time $t$ compared with its phase at time $t_0$. Fundamental to the phase behavior of a system of spins are the equations

$$\omega = \gamma B_L \qquad (3)$$

and

$$\phi = \omega t = \gamma B_L t \qquad (4)$$

The first of these equations states that the precessional frequency ($\omega$) of each nucleus is proportional to the local magnetic field strength, $B_L$. The proportionality constant $\gamma$ is the magnetogyric ratio. The second equation states that the phase ($\phi$) developed in a time $t$ is simply the product of frequency and the time given for the phase to evolve. In Fig. 1, the runners become dispersed because some move faster than others. Similarly, the nuclei in a system of spins become

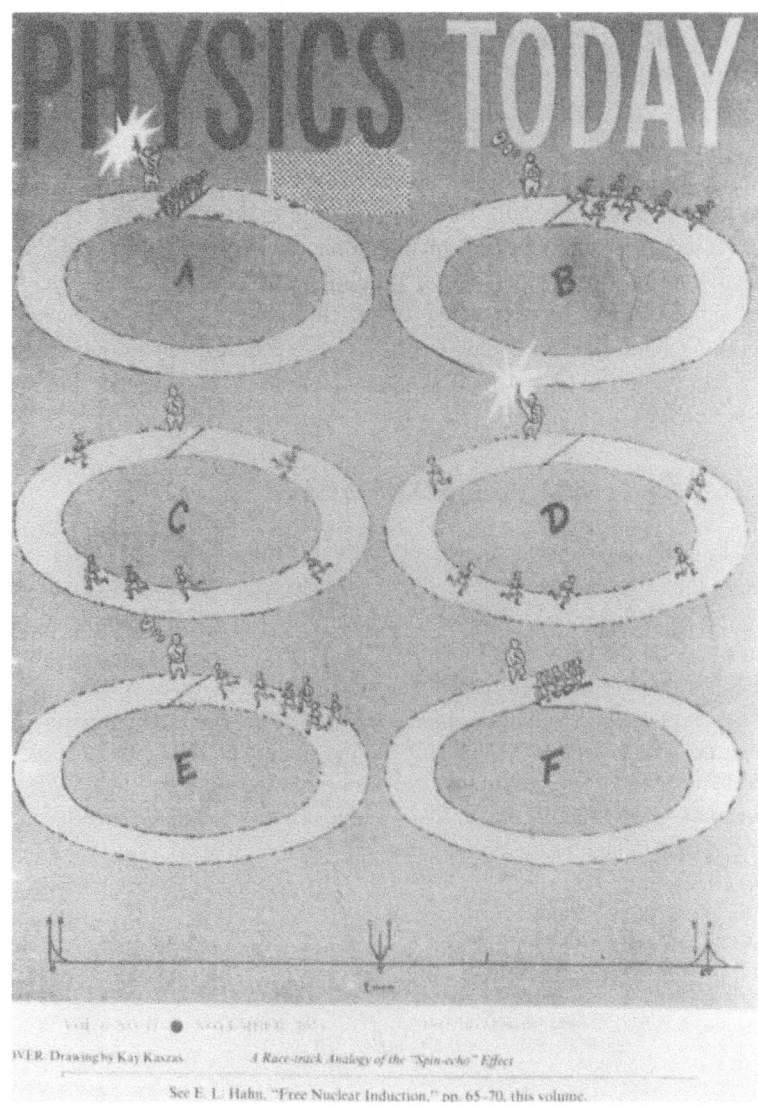

IVER. Drawing by Kay Kaszas          *A Race-track Analogy of the "Spin-echo" Effect*

See E. L. Hahn, "Free Nuclear Induction," pp. 65–70, this volume.

phase-dispersed if their precessional frequencies are not identical. Equation (4) states that this occurs when the magnetic field is spatially inhomogeneous. The phase evolution for those nuclei experiencing a larger field is greater than for those at a lower field owing to their larger precessional frequency. In the spin-echo experiment, a 180° radio-frequency pulse is applied at a time $\tau$ after the 90° pulse and has the effect of reversing the direction of precession. This is illustrated in the cartoon by a reversal of the direction in which the race is run. If the race continues and each runner maintains a constant speed, then every runner will cross the starting line at time $2\tau$. Similarly, in the spin-echo experiment, if every nucleus maintains a constant rate of precession, then the phase differences that develop prior to the 180° pulse are lost during the period following the 180° pulse, with the result that all the spins finish in phase at a time $2\tau$. The condition that all spins precess at a constant rate requires that the local magnetic field seen by each nucleus remains unchanged during the evolution period. In an inhomogeneous field, this requires that the positions of all the nuclei remain fixed; usually, however, spins undergo a significant diffusive displacement during the evolution period, with the result that the precessional frequency is constantly changing. Thus, the precise refocusing that occurs at time $2\tau$ in the case of stationary spins does not occur if the spins are free to diffuse. The refocusing is imperfect, and consequently the intensity of the echo is reduced. This is the basis of the spin-echo method for measuring diffusion. The magnitude of the echo signal attenuation that occurs due to diffusion in a constant magnetic field gradient $G$ for a period of time $2\tau$ is given by

$$I(G) = I(0) \exp(-\frac{2}{3} \gamma^2 G^2 D \tau^3) \tag{5}$$

where $I(0)$ is the echo intensity in the absence of a gradient (Carr and Purcell, 1954).

The effect of diffusion can be accentuated by the deliberate application of a

---

FIGURE 1. The behavior of a population of spins in a spin-echo experiment. This cartoon is taken from the front cover of the issue of *Physics Today* (November 1953) in which Hahn published his famous paper on the spin echo (Hahn, 1953). Each runner represents a spin packet. Immediately following the creation of transverse magnetization, all spin packets are in phase (A), but phase coherence is lost with time due to magnetic field inhomogeneity, as represented by the dispersal of the runners in B and C. A 180° radio-frequency pulse is applied at a time $\tau$, which has the effect of reversing the direction of precession. This is illustrated in the cartoon by the second firing of the starter's gun and a reversal of the direction in which the race is run (D). If the race continues and each runner maintains a constant speed, then every runner will cross the starting line at time $2\tau$ (F). Similarly, in the spin-echo experiment, if every nucleus maintains a constant rate of precession, then the phase differences that develop prior to the 180° pulse are lost during the period following the 180° pulse, with the result that all the spins finish in phase at a time $2\tau$. [Reproduced, with permission, from Hahn (1953).]

relatively large and time-independent magnetic field gradient, and this formed the basis of a number of studies in which very precise diffusion coefficients were obtained.

## 2.2. The Pulsed-Field Gradient Spin-Echo Method

Instrumental limitations (transmitter and receiver bandwidth constraints) place an upper limit on the magnitude of the magnetic field gradient that can be applied when the constant-gradient spin-echo method is used to measure diffusion. This, in turn, places a lower limit on the range of diffusion coefficients that can be measured using this method. McCall *et al.* (1963) suggested that, given the instrumentation that was in use at that time, a practical lower limit was $10^{-7}$ $cm^2$ $sec^{-1}$. They noted, however, that this limit might be lowered if, instead of a constant magnetic field gradient, an "alternating" field gradient were used, the magnitude of which is small, or zero, during the radio-frequency sections of the sequence and is increased during the intervening periods. Stejskal and Tanner (1965) were among the first to use the pulsed-field gradient spin-echo method; the sequence is shown in Fig. 2. These authors also derived the much-quoted result relating the pulsed-field gradient echo attenuation to the diffusion coefficient, namely,

$$\ln[I(G)/I(0)] = -\gamma^2 D \delta^2 (\Delta - \frac{1}{3} \delta) G^2 \tag{6}$$

where $\Delta$ is the separation between the leading edges of the two pulsed gradients, and $\delta$ is their duration. Using this method, diffusion coefficients may be measured with precision down to $10^{-10}$ $cm^2$ $sec^{-1}$.

## 2.3. Restricted Diffusion in Colloidal and Macromolecular Solutions

Soon after the introduction of the pulsed-field gradient method for measuring diffusion, researchers turned their attention to the problem of restricted diffusion. As stated above, diffusion measurements have been central to the study of colloidal systems, but the theory that led to the Stejskal–Tanner relationship applies only to free diffusion and breaks down when applied to colloids. During the 1960s, a number of theoreticians worked on the restricted-diffusion problem, including Tanner and Stejskal, who, in 1968, published the results of a particularly important study in which they derived expressions for the pulsed-field gradient spin-echo signal attenuation that occurs in a variety of restricted systems (Tanner and Stejskal, 1968). They also compared their theoretical results with those obtained by experiment. It should be noted that unlike the free-diffusion problem, which gives rise to Eq. (6) and which places no constraints on the

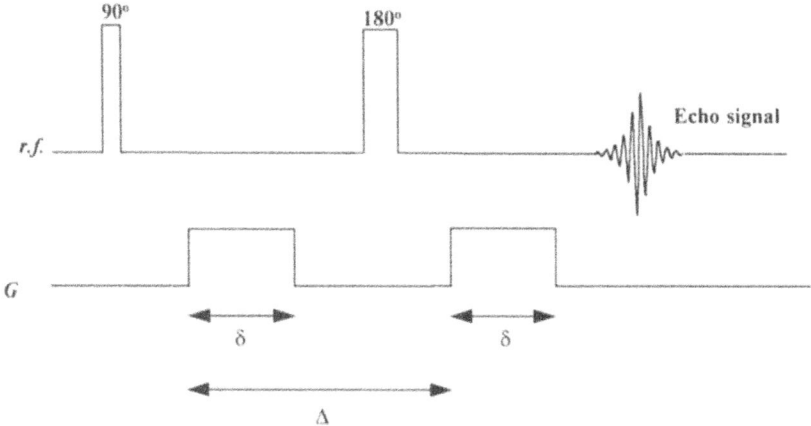

FIGURE 2. The pulsed-field gradient spin-echo sequence. The pulsed-field gradient spin-echo sequence consists of two gradient pulses placed one on each side of the 180° refocusing radio-frequency pulse.

values of $\Delta$ and $\delta$ (except, of course, that by definition $\Delta \geq \delta$), the restricted diffusion problem was found to be generally intractable. Consequently, Tanner and Stejskal obtained solutions that were applicable only to the special case in which $\delta$ is sufficiently short that the diffusive displacement occurring during this interval is small compared with that which occurs during the time $\Delta$. Subsequent publications gave analytical solutions with the short-$\delta$ constraint relaxed, Murday and Cotts (1968) providing the solution for a spherical geometry and Neuman (1974) extending this to the cylindrical and planar cases. These theoretical papers have acquired new significance with the advent of diffusion-weighted imaging (DWI).

## 2.4. Diffusion in Biological Media: Diffusion and Perfusion Imaging

In the mid 1980s it was shown that through the incorporation of pulsed-field gradients into a conventional imaging sequence, images can be obtained in which the intensity is weighted by the diffusion characteristics of the sample (Taylor and Bushell, 1985; Le Bihan et al., 1986). Figure 3 shows a typical DWI spin-echo sequence; the technique was named "intravoxel incoherent motions" (IVIM) imaging. Among the observations made during early DWI studies of the brain was that white matter exhibits a striking hyperintensity, the magnitude of which depends on the orientation of the sensitizing gradients (Moseley et al., 1990a,b; Doran et al., 1990), leading to the suggestion that DWI would become an important diagnostic tool for investigating white-matter disorders. A role for DWI in the early detection of ischemia was also demonstrated (Moseley et al.,

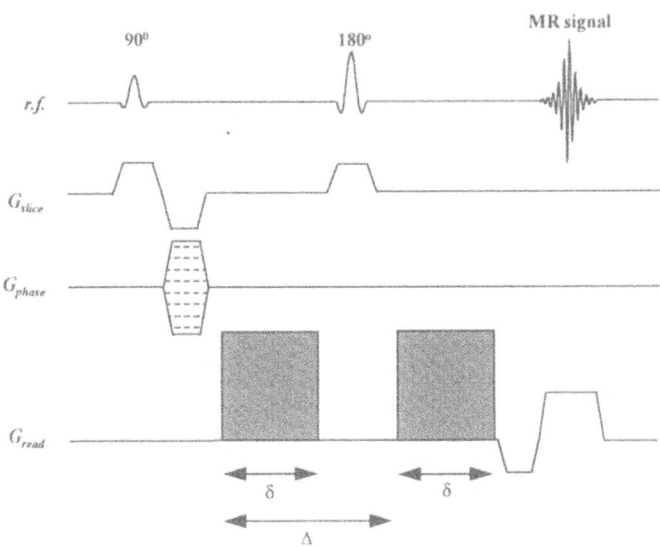

FIGURE 3. A typical diffusion-weighted spin-echo imaging sequence. A conventional spin-echo imaging sequence can be modified to acquire diffusion-weighted images by incorporating a pair of diffusion-sensitizing gradients into any one of the imaging axes. The slice-select gradient determines the orientation and position of the imaging plane, while the phase-encode and read gradients provide spatial encoding within this plane. In this example, the diffusion-sensitizing gradients (shaded bars) have been placed on the read axis.

1990b). These observations are discussed in Section 4, together with the DWI work that followed these initial studies.

The addition of imaging gradients to the pulsed-field gradient spin-echo sequence causes some complications, and Eq. (6) must be modified to account for the additional diffusion weighting that arises from the imaging gradients. This modification must take account of the direct diffusion weighting of the imaging gradients, as well as the indirect weighting that arises from so-called cross terms, involving products between the imaging gradients and the pulsed diffusion-encoding gradients. Although it is possible to write analytical expressions for these effects (see, e.g., Neeman *et al.*, 1990), an easier approach is to evaluate the "overall" diffusion weighting using numerical integration. (Any introductory text on numerical methods will include a description of one or more methods for performing numerical integration.) To this end, the attenuation caused by diffusion in the presence of magnetic field gradients can be expressed in the more general form.

$$I(G)/I(O) = \exp(-bD) \tag{7}$$

where $b$, which has become known as the $b$-factor, is a function of all of the gradients that are applied between the radio-frequency excitation of the spin system and data acquisition. It is given by

$$b = \int |k|^2 dt \tag{8}$$

and

$$k = \gamma \int G(t)dt \tag{9}$$

where $G(t)$ is the time-dependent gradient amplitude, which includes both the standard imaging gradients and the diffusion-encoding gradients. A failure to account properly for the effect of imaging gradients in the evaluation of the $b$-factor leads to biased diffusion estimates. Often, however, the objective is restricted to the generation of diffusion-weighted contrast as opposed to the measurement of precise diffusion coefficients, in which case $b$-factor calculations do not arise.

Among the original applications suggested for IVIM imaging was perfusion mapping. If realized, this application would have major clinical implications. The idea is based on the observation that the movement of blood through the randomly oriented capillary segments of the brain can be modeled as a pseudo-random process. Consequently, it was suggested (Le Bihan et al., 1988) that central nervous system (CNS) tissue can be modeled as a two-compartment system, the extracellular and intracellular spaces constituting one compartment, and the capillary bed the other. Calculations showed that, according to the model and given the known capillary geometry of the brain, the capillary compartment will yield a pseudo-diffusion coefficient that is an order of magnitude greater than the diffusion coefficient of the extracapillary water. IVIM imaging measurements should, therefore, yield values for two diffusion coefficients, i.e., the pseudo-diffusion coefficient relating to the capillary network and the diffusion coefficient of water in the combined "stationary extracellular + intracellular" pools, together with the relative volumes of the two pools. Ahn et al. (1987) conducted a theoretical and experimental study of capillary flow and suggested that the coherent movement of blood within a population of randomly oriented capillary segments can be distinguished from random Brownian motion through the use of suitable imaging sequences. They demonstrated that this difference can be exploited to obtain capillary density maps [see Fig. 7 of Ahn et al. (1987)].

Turner and Keller (1991) showed that, given a model for the geometry of the capillary network for the tissue in question, it is possible to convert the data obtained by IVIM imaging into perfusion rate estimates (having units of milliliters per minute per gram). In this context, Henkelman (1990) pointed out that the IVIM technique detects something that is fundamentally different from the perfusion that is measured by the classical isotope deposition and wash-out techniques. Classical perfusion is concerned with "terminal deposition and terminal pickup" as distinct from flow or transport, whereas the IVIM method is, in contrast, sensitive only to flow. Detailed information on the structure of the vascular system in the tissue in question is required to convert the flow rates obtained from the IVIM method into the equivalent of classical perfusion.

Considerable effort has been made to develop clinical perfusion imaging, and a number of variants of the initial method have been tried, among which steady-state methods have received particular attention. The successful application of the IVIM method for measuring perfusion remains, however, to be demonstrated. Indeed, statistical considerations indicate that it is barely feasible (Pekar *et al.*, 1992; King *et al.*, 1992). Nevertheless, the original pulsed-field gradient spin-echo imaging study that was performed by Le Bihan and his coworkers provided the stimulus for numerous subsequent DWI investigations, and these, in turn, have led to several important clinical applications.

Following the demonstration by Le Bihan and co-workers that pulsed-field gradient imaging sequences could be used to perform clinical diffusion imaging (Le Bihan *et al.*, 1986, 1988), Moseley *et al.* (1990b) published an important paper in which they showed that DWI might be used for the early detection of cerebral ischemia. This was based on the demonstration that DW images obtained from cats exhibited a regional hyperintensity as early as 45 min after the onset of an ischemic episode (Moseley *et al.*, 1990b). In the same study, it was shown that white matter exhibits an intensity that is dependent on the orientation of the applied pulsed gradients with respect to the neuronal tracts. This so-called anisotropic behavior is attributed to the presence of structural material which restricts the diffusion of water in a direction perpendicular to the white-matter tracts but allows water to diffuse relatively freely in the parallel direction. The appearance of this paper in the literature was followed by a period of intense research activity during which these observations were confirmed and extended to a clinical setting. For example, Doran *et al.* (1990) confirmed that clinical DWI is feasible and that white-matter abnormalities could be observed in a variety of diseases. These observations were followed by studies of myelination in the normal and pathological neonatal brain (Rutherford *et al.*, 1991; Sakuma *et al.*, 1991), but the value of DWI in the investigation of demyelinating diseases remains to be demonstrated. On the other hand, the role of DWI in the detection of ischemic tissue is established (van Bruggen *et al.*, 1994; see Section 4 of this chapter).

## 2.5. Diffusion-Weighted Imaging Techniques

Numerous modifications of the original pulsed-field gradient spin-echo imaging sequence have been suggested since its first application to biomedical imaging. The traditional method, in which two gradient pulses are placed at either side of a 180° radio-frequency pulse, can be modified by using a bipolar gradient pair or trains of bipolar pulses. This approach can be adopted to achieve short diffusion times and has the added advantage of being less susceptible to the effects of eddy currents, because the eddy currents induced by one-half of the bipolar pair are largely canceled by the polarity reversal. Furthermore, the effects of spatial variation in the magnetic susceptibility and the resulting internal gradients are reduced by using bipolar gradients. It should be noted, however, that a number of alternative and more sophisticated methods have been devised for dealing with the effects of internal field gradients, similar principles being applied to both imaging and nonimaging applications (Williams et al., 1978; Karlicek and Lowe, 1980; Latour et al., 1993; Lian et al., 1994).

In applications that require long diffusion times, a diffusion-weighted stimulated-echo sequence can be used (Tanner, 1970). Stimulated-echo sequences have been used to obtain diffusion-weighted images (Merboldt et al., 1985) as well as diffusion data on various metabolites (Merboldt et al., 1993). Horsfield et al. (1994) have used the stimulated echo to measure the diffusion of water in selected volumes. The basic stimulated-echo sequence consists of three 90° radio-frequency pulses which, as shown in Fig. 4, divide the spin evolution time into two so-called TE periods and a central TM period. The essential feature of this scheme is that the magnetization created during the first TE period is transformed into longitudinal magnetization by the second radio-frequency pulse and is therefore insensitive to $T_2$ relaxation during the TM period. Long diffusion times can therefore be obtained by increasing the duration of the TM period, the upper limit of which is determined by the $T_1$ of the sample. $T_1$ is typically of the order of hundreds of milliseconds in biological tissue, whereas $T_2$ values are of the order of tens of milliseconds. A useful application for the DW stimulated-echo technique is in a clinical environment, because moderate diffusion weighting can be achieved despite the limited gradient capability of most clinical imaging systems.

Diffusion weighting renders magnetic resonance imaging (MRI) extremely sensitive to motion, and early attempts to measure regional variation in the apparent diffusion coefficient of water in human brain were unsuccessful owing to small but uncontrollable movement (Turner et al., 1990a), even with careful immobilization of the head. The reason for this sensitivity to motion is that conventional MRI is based on repeated acquisition of the NMR signal. Each acquisition requires separate diffusion encoding and is therefore subject to varying amounts of motion-induced phase error. Rapid imaging techniques, on the other hand, employ a single preparation period; the large diffusion-encoding gradients are incorporated into this period. This removes the diffusion-gradient-induced phase inconsistencies that inevitably occur, due to motion, between the separate phase-encoding acquisitions of a conventional pulsed-field gradient imaging sequence. A number of different approaches have been employed, includ-

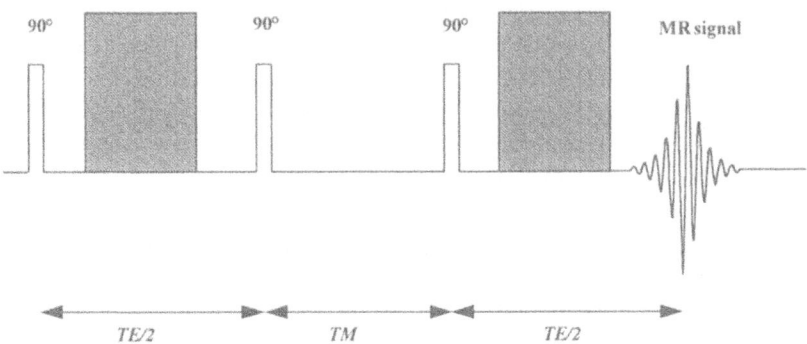

*FIGURE 4.* Diffusion-weighted stimulated-echo sequence. This figure shows a basic diffusion-weighted stimulated-echo (STEAM) sequence that can be made volume-selective by adding appropriate slice-selecting gradients or which can be incorporated into an imaging sequence. The stimulated-echo sequence consists of three 90° radio-frequency pulses which divide the spin evolution time into two so-called TE periods and a central TM period.

ing those based upon the echo-planar imaging (EPI) technique (Stehling *et al.,* 1991; Turner *et al.,* 1990b) and other rapid imaging modalities (Le Bihan, 1988; Le Bihan *et al.,* 1989; Wu and Buxton, 1990; Merboldt *et al.,* 1989; Niendorf *et al.,* 1994).

An entirely different approach was used by Chenevert *et al.* (1990) in which one-dimensional perfusion/diffusion profiles were obtained by using two orthogonal slice-selective radio-frequency pulses to excite a column of tissue prior to frequency encoding along the column. The absence of phase encoding in this volume-limited acquisition renders the method relatively insensitive to the effects of bulk tissue motion. Although this method has been used to obtain quantitative diffusion data from the human brain, high spatial resolution is limited to óne dimension.

## 2.6. Hardware Requirements for Diffusion Imaging

In order to develop useful contrast in a diffusion-weighted image, it is necessary to use a *b*-factor of a few hundred seconds per square millimeter. Too large a *b*-factor results in a less than optimum contrast-to-noise ratio. The *b*-factor required for optimum contrast-to noise must, of course, depend on the tissue and/or pathology. [A useful definition and treatment of the contrast-to-noise ratio in DWI is given in Appendix B of Prasad and Nalcioglu (1991).] On a clinical system it may not be possible to obtain the optimum *b*-factor, given the limited gradient strengths. One approach to achieving larger gradient amplitudes is the use of gradient coils that fit closely around the object (Turner *et al.,* 1990a; Hajnal *et al.,* 1992). Another issue relating to the rapid imaging techniques discussed above is the requirement for fast gradient switching. This rapid switching induces eddy currents in adjacent conducting structures, which, in turn, generate slowly decaying field gradients and cause image artifacts and errors in the diffusion parameters; this can be alleviated by using actively shielded gradients (Turner and Bowley, 1986; Mansfield and Chapman, 1986).

## 2.7. q-Space Imaging

An alternative approach to investigating the diffusivity of water is provided by a variant of diffusion-weighted imaging called $q$-space imaging. This is a technique that provides direct and unequivocal information about both coherent and incoherent water displacement. Information about incoherent water displacement is obtained in the form of water displacement profiles, i.e., unidirectional displacement probability density distributions. This technique was originally employed in materials science research, but it has recently been used in a study of normal and ischemic brain (King *et al.*, 1994). Cory and Garroway (1990) have shown that, given a homogeneous sample consisting of relatively impermeable cells of known geometry (e.g., approximately spherical cells), then a well-defined mathematical relationship exists between the water displacement profile and cell shape. It follows that, under these conditions, a direct estimate of the cell dimensions can be obtained. The application of this technique to ischemia may well lead to new insights into the underlying causes of the intensity changes that are observed in DWI. It should be noted, however, that $q$-space imaging requires magnetic field gradients that are several orders of magnitude greater than those available on clinical or most experimental imaging systems, and the technique is restricted to experimental systems with dedicated hardware.

## 3. ANISOTROPIC DIFFUSION AND THE DIFFUSION TENSOR

During the decade that has passed since Le Bihan *et al.* (1986) published their pioneering work on IVIM imaging, the technique has been used mainly for obtaining qualitative diagnostic information in the form of images intended for visual inspection and interpretation. In the majority of studies, quantification has been restricted to the calculation of so-called apparent diffusion coefficients (ADCs). This parameter is calculated by using the Stejskal–Tanner equation (Eq. 6), recognizing that the diffusion coefficients are biased since the equation does not generally apply to restricted systems. The last few years have seen the start of a new and more quantitative approach. A brief description of these methods is given in this section.

In order to describe some recently developed methods for quantifying the information obtained from diffusion imaging, it is necessary to include a brief account of the use of tensors in the mathematical description of anisotropic systems. Most of the introduction to this chapter was concerned with so-called isotropic diffusion, i.e., diffusion that exhibits no directional dependence. The exception is restricted diffusion, which is often anisotropic, due to an underlying anisotropy in the diffusion barriers. Thus, diffusion anisotropy arises when displacement along one direction occurs more readily than in some other direction. This behavior is to be expected in some biological tissues as a result of the macroscopic structure of the tissue. Both muscle and white matter, for example, have longitudinal structures that are expected to affect the diffusion of tissue water, with the result that it will be anisotropic. The diffusion tensor is a mathematical entity (a second-rank tensor) for dealing with anisotropic diffusion. The isotropic diffusion coefficient ($D$) that appears in the equations in Section 2 is replaced by the diffusion tensor, to express the fact that diffusive flux is orientation-dependent. Although an understanding of tensor algebra requires a knowledge of matrix and vector mathematics, the underlying principles are intuitive. In order to deal with the condition that flux has a directional dependence, the flux is written as an ordered set of three values, $J_x$, $J_y$, and $J_z$, which represent the flux in three mutually independent directions $x$, $y$, and $z$. This ordered set of numbers is written as a column, to form a so-called column vector:

$$\begin{bmatrix} J_x \\ J_y \\ J_z \end{bmatrix}.$$

A tensor is an array of numbers that is used for dealing with situations in which the response to some input has a directional dependency. (In the context of ordinary physical three-dimensional space, the second-rank tensors with which this section is concerned are $3 \times 3$ arrays of numbers.) Some aspects of tensor algebra are, we feel, rather more intuitive when dealing with the tensors that arise in the treatment of other physical problems. In order to give some insight into tensor mathematics, it is helpful, therefore, to digress and to consider one of these intuitively simpler examples, namely, the susceptibility tensor.

The magnetization ($M$) of an isotropic material is related to the applied field ($B$) by the relationship

$$M = \chi B \tag{10}$$

where $\chi$ is the magnetic susceptibility and $M$ is the response to the stimulus $B$. In an anisotropic sample the magnetization is, for some arbitrary orientation of the magnetic field, given by

$$\begin{bmatrix} M_x \\ M_y \\ M_z \end{bmatrix} = \begin{bmatrix} \chi_{xx} & \chi_{xy} & \chi_{xz} \\ \chi_{yx} & \chi_{yy} & \chi_{yz} \\ \chi_{zx} & \chi_{zy} & \chi_{zz} \end{bmatrix} \begin{bmatrix} B_x \\ B_y \\ B_z \end{bmatrix} \tag{11}$$

where the right-hand side is a multiplication of a matrix (the susceptibility tensor) by a vector. We will not concern ourselves with the rules for performing this operation but merely state that Eq. (11) can be expressed as three separate equations, one each for $M_x$, $M_y$, and $M_z$. To take an example, the equation for $M_x$ is

$$M_x = \chi_{xx}B_x + \chi_{xy}B_y + \chi_{xz}B_z \qquad (12)$$

The important points about this equation are that there is a coupling of the orthogonal (i.e., mutually perpendicular) components of $M$ and $B$ and that the component of the response in any one direction depends on the magnitude of the field in all three mutually perpendicular directions. In general, the magnetization and magnetic field vectors are not parallel. A schematic illustration of the behavior of a diamagnetic sample is shown in Fig. 5. In the special situation in which the susceptibility tensor is specified using so-called principal axes, i.e., axes that coincide with the axes of (structural) symmetry, the off-diagonal tensor elements are zero, and the expressions for $M_i$, $i = x$, $y$, or $z$, are simplified. For example, the equation for $M_x$ becomes

$$M_x = \chi_{xx}B_x \qquad (13)$$

with similar expressions for $M_y$ and $M_z$. In an isotropic system $\chi_{xx} = \chi_{yy} = \chi_{zz}$, and consequently $M$ and $B$ are parallel or antiparallel, depending on the sign of $\chi$. When dealing with anisotropic systems, only in the special case in which $B$ is

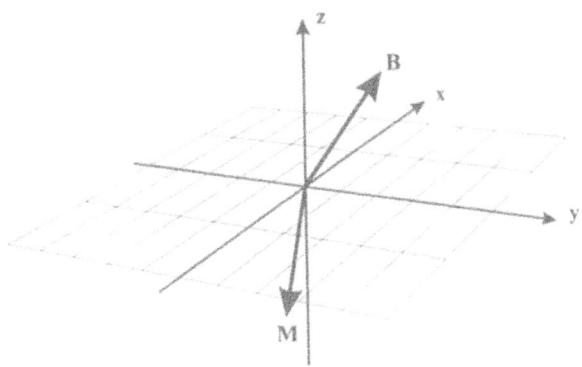

FIGURE 5. A diagram of diamagnetic susceptibility, showing the principal axes, $x$, $y$, and $z$. Except for some special cases, the magnetization vector ($M$) is not generally aligned with the applied magnetic field ($B$), unless $B$ lies along one of the principal axes.

applied along one of the principal axes are the stimulus and response vectors ($B$ and $M$, respectively) aligned.

The diffusion tensor behaves in an identical manner, and we can write an expression analogous to Eq. (11), namely,

$$\begin{bmatrix} J_x \\ J_y \\ J_z \end{bmatrix} = \begin{bmatrix} D_{xx} & D_{xy} & D_{xz} \\ D_{yx} & D_{yy} & D_{yz} \\ D_{zx} & D_{zy} & D_{zz} \end{bmatrix} \begin{bmatrix} \partial C/\partial x \\ \partial C/\partial y \\ \partial C/\partial z \end{bmatrix}_t \qquad (14)$$

This equation indicates that, in general, the flux and concentration vectors are not parallel. In the case of pulsed-field gradient MR, the concentration in question is that of spins possessing any given amount of evolved phase. Note that the concentration gradient vector is time-dependent. Its direction is initially determined by the direction of the applied gradient, while its evolution depends on the relative orientations of the applied gradient and the symmetry axes of the system.

A number of important issues have arisen from studies in which this more rigorous treatment of diffusion has been adopted. Among these is the observation that the usual practice in which the off-diagonal tensor elements are ignored can lead to poor estimates of the true diffusivity (Basser et al., 1994b). Secondly, the diffusion tensor approach has led to new and exciting developments in CNS imaging. These are discussed in Section 4.3.

# 4. DIFFUSION-WEIGHTED IMAGING AND CEREBRAL PATHOLOGY

The introduction of DW magnetic resonance imaging has had a major impact on the study and diagnosis of a number of cerebral dysfunctions, in particular, cerebral ischemia. Moseley et al. (1990b) were the first to demonstrate that ischemic tissue resulting from large vessel occlusion in the cat brain could be highlighted by DWI. The ischemic tissue appeared relatively hyperintense as early as 45 minutes following the induction of ischemia, whereas conventional imaging techniques failed to demonstrate any significant tissue damage up to 2–3 hours after occlusion. The ability of DWI to highlight ischemic tissue is exemplified in Fig. 6, which shows the striking difference between diffusion-weighted and $T_2$-weighted images in early cerebral ischemia. Conventional MRI, based upon changes in the relaxation parameters $T_1$ and $T_2$ has failed to demonstrate ischemic pathology with any reliability during the acute phase of stroke, i.e., prior to disruption of the blood–brain barrier. The precise mechanisms responsible for the changes seen with DWI during cerebral ischemia and related pathologies remain uncertain but, despite this, the capability of DWI to deline-

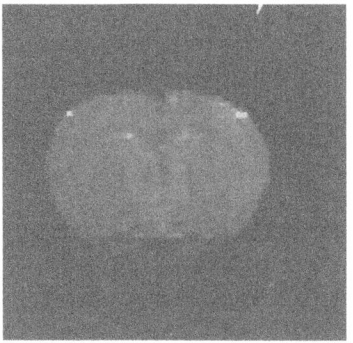 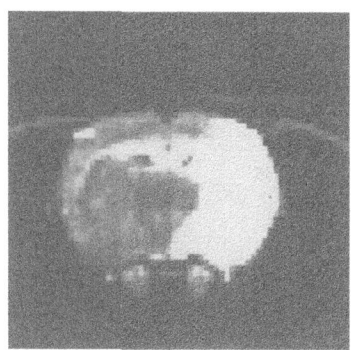

FIGURE 6. Diffusion-weighted imaging of cerebral ischemia. Rat brain $T_2$- and diffusion-weighted MR images recorded from a 2-mm coronal slice, 1.5 hr following occlusion of the middle cerebral artery (MCA) using the intraluminal filament technique. Ischemic tissue in the territory supplied by the MCA is highlighted on the DW image (*right*) but is not detectable in the equivalent $T_2$-weighted image (*left*).

ate ischemic tissue in the acute stage has tremendous potential for monitoring therapy.

## 4.1. Sensitivity and Specificity to Pathology

The seminal paper of Moseley *et al.* (1990b) clearly demonstrated that DWI is sensitive to the changes that occur during the acute phase of cerebral ischemia. It is established that the severity of cerebral ischemia is critically dependent upon the degree of perfusion deficit. As the cerebral blood flow (CBF) falls, the brain compensates by increasing oxygen and glucose extraction. With a further reduction in flow, these physiological compensatory mechanisms are insufficient to support neuronal electrical activity, as indicated by a suppression of electroencephalogram (EEG) activity, until a critical threshold of CBF is reached below which ATP depletion results in a disruption of the actively maintained cellular ion homeostasis. Large amounts of intracellular potassium are lost to the extracellular space, while sodium and osmotically obliged water move into the cell, resulting in cell swelling (a stage often referred to as cellular or cytotoxic edema). In a model of global cerebral ischemia produced by occlusion of the common carotid arteries (CCA) in the gerbil (see Chapter 5), Busza *et al.* (1992) were able to demonstrate that DWI signal enhancement only occurred when the cerebral blood flow fell below 15–20 ml $(100 \text{ g})^{-1}$ min$^{-1}$ (Fig. 7). This is similar to the flow threshold for the maintenance of tissue high-energy metabolites (Allen *et al.*, 1988). This compares well with the results of conventional bio-

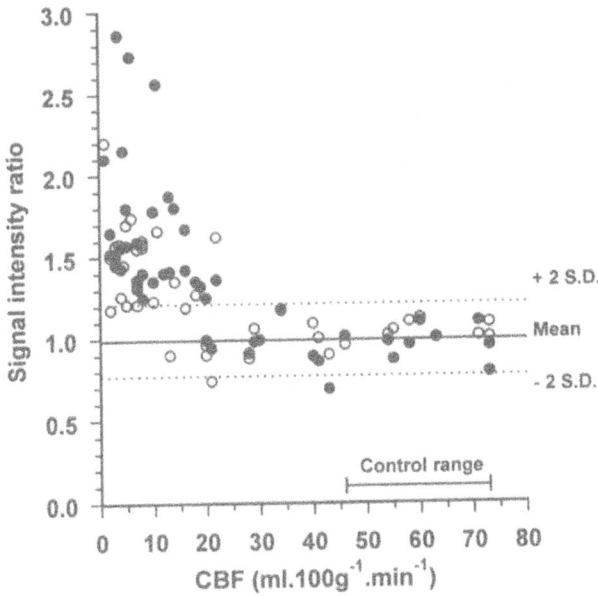

FIGURE 7. The relationship between diffusion-weighted imaging and cerebral blood flow (CBF) following partial occlusion of the common carotid arteries in the gerbil. Signal intensity ratios (SIR) were determined with respect to preischemia (control) images for the thalamus (○) and the cortex (●) in eight gerbils. The solid and dotted lines represent the mean ± 2 standard deviation calculated using data obtained when the CBF was 30 ml $(100 \text{ g})^{-1}$ $\text{min}^{-1}$ or greater. No significant increase in signal intensity occurred until the CBF fell below 15–20 ml $(100 \text{ g})^{-1}$ $\text{min}^{-1}$.

chemical studies performed on the baboon, where changes in lactate and ATP only occurred when the CBF had fallen below 20 ml $(100 \text{ g})^{-1}$ $\text{min}^{-1}$ (Obrenovitch *et al.*, 1988). Furthermore, during severe ischemia, when the CBF is reduced to below approximately 8 ml $(100 \text{ g})^{-1}$ $\text{min}^{-1}$, DWI signal enhancement occurred after a delay of approximately 2 minutes and increased gradually thereafter, consistent with the time course of ATP depletion and the increase in extracellular potassium (Hansen, 1985). These studies suggested that the DWI changes were a consequence of a redistribution of tissue water following energy failure. In focal cerebral ischemia, the flow threshold for the suppression of energy metabolism is less defined, and in penumbral tissue it is believed to increase with time following the onset of the insult. Autoradiographic techniques for measuring CBF, combined with bioluminescence and fluoroscopic methods for quantifying ATP, glucose, lactate, and pH, have been used to study the spatial relationship between DW image changes, brain blood flow, and metabolism following occlusion of the middle cerebral artery (MCA) (Kohno *et al.*, 1995).

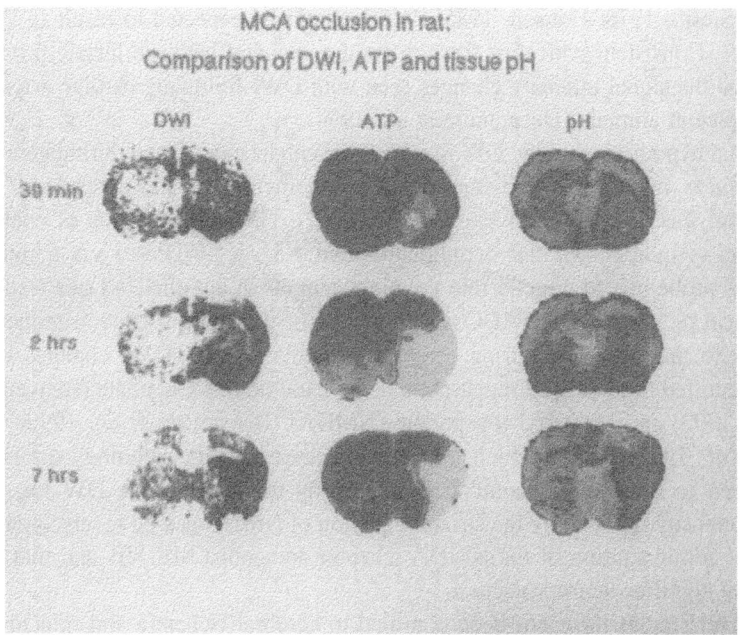

FIGURE 8. Middle cerebral artery (MCA) occlusion in the rat brain: diffusion-weighted imaging and metabolite maps. From left to right, diffusion-weighted ($b$ = 1500 sec mm$^{-2}$), ATP biolumines-cence, and tissue pH fluorescence images of rat brain obtained at 30 min (*top*), 2 hr (*middle*), and 7 hr (*bottom*) after MCA occlusion (the color coding is in the order red–yellow–green–blue, with high ATP bioluminescence and acidosis in red and low ATP bioluminescence and alkalosis in blue). At 30 min, the hyperintense region in the DW image corresponds to the region of severe tissue acidosis (orange–red), which is greater than the area of ATP depletion. These differences diminish with time, and the three regions eventually become indistinguishable. Details can be found in Kohno *et al.* (1995). (Figure kindly provided by Dr. M. Hoehn-Berlage and colleagues.) (A color reproduction of this figure appears on the insert found after p. 198).

In contrast to the diffusion-weighted changes observed in global ischemia, Kohno *et al.* reported that during the early phase of cerebral ischemia (within 30 minutes of the occlusion) the area of hyperintensity seen on the DW images was significantly larger than the region of ATP depletion, although it matched the area exhibiting tissue acidosis (see Fig. 8). The difference became progressively smaller with the evolution of the lesion, such that by 7 hours the area of tissue damage, as indicated by DWI, was identical to the region of ATP depletion and histological infarction. If, in fact, DWI changes result from an alteration in the compartmentation of tissue water, then the latter observations indicate that water redistribution occurs prior to a loss of high-energy metabolites. These observa-tions are not necessarily contradictory, since the anaerobic production of metabolites, including lactate, can also give rise to a redistribution of water (Staub *et al.*, 1993; Siesjö, 1978; Hansen, 1985), which might be expected to result in DWI changes. Consistent with this, there is good agreement between lactate production and the signal intensity changes seen with DWI following cardiac arrest in experimental animals (Decanniere *et al.*, 1994).

The hypothesis that the DW image signal enhancement seen during cerebral ischemia is related to the disruption of ion homeostasis, and a resulting cell swelling, was tested by Benveniste *et al.* (1992). The administration of ouabain (a specific inhibitor of the membrane-bound $Na^+/K^+$-ATPase) via a microdialysis probe placed directly into the hippocampus of anesthetized rats resulted in a local reduction in the ADC to about 33% of the control value, a reduction similar to that observed during cerebral ischemia. Similar DWI changes have been reported following administration of the excitotoxins glutamate (Benveniste *et al.*, 1992) and *N*-methyl-D-aspartate (NMDA) (Benveniste *et al.*, 1992; Verheul *et al.*, 1993), both of which are known to cause an acute pathology similar to that seen in cerebral ischemia. It is interesting to note that the DW imaging hyperintensity initiated by the striatal injection of NMDA can be reversed by the prompt administration of the NMDA receptor antagonist MK-801 and that this reversal signifies neuroprotection.

DWI studies have not been confined to cerebral ischemia and related pathologies. Zhong *et al.* (1993) were able to demonstrate changes in DW image signal intensity associated with seizure activity in the rat brain. Bicuculline, a GABA antagonist, can be administered intraperitoneally to induce status epilepticus in experimental animals, which is believed to be a suitable model for acute epilepsy in humans. Immediately following the administration of bicuculline, a 14–18% decline in the ADC was seen, which, under the experimental conditions used by Zhong *et al.*, was approximately 50% of the decline seen upon total cessation of CBF. The alterations in CBF and metabolism seen during bicuculline-induced seizures are very different from those associated with cerebral ischemia (a marked increase in blood flow and a relative preservation of ATP occurs during seizure), although some of the resulting pathologies may be similar. While it is not possible to identify unequivocally the underlying pathology responsible for the DWI changes seen during status epilepticus, these data serve to illustrate the diversity and scope for DWI in the study of cerebral pathology. An interesting aside, but one of direct relevance to epilepsy research, is that DWI has also been used to demonstrate the pathology resulting from chronic exposure to the drug vigabatrin. Vigabatrin, which inhibits GABA transaminase and raises cerebral GABA, has been shown to be clinically effective in the treatment of intractable complex partial seizures. Animal experiments have shown, however, that prolonged exposure to this drug results in intramyelinic edema and the formation of microvacuoles in cerebellar white matter (Gibson *et al.*, 1990). Pathological changes associated with exposure to vigabatrin have been shown using DWI in the rat brain (Preece *et al.*, 1994; Fig. 9). It is interesting to note

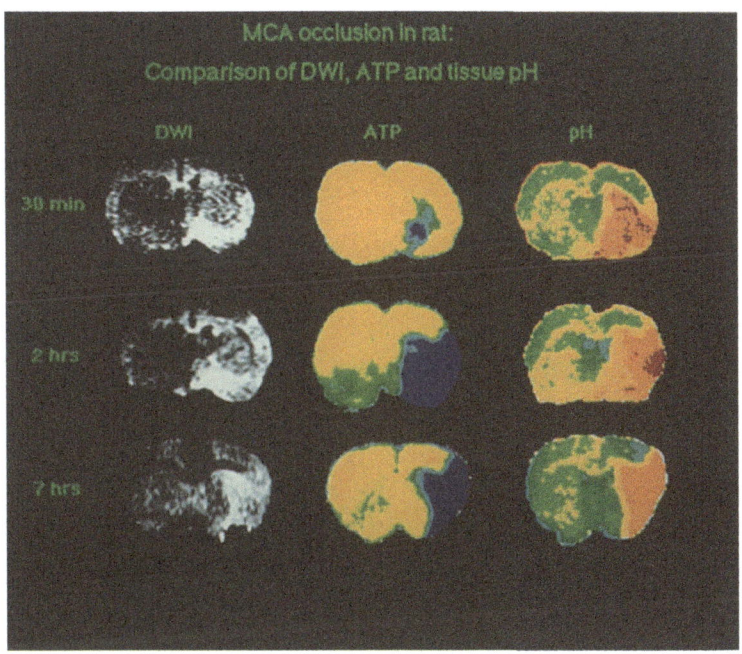

*FIGURE 8.* Middle cerebral artery (MCA) occlusion in the rat brain: diffusion-weighted imaging and metabolite maps. From left to right, diffusion-weighted ($b = 1500$ sec mm$^{-2}$), ATP bioluminescence, and tissue pH fluorescence images of rat brain obtained at 30 min (*top*), 2 hr (*middle*), and 7 hr (*bottom*) after MCA occlusion (the color coding is in the order red–yellow–green–blue, with high ATP bioluminescence and acidosis in red and low ATP bioluminescence and alkalosis in blue). At 30 min, the hyperintense region in the DW image corresponds to the region of severe tissue acidosis (orange–red), which is greater than the area of ATP depletion. These differences diminish with time, and the three regions eventually become indistinguishable. Details can be found in Kohno *et al.* (1995). (Figure kindly provided by Dr. M. Hoehn-Berlage and colleagues.) (A black and white reproduction of this figure appears on p. 197.)

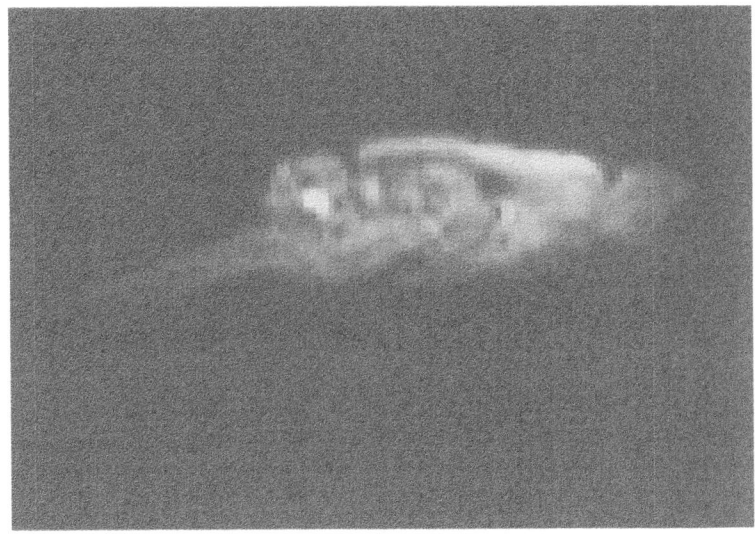

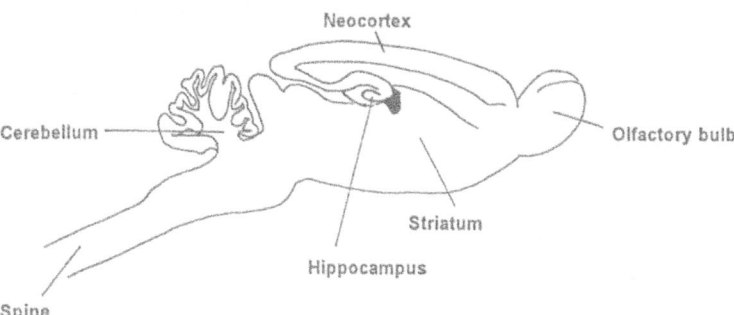

FIGURE 9. Diffusion-weighted imaging of vigabatrin-induced lesions in the rat brain. A diffusion-weighted magnetic resonance image recorded from a 2-mm-thick sagittal section through the brain of a rat that had been fed a diet containing the antiepilepsy drug vigabatrin. A striking increase in signal intensity is seen in the white matter of the cerebellum. Postmortem investigations demonstrated the presence of intramyelinic edema in this region. Details can be found in Preece *et al.* (1994). (Figure kindly provided by Dr. N. E. Preece and colleagues.)

that upon the withdrawal of the drug, the observed diffusion-weighted hyperintensity eventually returns to normal, although the intramyelinic edema persists in the cerebellar Purkinje fibers (Preece *et al.*, 1994).

A number of other experimental conditions lead to a loss of cell volume regulation and a subsequent cell swelling. The spreading depression phenomenon

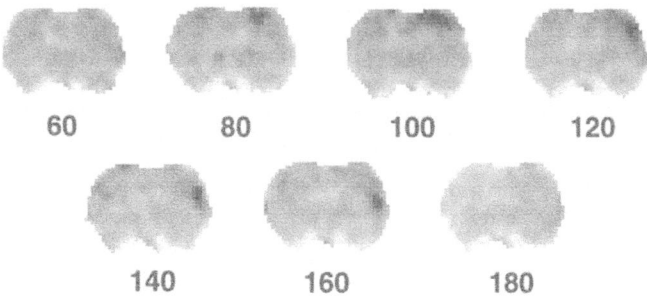

60        80        100        120

140       160       180

## Time After Application (seconds)

FIGURE 10. Diffusion imaging of cortical spreading depression in the rat brain. Following the topical application of KCl, a zone of decreased diffusion coefficient, approximately 2 mm in size, is clearly seen moving away from the site of initiation toward the lower cortex. Details can be found in Latour *et al.* (1994). (Figure kindly provided by Dr. L. L. Latour.)

is one example. Spreading depression (SD) was first described by Leao (1944a), who observed a slowly moving depression of cortical activity following cortical stimulation of the brain of an anesthetized rabbit. The changes in cortical interstitial space during SD are similar to those in anoxia and are related to ion movement (Hansen and Olsen, 1980). Spreading depression can be initiated readily in the cerebral cortex of a rat, forming a wave that spreads from the site of origin to invade the entire ipsilateral cortex, affecting each area profoundly for about 2 min and propagating at a characteristic velocity of about 3–6 mm min$^{-1}$. MRI has been used to demonstrate propagating waves of SD in the brain of anesthetized rats using a $T_2$*-weighted imaging protocol (Gardner-Medwin *et al.*, 1994); an increase in signal was observed in this study and was attributed to the increase in tissue oxygenation and to the arterialization of venous blood, both of which are thought to occur in SD (Leao, 1944b; Back *et al.*, 1993). More recently, pulsed-field gradient MRI techniques have been employed to demonstrate changes in the diffusivity of water during an episode of SD (Latour *et al.*, 1994). Diffusion maps (Fig. 10) show an area of reduced diffusivity approximately 2 mm in size, propagating away from the site of initiation at approximately 3 mm min$^{-1}$, consistent with the known physiology of SD. It is not known whether SD occurs in humans, but it has been purported to be involved in a number of neurological disorders, including migraine, stroke, and trauma (Hansen and Lauritzen, 1984).

It is clear that DWI can be used to demonstrate the presence of altered pathophysiologies as diverse as spreading depression, ischemic depolarization, excitotoxicity and epilepsy. A reduction in the interstitial space is common to all of these pathologies. Although the accurate determination of interstitial volume fraction (IVF) is difficult, exogenous markers have been used to determine interstitial diffusion characteristics and size. Ion-selective microelectrodes placed into the cortex facilitate the continuous monitoring of the extracellular concentrations of markers such as choline or trimethyltris(hydroxymethyl)methyl ammonium chloride. These markers have a low cellular permeability, and thus their concentrations can be used as an index of the extracellular water volume. Using this technique, it has been shown that the extracellular space decreases by approximately 50% during both SD and ischemia (Hansen and Olsen, 1980).

Changes in the extracellular volume fraction can also be estimated *in situ* by measuring the electrical impedance of the brain. Electrical conductance occurs mainly via the extracellular space, and changes in the extracellular volume fraction are therefore accompanied by conductance changes. Although no simultaneous measurements of electrical impedance and DWI changes have been performed to date, parallel experiments have demonstrated a remarkable correlation between the evolution of DW image hyperintensity and electrical impedance changes following the intrastriatal injection of the neurotoxin NMDA and the subsequent recovery that occurs after MK-801 treatment (Verheul *et al.*, 1994). Changes in the extracellular space, as determined by electrical impedance, have been used as an indicator of potential recovery from an ischemic insult (Hossmann, 1971). For example, it has been shown that animals that do not recover from a prolonged period of cerebral ischemia undergo a transient recovery of the extracellular space before a secondary and sustained decline. A similar pattern of partial recovery followed by a subsequent decline in the ADC has been observed in several related pathological conditions (Hossmann *et al.*, 1994). These data not only provide further support for the hypothesis that the ADC can be used as an index of extracellular fluid volume but also, and perhaps more importantly, they demonstrated that a full recovery of the ADC following prolonged cerebrocirculatory arrest is a reliable indicator of the restoration of metabolism and function.

## 4.2. Diffusion-Weighted Imaging and the Therapeutic Window

The notion of a therapeutic window arose from studies in which the size of the infarct that developed in transient ischemia was compared with measurements made after permanent occlusion. It was shown that infarct size could be markedly reduced if the blood flow were restored after brief periods of focal ischemia. This therapeutic window can be regarded as the time during which ischemic tissue may be salvaged with appropriate physiological or pharmacological intervention. It is widely recognized that a reversal of the impending infarction can only be achieved within a limited period; for example, in the case of a focal ischemia, irreversible damage occurs within 2–4 hours of stroke onset. Efficacy of therapy is therefore crucially dependent upon early detection. Conventional histopathological techniques for defining the extent of infarction are restricting not only by their invasive nature but also because long survival times are required (24–48 hours) for full morphological expression. DWI provides a noninvasive and immediate alternative.

Much interest is being shown in pharmacological methods for salvaging ischemic tissue and minimizing the effects of ischemia, and an understanding of the physiology and biochemistry of tissue recovery is necessary if this goal is to be achieved. A cerebral ischemic insult may have consequences far beyond the immediate effects of impaired blood supply. While the effects of ischemia on tissue metabolism and homeostasis are largely well understood and predictable, insight into tissue recovery is often confounded by the interrelationship of many factors, including the severity of the perfusion deficit and its duration. Overall, however, it is generally agreed that prompt restoration of the cerebral circulation is the single most effective way of minimizing ischemic damage.

Successful restoration of cerebral blood flow after short periods of ischemia results in a relatively rapid reversal of many of the changes that develop when the cerebral blood flow is interrupted. This includes recovery of energy metabolism, reactivation of the ion-exchange pumps, and normalization of water balance and ion homeostasis. Thus, the fluid imbalance caused by ischemia is largely reversible under favorable conditions.

In addition to the therapeutic window, we should also consider the ischemic penumbra. The concept of the ischemic penumbra originally arose from attempts to define the existence of two different blood-flow thresholds, one for the cessation of electrical activity and one for the maintenance of transmembrane ion gradients (Astrup et al., 1981). The term now has a broader meaning, one that encompasses the potentially reversible perifocal region that may be recruited into the infarction process (Memezawa et al., 1992).

Reversibility of diffusion-weighted hyperintensity after cerebral ischemia was first reported by Mintorovitch et al. (1991). Using a rat model of temporary MCA occlusion, they were able to demonstrate that after a 33-min ischemic insult, diffusion-weighted signal intensity returned to control levels within 2 hr after the restoration of the cerebral circulation. This was an important finding because, given that it is now established that the development of diffusion-weighted hyperintensity in ischemia is related to a perturbation of energy metabolism, it indicated that it may be possible to follow the physiological recovery of cerebral tissue following a stroke. Other important aspects to consider are the probable development and progression of the infarct and whether penumbral, i.e., potentially salvageable, tissue can be imaged. To this end, comparisons have been made between the size of the diffusion-weighted image hyperintense regions and the final size of the lesion, as determined by subsequent histology. For instance, after temporary MCA occlusion in the rat, areas of DWI hyperintensity measured 30 min after recirculation correspond to infarcted areas at postmortem examination (Minematsu et al., 1992). This was shown to depend on the duration of the ischemia, since after a 2-hr insult the postischemic hyperintense area was significantly greater than that observed after a 1-hr insult. Further confirmation of the sensitivity of DWI in depicting an ischemic lesion was provided by the fact that the hyperintense region correlated with infarcted tissue volume at 24 hr, as determined by 2,3,5-triphenyltetrazolium chloride (TTC) staining.

Attempts have been made to quantify the relationship between DWI changes and the underlying pathophysiology. There have been several studies aimed at establishing the ADC threshold below which irreversible ischemic damage occurs (Helpern *et al.*, 1993). Dardzinski *et al.* (1993) have suggested that brain tissue exhibiting an ADC greater than $0.55 \times 10^{-5}$ mm$^2$ sec$^{-1}$ at 2 hr post permanent MCA occlusion will not become infarcted, whereas areas of tissue with ADC values below this will be uniformly infarcted. Furthermore, regions where the ADC was below $0.45 \times 10^{-5}$ mm$^2$ sec$^{-1}$ became enlarged during a 2-hr ischemic observation period. In another study, Hasegawa *et al.* (1994) showed that a recovery of the ADC occurred only in those ischemic regions where it had decreased by less than $0.25 \times 10^{-5}$ mm$^2$ sec$^{-1}$. In relation to these studies, it should be noted that the ADC obtained is dependent upon the experimental conditions employed to make the measurement and that the values published by different researchers are not necessarily comparable.

Hossmann *et al.* (1994) reported that the recovery of the ADC following resuscitation after complete cerebrocirculatory arrest for 1 hr in cats depended on the postischemic reperfusion pressure. Furthermore, the ADC recovery correlated closely with tissue pH and metabolic recovery, in that those animals that demonstrated no ADC recovery exhibited a global depletion of ATP and glucose, together with severe lactic acidosis. Conversely, animals that showed a recovery of the ADC exhibited a replenishment of ATP and glucose and a substantial reversal of lactic acidosis. A reversal of the ischemic changes is thus at least partially possible with prompt recirculation. This is not always possible in a clinical context, and an alternative strategy is required. Pharmacological neuroprotection is one area of investigation which has been pursued.

A variety of neuroprotective agents have been investigated in an attempt to identify a treatment capable of inhibiting or even reversing the physiological decline of nervous tissue that has suffered an ischemic insult. DWI is ideally suited to this type of study. The NMDA antagonist MK-801 has been extensively studied and has become the standard against which other agents are compared. Lo *et al.* (1994) have shown that DWI changes in rat brain following permanent occlusion of the MCA can be reversed, at least partially, by the administration of MK-801 60 minutes after the onset of ischemia. A convincing demonstration of the MK-801-induced reversal of the DWI changes was given by Verheul *et al.* (1994), who showed that the DW image hyperintensity resulting from the intrastriatal injection of the neurotoxin NMDA could be completely reversed if MK-801 was administered within 30 minutes. Furthermore, this reversal was consistent with neuronal protection as determined from subsequent histology.

Delayed treatment (1–6 hr) resulted in incomplete reversal of the hyperintensity and eventual tissue necrosis.

## 4.3. Trace Imaging, Fiber Orientation

As stated above, the ADC has become widely used as a crude parameter for quantifying DW images. Recently, however, a number of research centers have developed improved methods of quantification (see, e.g., Basser *et al.*, 1994a). These methods explicitly account for diffusion anisotropy, a feature which is lacking in the ADC approach. One method is based on the calculation of "diffusion ellipsoids" (Basser *et al.*, 1994a). These are surface plots of the mean translational displacement at a given diffusion time. Isotropic diffusion yields spheres, whereas anisotropic diffusion yields ellipsoids that are elongated in the direction of least restriction. Diffusion ellipsoid maps of the CNS consist of regions in which the ellipsoids are approximately spherical, corresponding to gray matter and the ventricular system, while white-matter tracts are revealed by ellipsoids, the long axes of which are oriented parallel to the fibers. A diffusion ellipsoid map is shown in Fig. 11; the lateral ventricles and the corpus callosum are clearly distinguished.

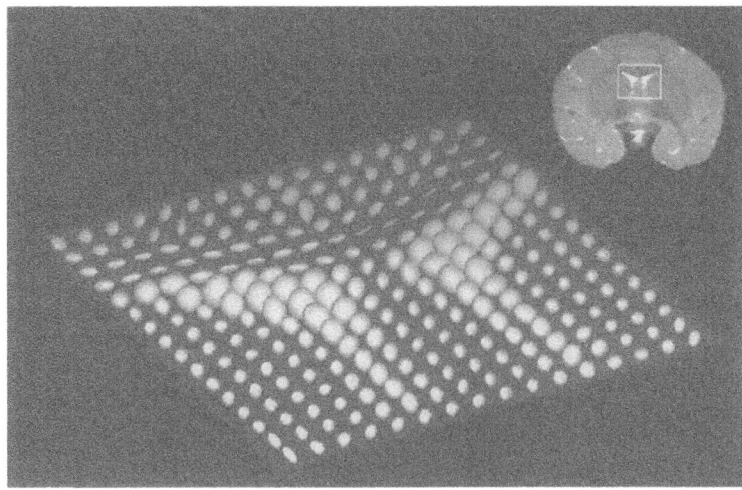

*FIGURE 11.* Diffusion ellipsoid map. This figure shows diffusion ellipsoids, calculated on a pixel-by-pixel basis, in the region of interest indicated on the inserted coronal image of a monkey brain. The most striking features are the two lateral ventricles, which appear as regions of high intensity on the inserted image and which give rise to large spheres in the map, reflecting the isotropic and free diffusion of water in the cerebrospinal fluid. The corpus callosum, which lies immediately above the lateral ventricles and appears hypointense on the inserted image, gives rise to prolate ellipsoids orientated along the fibers. (Figure kindly provided by Dr. C. Pierpaoli and colleagues.)

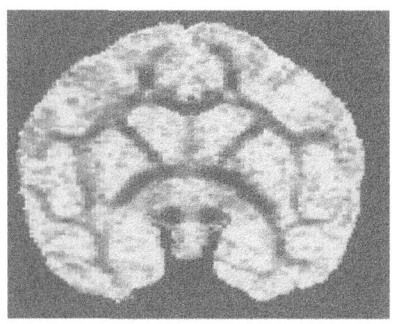

FIGURE 12. Volume ratio image of monkey brain. This image was constructed from a pixel-by-pixel calculation of the volume ratio. This parameter is defined in the text. Major white-matter tracts are clearly seen as regions of reduced signal intensity. (Figure kindly provided by Dr. C. Pierpaoli and colleagues.)

An alternative approach is based on the calculation of scalar invariants. A scalar invariant is a number that is independent of orientation. In the context of diffusion mapping, a scalar invariant is independent of the relative orientations of the object and the gradient axes. Among the possible scalar invariants is the diffusion tensor trace, which is the sum of the diagonal elements. The use of trace images in acute stroke has been examined by van Gelderen *et al.* (1994), who have suggested that trace measurements provide a more accurate delineation of the ischemic area than is obtained with unidirectional diffusion-weighted images because, in the latter, contrast between normal and affected tissue can be obscured by orientationally dependent gray–white matter contrast. This orientation-dependent contrast is not present in the trace image, with the result that little gray–white matter contrast remains, and affected tissue appears more prominent.

Figure 12 shows the image obtained using another scalar invariant, one that has been called the "volume ratio." Among the structures seen in this image is the ventral internal capsule, a neuronal tract which has an oblique orientation with respect to the sensitizing gradient. Pierpaoli *et al.* (1994) showed that this feature is not so clearly displayed in other types of diffusion images. The volume ratio[+] is the square of the volume of the diffusion ellipsoid divided by the square of the volume of a sphere having a radius equal to $[\text{Trace}(D_{\text{eff}})/3]^{1/2}$.

## 5. DIFFUSION-WEIGHTED SPECTROSCOPY

The discussion so far has been concerned with the measurement of the self-diffusion of water and its application to MR imaging. One obvious extension is to

[+]The formal definition of the volume ratio (VR) is

$$VR = \frac{27\text{Det}(D_{\text{eff}})}{[\text{Trace}(D_{\text{eff}})]^3}$$

where $\text{Det}(D_{\text{eff}})$ is the determinant of the effective diffusion tensor.

explore the diffusion characteristics of cerebral metabolites using diffusion-weighted MR spectroscopy. It has been realized for some time that differences among the properties of various metabolites can be used to probe certain cellular functions. These characteristics include membrane transport, metabolic compartmentation, and metabolic flux. Exploiting differences in metabolite diffusivity could be a useful adjunct to these techniques. Although diffusion spectroscopy has not been fully exploited, a number of interesting applications have been reported. For instance, diffusion-weighted proton spectra have been obtained *in vivo* from anesthetized rat brain (Merboldt *et al.*, 1993) and human brain (Posse *et al.*, 1993). In these experiments, localization and diffusion weighting were achieved using a modification of the stimulated-echo acquisition mode (STEAM) sequence, as outlined in Section 2.5. Apparent diffusion coefficients were calculated for the major intracellular metabolites detectable by $^1H$ MR spectroscopy. These studies were, however, restricted to normal brain tissue, and it remains to be seen whether this approach can provide useful information about disease.

The application of diffusion-weighted spectroscopy to systems *in vitro* has required the development of techniques that enable viable cells to be maintained at high densities by continuous perfusion. For example, cells can be embedded in agarose gel or packed into hollow fiber bioreactors and subsequently perfused. A major problem associated with using these techniques, especially for $^1H$ MR spectroscopy, is the high concentration of metabolites and of water in the perfusate, because these dominate the MR spectrum. It has been shown that an adequate suppression of extracellular water and metabolites can be achieved by applying diffusion weighting. For example, complete suppression of extracellular signals was demonstrated by van Zijl *et al.* (1991), who obtained well-resolved $^1H$ MR spectra reflecting the metabolic composition of the intracellular compartment.

## 6. CONCLUSION

We have shown in this chapter that DWI can be used to advantage in both clinical medicine and biomedical research. The former is vividly demonstrated in Fig. 13, which shows an example in which diffusion- and $T_2$-weighted imaging have been combined in order to differentiate between acute and chronic pathologies. A case might be made for the routine use of DWI in the examination of stroke patients. Several other applications have been mentioned, including the study of myelination in neonates. A future role for diffusion tensor imaging in the study of demyelinating disorders is expected.

Particular emphasis is being placed on the application of DWI to drug development and for evaluating therapy, in both experimental research and in the management of stroke patients. Despite the absence of a complete understanding

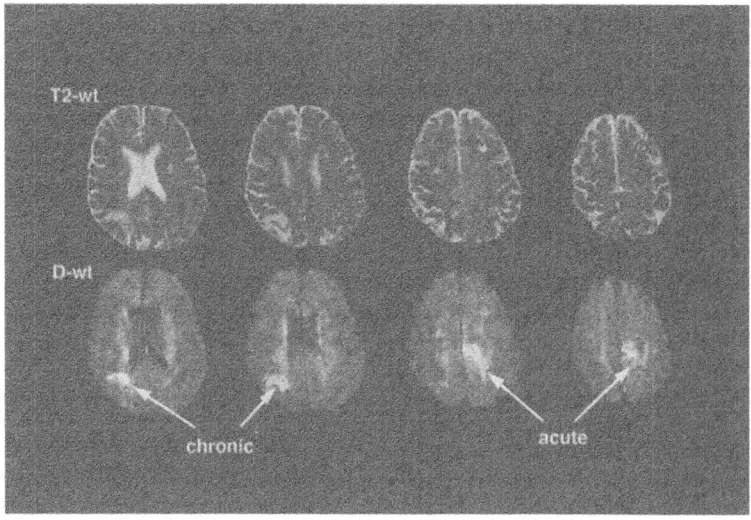

**FIGURE 13.** Diffusion- and $T_2$-weighted images of a human stroke. $T_2$-weighted (top row, TE 110 msec) and diffusion-weighted (bottom row, $b$ = 550 sec mm$^{-2}$) images (four slices) recorded from a 64-year-old male, approximately 3.5 hr following the onset of a stroke. The appearance of hyperintensity in the diffusion-weighted scan in the absence of any detectable changes in the $T_2$-weighted image is indicative of an acute insult. A chronic lesion is also visible as a hyperintense region in both the diffusion- and $T_2$-weighted images. (Figure kindly provided by Dr. A. De Crespigny.)

of the mechanisms underlying the changes seen in the diffusion-weighted image, the role of DWI in neuroimaging is now established.

## REFERENCES

Ahn, C. D., Lee, S. Y., Nalcioglu, O., and Cho, Z. H., 1987, The effects of random directional distributed flow in nuclear magnetic resonance imaging, *Med. Phys.* **14**:43–48.

Allen, K., Busza, A. L., Crockard, H. A., Frackowiak, R. S., Gadian, D. G., Proctor, E., Ross Russell, R., and Williams, S. R., 1988, Acute cerebral ischaemia: Concurrent changes in cerebral blood flow, energy metabolites, pH, and lactate measured with hydrogen clearance and $^{31}$P and $^1$H nuclear magnetic resonance spectroscopy: III Changes following ischaemia, *J. Cereb. Blood Flow Metab.* **8**:816–821.

Astrup, J., Siesjö, B. K., and Symon, L., 1981, Thresholds in cerebral ischemia—the ischemic penumbra, *Stroke* **12**:723–725.

Back, T., Kohno, K., Radermacher, B., Fischer, M., and Hossmann, K.-A., 1993, Characterisation of tissue oxygenation and spreading depression-like DC shifts in the penumbra of focal ischemic rats, *J. Cereb. Blood Flow Metab.* **13**(Suppl. 1):S83.

Basser, P. J., Mattiello, J., and Le Bihan, D., 1994a, Estimation of the effective self-diffusion tensor from the NMR spin echo, *J. Magn. Reson. B* **103**:247–254.

Basser, P. J., Mattiello, J., and Le Bihan, D., 1994b, MR diffusion tensor spectroscopy and imaging, *Biophys. J.* **66:**259–267.

Benveniste, H., Hedlund, L. W., and Johnson, G. A., 1992, Mechanism of detection of acute cerebral ischemia in rats by diffusion-weighted magnetic resonance microscopy, *Stroke* **23:**746–754.

Busza, A. L., Allen, K. L., King, M. D., van Bruggen, N., Williams, S. R., and Gadian, D. G., 1992, Diffusion-weighted imaging studies of cerebral ischemia in gerbils. Potential relevance to energy failure, *Stroke* **23:**1602–1612.

Carr, H. Y., and Purcell, E. M., 1954, Effects of diffusion on free precession in nuclear magnetic resonance experiments, *Phys. Rev.* **94:**630–638.

Chenevert, T. L., Brunberg, J. A., and Pipe, J. G., 1990, Anisotropic diffusion in human white matter: Demonstration with MR techniques *in vivo, Radiology* **177:**401–405.

Cory, D. G., and Garroway, A. N., 1990, Measurement of translational displacement probabilities by NMR: An indicator of compartmentation, *Magn. Reson. Med.* **14:**435–444.

Dardzinski, B. J., Sotak, C. H., Fisher, M., Hasegawa, Y., Li, L., and Minematsu, K., 1993, Apparent diffusion coefficient mapping of experimental focal cerebral ischemia using diffusion-weighted echo-planar imaging, *Magn. Reson. Med.* **30:**318–325.

Decanniere, C., Eleff, S., Sugimoto, H., Davis, D., and van Zijl, P., 1994, Rapid changes in lactate, purine nucleotides, and average water diffusion constant during global ischemia in cat brain, *in "Proceedings of the 2nd Annual Meeting of the Society for Magnetic Resonance, San Francisco."* 235. (Abstract).

Doran, M., Hajnal, J. V., van Bruggen, N., King, M. D., Young, I. R., and Bydder, G. M., 1990, Normal and abnormal white matter tracts shown by MR imaging using directional diffusion weighted sequences, *J. Comput. Assist. Tomogr.* **14:**865–873.

Gardner-Medwin, A. R., van Bruggen, N., Williams, S. R., and Ahier, R. G., 1994, Magnetic resonance imaging of propagating waves of spreading depression in the anaesthetised rat, *J. Cereb. Blood Flow Metab.* **14:**7–11.

Gibson, J. P., Yarrington, J. T., Loudy, D. E., Gerbig, C. G., Hurst, G. H., and Newberne, J. W., 1990, Chronic toxicity studies with vigabatrin, a GABA-transaminase inhibitor, *Toxicol. Pathol.***18:**225–238.

Hahn, E. L., 1953, Free nuclear induction, *Phys. Today* **6:**4–9.

Hajnal, J. V., Doran, M., and Bydder, G. M., 1992, Clinical diffusion imaging, *in "Magnetic Resonance Imaging"* (D. D. Stark and W. G. Bradley, eds.), pp. 1081–1090, Mosby Year Book, St. Louis.

Hansen, A. J., 1985, Effect of anoxia on ion distribution in the brain. *Physiol. Rev.* **65:**101–148.

Hansen, A. J., and Lauritzen, M., 1984, The role of spreading depression in acute brain disorders, *An. Acad. Bras. Cienc.* **56:**447–479.

Hansen, A. J., and Olsen, C. E., 1980, Brain extracellular space during spreading depression and ischemia, *Acta Physiol. Scand.* **108:**355–365.

Hasegawa, Y., Fisher, M., Latour, L. L., Dardzinski, B. J., and Sotak, C. H., 1994, MRI diffusion mapping of reversible and irreversible ischemic injury in focal brain ischemia, *Neurology* **44:**1484–1490.

Helpern, J. A., Dereski, M. O., Knight, R. A., Ordidge, R. J., Chopp, M., and Qing, Z. X., 1993, Histopathological correlations of nuclear magnetic resonance imaging parameters in experimental cerebral ischemia, *Magn. Reson. Imaging* **11:**241–246.

Henkelman, R. M., 1990, Does IVIM measure classical perfusion? *Magn. Reson. Med.* **16:**470–475.

Horsfield, M. A., Barker, G. J., and McDonald, W. I., 1994, Self-diffusion in CNS tissue by volume-selective proton NMR, *Magn. Reson. Med.* **31:**637–644.

Hossmann, K.-A., 1971, Cortical steady potential, impedance and excitability changes during and after total ischemia of cat brain, *Exp. Neurol.***32:**163–175.

Hossmann, K.-A., Fischer, M., Bockhorst, K., and Hoehn-Berlage, M., 1994, NMR imaging of the apparent diffusion coefficient (ADC) for the evaluation of metabolic suppression and recovery after prolonged cerebral ischemia, *J. Cereb. Blood Flow Metab.* **14**:723–731.

Karlicek, R. F., and Lowe, I. J., 1980, A modified pulsed gradient technique for measuring diffusion in the presence of large background gradients, *J. Magn. Reson.* **37**:75–91.

King, M. D., van Bruggen, N., Busza, A. L., Houseman, J., Williams, S. R., and Gadian, D. G., 1992, Perfusion and diffusion MR imaging, *Magn. Reson. Med.* **24**:288–301.

King, M. D., Houseman, J., Roussel, S. A., van Bruggen, N., Williams, S. R., and Gadian, D. G., 1994, *q*-Space imaging of the brain, *Magn. Reson. Med.* **32**:707–713.

Kohno, K., Hoehn-Berlage, M., Mies, G., Back, T., and Hossmann, K.-A., 1995, Relationship between diffusion-weighted MR images, cerebral blood flow, and energy state in experimental brain infarction, *Magn. Reson. Imaging* **13**:73–80.

Latour, L. L., Li, L., and Sotak, C. H., 1993, Improved PFG stimulated-echo method for the measurement of diffusion in inhomogeneous fields, *J. Magn. Reson. B* **101**:72–77.

Latour, L. L., Hasegawa, Y., Formato, J. E., Fisher, M., and Sotak, C. H., 1994, Spreading waves of decreased diffusion coefficient after cortical stimulation in the rat brain, *Magn. Reson. Med.* **32**:189–198.

Leao, A. A. P., 1944a, Spreading depression of activity in the cerebral cortex. *J. Neurophysiol.* **7**:359–390.

Leao, A. A. P., 1944b, Pial circulation and spreading depression of activity in the cerebral cortex, *J. Neurophysiol.* **7**:391–396.

Le Bihan, D., 1988, Intravoxel incoherent motion imaging using steady-state free precession, *Magn. Reson. Med.* **7**:346–351.

Le Bihan, D., Breton, E., Lallemand, D., Grenier, P., Cabanis, E., and Laval-Jeantet, M., 1986, MR imaging of intravoxel incoherent motions: Application to diffusion and perfusion in neurologic disorders, *Radiology* **161**:401–407.

Le Bihan, D., Breton, E., Lallemand, D., Aubin, M.-L., Vignaud, J., and Laval-Jeantet, M., 1988, Separation of diffusion and perfusion in intravoxel incoherent motion MR imaging. *Radiology* **168**:497–505.

Le Bihan, D., Turner, R., and MacFall, J. R., 1989, Effects of intravoxel incoherent motions (IVIM) in steady-state free precession (SSFP) imaging: Application to molecular diffusion imaging, *Magn. Reson. Med.* **10**:324–337.

Lian, J., Williams, D. S., and Lowe, I. J., 1994, Magnetic resonance imaging of diffusion in the presence of background gradients and imaging of background gradients, *J. Magn. Reson. A* **106**:65–74.

Lo, E. H., Matsumoto, K., Pierce, A. R., Garrido, L., and Luttinger, D., 1994, Pharmacologic reversal of acute changes in diffusion-weighted magnetic resonance imaging in focal cerebral ischemia. *J. Cereb. Blood Flow Metab.* **14**:597–603.

Mansfield, P., and Chapman, B., 1986, Active magnetic screening of gradient coils in NMR imaging, *J. Magn. Reson.* **66**:573–576.

McCall, D. W., Douglass, D. C., and Anderson, E. W., 1963, Self-diffusion studies by means of nuclear magnetic resonance spin-echo techniques, *Ber. Bunsenges Phys. Chem.* **67**:336–340.

Memezawa, H., Smith, M. L., and Siesjö, B. K., 1992, Penumbral tissues salvaged by reperfusion following middle cerebral artery occlusion in rats, *Stroke* **23**:552–559.

Merboldt, K.-D., Hänicke, W., and Frahm, J., 1985, Self-diffusion NMR imaging using stimulated echoes, *J. Magn. Reson.* **64**:479–486.

Merboldt, K.-D., Bruhn, H., Frahm, J., Gyngell, M. L., Hänicke, W., and Deimling, M., 1989, MRI of "diffusion" in the human brain: New results using a modified CE-FAST sequence, *Magn. Reson. Med.* **9**:423–429.

Merboldt, K.-D., Hörstermann, D., Hänicke, W., Bruhn, H., and Frahm, J., 1993, Molecular self-diffusion of intracellular metabolites in rat-brain *in vivo* investigated by localized proton NMR diffusion spectroscopy, *Magn. Reson. Med.* **29**:125–129.

Minematsu, K., Li, L. M., Sotak, C. H., Davis, M. A., and Fisher, M., 1992, Reversible focal ischemic injury demonstrated by diffusion-weighted magnetic resonance imaging in rats, *Stroke* **23:**1304–1311.

Mintorovitch, J., Moseley, M. E., Chileuitt, L., Shimizu, H., Cohen, Y., and Weinstein, P. R., 1991, Comparison of diffusion- and T2-weighted MRI for the early detection of cerebral ischemia and reperfusion in rats, *Magn. Reson. Med.* **18:**39–50.

Moseley, M. E., Cohen, Y., Kucharczyk, J., Mintorovitch, J., Asgari, H. S., Wendland, M. F., Tsuruda, J., and Norman, D., 1990a, Diffusion-weighted MR imaging of anisotropic water diffusion in cat central nervous system, *Radiology* **176:**439–445.

Moseley, M. E., Cohen, Y., Mintorovitch, J., Chileuitt, L., Shimizu, H., Kucharczyk, J., Wendland, M. F., and Weinstein, P. R., 1990b, Early detection of regional cerebral ischemia in cats: Comparison of diffusion- and T2-weighted MRI and spectroscopy, *Magn. Reson. Med.* **14:**330–346.

Murday, J. S., and Cotts, R. M., 1968, Self-diffusion coefficient of liquid lithium, *J. Chem. Phys.* **48:**4938–4945.

Neeman, M., Freyer, J. P., and Sillerud, L. O., 1990, Pulsed-gradient spin-echo diffusion studies in NMR imaging. Effects of the imaging gradients on the determination of diffusion coefficients, *J. Magn. Reson.* **90:**303–312.

Neuman, C. H., 1974, Spin echo of spins diffusing in bounded medium, *J. Chem. Phys.* **60:**4508–4511.

Niendorf, T., Norris, D. G., and Leibfritz, D., 1994, Detection of apparent restricted diffusion in healthy rat brain at short diffusion times, *Magn. Reson. Med.* **32:**672–677.

Obrenovitch, T. P., Garofalo, O., Harris, R. J., Bordi, L., Ono, M., Momma, F., Bachelard, H. S., and Symon, L., 1988, Brain tissue concentrations of ATP, phosphocreatine, lactate, and tissue pH in relation to reduced cerebral blood flow following experimental acute middle cerebral artery occlusion, *J. Cereb. Blood Flow Metab.* **8:**866–874.

Pekar, J., Moonen, C. T. W., and van Zijl, P. C. M., 1992, On the precision of diffusion/perfusion imaging by gradient sensitization, *Magn. Reson. Med.* **23:**122–129.

Pierpaoli, C., Linfante, I., Mattiello, J., Di Chiro, G., Le Bihan, D., and Basser, P. J., 1994, Diffusion tensor imaging of brain white matter anisotropy, *in* "Proceedings of the 2nd Annual Meeting of the Society for Magnetic Resonance, San Francisco," p. 1038 (Abstract).

Posse, S., Cuenod, C. A., and Le Bihan, D., 1993, Human brain: Proton diffusion MR spectroscopy, *Radiology* **188:**719–725.

Prasad, P. V., and Nalcioglu, O., 1991, A modified pulse sequence for *in vivo* diffusion imaging with reduced motion artifacts, *Magn. Reson. Med.* **18:**116–131.

Preece, N. E., Butler, W. H., Cannon, D. J., Houseman, J., Weller, R. O., and Williams, S. R., 1994, The development of vigabatrin-induced lesions in the rat brain studied by quantitative T2 measurements and diffusion-weighted imaging, *in* Proceedings of the 2nd Annual Meeting of the Society for Magnetic Resonance, San Francisco," p. 1378 (Abstract).

Rutherford, M. A., Cowan, F. M., Manzur, A. Y., Dubowitz, L. M. S., Pennock, J. M., Hajnal, J. V., Young, I. R., and Bydder, G. M., 1991, MR imaging of anisotropically restricted diffusion in the brain of neonates and infants, *J. Comput. Assist. Tomogr.* **15:**188–198.

Sakuma, H., Nomura, Y., Takeda, K., Tagami, T., Nakagawa, T., Tamagawa, Y., Ishii, Y., and Tsukamoto, T., 1991, Adult and neonatal human brain: Diffusional anisotropy and myelination with diffusion-weighted MR imaging, *Radiology* **180:**229–233.

Siesjö, B. K., 1978, "Brain Energy Metabolism," Wiley, Chichester.

Staub, F., Mackert, B., Kempski, O., Peters, J., and Baethmann, A., 1993, Swelling and death of neuronal cells by lactic acid, *J. Neurol. Sci.* **119:**79–84.

Stehling, M. K., Turner, R., and Mansfield, P., 1991, Echo-planar imaging: Magnetic resonance imaging in a fraction of a second, *Science* **254:**43–50.

Stejskal, E. O., and Tanner, J. E., 1965, Spin diffusion measurements: Spin echoes in the presence of a time-dependent field gradient, *J. Chem. Phys.* **42**:288–292.

Tanner, J. E., 1970, Use of the stimulated echo in NMR diffusion studies, *J. Chem. Phys.* **52**:2523–2526.

Tanner, J. E., and Stejskal, E. O., 1968, Restricted self-diffusion of protons in colloidal systems by the pulsed-gradient, spin-echo method, *J. Chem. Phys.* **49**:1768–1777.

Taylor, D. G., and Bushell, M. C., 1985, The spatial mapping of translational diffusion coefficients by the NMR imaging technique, *Phys. Med. Biol.* **30**:345–349.

Turner, R., and Bowley, R. M., 1986, Passive screening of switched magnetic field gradients, *J. Phys. E: Sci. Instrum.* **19**:876–879.

Turner, R., and Keller, P. J., 1991, Angiography and perfusion measurements by NMR. *Prog. NMR Spectrosc.* **23**:93–133.

Turner, R., Le Bihan, D., Maier, J., Vavrek, R., Hedges, L. K., and Pekar, J., 1990a, Echo-planar imaging of intravoxel incoherent motion, *Radiology* **177**:407–414.

Turner, R., von Kienlin, M., Moonen, C. T. W., and van Zijl, P. C. M., 1990b, Single-shot localized echo-planar imaging (STEAM-EPI) at 4.7 Tesla, *Magn. Reson. Med.* **14**:401–408.

van Bruggen, N., Roberts, T. P. L., and Cremer, J. E., 1994, The application of magnetic resonance imaging to the study of experimental cerebral ischaemia, *Cerebrovasc. Brain Metab. Rev.* **6**:180–210.

van Gelderen, P., de Vleeschouwer, M. H. M., Despres, D., Pekar, J., van Zijl, P. C. M., and Moonen, C. T. W., 1994, Water diffusion and acute stroke, *Magn. Reson. Med.* **31**:154–163.

van Zijl, P. C. M., Moonen, C. T. W., Faustino, P., Pekar, J., Kaplan, O., and Cohen, J. S., 1991, Complete separation of intracellular and extracellular information in NMR spectra of perfused cells by diffusion-weighted spectroscopy, *Proc. Natl. Acad. Sci. USA* **88**:3228–3232.

Verheul, H. B., Balazs, R., Berkelbach van der Sprenkel, J. W., Tulleken, C. A. F., Nicolay, K., and van Lookeren Campagne, M., 1993, Temporal evolution of NMDA induced excitoxicity in the neonatal rat brain measured with ¹H nuclear magnetic resonance imaging, *Brain Res.* **618**:203–212.

Verheul, H. B., Balazs, R., Berkelbach van der Sprenkel, J. W., Tulleken, C. A. F., Nicolay, K., Tamminga, K. S., and van Lookeren Campagne, M., 1994, Comparison of diffusion-weighted MRI with changes in cell volume in a rat model of brain injury, *NMR Biomed.* **7**:96–100.

Williams, W. D., Seymour, E. F. W., and Cotts, R. M., 1978, A pulsed-gradient multiple-spin-echo NMR technique for measuring diffusion in the presence of background magnetic field gradients, *J. Magn. Reson.* **31**:271–282.

Wu, E. X., and Buxton, R. B., 1990, Effect of diffusion on the steady-state magnetization with pulsed field gradients, *J. Magn. Reson.* **90**:243–253.

Zhong, J., Petroff, A. C., Prichard, J. W., and Gore, J. C., 1993, Changes in water diffusion and relaxation properties of rat cerebrum during status epilepticus, *Magn. Reson. Med.* **30**:241–246.

# HIGH-SPEED ECHO-PLANAR IMAGING AND ITS APPLICATION TO NEUROLOGY

## PENNY GOWLAND and PETER MANSFIELD

## SUMMARY

This chapter describes the theory of echo-planar imaging (EPI) and technical issues in its practical implementation. It also explains the developments of EPI which enable it to be used to make quantitative measurements of parameters such as magnetic resonance relaxation times, chemical shift, or flow. Finally, the main areas of application of EPI are discussed.

## INTRODUCTION

The phenomenon of nuclear magnetic resonance, or simply magnetic resonance (MR) as is nowadays preferred, was first applied to imaging in 1973 (Mansfield

PENNY GOWLAND and PETER MANSFIELD • The Magnetic Resonance Centre, Department of Physics, University of Nottingham, Nottingham, NG7 2RD, United Kingdom.

Magnetic Resonance Spectroscopy and Imaging in Neurochemistry, Volume 8 of Advances in Neurochemistry, edited by Bachelard, Plenum Press, New York, 1997.

and Grannell, 1973; Lauterbur, 1973), and in the early days many strategies for producing images were proposed. Among these, one method—spin warp imaging (Edelstein *et al.*, 1980)—proved robust enough to be commercially developed for medical imaging using the technology available at the time. Unfortunately, conventional spin warp imaging takes several seconds to minutes to produce an image and is therefore inherently prone to motion artifacts arising from cardiovascular, respiratory, peristaltic, and other involuntary and voluntary movements. As early as 1977, an alternative high-speed imaging technique, echo-planar imaging (EPI), was proposed (Mansfield, 1977). With this technique, imaging times could be reduced to a "snapshot" (30–130 msec), effectively freezing all movements *in vivo*. The hardware specifications to implement EPI were severe, and major technical breakthroughs were required to make the technique robust enough for clinical use (Turner *et al.*, 1988). EPI has only become generally available over the last five years and has recently gained widespread acceptance because at high field strengths it can be used to rapidly study brain activation with high resolution and high sensitivity (Belliveau *et al.*, 1991).

This chapter aims to explain the method of collecting an MR image in a snapshot, the general applications of high-speed imaging and specific clinical applications of the technique.


## 2. GENERAL APPLICATIONS

People who are used to conventional magnetic resonance imaging (MRI) are generally stunned when they first see an EPI image acquired. Each image is truly acquired in a snapshot, after minimal setting-up time on the subject. Data flood in, rapidly filling up all available storage space. This snapshot capability has many advantages, and there are three main impetuses driving the development of high-speed MR imaging techniques.

The first, and probably main, advantage of high-speed imaging is that it opens up the possibility of studying a whole range of dynamic systems. Only ultrasound offers the same safe fluoroscopic capability which allows extended and repeated scanning of patients and volunteers. However, ultrasound suffers from low image quality and physical problems with scanning beyond bone or air spaces, a particular disadvantage in the brain. When imaging dynamically, EPI can run at a rate of 0.3–3 scans per second, depending on the $T_1$ of the region being scanned; this can be increased to more than 8 scans per second in the heart, where the rapid inflow of blood refreshes the longitudinal magnetization. Using a low-angle excitation pulse, repeated scans have been obtained at 20 frames per second. Alternatively, multislice three-dimensional data sets can be acquired at the rate of more than 20 frames per second (each frame comprising 128 x 128 pixels.) The dynamic capability of EPI has been applied to the study of brain activation and cerebrospinal fluid (CSF) flow, the modeling of contrast agent uptake in tumors, the study of gut motility, and the study of cardiac abnormalities.

The second advantage of ultra-high-speed imaging is that it makes parametric imaging feasible *in vivo*. The signal intensity in MR images is controlled by many biophysical factors such as MR relaxation times (which relate largely to the binding of water molecules in a sample), the self-diffusion of water molecules, incoherent blood flow (that is, capillary flow within a voxel), coherent blood flow in large vessels, and magnetic susceptibility differences, in particular, the susceptibility differences between oxyhemoglobin and deoxyhemoglobin, the breakdown products of hemoglobin and tissue, and the susceptibility differences between blood vessels containing contrast agents and the surrounding tissues. An important feature of MRI is that, in principle, it is possible to measure these parameters *in vivo*. This generally involves magnetically preparing the sample before collecting the image, which always increases the imaging time by a certain fraction. Several images are then acquired with different preparation conditions. Therefore, it is generally not feasible to produce quantitative images with conventional MRI; an accurate measurement of $T_1$ may take eight minutes to map a single slice. This must be added to the time taken to collect anatomical images and to select the optimum imaging plane. Using high-speed MRI, $T_1$ maps of a single slice can now be acquired in two seconds.

The third advantage is financial: the capital investment tied up in a whole-body scanner means that high patient throughput is essential. At Nottingham we can complete a scan of a fetus to measure the volume of its internal organs in three minutes (the time between the patient lying down and standing up again). This increased throughput is not only due to the rapid time for data acquisition, but also to the interactive nature of EP imaging. As images are obtained in a snapshot, the operator has almost immediate feedback when setting up imaging parameters, allowing both rapid image optimization and immediate confirmation that the required image has been obtained.

## 3. ECHO-PLANAR IMAGING: THEORY AND PRACTICE

### 3.1. The EPI Sequence

In order to understand EPI, one must first consider NMR and its application to imaging. The phenomenon of NMR arises when a sample containing NMR-sensitive nuclei (spins) is placed in a static magnetic field. If a radio-frequency (RF) magnetic field is applied to the sample, the spin system will resonate and produce an RF signal at the Larmor frequency. The Larmor frequency depends on the nucleus under consideration and is directly proportional to the magnitude of the applied magnetic field, $B_0$ (unit: tesla, T):

$$\omega_o = \gamma B_o \tag{1}$$

where $\omega_o$ is the Larmor angular frequency (in megahertz) and $\gamma$ is the gyromagnetic ratio for the nucleus under consideration (this has a value of 42.57 MHz/T for protons). The applied static magnetic field is generally constant and extremely homogeneous (e.g., varying by less than 3 ppm over a 30-cm$^3$ sphere at the center of the magnet), but nearly all methods of producing an MR image involve superimposing a spatially varying linear magnetic field gradient on this field, thereby making the Larmor frequency of the spins spatially dependent. For a gradient along the $x$ direction ($G_x$ in units of T/m), the magnitude of the magnetic field at a given position $x$ can be described by

$$B(x) = B_o + xG_x \tag{2}$$

The total signal detected will therefore be the sum of all signal amplitudes contributing at different frequencies. The signal amplitude at each frequency depends primarily on the number of spins at the corresponding position in the magnetic field.

In the absence of a magnetic field gradient, all the spins will precess at approximately the same angular frequency, $\omega$, and are jointly known as an isochromat. The signal from an isochromat as a function of time is given by

$$S(t) = S_o e^{i\omega t} e^{-t/T_2} \tag{3}$$

where $\omega = \omega_o$ at resonance, $S_o$ is the magnitude of the signal (a function of the MR parameters of the isochromat and the exact pulse sequence), and $T_2$ is the transverse relaxation time that describes the exponential decay of the MR signal. However, the signal is detected via phase-sensitive detectors that compare the signal to a reference frequency that is also the carrier frequency of the transmitted RF excitation pulse. These give an audio-frequency output, the frequency of which depends on the difference in frequency between the signal and the reference frequency and the phase of which depends on the phase difference between the signal and the reference frequency as well as the sign of the difference in frequency. By using two-phase-sensitive detectors with reference frequencies 90° out of phase, it is possible to determine the sign of the difference in frequency between the signal and reference, and also to achieve a $\sqrt{2}$ improvement in signal-to-noise ratio. Usually, the term Larmor frequency is used to describe the reference frequency, which is generally set at the average resonance frequency across the whole sample. This frequency depends on intrinsic and induced inhomogeneities in the applied field and will be given the symbol $\omega_L$. The signal detected is given by

$$S(t) = S_o e^{i(\omega t + \phi)} e^{-t/T_2} \tag{4}$$

where $\omega$ ($\omega = \omega_o - \omega_L$) is the difference between the Larmor frequency and the frequency of the received signal, and $\phi$ is the phase of the received signal. Equation (4) describes the free induction decay as it is usually observed.

Let the concentration of spins (spin density) projected along the applied gradient direction (the $x$- axis) be described by $\rho(x)$; then the resonant frequency of the spins will depend on their position [according to Eqs. (1) and (2)], and the total signal obtained from the phase-sensitive detector will be given by

$$S(t) = \int_{-x}^{+x} \rho(x) \, e^{i(\gamma G_x x)t} \, e^{-t/T_2} dx \tag{5}$$

where $+X$ and $-X$ are the maximum and minimum extents of the spins along the gradient axis in the magnet bore. Equation (5) indicates that, because the frequency of precession depends on position, the spins will start to dephase with respect to one another, leading to signal attenuation caused by a decrease in the net transverse magnetization.

Normally, MRI is performed on only a single slice of the sample, which is defined using selective excitation (Garroway *et al.*, 1974; Mansfield *et al.*, 1979); conventionally the slice is defined in the $z$ direction. For a slice of thickness $\Delta z$ at position $Z_0$ in an extended three-dimensional object, $\rho(x)$ is given by:

$$\rho(x) = \int_{Z_0}^{Z_0 + \Delta z} \int_{-Y}^{+Y} \rho(x, y, z) \, dy \, dz \tag{6}$$

(i.e., $\rho(x)$ is the projection of a slice of the object onto the $x$ axis). In general, the direction and amplitude of the gradient may be varied, and so Eq. (5) can be generalised as

$$S(t) = \int \rho(\mathbf{r}) e^{i\gamma \int_0^t \mathbf{r} \cdot \mathbf{G}(t') dt'} \, e^{-t/T_2} d\mathbf{r}. \tag{7}$$

If the reciprocal space vector $\mathbf{k}$ is defined as

$$\mathbf{k} = \gamma \int_0^t \mathbf{G}(t') dt' \tag{8}$$

then Eq. (7) may be rewritten as

$$S(\mathbf{k}) = \int \rho(\mathbf{r}) \, e^{i\mathbf{r} \cdot \mathbf{k}} e^{-t/T_2} \, d\mathbf{r}. \tag{9}$$

Equation (9) therefore gives the MR signal expressed in terms of spatial frequencies. The spectrum of the spatial frequencies required to describe a one-dimensional projection of the object corresponds to the Fourier transform of the projection. The spatial frequency domain containing the two-dimensional spectrum of the two-dimensional object is usually known as $k$ space. MR imaging is often explained in an analogy with diffraction in optics (Mansfield and Grannell, 1973; Ljunggren, 1983). The Fourier transform of the object is formally the same as the diffraction pattern that would be obtained from it in plane-wave scattering. The spatial resolution is "diffraction limited" by a pseudo-wavelength given by $\lambda = 2\pi/k$. The signal described in Eq. (9) is the free induction decay obtained in the presence of a gradient; neglecting $T_2$ decay, it can be seen that this corresponds to a single line across $k$ space.

By varying the direction in which the gradient is applied (and hence r), it is possible to produce different scans across $k$ space. MRI involves devising schemes to adequately scan the whole of $k$ space. An image can be obtained by taking the two-dimensional Fourier transform of the "diffraction pattern" thus obtained.

One of the most easily understood schemes for imaging is projection reconstruction (Lauterbur, 1973), which involves rotating either r (the direction in which the gradient was applied) or the object itself to obtain radial lines across $k$ space. Rotation of r is achieved by applying two orthogonal gradients simultaneously and varying their relative amplitudes for each scan. The technique has the advantage of scanning with particularly short echo times, but $k$ space is sampled unevenly.

To scan $k$ space on a rectangular grid, it is necessary to be able to move the trajectory away from the center of $k$ space before scanning. This is achieved by applying a gradient pulse before sampling begins, which dephases the spins along the direction of this gradient. Sampling occurs as the gradient is reversed, which causes the phase evolution of the spins to be reversed, thereby moving the trajectory back across $k$ space. A gradient echo is formed when the time integral of the gradient waveform is zero. This occurs when the $k$-space trajectory crosses a $k$-space axis (Fig. 1). Such a line across $k$ space corresponds to the Fourier transform of the projection of the object in that direction.

To obtain similar horizontal scans, along lines parallel to $k_y = 0$ (Fig. 2), both $y$ and $x$ preparation gradients must also be applied to cause excursion along $k_y$ (OR in Fig. 2) and $k_x$ (RS) before $k_x$ is refocused (ST). The gradient echo thus formed will be attenuated because of signal dephasing due to the unrefocused $y$ gradient. This must be repeated for different amplitudes of $G_y$ preparation pulses so that the phase dispersion is incremented to scan many lines across $k$ space. On Fig. 2 consider the vertical line $AB$ drawn through the final complete data set, parallel to $k_x = 0$. The phase varies along this line in a systematic way, similar to the way in which it would have evolved during the application of a constant $y$ gradient—that is, the data along such a vertical line corresponds to the Fourier transform of the projection of the object in the $y$ direction.

In conventional spin warp imaging (Edelstein et al., 1980), a single "hori-

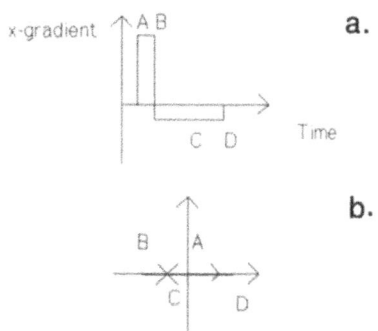

*FIGURE 1.* The variation in *x* gradient as a function of time (a) and the corresponding *k*-space trajectory (b). At point A, no gradient has been applied, and the *k*-space trajectory is at the origin. At point B, the integral under the gradient waveform is maximum, and the system has moved a maximum distance away from the center of *k* space. At point C, the integral under the gradient is zero. At point D the system has moved away from the centre of *k* space again.

zontal" line across *k* space is collected following each RF pulse. Many such scans are required to sample the whole of *k* space, each following the application of a preparation gradient pulse along *y* of varying amplitude. Therefore, the data acquisition takes a time $n \times TR$, where *n* is the matrix size of the image in the phase-encoding direction and TR is the chosen repetition time. Thus, the total imaging time is typically between a few seconds and a few minutes.

With EPI (Mansfield, 1977), the whole of *k* space is sampled in a single scan following a single RF pulse. Various trajectories have been suggested, but the most popular is a rectangular sampling, which requires the application of two preparation dephasing lobes followed by a switched *y* gradient to sweep horizontally across *k* space (in the *ky* direction) interspersed with a blipped *x* gradient to

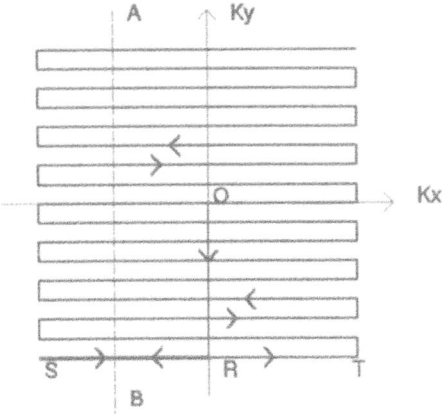

*FIGURE 2.* A scan across many lines of *k* space. A *y* gradient must be applied before each span across $k_x$ to move the *k*-space trajectory to a different position on the *y* axis.

step up through the $k_x$ direction (Howsemann *et al.*, 1988) (Fig. 2). Other trajectories employed include the original EPI sequence that used a constant broadening gradient to zigzag through $k$ space (Mansfield, 1977), a spiral scan that has some advantages in terms of signal attenuation from $T_2$ (Macovski, 1985), and the use of spin echoes rather than gradient echoes (Mansfield, 1977; Hennig *et al.*, 1986).

The EPI technique can be used to collect a volume data set, either by combining a conventional imaging sequence with EPI or by extending the snapshot EPI technique to three dimensions directly in echo-volumar imaging (EVI) (Mansfield *et al.*, 1989).

## 3.2. Technical Issues in EPI

Following Eq. (9), it was stated that $T_2$ decay would be neglected. However, during the sweep through $k$ space, the MR signal inevitably decays with spin–spin, and other $T_2^*$, relaxation processes which is the decay due to magnetic field homogeneity. It is this $T_2^*$ decay that places many constraints on the EPI imaging technique (Ordidge *et al.*, 1989). Furthermore, the spatial resolution of the images, $\delta x$, is determined by the maximum excursion in $k$ space as

$$\delta x = 2\pi/\Delta k_x \tag{10}$$

Substituting $\Delta k_x$ from Eq. (8), we see that the greater the gradient amplitude is and/or the longer it is applied for, the higher the spatial resolution. As a consequence, large gradients must be switched very rapidly. On the 0.5-T system at Nottingham, a switched gradient of about 10 m T/m is oscillated at 500 Hz. Generally, a trapezoidal waveform is used, with a rise time of 100 μsec.

At 3.0 T, higher frequency switched gradients are required to reduce the susceptibility artifacts. Typically, we use 1.7 kHz as the switching frequency for head imaging, but the choice of waveform depends largely on the gradient drives available. Whatever waveform is used, the sampling scheme must be adjusted to achieve linear sampling in $k$ space. If the gradient amplitude varies during sampling, an irregular sampling scheme is required. High voltages will be required to switch the gradient currents unless the coils have low inductance. Another consequence of rapid data sampling is that the receiver circuit requires a high bandwidth, which can introduce noise into the images. The long data acquisition period leads to a narrow bandwidth per pixel of about 8 Hz. Because of a proton chemical shift difference of about 72 Hz at 0.52 T, a large misregistration between fat and water images is produced, corresponding to about 9 points in the blipped, phase-encoding direction. If necessary, the fat or water signal can be suppressed prior to imaging. Susceptibility gradients can give rise to similar shifts and also to $T_2^*$ signal decay, if a large enough phase dispersion is acquired across a pixel during signal readout, and which is particularly significant in the long echo-planar experiment. In spin-echo EPI, a 90°–180° pulse sequence is applied before the imaging gradients, and this can be used to refocus the spin magnetization, thereby reducing the signal attenuation at the center of $k$ space. Strictly speaking, in order to maximize the spatial resolution for modulus images, EPI should be performed with symmetrical coverage of $k$ space; in practice, the first quarter of $k$ space is often not acquired. In this case, data collection starts at $T/4$ before the center of $k$ space, where $T$ is the theoretically correct acquisition time for a particular spatial resolution, thereby reducing signal attenuation due to $T_2^*$. Only just over half of $k$ space need be acquired if half-Fourier reconstruction can be performed to regain spatial resolution; this is at the cost of image signal-to-noise ratio (Margosian, 1985). At high field strengths, where the intrinsic MR signal is greater, this approach is often necessary to reduce signal losses due to $T_2^*$. To further reduce this problem, the time for sampling each line of $k$ space is reduced by using larger gradient amplitudes to maintain the same resolution, again increasing the experimental bandwidth.

The nature of the Modulus Blipped Echo-Planar Single-Pulse Technique (MBEST) EPI trajectory causes alternate lines of data to be collected under gradients of opposite polarity; therefore, alternate lines of data have to be reversed to fill the data matrix correctly (Mansfield, 1977; Johnson and Hutchinson, 1985). If there are any sequence errors that cause the echoes to be misaligned with respect to the sampling comb, reordering the data in this way can give rise to ghost artifacts (the Nyquist ghosts). Three major sources of such errors are nonideal gradient waveforms, eddy currents induced in the RF screen or coil, or delays in the receiver bandwidth filter. Various methods have been proposed to overcome them (Ordidge *et al.*, 1989; Bruder *et al.*, 1992; King *et al.*, 1995).

To reduce the peak voltage required to switch the gradient currents, a resonant sinusoidal or trapezoidal drive can be used. This involves connecting a capacitor network in series with the coil to produce a circuit that resonates at the desired number of frequency modes (Harvey and Mansfield, 1994).

## 3.3. The Safety of EPI

There are several aspects of MRI that could lead to hazards to health; each will be considered in turn. EPI is safer than standard MRI in terms of RF exposure, as only a single RF pulse must be applied for each image. If any risks were ever established to be associated with exposure to static magnetic fields, EPI would again be safer as the patient exposure is shorter due to more rapid throughput. Only the rate of switching of the magnetic field gradients for EPI gives cause for concern when EPI is compared with standard MRI. It is established that rapidly varied magnetic fields ($>60$ T/sec$^{-1}$) can give rise to sensations in volunteers due to neural stimulation by the eddy currents induced in the body. The magnitude of these fields depends on the design of the gradient coil, the relative geometry of the coil and the patient, and the exact gradient waveform. However, the mechanism is not fully explained, and it has recently been suggested that it is the maximum change in the field, rather than the rate of change of the field, that is significant in relation to neural stimulation (Mansfield and Harvey, 1993). Currently, EPI is carried out well below established limits for gradient switching, but this phenomenon could be a problem in the future if the experimental time is reduced.

## 3.4. Real-Time "Movie" Imaging

As EPI can produce snapshot images (40 msec or less), it has the potential to acquire images at a rate of more than 20 frames per second, allowing the monitoring of biodynamic processes in real time. For instance, the movement of the cardiac wall can be captured in situations in which cardiac gating would be inappropriate (Chapman *et al.*, 1987); similarly, movements of the gut wall can be captured (Stehling *et al.*, 1989b). The constraints on the rate of image acquisition are both technical and physical. The technical constraints arise mainly from the maximum power output of the gradient amplifiers and consequently will depend on the efficiency of the gradient coils. This is not a serious limitation if only short bursts of rapid imaging are required over a period of a few seconds. The physical constraint is that saturation of the longitudinal magnetization by rapidly repeated RF pulses will cause a reduction in the signal-to-noise ratio of the images. This effect can be mitigated by setting the pulse angle to be equal to the Ernst angle:

$$\cos \theta = \frac{TR}{T_1} \tag{11}$$

Alternatively, if a short burst of $n$ images is required, the flip angle of the $i$th RF pulse, $\theta_i$, can be chosen to satisfy the condition

$$\tan \theta_{i-1} = \sin \theta_i \tag{12}$$

where $\theta_{i-1}$ is the flip angle of the preceding RF pulse. The signal after the $n$th pulse is given by

$$S_n = S_0/\sqrt{n} \tag{13}$$

where $S_0$ is the equilibrium signal. This is applicable when $T_1 > > T_2$ and the total time to acquire $n$ images is less then $T_1$ (Mansfield, 1984).

Finally, if the $T_2$ of the sample is long, transverse magnetization will persist between RF pulses as stimulated and spin echoes, which can lead to image artifacts. Phase scrambling gradient pulses applied at the end of the imaging gradient sequence will destroy the coherence of this magnetization. However, additional time must be allowed between imaging modules for the application of the scrambling gradients and associated eddy currents or amplifier drift.

Once a set of images of a periodically moving object has been acquired, the signal-to-noise ratio of the images can be enhanced, at no cost in image resolution, by Fourier filtering the data set (Doyle and Mansfield, 1986).

Real-time EPI is used to optimize images and has been applied to monitor uptake of contrast agents, to observed fetal movement, to monitor gut motility and to assess cardiac movement and is also now used in interventional procedures. However, we emphasize that there are many more potential applications of EPI. In its real-time capability, EPI is only rivaled by ultrasound scanning; EPI is best in terms of image quality, whereas ultrasound wins on price.

## 3.5. State-of-the-Art EPI

For reasons discussed above, EPI is a technique that has a relatively low signal-to-noise ratio, but this can be improved by increasing the field strength, which allows an increase in image quality in the form of either improved resolution or improved signal-to-noise ratio. It also allows an increase in bandwidth for a given $Q$ (quality factor) of the RF coil. Unfortunately, increasing the field strength also increases the susceptibility field shift and can lead to distortion of the image and signal attenuation at boundaries between soft tissue and air or bone. However, this also gives rise to an increase in the contrast between vessels containing oxyhemoglobin and deoxyhemoglobin and it is this increase in contrast that is exploited in brain activation imaging using high-field EPI (see Chapter 9). If a resonant gradient drive is used instead of a switched trapezoidal gradient to improve the efficiency of the coil/amplifier system, the switched gradient can be oscillated at a faster rate, thereby reducing the length of the data acquisition and hence reducing the susceptibility dephasing and distortion. The current state-of-the-art echo-planar images have been obtained from an EPI machine built in-house, based around a highly homogeneous 3.0-T magnet (<0.5 ppm over 35 cm). Using a slice thickness of 2.5 mm, $256 \times 256$ pixel images can be acquired with an in-plane resolution of 0.75 mm in 140 msec (Mansfield *et al.*, 1994). Multislice data sets can be collected at a rate of 10 images per second with a slice thickness of 5.0 mm and resultant $128 \times 128$ pixel images have an in-plane resolution of 1.5 mm. Fifty frames a second can be repeated continuously.

## 4. DEVELOPMENTS OF ECHO-PLANAR IMAGING

The nature of the MR phenomenon is such that the signal intensity in MR images depends on many features of the tissue. The most important of these are the MR relaxation times ($T_1$ and $T_2$), which in turn depend chiefly on the exchange of free water molecules with those bound to macromolecules. Another important parameter controlling signal intensity in EPI is locally induced magnetic field inhomogeneity arising from susceptibility and characterized in the time domain by $T_2^*$. Such inhomogeneity reflects constant local variations in the applied magnetic field due to variations in the magnetic susceptibility of tissues (especially at soft tissue/air or soft tissue/bone interfaces). Similar effects can also be produced by paramagnetic centers in a contrast agent or by small susceptibility changes in blood arising from differences in blood oxygenation. It is possible to produce maps of these different parameters by comparing images collected with different weightings to a chosen feature. As many images may need to be acquired, this can be very time-consuming. However, by using EPI it becomes possible to produce parametric maps in a short enough time to be clinically useful.

### 4.1. $T_1$ Mapping Using EPI

Many of the methods for measuring $T_1$ with standard MRI can also be applied to EPI. However, accurate sequences such as the standard inversion recovery (IR) technique that are too slow for use with standard MRI *in vivo* can be used conveniently with EPI. In the IR–EPI sequence, the spins are inverted initially and then the longitudinal magnetization is read out using a standard EPI sequence after a variable inversion time TI (Fig. 3). The magnetization is usually allowed to recover fully before the next inversion pulse is applied and consequently it takes about 20 sec to make an 8-point $T_1$ measurement with this sequence. Using this method, for a typical signal-to-noise ratio of 40, the relative error in $T_1$ due to the noise in the image was calculated to be 25% over the normal biological range of $T_1$ (Stehling *et al.,* 1990a).

If multislice $T_1$ measurements are required, the efficiency of this sequence can be greatly increased by applying a nonselective inversion pulse succeeded by multiple EPI readout modules, each scanning a different slice. These modules are applied as quickly as possible, the only limiting factors being the combined duration of the EPI experiment and the gradient amplifier settling time, in prac-

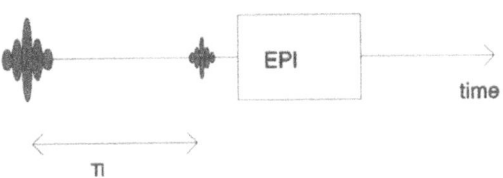

*FIGURE 3.* The inversion recovery sequence. The first pulse is an inversion pulse, and the second is a 90° (readout) pulse.

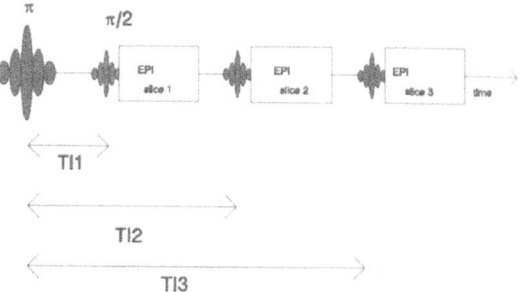

*FIGURE 4.* Multislice inversion recovery. Following a single inversion pulse, a series of readout pulses are applied, each sampling the signal from a different slice. By repeating the sequence and cycling the order in which the slices are scanned, a whole recovery curve can be built up for each slice.

tice about 180 msec. The experiment is repeated with the order of slice acquisition cycled so that eventually each slice has been sampled at each inversion time. In principle, 8 slices can be scanned in about 8 sec in this way. By reordering and fitting the data, $T_1$ maps can be produced (Fig. 4; Ordidge *et al.*, 1990).

If extreme rapid single-slice measurements of $T_1$ are required, for instance, in regions of the body that are moving unpredictably, such as the fetus, or in studies of the uptake of contrast agents, then the Look–Locher (LL) sequence for measuring $T_1$ (Look and Locher, 1970) can be combined with the echo-planar imaging sequence, giving the LL–EPI sequence. In the Look–Locher sequence a single recovery curve is sampled with multiple low-flip-angle readout pulses (Fig. 5). With this LL–EPI sequence, accurate measurements of $T_1$ in a single slice can be made in 2 sec (Gowland and Mansfield, 1993). In this sequence a single inversion pulse is followed by multiple readout modules, each consisting of a low-flip-angle pulse and the echo-planar imaging sequence. The modified recovery curve obtained can be fitted to calculate a $T_1$ map. The mean systematic

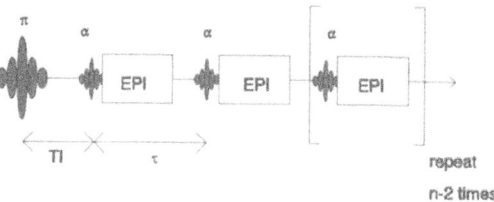

*FIGURE 5.* The Lock–Locher echo-planar imaging (LL-EPI) sequence. Following a single inversion pulse, the signal from a slice is sampled many times, allowing $T_1$ to be measured.

error obtained with this sequence over a $T_1$ range of approximately 100–1300 msec was 7% (Gowland and Mansfield, 1993).

If there is perfusion into a voxel, then the spins within the voxel will continually be being replaced with new spins flowing in. Therefore, if the longitudinal magnetization within the voxel is inverted, its apparent rate of recovery will be increased by the equilibrium magnetization flowing in. Conversely, if the magnetization that is flowing in had also been inverted at the same time, then the $T_1$ measured in the slice will be unaffected by perfusion. By comparing $T_1$ values measured in a slice when the inflowing magnetization is inverted with those when it is at equilibrium, the rate of perfusion can be calculated (Detre *et al.*, 1992). These principles have recently been used to monitor perfusion, quantitatively in animals and qualitatively in humans, using EPI (a technique known as EPISTAR) (Edelman *et al.*, 1994). Both sequences involve saturating or inverting the magnetization in an artery feeding the capillaries in the voxel, or in an adjacent slice to the voxel, and then observing the inflow of this blood into the slice of interest.

## 4.2. Diffusion Mapping Using EPI

It has been established that following stroke there is an acute decrease in the apparent diffusion coefficient of water followed by a later increase in the apparent diffusion coefficient (Moseley *et al.*, 1990). The MR signal can be made sensitive to the diffusion of water molecules by the application of gradient pulses during the RF pulse sequence (Fig. 6; Stetskal and Tanner, 1965). For stationary spins, the transverse magnetization created by the 90° pulse is dephased when the gradient is first applied, flipped by the 180° pulse, and then rephased when the gradient is applied a second time. However, diffusing spins will have moved to a different position in the gradient by the time of the second application and are therefore not totally rephased. The amount of phase gained by each spin at the end of the sequence depends on how far it has diffused in the direction of the gradient, and, as diffusion is a random process, this implies that each spin will have gained a randomly varying phase. Therefore, the total transverse magnetization (which is the sum of the transverse magnetization available from each spin) will be attenuated by an amount that depends on the average distance diffused by the molecules and hence on the diffusion coefficient. This sequence can be performed using conventional MRI, but it is very sensitive to motion artifacts, and, as with $T_1$ measurements, the time required to collect sufficient images, each with a different diffusion weighting, to produce a diffusion coefficient map is prohibitively long. Therefore, diffusion measurements are made ideally using EPI (Turner and Le Bihan, 1990). $T_2$ attenuation of the signal is always a problem with EPI and can be very severe in diffusion measurements where $T_2$ decay also occurs during the diffusion encoding period. This effect can be minimized by using a spin-echo diffusion sequence as described above. If $T_1$ is long enough, a stimulated echo sequence can also be used to reduce $T_2$ decay.

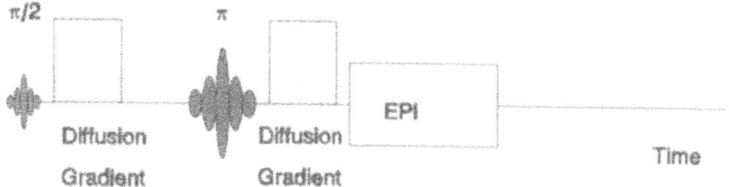

FIGURE 6. The echo-planar imaging (EPI) diffusion sequence.

It is often assumed that diffusion measurements by EPI will be adversely affected by contributions to diffusion weighting from the large imaging gradients, but there are several reasons why this need not be a problem. If the diffusion gradient is parallel to the switched EPI gradient, there is a negligible contribution to the diffusion encoding from the imaging gradients as they are rapidly rephased at each gradient echo; this is not the case for the blipped gradient, which is refocused at the center of $k$ space. However, it is simple to eliminate cross terms between the imaging and diffusion gradients by ensuring that the net phase accumulated by spins under the imaging gradient is zero at the time that the diffusion gradient in that direction is applied (and vice versa). In that case, as the amplitude of the diffusion gradients is varied, the imaging gradient will only have a constant additive effect on the value on diffusion weighting, and this effect can be ignored as the diffusion coefficient is found as the slope of the plot of signal amplitude against diffusion weighting (Neeman et al., 1990; Hong and Dixon, 1992). In practice, this simply means completing the refocusing of the slice selection gradient before the application of diffusion gradients and applying the preparation switched and blipped gradients after the completion of the diffusion gradient.

## 4.3. Flow

EPI is particularly useful for measuring flow in vivo as the high-speed nature of the technique is especially valuable when the flow rate is changing (e.g., during the cardiac cycle). A rapid technique like EPI can obtain a snapshot of flow velocities, in contrast to slower techniques that can be used either in gated mode, which is very slow and relies on a regular heart rhythm, or to obtain an average velocity.

A preparation pulse sequence similar to that used for the measurement of diffusion (Fig. 6) has also been used to measure flow (Firmin et al., 1989). In this study the slice-select and phase-encoding gradients were gradient moment nulled to the center of $k$ space to prevent signal dephasing within a voxel. At least two measurements were required, the first obtained with the additional flow-encoding gradients applied and the second obtained without them; the resulting phase maps were subtracted, yielding a map in which the phase $\phi$ is proportional to velocity, $v$, and is given by

$$\phi = 2\gamma G\tau^2 v \qquad (14)$$

where $\tau$ is the time of each application of the gradient.

An alternative approach is that of Guilfoyle *et al.* (1991). Velocity sensitization is achieved in the same way, but in this case an additional 90° pulse is included at the end of the sequence. Two images are collected, with the phase of this final 90° pulse shifted by 90° with respect to the first pulse. In the first image (phase of both 90° pulses the same), the signal in the final image is given by $M_{stat} + M_v \cos \phi$ (where $M_{stat}$ is the magnitude of static magnetization and $M_v$ is the magnitude of moving magnetization and where it is assumed that each pixel contains either static spins or spins moving with a constant velocity, but not both). In the second image, the signal is given by $M_v \sin \phi$, and therefore this is just an image of the moving spins, allowing quick and simple angiography. However, both images can be combined to quantify flow accurately. An advantage of this technique is that data are evaluated from magnitude images, thereby minimizing sensitivity to phase errors. As the flow-encoding gradients are present for both images, there is less sensitivity to eddy current effects. Once an initial pair of images has been collected, allowing a calculation of $M_v$, it is possible, in principle, to monitor dynamic changes in velocity from a single image using only the $M_v \sin \phi$ component.

It is possible to extend all these methods to measure the velocity distribution within a voxel, by systematically incrementing the amplitude of the velocity-encoding gradients, reconstructing each image, and then Fourier transforming along the third dimension. This has two benefits: it prevents partial-voluming of flow information where this may be relevant, and the width of the peaks in the velocity dimension gives information about diffusion (or perfusion) in the voxel (Callaghan & Xia, 1991). To be able to apply this technique *in vivo* in a reasonable imaging time, EPI must be used.

When imaging flow, it must be remembered that all MR imaging techniques are susceptible to artifacts in the presence of flow, and although gradient moment nulling can reduce some effects, it is rarely totally successful *in vivo* because the movement is frequently turbulent and it is not practically possible to null above the first-order gradient moment within a reasonable echo time. EPI is not immune to these problems. The severe motion artifacts that can be observed in spin warp imaging, caused by unpredictable random variations in position between phase-encoding steps, are not observed in EPI as the motion during the course of the experiment is usually insignificant. However, high rates of flow can lead to image artifacts. Flow along the slice-select axis leads to signal loss, which is easily compensated by gradient moment nulling of this gradient. The switched read gradient exhibits even echo refocusing for constant flow, but the dephasing on odd echoes gives rise to ghosts in the image. This can be overcome by rearranging the sequence to collect data only on even echoes. Blurring due to quadratic phase accumulation (i.e., acceleration effects) of moving spins during the read gradient is insignificant in the times used in EPI. The blipped phase-encoding gradient is analogous to the read gradient in spin warp imaging. Motion during this gradient leads firstly to constant phase increments, which cause signal attenuation in areas of varying flow rates and can be removed by simple gradient moment nulling to the center of $k$ space (Firmin *et al.*, 1989). Secondly, it causes quadratic phase increments, which leads to shifting and broadening of the point spread function. This can be overcome by design of phase-encoding gradients that encode with the position held by the spins at a single instant of time and compensate for quadratic phase increments, at the expense of greater gradient amplitude (Butts and Riederer, 1992; Duerk and Simonetti, 1991). In practice, these artifacts are not significant *in vivo* (where the maximum flow rate is 20 cm/sec in the aorta) for EPI implemented on a dedicated EPI scanner, with high gradient amplitudes and rapid gradient switches so that the phase gain during the acquisition due to the flow is small.

## 4.4. Chemical Shift Imaging

MR spectroscopy can be used to determine biochemical changes *in vivo*. Nowadays, chemical shift imaging (CSI) is used routinely to obtain this information with regional localization. Unfortunately, conventional CSI is very slow, because it is a three-dimensional technique, and $n \times n$ or $n \times m$ shots are required if $n$ pixels are required in both spatial directions and $m$ points are required in the chemical shift direction of the final data set. The time penalty can be greatly reduced or overcome by using EPI. Two approaches have been described: the first is an extremely rapid one-shot technique, known as echo-planar shift mapping (EPSM) (Mansfield, 1984; Guillfoyle and Mansfield, 1985; Posse *et al.*, 1995), and the second is a compromise, $n$-shot technique, with an acquisition time comparable to that of a conventional spin-warp imaging. The latter approach has been realized as projection reconstruction echo-planar (PREP) imaging (Doyle and Mansfield, 1987; Bowtell *et al.*, 1989) and phase-encoded echo-planar (PEEP) imaging (Guilfoyle *et al.*, 1989).

EPSM is a true one-shot technique, using two switched gradients (one switching $n$ times as fast as the other). Intrinsic broadening during the experiment is produced by the chemical shifts experienced by the nuclei. So far, EPSM has been attempted only at low field strengths, and as all the information must be collected in less than 150 msec before total loss of coherence of transverse magnetization ($T_2^*$ decay), the resulting chemical shift images can have only low spatial and chemical shift resolution. However, with current technology it is probable that the 24-mm spatial resolution and 8-ppm/point spectral resolution reported so far (Guifoyle and Mansfield, 1985) could be improved on.

In PREP, chemical shift encoding is provided by spin phase evolution during the echo-planar module (i.e., no broadening or blipped echo-planar gradient is applied), and spatial encoding is provided by a standard switched echo-planar gradient, the direction of which is altered with each experiment to allow an image to be produced by back-projection reconstruction. The problem with this technique is that the two imaging gradient coils must have identical time constants so that the switched gradient maintains a constant orientation throughout each experiment. The advantage is that there is no phase encoding period after slice selection leading to $T_2^*$ attenuation of the signal. Proton images have been acquired with this technique with a chemical shift resolution of 1.0 ppm and a spatial resolution of 0.6 mm in 77 sec for a single pass with 16 averages (Bowtell et al., 1989). In PEEP, spatial encoding is obtained by the application of a phase-encoding gradient pulse between the RF pulse and the rest of the imaging gradient together with a standard switched echo-planar imaging gradient. Spectral encoding is achieved as in PREP. The advantage of this technique over PREP is that $k$ space is sampled uniformly. In all these techniques, the spectral resolution is inversely proportional to the total sampling time of the echo train ($T_S$), and the number of points in each chemical shift spectrum is given by the number of times the imaging gradient is switched. As with all MR spectroscopy techniques, absolute quantification in terms of metabolite concentration is very difficult. PREP has also been applied to brain imaging (Posse et al., 1995).

## 4.5. Echo-Volumar Imaging

The ultimate method of high-speed MR imaging is echo-volumar imaging (EVI), in which a whole volume of data is acquired following a single thick-slice excitation pulse (Mansfield et al., 1989). This pushes the technique of MR to its physical limits. Over the last few years, the array size has been increased from $16^3$ voxels for the first head studies performed at 0.1 T (Mansfield et al., 1989) to $64 \times 32 \times 8$ voxels for whole-body studies at 0.5 T (Harvey and Mansfield, 1996). More recently, EVI results have been obtained at 3.0 T; these consist of volume arrays comprising $8 \times 64 \times 32$ and $8 \times 64 \times 64$ voxels collected in times ranging from 102 to 128 msec (Mansfield et al., 1995; Hykin et al., 1995).

## 5. CLINICAL APPLICATIONS IN NEUROLOGY

### 5.1. Functional Imaging

The recent explosion in the number of scanners in the world that are capable of performing EPI has largely been fueled by the discovery that EPI is fortunately extremely well suited to monitoring brain activation. This is because the intrinsically long echo time of EPI makes it very sensitive to changes in the local susceptibility of tissues. It appears that, following activation of a region of the brain, the local blood flow increases, overcompensating for the increased metabolism and leading to a local decrease in the amount of deoxyhemoglobin in the tissue. Deoxyhemoglobin is more paramagnetic than the oxygenated form of the molecule, and this leads to an increase in the local susceptibility dephasing, which is proportional to field strength. However, the resulting change in $T_2$ is only small and even at high field can only be detected with a long-echo-time technique. Furthermore, as transient effects are to be studied, a high-speed technique is required. The conventional method of studying brain activation has been positron-emission tomography (PET) studies of blood flow and metabolism using $^{15}O_2$ and $[^{18}F]$-FDG (fluorine-labeled glucose) tracers; however, PET images have a limited resolution of 5 mm, and, because of radiation dose considerations, PET studies cannot be easily repeated on the same subject.

Many results have been reported already for different activation paradigms (Society of Magnetic Resonance in Medicine, Book of Abstracts, 1992, 1993, 1994, and 1995). To fully realize the exciting potential of this technique and to carry the subject forward, the imaging methodology must be fully exploited to give the highest achievable quality images with high spatial and temporal resolution (Mansfield *et al.*, 1994). Using EVI, it has recently been possible to determine the temporal differences in functional activation between different regions of the brain (Hykin *et al.*, 1995). It will also be necessary to gain a better understanding of the mechanism of the signal change that occurs with brain activation, and this may be achieved by using alternative MR sequences that are sensitive to flow rather than blood oxygenation, such as EPISTAR. The application of this technique is discussed fully in Chapter 9.

## 5.2. Cerebrospinal Fluid Flow

The cerebrospinal fluid (CSF) slowly flows from its point of secretion in the choroid plexus, through the ventricular system, to its point of absorption in the arachnoid villi, which are associated with the major dural sinuses. More rapid pulsatile motion, controlled primarily by movement of the brain, which is driven by changes in intracranial blood volume associated with the respiratory and cardiac cycles, is superimposed upon this flow (Bergstrand *et al.*, 1985; Stehling *et al.*, 1991). There has been speculation about the origins of the CSF pump for many years, but recently the exact relationship between the movement of the brain and the flow of the CSF has been partly explained using MRI velocity measurements of brain tissue during the cardiac cycle (Poncelet *et al.*, 1992; Enzmann and Pelc, 1992). In general, the pulsatile increase in blood volume leads to compression of the ventricles by the brain parenchyma. The brain stem also moves downward, driving CSF out of the brain. The resulting bidirectional movement of the CSF allows mixing and prevents pulsatile pressure from being exerted on the brain tissues. The clinical interest in this subject arises from its relevance to the study of normal pressure hydrocephalus (NPH). Recent studies have suggested that abnormal brain movement occurs in NPH. It is hypothesized that this movement impedes the expulsion of CSF from the brain. This would lead to an increase in pressure in the periventricular brain, leading to ischemia (Feinberg, 1992).

To make useful measurements of brain velocity, velocities must be monitored in all three spatial dimensions, the results must be corrected for bulk movement of the head (Poncelet *et al.*, 1992), and high spatial, temporal, and velocity resolution are required (Feinberg, 1992). To fully describe normal brain motion and its relationship to CSF flow and vascular blood flow, brain tissue velocity measurements will be required in all three directions, across the entire brain and throughout the entire cardiac cycle, and will probably have to be combined with perfusion studies. Such studies will only be feasible if a one-shot technique like EPI is used. The use of EPI will also allow the study of transient movements in the brain in response to applied perturbations, such as rapid decelerations.

## 5.3. Perfusion Imaging in Cerebrovascular Disease

The changes in signal intensity in a tissue following the intravenous administration of a contrast agent can be modeled in terms of the nature of the capillary system of the tissue being examined (Stehling *et al.*, 1989a). One way of studying this is to monitor the $T_2^*$ changes that occur as the bolus of contrast agent makes its first pass through the capillary bed. The signal intensity drops and then recovers over about 20 sec. If this signal intensity–time curve is converted to concentration, then the area under the curve can be related to the cerebral blood volume. By deconvoluting the measured curve from the arterial input function, the mean transit time can also be estimated (Rosen *et al.*, 1990; Belliveau *et al.*, 1990). This technique has been applied to the measurement of brain perfusion and could be applied to study therapeutic regimes in arterial disease.

## 5.4. Contrast Agent Uptake to Characterize Brain Tumors

Most contrast agents are constrained to the capillaries in the normal brain, but they may leak into surrounding tissues in regions where the blood–brain barrier has been breached (for instance, in some tumors or in demyelinating disease). The changes in $T_1$ due to the inflow of paramagnetic contrast agents can be used to monitor tissue parameters such as cerebral blood volume (Stehling *et al.*, 1989a; Gowland *et al.*, 1992; Worthington *et al.*, 1991) and blood–brain barrier permeability (Stehling *et al.*, 1989a; Gowland *et al.*, 1992; Worthington *et*

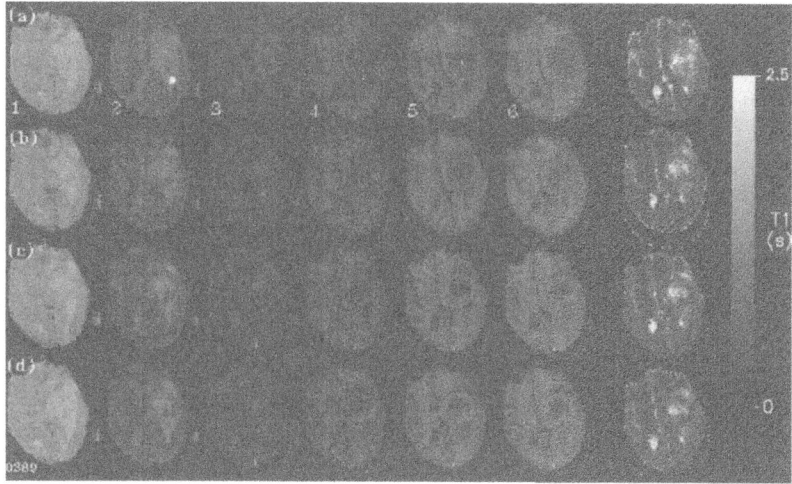

FIGURE 7. Contrast agent uptake images in a brain tumor. Each row contains six LL-EPI images (averages of four sets) and a corresponding $T_1$ map. The rows were collected before (a) and 30 sec (b), 132 sec (c), and 180 sec (d) after injection of Prohance contrast agent.

al., 1991; Tofts and Kermode, 1991; Larrson et al., 1990). When adequate temporal resolution is used, it is observed that the uptake curve obtained generally falls into two phases, an initial rapid phase and a later slow phase. The form of the rapid initial phase of the inflow is dominated by cerebral blood volume and tissue perfusion; only high-speed imaging gives adequate temporal resolution to monitor this, particularly if true quantitation is required, necessitating the dynamic measurement of $T_1$ at 3-sec intervals or less (Freeman et al., 1994). Figure 7 shows a set of $T_1$ maps collected after injection of a gadolinium-based contrast agent, Prohance. The form of the slow phase depends on blood–brain barrier permeability. The information available from both phases has been applied to the problem of characterization of tumors (Gowland et al., 1992; Worthington et al., 1991).

To accurately model the uptake curve, the arterial input function must be known with high temporal resolution. With EPI it is possible to monitor this directly by measuring the signal in a vessel in the same field of view as the region of interest.

## 5.5. Diffusion Measurements in Stroke

Molecular diffusion can be used to obtain information about the structures of tissues. In particular, it has been found that diffusion in the brain is anisotropic, with the diffusion coefficient being up to three times greater along the direction of the nerve fibers compared with that across them (Hajnal *et al.*, 1991). This can be studied by altering the direction in which the diffusion gradient is applied.

Most excitingly, it has been observed that the diffusion coefficient changes rapidly following ischemic injury. This occurs prior to changes in any other MR image parameters, such as $T_2$, which is known to change later after ischemic injury. It is expected that this will be very useful in the evaluation of stroke and in determining whether therapy is appropriate (Knight *et al.*, 1994).

Only with EPI can diffusion be measured accurately without errors due to movement, which is a problem even in the brain. Careful design of the sequence can ensure that the imaging gradients will not contribute to the diffusion weighting in such a way that it will introduce errors into the measurement of the diffusion coefficient.

## 5.6. Multiple Sclerosis

MRI is very successful at demonstrating multiple sclerosis plaques, and contrast-agent-enhanced MRI can determine which plaques are active at a given time. However, there is currently very little correlation between the apparent lesion load in the brain and the clinical condition of the patient. Recent studies looking at the relaxation times of the normal-appearing white matter suggest that much of the lesion load may exist as areas of disease whose size is of the order of the volume of a pixel only (Barbosa *et al.*, 1994). The functional relevance of these minute lesions remains to be determined, but it is clear that to study this further it is essential to have echo-planar methods to produce relaxation time maps of the whole brain in a reasonable time.

It has recently been demonstrated that within multiple sclerosis plaques there is a general increase in the diffusion coefficient and an increase in the anisotropy of the molecular diffusion in the brain (Graham *et al.*, 1995). The significance of this result, that is, whether changes in diffusion coefficient will relate better to the functional deficit of the patient, has not been established yet.

The blood–brain barrier is often broken down in multiple sclerosis plaques. Consequently, contrast agents are administered during MRI scans to study the number of enhancing lesions the patient has. This technique can be refined by collecting sequential scans to monitor the rate of uptake of contrast agent and hence to measure the blood–brain barrier permeability (Tofts and Kermode, 1991). Although the rate of uptake into plaques is usually slow enough that the uptake does not need to be studied by EPI, various refinements to the technique, such as monitoring the arterial input function, may require the use of EPI. This application is discussed more fully in Chapter 10.

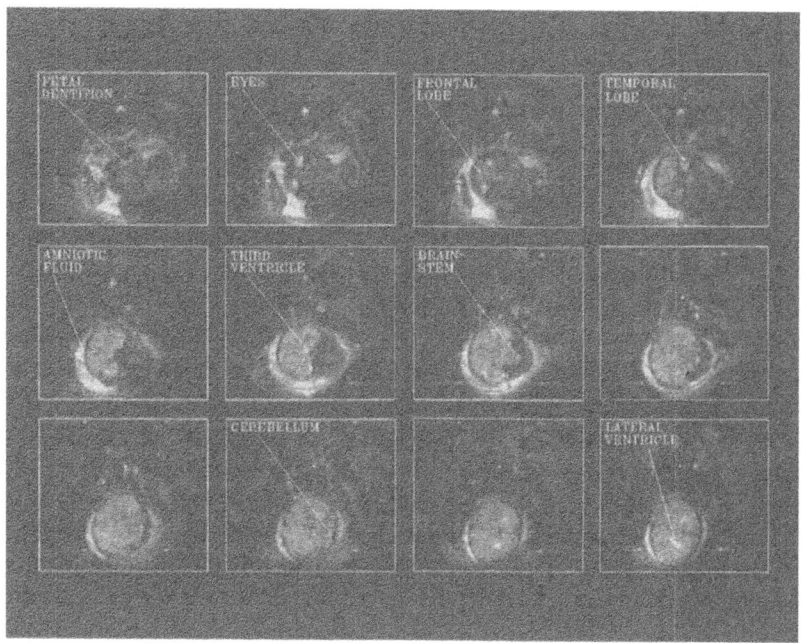

*FIGURE 8.* Brain of a 36-week fetus, showing ventricles.

## 5.7. Fetal Imaging

High-speed imaging is essential for imaging fetuses because of the random and unpredictable nature of their movements. EPI has been used to study fetal morphology in normal and pathological cases (Mansfield *et al.*, 1990; Stehling *et al.*, 1990b). The fetal brain gives high signal intensity on standard MBEST images because of the high water content of the unmyelinated tissue. The brain is clearly differentiated from amniotic fluid by the lack of signal from the skull, and the two hemispheres can be distinguished (Fig. 8). The base of the skull and eye orbits can be seen. The cerebrospinal fluid space in the spinal canal is always clearly seen. Recently, fetal growth has been studied by EPI (Baker *et al.*, 1994a,b). In this work the volume of the fetus has been estimated by planimetry of the images, and this has been used to predict fetal birth weight, to study fetal lung development and to measure organ growth rates in normal pregnancies and those suffering from intrauterine growth retardation (IUGR). The latter study confirmed the previous finding that in IUGR the brain is spared and grows normally whereas the liver size is significantly reduced. Liver size was reported to be a better indicator than total fetal size alone of whether a baby would ultimately be deemed to have suffered IUGR (Baker *et al.,* 1994a).

## 5.8. Interventional Imaging

With new magnet designs, interventional MRI is just starting to reach the stage of clinical utility. However, to fully realize its potential, true real-time EPI will be required (Gowland and Mansfield, 1991). The ability of EPI to image oblique slices, which can be steered automatically by the surgeon (in a similar fashion to real-time ultrasound), combined with the excellent resolution and controllable soft-tissue contrast of EPI and its safety for both patient and operator, will make MRI the method of choice for many interventional procedures. MRI will be used increasingly to guide biopsies and surgery and to monitor drug delivery in real time, without the complications of stereotactic techniques. Furthermore, changes in the MR properties of the tissues will be used to monitor changes in the tissues in response to therapy in real time; for instance, both the diffusion coefficient and $T_1$ are dependent on the temperature, and changes in these parameters could be used to monitor response to hyperthermia and laser therapies.

## REFERENCES

Baker, P. N., Johnson, I. R., Gowland, P. A., Hykin, J., Adams, V., Mansfield, P., and Worthington, B., 1994a, Measurement of fetal liver, brain and placental volumes with echo-planar magnetic resonance imaging, *Br. J. Obstet. Gynaecol.* **102:**35–39.

Baker, P. N., Johnson, I. R., Gowland, P. A., Hykin, J., Harvey, P. R., Freeman, A., Adams, V., Worthington, B. S., and Mansfield, P., 1994b, Fetal weight estimation using echo-planar magnetic resonance imaging, *Lancet* **343:**644–645.

Barbosa, S., Blumhardt, L. D., Roberts, N., Lock, T., and Edwards, R. H. T., 1994, Magnetic resonance relaxation time mapping in multiple sclerosis: normal appearing white matter and the invisible lesion load, *Magn. Reson. Imaging* **12:**33–42.

Belliveau, J. W., Rosen, B. R., Kantor, H. L., Rzedzian, R. R., Kennedy, D. N., McKinstry, R. C., Vevea, J. M., Cohen, M. S., Pykett, I. L., and Brady, T. J., 1990, Functional cerebral imaging by susceptibility-contrast NMR, *Magn. Reson. Med.* **14,** 3:538–546.

Belliveau, J. W., Kennedy, D. N., McKinstry, R. C., Buchbinder, B. R., Weisskoff, R. M., Cohen, M. S., Vevea, J. M., Brady, T. J., and Rosen, B. R., 1991, Functional mapping of the human visual cortex by magnetic resonance imaging, *Science* **254:**716–719.

Bergstrand, G., Berstrom, M., Nordell, B., *et al.,* 1985, Cardiac gated MR imaging of cerebrospinal fluid flow, *J. Comput. Assist. Tomogr.* **9:**1003.

Bowtell, R., Cawley, M. G., and Mansfield, P., 1989, Proton chemical-shift mapping using PREP, *J. Magn. Reson.* **82:**634–639.

Bruder, H., Fischer, H., Reinfelder, H. E., and Schmitt, F., 1992, Image reconstruction for echo-planar imaging with non-equidistant $k$-space sampling, *Magn. Reson. Med.* **23:**311–323.

Butts, K., and Riederer, S. J., 1992, Analysis of flow effects in echo-planar imaging, *J. Magn. Reson. Imaging* **2:**285–293.

Callaghan, P. T., and Xia, Y., 1991, Velocity and diffusion imaging in dynamic NMR microscopy, *J. Magn. Reson.* **91**:326.

Chapman, B., and Mansfield, P., 1986, Double active magnetic screening of coils in NMR, *J. Phys. D: Appl. Phys.* **19**:L129–L131.

Chapman, B., Turner, R., Ordidge, R. J., Doyle, M., Cawley, M., Coxon, R., Glover, P., and Mansfield, P., 1987, Real-time movie imaging from a single cardiac cycle by NMR, *Magn. Reson. Med.* **52**:246–254.

Detre, J. A., Leigh, J. S., Williams, D. S., and Koretsky, A. P., 1992, Perfusion imaging, *Magn. Reson. Med.* **23**:37–45.

Doyle, M., and Mansfield, P., 1986, Real-time movie enhancement in NMR, *J. Phys. E: Sci. Instrum.* **19**:439–444.

Doyle, M., and Mansfield, P., 1987, Chemical shift imaging: A hybrid approach, *Magn. Reson. Med.* **5**:255–261.

Duerk, J. L., and Simonetti, M. S., 1991, Theoretical aspects of motion sensitivity and compensation in echo-planar imaging, *J. Magn. Reson. Imaging* **1**:643–650.

Edelman, R. R., Siewert, B., Darby, D. G., Thangaraj, V., Nobre, A. C., Mesulam, M. M., and Warach, S., 1994, Qualitative mapping of cerebral blood flow and functional localization with echo-planar MR imaging and signal targeting with alternating radio frequency, *Radiology* **192**:513–520.

Edelstein, W. A., Hutchinson, J. M. S., Johnson, G., and Redpath, T., 1980, Spin warp NMR imaging and applications to human whole body imaging, *Phys. Med. Biol.* **25**:751–756.

Enzmann, D. R., and Pelc, N. J., 1992, Brain motion: Measurement with phase contrast MR imaging, *Radiology* **185**:653–660.

Feinberg, D., 1992, Modern concepts of brain motion and cerebrospinal fluid flow, *Radiology* **185**:630–632.

Firmin, D. N., Klipstein, R. H., Houndsfield, G. L., Paley, M. P., and Longmore, D. B., 1989, Echo-planar high resolution flow velocity mapping, *Magn. Reson. Med.* **12**:316–327.

Freeman, A., Gowland, P., Jellinek, K., Wilock, D., Firth, J., Worthington, B., Mansfield, P., and Radcliffe, G., 1994, Dynamic studies of gadolinium update in brain tumours using LL-EPI, *Magma* **2**:409–413.

Garroway, A., Grannell, R. K., and Mansfield, P., 1974, Image formation in NMR by a selective irradiative process, *J. Phys. C: Solid State Phys.* **7**:L453–L462.

Gowland, P., and Mansfield, P., 1991, Real time interventional echo-planar imaging, *in* "Proceedings of the 8th Annual Congress of the European Society of Magnetic Resonance in Medicine, Zurich," p. 263.

Gowland, P. A., and Mansfield, P., 1993, Accurate measurement of $T_1$ in vivo in less than 3 seconds using echo-planar imaging, *Magn. Reson. Med.* **30**:351–354.

Gowland, P., Mansfield, P., Bullock, P., Stehling, M., Worthington, B., and Firth, J., 1992, Dynamic studies of Gd-DTPA uptake in brain tumors using echo-planar imaging, *Magn. Reson. Med.* **22**:241.

Graham, G. D., Zhong, J., Guarnaccia, J. B., and Gore, J. C., 1995, Echo-planar imaging of water directional diffusion and diffusion anisotropy within multiple sclerosis plaques, *in* "Proceedings of the Third Annual Meeting of the Society for Magnetic Resonance," p. 278.

Guilfoyle, D. N., and Mansfield, P., 1985, Chemical shift imaging, *Magn. Reson. Med.* **2**:478–489.

Guilfoyle, D. N., Blamire, A., Ordidge, R. J., and Mansfield, P., 1989, PEEP—a rapid chemical-shift imaging method, *Magn. Reson. Med.* **10**:282–287.

Guilfoyle, D. N., Gibbs, P., Ordidge, R. J., and Mansfield, P., 1991, Real-time flow measurements using echo-planar imaging, *Magn. Reson. Med.* **18**:1.

Hajnal, J. V., Doran, M., Hall, A., Collins, A. G., Cartridge, A., Pennock, J. M., Young, I. R., and

Bydder, G. M., 1991, MR imaging of anisotropically restricted diffusion of water in the nervous system: Technical, anatomic and pathologic considerations, *J. Comput. Assist. Tomogr.* **15**:1–18.

Harvey, P. R., and Mansfield, P., 1994, Resonant trapezondal gradient generation for use in echo planar imaging. *Magn. Reson. Imaging* **12**:93–100.

Harvey, P. R., and Mansfield, P., 1996, Echo volumar imaging (EVI) at 0.5T: First whole body volunteer studies, *Magn. Reson. Med.* **35**:80–88.

Hennig, J., Nauerth, A., and Friedburg, H., 1986, RARE imaging: A fast imaging method for clinical MR, *Magn. Reson. Med.* **3**:823.

Hong, X., and Dixon, T., 1992, Measuring diffusion in inhomogeneous systems in imaging mode using antisymmetric sensitizing gradients, *J. Magn. Reson.*

Howseman, A. M., Stehling, M. K., Chapman, B., Coxon, R., Turner, R., Ordidge, R. J., Cawley, M. G., Glover, P., Mansfield, P., and Couplans, R. E., 1988, Improvements in snap-shot nuclear magnetic resonance imaging, *Br. J. Radiol.* **61**:822–828.

Hykin, J., Coxon, R., Bowtell, R., Glover, P., and Mansfield, P., 1995, Temporal differences in functional activation studies between separate regions of the brain investigated with single shot echo volumar imaging at 3.0 T, Third Annual Meeting of the Society of Magnetic Resonance, Nice, p. 451.

Johnson, G., and Hutchinson, J. M. S., 1985, The limitations of NMR recalled-echo imaging techniques, *J. Magn. Reson.* **63**:14–30.

King, K. F., Crawford, C. R., and Maier, J. K., 1995, Correction for filter induced ghosts in echoplanar imaging. Third Annual Meeting of the Society of Magnetic Resonance, Nice, p. 105.

Knight, R. A., Dereski, M. O., Helpern, J. A., Ordidge, R. J., and Chopp, M., 1994, *Stroke* **25**:1252.

Larrson, H. B. W., Stubgoard, M., Fredriksen, J. L., Henson, M., Henrikson, O., and Paulson, O. B., 1990, Quantitation of blood brain barrier defect by magnetic resonance imaging and GdDBA in patients with multiple sclerosis and brain tumours, *Magn. Reson. Med.* **16**:112–131.

Lauterbur, P. C., 1973, Image formation by induced local interactions: Examples employing nuclear magnetic resonance, *Nature (London)* **242**:190–191.

Ljunggren, S., 1983, A simple graphical representation of Fourier based imaging methods, *J. Magn. Reson.* **54**:338–343.

Look, D. C., and Locher, D. R., 1970, *Rev. Sci. Instrum.* **41**:250.1.

Macovski, A., 1985, Volumetric NMR imaging with time varying gradients, *Magn. Reson. Med.* **2**:29.

Mansfield, P., 1977, Multi-planar image formation using NMR spin echoes, *J. Phys. C: Solid State Phys.* **10**:L55–L58.

Mansfield, P., 1984, Spatial mapping of the chemical shift in NMR, *Magn. Reson. Med.* **1**:370–386.

Mansfield, P., and Grannell, P. K., 1973, NMR 'diffraction' in solids?, *J. Phys. C: Solid State Phys.* **6**:L422.

Mansfield, P., Maudsley, A. A., Morris, P. G., and Pykett, T. L., 1979, Selective pulses in NMR imaging: A reply to a criticism, *J. Magn. Reson.* **33**:261–274.

Mansfield, P., and Harvey P. R., 1993, Limits to neural stimulation in echo planar imaging, *Magn. Reson. Med.* **29**:746–758.

Mansfield, P., Howseman, A., and Ordidge, R., 1989, Volumar imaging using NMR spin echoes, echo-volumar imaging at 0.1 T, *J. Phys. E: Sci. Instrum.* **22**:324–330.

Mansfield, P., Stehling, M. K., Ordidge, R. J., Coxon, R., Chapman, B., Blamire, A., Gibbs, P., Johnson, I. R., Symonds, E. M., Worthington, B. S., and Coupland, R. E., 1990, Echo-planar imaging of the human fetus *in utero* at 0.5T, *Br. J. Radiol.* **63**:833.

Mansfield, P., Coxon, R., and Glover, P., 1994, Echo-planar imaging of the brain at 3.0T: First normal volunteer results, *J. Comput. Assist. Tomogr.* **18**:339–343.

Mansfield, P., Coxon, R., and Hykin, J., 1995, Echo-volumar imaging (EVI) of the brain at 3.0 T: First normal volunteer and functional imaging results, *J. Comput. Assist. Tomogr.* **19**:847–852.

Margosian, P., Faster, 1985, MR imaging: Imaging with half the data. Proceedings of fourth annual meeting of Soc. Magn. Res. Med., 1024.

Moseley, M. E., Cohen, Y., Mintorovitch, J., Chilevitt, L., Shimizu, H., Kucharczyk, J., Wendland, M. F., and Weinstein, P. R., 1990, Early detection of regional cerebral ischemia in cats—comparison of diffusion weighted and $T_2$ weighted MRI and spectroscopy, *Magn. Reson. Med.* **14**:330.

Neeman, M., Freyer, J. P., and Sillerud, L. O., 1990, Pulsed-gradient spin-echo diffusion studies in NMR imaging. Effects of the imaging gradients on the determination of diffusion coefficients, *J. Magn. Reson.* **90**:303–312.

Ordidge, R. J., Howseman, A., Coxon, R., Turner, R., Chapman, B., Glover, P., Stehling, M., and Mansfield, P., 1989, Snapshot imaging at 0.5T using echo-planar techniques, *Magn. Reson. Med.* **10**:227–240.

Ordidge, R. J., Gibbs, P., Chapman, B., Stehling, M. K., and Mansfield, P., 1990, High-speed multislice $T_1$ mapping using inversion-recovery echo-planar imaging, *Magn. Reson. Med.* **16**:238–245.

Poncelet, B. P., Wedeen, V. J., Weisskoff, R. M., and Cohen, M. S., 1992, Brain parenchyma motion: Measurement with cine echo-planar MR imaging, *Radiology* **185**:645–651.

Posse, S., Tedeschi, G., Risinger, R., Ogg, R., and Le Bihan, D., 1995, High speed $^1$H spectroscopic imaging in human brain by echo planar spatial-spectral encoding, *Magn. Reson. Med.* **33**:34–40.

Rosen, B. R., Belliveau, J. W., Vevea, J. M., and Brady, T. J., 1990, Perfusion imaging with NMR contrast agents, *Magn. Reson. Med.* **14**:249–265.

Stehling, M. K., Bullock, P., Firth, J. L., Blamire, A. M., Ordidge, R. J., Coxon, R., Gibbs, P. J., and Mansfield, P., 1989a, Gd DTPA realtime studies of the brain with EPI: A dynamic approach to perfusion and blood brain barrier assessment, *In* "Proceedings of the 8th Annual Meeting of the Society for Magnetic Resonance in Medicine," p. 358.

Stehling, M. K., Evans, D. F., Lamont, G., Ordidge, R. J., Howseman, A. M., Chapman, B., Coxon, R., Mansfield, P., Hardcastle, J. D., and Coupland, R. E., 1989b, Gastrointestinal tract: Dynamic MR studies with echo-planar imaging, *Radiology* **1171**:41–46.

Stehling, M. K., Ordidge, R. J., Coxon, R., and Mansfield, P., 1990a, Inversion-recovery echo-planar imaging (IR-EPI) at 0.5T, *Magn. Reson. Med.* **13**:514–517.

Stehling, M. K., Mansfield, P., Ordidge, R. J., Coxon, R., Chapman, B., Blamire, A., Gibbs, P., Johnson, I. R., Symonds, E. M., Worthington, B. S., and Coupland, R. E., 1990b, Echo-planar imaging of the human fetus *in utero, Magn. Reson. Med.* **13**:314–318.

Stehling, M. K., Firth, J. L., Worthington, B. S., Guilfoyle, D. N., Ordidge, R. J., Coxon, R., Blamire, A. M., Gibbs, P., Bullock, P., and Mansfield, P., 1991, Observation of cerebrospinal fluid flow with echo-planar magnetic resonance imaging, *Br. J. Radiol.* **64**:89–97.

Stetskal, E. O., and Tanner, J. E., 1965, Spin diffusion measurements: Spin echoes in the presence of a time-dependent field echo, *J. Chem. Phys.* **42**:288–292.

Tofts, P. S., and Kermode, C., 1991, Measurement of blood brain barrier permeability and leakage during dynamic MR imaging. I. Fundamental concepts: *Magn. Reson. Med.* **17**, **2**:357–367.

Turner, R., and Le Bihan, D., 1990, Single-shot diffusion imaging at 2.0 Tesla. *J. Magn. Reson.* **86**:445–452.

Turner, R. Chapman, B., Howseman, A. M., Ordidge, R. J., Coxon, R., Glover, P., and Mansfield, P., 1988, Snapshot magnetic resonance imaging at 0.1T using double screened gradients, *J. Magn. Reson.* **80**:248–258.

Worthington, B. S., Bullock, P., Stehling, M., Gowland, P. A., Firth, J. L., and Mansfield, P., 1991, Clinical experience with contrast enhanced echo planar imaging of the brain, *Magn. Reson. Med.* **22**, **2**:255–258.

CHAPTER 9

# BRAIN ACTIVATION STUDIES USING MAGNETIC RESONANCE IMAGING

STEVE C. R. WILLIAMS, ANDREW SIMMONS, CHRIS M. ANDREW, MICK J. BRAMMER, ED T. BULLMORE, and SOPHIA RABE-HESKETH

## SUMMARY

The principles underlying functional magnetic resonance imaging (FMRI) are described, with examples of the various approaches that are employed. These include the use of exogenous (invasive) contrast agents such as gadolinium-

*STEVE C. R. WILLIAMS, ANDREW SIMMONS, and CHRIS M. ANDREW • Department of Clinical Neurosciences, Institute of Psychiatry, De Crespigny Park, London, SE5 8AF, United Kingdom.*
*MICK J. BRAMMER • Departments of Neuroscience and Biostatistics & Computing, Institute of Psychiatry, De Crespigny Park, London, SE5 8AF, United Kingdom.*
*ED T. BULLMORE • Departments of Psychological Medicine and Biostatistics & Computing, Institute of Psychiatry, De Crespigny Park, London, SE5 8AF, United Kingdom.*
*SOPHIA RABE-HESKETH • Department of Biostatistics & Computing, Institute of Psychiatry, De Crespigny Park, London, SE5 8AF, United Kingdom.*

*Magnetic Resonance Spectroscopy and Imaging in Neurochemistry,* Volume 8 of *Advances in Neurochemistry,* edited by Bachelard, Plenum Press, New York, 1997.

DTPA and noninvasive endogenous contrast methods (BOLD magnetic resonance imaging, BOLD magnetic resonance spectroscopy, and arterial spin labeling). The advantages and disadvantages of each are discussed, and technical aspects of the implementation are reviewed. Illustrations of the applications of BOLD magnetic resonance imaging being performed in the authors' MR Unit include investigations of motor tasks, language dominance, verbal working memory, covert verbal fluency, and hallucinations.

## 1. INTRODUCTION

Since the first human brain magnetic resonance (MR) images were produced in the late 1970s, the major focus of the technique has been on probing the anatomy, morphology, and pathology of the central nervous system (CNS). Because the method can be made highly sensitive to flow, many groups, even at this early stage, also postulated extending the potential of MRI to interrogate brain physiology. Observing such changes was, however, limited to large vessels, and detection of "changes in blood supply in accordance with local variations of functional activity" (Roy and Sherrington, 1890) was prohibited by the need to acquire data for a minimum of several minutes.

Brain tasks are, by definition, often short-lived, and the brain deals with many simultaneous sensory inputs which may vary rapidly with time. Hence, in order to achieve MR images reflecting brain function, one must perform extremely rapid MR image acquisition. Conventional MR systems have, in the past, taken several minutes to acquire information from any one anatomical location. To monitor the effect of rapid bursts of neuronal activity in functional magnetic resonance imaging (FMRI), the temporal resolution has to be improved to a maximum collection time of a few seconds. Furthermore, information is, ideally, captured during alternating periods of activity and rest states. Thus, the scanning sequence has to be repeated many times. Fast MRI with conventional hardware has been achieved using fast gradient-echo techniques with such acronyms as FLASH, SPGR, FISP, PSIF, AND SSFP (for a review of fast imaging methods, see Wehrli, 1990).

These gradient-echo methods use radio-frequency pulses which are optimized for experiments with short repetition times (TR) (typically, between 20 and 50 msec) using flip angles of less than 90°. This can result in a total acquisition time of the order of seconds for an image at an in-plane spatial resolution of about 2 mm. The three major disadvantages of these conventional methods for FMRI experiments are (1) the limited slice coverage within a time frame of a few seconds, (2) the serious loss of signal-to-noise ratio due to the necessary reduction of flip angle when compared with a 90° excitation, long-TR scan that "samples" all the available magnetization, and (3) due to the short TR used, a potentially greater contribution from draining vessels at a considerable distance downstream from the activated brain region (Lai *et al.*, 1993).

Currently, we believe the best method available for collecting functional MR data is the less conventional echo-planar imaging (EPI) technique, developed at the University of Nottingham (see Chapter 8). EPI is an ultrafast, snapshot MR imaging method that allows capture of a full, single-slice image after the application of only one radio-frequency pulse in a total acquisition time of less than 100 msec. In order to acquire data in such a short time, there are more demanding hardware requirements. These include very rapidly switching magnetic field gradients, controlled by powerful current amplifiers, and wide-receiver-bandwidth technology to allow an extremely high signal sampling rate. Until recently, there were only 12 sites worldwide that had the capability of producing EP images of the brain that would be of sufficient quality to be regarded as adequate for functional studies, but, since late 1995, the major MR system manufacturers have begun to produce both new systems and upgrades for older machines that have the ability to collect single-shot EP images of the human brain. It is expected that approximately 500 sites worldwide will be installed with EPI capability by late 1997.

EPI does, however, have two major disadvantages, namely, (1) the need for a wide receiver bandwidth, which leads to a reduction in signal-to-noise ratio, and (2) the requirement to sample data for long acquisition times after each excitation, which can lead to increased image distortion due to a higher sensitivity to magnetic field inhomogeneity. Despite the implementation of field mapping and subsequent correction algorithms, the configuration of the object itself introduces magnetic field variations due to differences in magnetic susceptibility between air, bone, and the tissues of interest. At the higher magnetic field strengths necessary for functional MRI, signals from such boundaries, for example, the base of the frontal lobes above the paranasal sinuses and near the cerebellum, can be grossly reduced, increased, or displaced in EP images.

## 2. MR METHODS FOR PROBING BRAIN FUNCTION

### 2.1. Exogenous Contrast Agents

Early developments in functional imaging with MR were hampered by problems of dynamic range. Only some 5% of the total water protons are located in blood flowing through the small cortical vessels from which one might expect a change in signal intensity during increased neuronal activity. Numerous methods to minimize the dynamic range problem have been reported; these include the use of freely diffusible exogenous tracers such as fluorinated agents (Pekar *et al.*, 1994; Ewing *et al.*, 1990; Barranco *et al.*, 1989; Eleff *et al.*, 1988), $D_2O$ (Detre *et al.*, 1990), and $H_2{}^{17}O$ (Pekar *et al.*, 1991), but in most cases low MR sensitivity has resulted in long acquisition times (several minutes) and low spatial resolution (typically, approximately 1 cm³) with these techniques, which have limited their use (Turner and Keller, 1991). The first major breakthrough in functional neuroimaging came with the advent of intravascular paramagnetic contrast agents [e.g., gadolinium diethylenetriaminepentaacetic acid (G-DTPA)]. These lipophobic agents have previously received clinical approval for use in determining the integrity of the blood–brain barrier in, for example, patients with suspected multiple sclerosis, and they have also been used in clinical research for probing the vascularity of certain CNS tumors. For brain activation studies, the exogenous agent is administered intravenously as a rapid bolus (<8 sec), which causes an immediate drop in the MR signal (as much as 50% in well-perfused gray matter) in those tissue areas fed by this "labeled" blood.

If a direct linear relationship is assumed between the fall in MR signal and the blood concentration of the agent (Fisel *et al.*, 1991), then, by integrating the MR signal change in a series of brain images acquired over the time course of the initial wash-in/wash-out of the contrast agent and applying first-pass kinetic modeling, maps of both relative cerebral blood volume (CBV) and mean tracer transit time can be calculated.

The NMR group at Massachusetts General Hospital, Boston, investigated brain function with this exogenous agent during a photic stimulation study (Belliveau *et al.*, 1991). Seven normal subjects were injected twice with an intravenous bolus of Gd-DTPA: once in darkness and once during photic stimulation by light-emitting diodes (LED) flashing at 8 Hz in lightproof goggles placed over the subjects' eyes. Typically 60 EP images from a near-axial, 10-mm slice along the calcarine fissure were acquired over some 45 seconds after each injection. CBV maps were then generated, and a "difference map" was created by subtracting the "OFF" from the "ON" CBV maps. These "difference" data yielded highly significant changes in blood volume within the gray matter of the visual cortex for all subjects. This experiment was not, however, ideal because of the requirement to use repeated, rapid bolus administration of an exogenous agent. There are often problems with provision of a discrete bolus as it passes through the thoracic cavity to the brain, and, if one is to compare functional imaging modalities, MR begins to lose one of the major advantages that this technique offers, in that bolus tracking renders the method invasive. The technique does, however, allow the potential for further quantification because analysis of signal from a slice perpendicular to the major feeding arteries when imaged simultaneously would allow the measurement of the arterial input function (AIF) (Rempp *et al.*, 1994) and, on subsequent deconvolution, yield a measure of cerebral blood flow (CBF).

## 2.2. Endogenous Contrast Methods

### 2.2.1. BOLD Magnetic Resonance Imaging

The dynamic, bolus tracking technique described above has recently been superseded by a method that allows mapping of hemodynamic changes in the brain without the use of any invasive exogenous contrast agent. This method relies on a natural change in the magnetic properties of the blood during neuronal activity. As early as 1936, Pauling and Coryell noted that the magnetic susceptibility of hemoglobin and deoxyhemoglobin differed slightly (Pauling and Coryell, 1936), and Thulborn et al. (1982) demonstrated that the magnetic state of hemoglobin in red cells is strongly dependent upon oxygen saturation. Deoxygenated blood is considerably more paramagnetic than oxygenated blood (0.2 ppm difference; Weisskoff and Kiihne, 1992). Therefore, the MR signal decay rate of deoxygenated blood is more rapid than that of its oxygenated counterpart.

The effects of oxygen on the relaxation time $T_2^*$ were first reported in MR images of anesthetized rats by Ogawa et al. (1990). They noted that cortical blood vessels appeared visibly larger in gradient-echo images obtained at high magnetic field strengths as blood oxygenation was lowered by using different anesthetics. They interpreted this change to be due to deoxygenated blood creating local magnetic field inhomogeneities, and thus signal loss, in a fashion comparable to the previously described intravascular paramagnetic contrast agents. Ogawa et al. termed their experiment "BOLD"—blood oxygen level dependent imaging. These results were elaborated upon by Turner et al. (1991), who used gradient-echo EP imaging to observe the time course of these oxygenation changes while an animal breathed an oxygen-deprived, nitrogen atmosphere. The spatial resolution was insufficient to depict individual vessels, but the drop in MR signal during anoxia within regions of well-perfused gray matter was readily observed.

These initial results gave Kwong and colleagues at Massachusetts General Hospital the necessary fillip to attempt the same photic stimulation study described earlier, but now using a simple gradient-echo EPI method and probing MR signal changes due solely to the endogenous BOLD mechanism (Kwong et al., 1992). An echo time (TE) of 40 msec and a repetition time (TR) of 3 sec were used to obtain a 1-cm-thick axial image through the primary visual cortex at an in-plane resolution of 1.5 mm. Subjects experienced short, 30-sec epochs of stimulus from flashing LED goggles interspersed with equal periods of darkness. When the sum of the "off-state" images was subtracted from those acquired during photic stimulation (the "on-state"), the resulting difference image depicted a signal increase of some 3% in the primary visual cortex. The MR signal increase reflects a reduction in deoxyhemoglobin concentration following the hyperoxemia that is associated with increased neuronal activity. Therefore, the

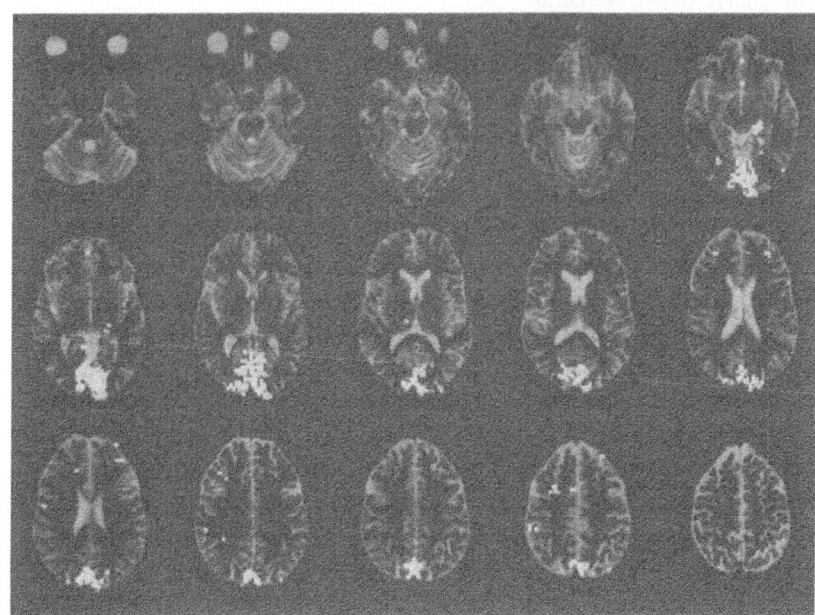

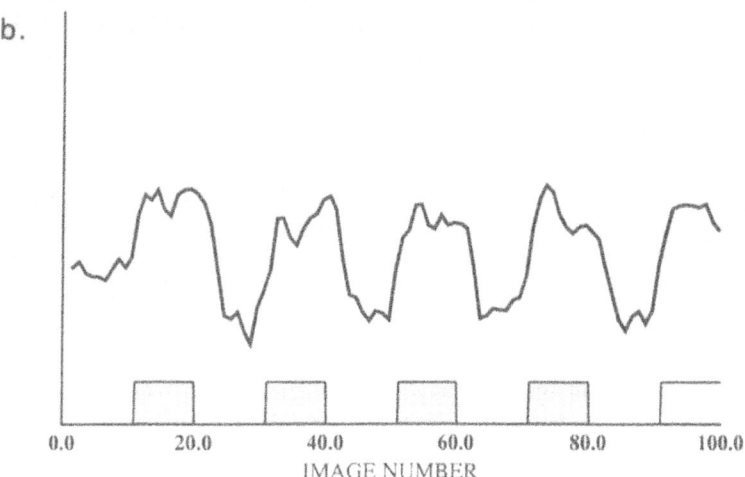

IMAGE NUMBER

*FIGURE 1.* (a) Functional MR signal response to periodic photic stimulation applied by LED goggles flashing at 8 Hz for 30-sec intervals. Pixels that were significantly different between the resting (OFF) state and the stimulus (ON) epochs are depicted in white and are mapped onto a near-axial, high-resolution echo-planar image series (TE = 80 msec, TR = 16,000 msec, TI = 180 msec, 8 Signal averages, 3-mm slice thickness, 1.5-mm in-plane resolution). (b) Conventional boxcar input function for our periodic stimulus presentation—10 images "OFF"/10 images "ON" (*bottom*)—and a typical BOLD response from a region of interest within the visual cortex (*top*). A 3–4% MR signal increase is normally observed during such a stimulus at 1.5 T.

interpretation is that the brain receives a greater supply of oxygen than required (Frostig *et al.,* 1990). This method has now become one of the benchmark experiments for new groups endeavoring to obtain functional MR images from their EPI systems. Comparable results from a similar photic stimulation experiment on our 1.5-T system at our institute in London are shown in Fig. 1. One feature that is apparent from these data is the strong but delayed correlation between the input stimulus and the MR response. The rise to maximum BOLD effect can be delayed by 5–8 sec, reflecting the hemodynamic origin of the change of signal, but significant signal intensity increases can be observed within a few seconds after the onset of the stimulus.

The University of Nottingham group that originally developed EPI at 0.5 T (Chapter 8) is now operating at 3.0 T, obtaining snapshot functional images within 50–60 msec, with in-planar resolution of $0.75 \times 0.75$ mm$^2$. In their system the delay between input stimulus and maximum BOLD effect is about 3 sec (Mansfield *et al.,* 1994a). In order to ensure that the BOLD signal change observed is primarily derived from small vessels surrounded by activated brain tissue, an MR angiogram is often performed so that all signals that can be attributed to major vessels are eliminated (Frahm *et al.,* 1994). Spin-echo echo-planar images (SE-EPI) have been shown to be more specific for probing BOLD effects in small capillary vessels than gradient-echo echo-planar images (GE-EPI), which display equal sensitivity to all vessel sizes (Weisskoff *et al.,* 1993). Therefore, one would predict that asymmetric SE-EPI methods would be more commonly used in practice than GE-EPI. Empirically, however, the requirement for repeated whole-head coverage in functional neuroimaging studies necessitates slow radio-frequency power deposition and minimal cross stalk between contiguous slices. Introduction of an imperfect 180° refocusing pulse in spin-echo sequences can be more deleterious than an optimally calibrated, long-TR (ca. 3000 msec) gradient-echo EPI sequence where flow effects are minimized. There are also concerns about the loss of BOLD sensitivity when an asymmetric spin-echo method is used. Therefore, whereas 3-T and 4-T instruments are available (Chapter 8), most groups are able to use gradient-echo EPI methods at a modest magnetic field strength of 1.5 T when probing more subtle functional responses, while performing multislice experiments.

A novel form of functional imaging (echo-volumar imaging, EVI) allows the whole volume of the brain to be imaged within 100 msec; this advantage is offset by the lower resolution at 3 T of 3.0 mm$^2$, compared with the 0.75 mm$^2$ obtained using EPI at 3 T (Mansfield *et al.,* 1995a).

## 2.2.2. *BOLD Magnetic Resonance Spectroscopy*

A variation of the BOLD sensitive FMRI method is functional magnetic resonance spectroscopy (FMRS), where spatial resolution is sacrificed for much higher temporal resolution (millisecond range) from one selected volume of brain tissue (Ernst and Hennig, 1994; Hennig *et al.*, 1994). A transient shortening of the $T_2^*$ of water in an activated voxel has been observed within 1 sec after a brief visual stimulus, followed by the more commonly observed increase in $T_2^*$. The delayed increase in $T_2^*$ corresponds to the signal intensity rise seen in conventional FMRI and is due to a drop in local deoxygenated hemoglobin concentration. Hence, the early decrease of $T_2^*$ that precedes this rise can be attributed to an early transient deoxygenation of blood in the voxel of interest. Results from this FMRS method are, however, still very preliminary, and replication of these results by other groups using FMRI is currently under way (Menon *et al.*, 1995).

Care must be taken when interpreting BOLD contrast with different methods. It has been suggested that a change in, for example, blood arterial oxygenation, volume, and flow as well as hematocrit and tissue oxygen uptake may alter the extent of BOLD signal change (Turner and Jezzard, 1994), and detailed experimental studies to investigate this further are currently in progress. One encouraging observation has been the potential for increased BOLD contrast at higher magnetic field strengths. Turner *et al.* (1993) observed changes of up to 15% in the primary visual cortex when they repeated the photic stimulation described above at 4 T. This improvement is due to an increased signal-to-noise ratio at the higher field strength and the substantially greater change in magnetic susceptibility due to larger, local magnetic field gradients. Conversely, this also leads to a potential for both increased loss of signal and greater image distortion near air/bone/tissue boundaries at higher fields. Numerous groups are currently developing improved shimming software and hardware in order to optimize the magnetic field homogeneity across the whole brain prior to acquisition of the FMRI experiment (Blamire *et al.*, 1996; Reese *et al.*, 1993). It should be noted that both high-field magnets and higher order shim coils are currently nonstandard and expensive when compared with a 1.5-T system which can also serve as a conventional diagnostic instrument.

## 2.2.3. Arterial Spin Labeling

Another noninvasive MR method to quantify brain perfusion has recently been reported by Detre *et al.* (1992). This method relies on the saturation or inversion of blood flowing into the brain. Two images are collected, one with and one without radio-frequency "tagging" of upstream sources of flow. (Note that the "tag" is due to the spin, rather than an invasive isotope). In the first acquisition, saturated or inverted spins flow from the neck into the region of interest and exchange with local tissue water. This causes a change in the apparent spin–lattice relaxation time $(T_1)$ of the tissue at a rate dependent upon the delivery of labeled spins. Care must be taken with such methods to correct for signal changes caused by off-resonance, radio-frequency irradiation of the bound water pool, commonly termed "magnetization transfer effects." Therefore, a second, control image is usually collected by irradiating an equivalent-sized slab downstream from the desired image at an equal but opposite frequency from the slab used to "tag" inflowing spins. The difference between the two acquisitions results in an image in which the signal is proportional to the amount of perfusion through the imaged slice. Numerous variations on this common theme exist, such as EPIS-TAR (Edelman *et al.*, 1994) and FAIR (Kim, 1995). Since this "label" is freely diffusible, the potential signal change is no longer limited by the volume fraction of the vasculature. The relationship between $T_1$ and regional blood flow can be described by

$$\frac{1}{T_{1app}} = \frac{1}{T_1} + \frac{f}{\lambda}$$

where $T_{1app}$ is the observed spin–lattice relaxation time with flow effects included, $T_1$ is the true tissue longitudinal relaxation time in the absence of flow, $f$ is the flow in milliliters per gram per unit time, and $\lambda$ is the blood–brain partition coefficient of water (ca. 0.95 ml/g).

One potential advantage of these spin labeling techniques is that they use freely diffusible water, which will allow direct comparison with the contrast mechanism in $H_2^{15}O$ positron-emission tomography (PET), a well-established functional neuroimaging method. PET studies, however, use radioactive agents which are injected into the plasma space, whereas the arterial spin labeling method involves only radio-frequency excitation of water molecules both inside and outside blood cells. These flow-dependent FMRI methods using relative differences in the relaxation characteristics of inflowing water have been successfully employed to localize areas of neuronal activity during visual stimulation (Kwong *et al.*, 1992) and sensorimotor activation (Edelman *et al.*, 1994), but most experiments to date have only involved single-slice acquisitions, thus limiting the potential utility of such methods in their present form. Several groups are currently developing appropriate radio-frequency hardware that will allow continuous irradiation of the major feeding arteries in the neck via a dedicated surface coil during multislice data acquisition through the region of interest.

## 3. PRACTICAL CHALLENGES IN IMPLEMENTING FMRI STUDIES

### 3.1. General Considerations

The same contraindications must be considered for an FMRI study as would apply to a patient presenting for a standard clinical MR examination. Over and above such considerations of metallic implants, cardiac pacemakers, early pregnancy, etc., one must also consider acoustic noise. FMRI experiments using either conventional or EP imaging techniques are among the noisiest MR examinations. It is therefore important to carefully protect the subjects' ears prior to commencement of the experiment. Similarly, one must be realistic about the type of FMRI experiments involving auditory perception that can currently be achieved. Acoustic noise is of such concern that there is now considerable effort being devoted to the development of quiet gradient coils, which should overcome much of this disadvantage in the future (Mansfield *et al.,* 1994b, 1995b).

Many FMRI tasks involve the subject looking at either a video recording or a computer projection onto a screen at the end of the magnet bore. Therefore, good eyesight or nonmetallic glasses are often necessary. Contact lenses are considered a contraindication for MR, and many units insist they be removed. Task complexity is also an issue. It is important that all subjects practice the required task prior to their insertion in the magnet. Numerous hours can be lost trying to explain the paradigm "on the fly" while the subject is lying prone in the scanner. One must also be careful not to conduct overelaborate paradigm in normal volunteers that would be impossible in certain patient groups. For example, an unrehearsed complex finger tapping task would be very difficult for most chronic motor neuron disease (MND) patients to perform. There are, of course, certain cognitive studies, for example, involving memory tasks, that preclude rehearsal prior to scanning.

## 3.2. Quality Control

Since BOLD FMRI studies at 1.5 T rely on very small increases in signal intensity during activation, these changes may be masked by either artifacts or temporal system drifts. EPI is particularly susceptible to Nyquist ghosts caused by slight misregistration of alternate lines in $k$ space. Figure 2 illustrates such ghosts, which manifest themselves as a low-signal copy of the image shifted (or "wrapped around") by a distance equal to half the field of view. Appropriate calibration must be carried out on a daily or scan-by-scan basis to eliminate, or at least minimize, these effects. Temporal drifts in signal intensity can also occur due to gradient instabilities or changes in temperature of electronic circuitry. Phantom scans must be carried out on a daily basis to ensure temporal stability. We believe that a signal variation of no greater than 0.5% over a 5-min, 100-time-points examination of a standard sphere phantom using multislice EPI is acceptable for our cognitive FMRI studies. Hardware instability can often be addressed by regular servicing and daily quality assurance. Sophisticated paradigm designs can also help to compensate for linear system drift.

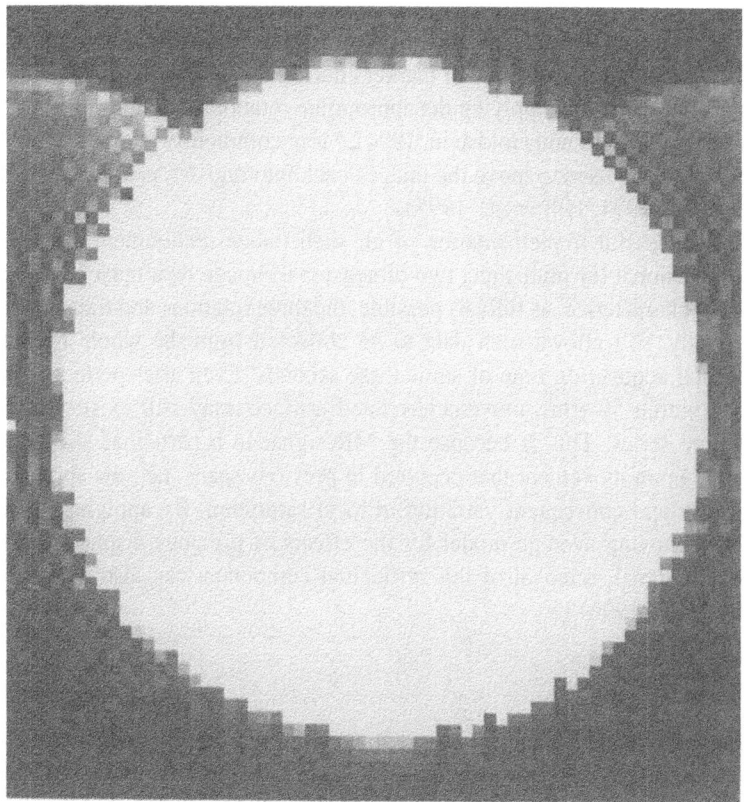

FIGURE 2. Gross Nyquist ghost artifacts created from a standard sphere phantom when collecting a full $k$-space EPI data set where alternate lines of $k$-space ($k_y$) are not perfectly aligned.

## 3.3. Subject Motion

The biggest practical concern in FMRI studies is subject motion. All manner of devices to minimize such movement, including American football helmets and bite-bars, have been tried, but we have found, empirically, that subjects move least if (1) they are made comfortable using minimal binding straps and cushions, (2) the paradigms are kept as short as possible, and (3) they are given the opportunity to rehearse the task beforehand.

Even in those cases in which rigid fixation of the head has been employed, the most willing subject will typically move by at least a millimeter or so during the course of the study. Testing with interpolated data sets has confirmed that even if this movement is less than a pixel on direct image subtraction, a noticeable difference, which may initially be interpreted as brain activation, is still observed. Image realignment must therefore be performed. Early work involved monitoring movements of fiducial markers during the course of an examination and then retrospectively applying the appropriate rotations and translations to the image series (Jezzard and Goldstein, 1994). More commonly, features within the images have been used to move the images back into register, *post hoc* (see, e.g., Tyszka *et al.*, 1994; Hill *et al.*, 1995a).

For successful implementation of all such image realignment algorithms, three-dimensional (or multislice, two-dimensional) image data must be acquired in order to characterize, as fully as possible, the three rotations and translations of a rigid body. EPI allows such data to be collected from the whole head in a typical total acquisition time of some three seconds. Even after perfect realignment (to within 50 μm), movement-related artifacts may still exist within the FMRI time series. This is because the MR signal in a particular slice is also dependent upon movement that occurred in previous scans, i.e., the spin excitation history and consequent variation in local saturation. By applying an autoregression moving-average model for the effects of previous displacements on the current signal, removal of this artifactual component can also be achieved (Friston *et al.*, 1996).

## 3.4. Image Analysis

Having realigned each image and applied the appropriate intensity transformations to remove motion-related artifacts from the time series, one can apply a variety of statistical methods to the data in order to extract areas of significant brain activation [see reviews by Rabe-Hesketh and Bullmore (1997) and Lange (1996)]. One must remember that at 1.5 T neuronal activation only induces small signal intensity changes which are often on a par with those fluctuations that can be attributed to, for example, respiration, blood flow, and cerebrospinal fluid motion. It is therefore necessary to test the statistical significance of any changes which appear to be associated with neuronal activation. Typically, significance tests are applied voxel by voxel, with some correction for multiple comparisons. Early approaches compared the distributions of signal intensities with an without neuronal activation using Kolmogorov–Smirnov tests or $t$-tests or involved similar correlations between the time series and the boxcar function describing neuronal activation. However, these methods assume that the effect of neuronal activation on the signal is instantaneous, whereas it has been shown that the hemodynamic response to activation is gradual, reaching its maximum only after a delay of about 3 to 6 seconds (see above). Hemodynamic response may be taken into account by regressing the observed time series on (or correlating it with) a reference vector representing the expected temporal response (Friston *et al.*, 1994; Bandettini *et al.*, 1993). The simplest approach is to assume that the shape of this expected response is the same in all activated regions and to estimate it, for example, from the mean response in a region known to be activated. However, it has been shown that the response, particularly the delay to a maximum BOLD response relative to the onset of the stimulus, differs significantly between regions (Bandettini *et al.*, 1995; Takahashi *et al.*, 1993). Methods of estimating the shape of the temporal response include nonlinear estimation of the point spread function which is assumed to yield the expected temporal response when convolved with the boxcar function (Lange and Zeger, 1997) and least-squares fitting of linear regression models (Bullmore *et al.*, 1996a; Worsley and Friston, 1995) for examples using sinusoidal terms (see below). These more recent approaches also tackle the problem of temporal correlations in the time series, which had previously been ignored.

In summary, data processing requires good models and sophisticated statistical techniques to cope with the low signal-to-noise ratio, the complexity of the hemodynamic response, and the presence of temporal correlations in the signal. Some promising methods have been proposed, but further research is required in this area. The variety of processing procedures and the dependence on both paradigm design and the details of the MR experiment currently make direct comparison between the results from different institutions difficult.

## 4. FMRI—HOW WE DO IT

Because FMRI research is still in its infancy, there is no standardized procedure for data acquisition and processing. Although this section does not purport to satisfy this goal fully, we have endeavored to summarize how FMRI studies are carried out at the Institute of Psychiatry and Maudsley Hospital, while bearing in mind our hardware limitations and patient profile.

### 4.1. Image Acquisition

Gradient-echo, echo-planar MR images are acquired using a 1.5-T GE Signa System (General Electric, Milwaukee, WI) retrofitted with Advanced NMR hardware (ANMR, Woburn, MA). A quadrature birdcage coil encompassing the whole head is used for RF transmission and reception. In each of 14 contiguous planes parallel to the intercommissural plane, 100 $T_2^*$-weighted MR images depicting BOLD contrast are acquired at a TE of 40 msec, a TR of 3000 msec, an in-plane resolution of 3 mm, and a slice thickness of 5 mm. Head movement is minimized by foam padding with the head coil and a restraining band placed across the forehead. In the same session, a 43-slice, high-resolution inversion recovery, gradient-echo, echo-planar image series of the whole brain is again acquired parallel to the intercommissural (AC–PC) plane with a TE of 40 msec, a

TI of 180 msec, a TR of 16 sec, an in-plane resolution of 1.5 mm, and a slice thickness of 3 mm (8 signal averages). This latter data set allows improved visualization of the anatomy while maintaining any geometric distortion inherent within our EPI methodology. These EP images allow direct superimposition of activated voxels from the time series without correction for geometric distortion, which is necessary when such functional maps are registered onto conventional MR images.

## 4.2. Movement Correction

Slight subject motion during our MR image acquisition can cause changes in $T_2$*-weighted signal intensity unrelated to changes in the oxyhemoglobin/deoxyhemoglobin ratio. The following procedure is currently adopted to estimate and correct the effects of motion prior to any further analysis of the images:

(i)"Base" images of the mean signal intensity over time are created by averaging the 100 "match" images acquired in each plane.

(ii)The sum of absolute differences in gray-scale values between the voxels of each match image and its corresponding base image is then computed.

(iii)A nonlinear search algorithm (Press *et al.*, 1992) is used to estimate the extent of translation and rotation in three dimensions which minimizes the total difference between all match and base images.

(iv)The match images are then realigned relative to the base image by tricubic spline interpolation.

(v)The realigned, $T_2$*-weighted time series are then regressed on the concomitant and lagged time series of estimated movement at each voxel (Friston *et al.*, 1996).

The residual time series resulting from the last stage of this procedure are uncorrelated with estimated rigid body motion in three dimensions.

## 4.3. Time Series Analysis and Hypothesis Testing

The power of periodic signal change at the (fundamental) OFF–ON frequency of our stimulus (or task) is estimated by iterated least-squares fitting of a sinusoidal regression model to the motion-corrected time series at each voxel for all images. The model allows the shape and delay of the hemodynamic response to be estimated at each voxel and takes account of temporal autocorrelations. The fundamental power quotient (FPQ; fundamental power divided by its standard error) is then estimated at each voxel and represented in a descriptive parametric map. Each observed FMRI time series is then randomly permuted 10 times, and the FPQ estimated after each permutation. This results in 10 parametric maps (for each subject at each plane) of FPQ estimated under the null hypothesis that FPQ is not determined by periodic stimulation (Bullmore *et al.,* 1996a). Equivalence class testing may then be used to determine the size and number of 8 connected voxel clusters in the observed and randomized images. The distribution of voxel cluster size in the randomized images is used to assign a probability under the null hypothesis to each voxel cluster in the observed image. For each voxel in the observed image, its probability under the null hypothesis in terms of time series activity is combined with its probability under the null hypothesis in terms of spatial clustering. Voxels with such a spatio-temporally combined probability of false positive activation of <0.0005 are regarded, by us, as activated in a resulting brain activation map (BAM) from a conventional sensory stimulus.

All parametric maps of FPQ are then registered in the standard space of Talairach and Tournoux (1993). This is done in two stages, using the realignment algorithms previously used for movement correction. First, the 14-slice set of FPQ maps derived from each subject is registered with that subject's high-resolution EPI data set; it is then registered and rescaled relative to a template image (previously obtained by averaging high-resolution EPI data sets acquired from four normal volunteers in Talairach space). Identical transformations are applied to the randomized FPQ maps obtained for each subject. After spatial normalization, the observed and randomized FPQ maps from each subject are identically smoothed with a Gaussian filter (SD = 3 mm or 1 voxel), to accommodate variability in gyral anatomy and error of voxel displacement during normalization. Generic activation is then robustly determined by computing the median value of FPQ at each voxel of the observed parametric maps and comparing it to a null distribution of median FPQ values computed from the randomized parametric maps. If the observed median FPQ exceeds the critical value of randomized median FPQ, for a test of size $\alpha = 2.5 \times 10^{-4}$, then that voxel is generically activated with probability of false positive activation equal to $\alpha$. At this level of significance testing, 5 voxels are expected to be "activated" by change over the whole median image.

Generically activated voxels are colored (typically red) and superimposed on (gray-scale) high-resolution EPI data, to create generic brain activation maps (GBAMs), in one of two ways:

(i)Activated voxels may then be registered in the space of an individual, surface-rendered, high-resolution EPI data set. This display allows us to locate generic activation in the context of an individual's sulco-gyral anatomy.

(ii)Activated voxels are overlaid on the template (high-resolution EPI) image in Talairach space, and the combined image is represented either as three orthogonal maximum intensity projections or as a series of oblique, near-axial slices in the AC–PC plane. These displays allow us to identify the standard stereotactic coordinates of each regional focus of generic activation.

## 5. APPLICATIONS

Due to the limited length of this chapter, the focus of this final section will be restricted to some applications of FMRI in our institution that have clinical potential.

### 5.1. Motor Tasks

We routinely conduct studies in which the patient performs finger tapping tasks using a visual cue which alternates between, for example, the words REST and LEFT every 30 seconds over a five-minute interval. The primary objective of these studies is to ascertain all the major, eloquent areas of the brain required to perform a simple interdigitation. Our typical patient subgroup for such FMRI studies consists of patients who are about to undergo neurosurgery, e.g., debulking a mass lesion overlying the motor cortex. Currently, we are comparing our FMRI findings with those from conventional electrode mats (Hill *et al.,* 1995b). This task is often repeated twice for both left and right hands in those patients who can tolerate a total of one hour in the magnet. Repetition of the experiment allows increased confidence in our interpretation of areas of neuronal activity by means of intrasubject averaging and reduced sensitivity to random movements by the patient. Figure 3 shows "group-averaged" data from a single subject who performed the same five-minute, complex finger tapping task of the left hand

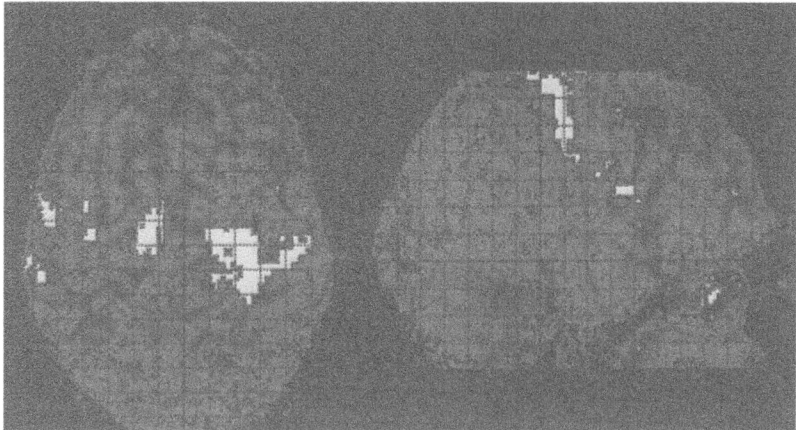

*FIGURE 3.* Significant areas of cortical activation in response to a finger tapping task of the left hand. Most of the activation is apparent in the primary motor areas (M1) of the contralateral, right hemisphere as viewed in both axial *(left)* and sagittal *(right)* planes.

four times during the course of a single session. The data have been presented in such a way as to highlight the area of the primary motor cortices involved in such a task. We envisage that such information allied with data from existing techniques will, in time, assist the neurosurgeon in planning the most appropriate means of intervention.

## 5.2. Language Dominance

Another FMRI study again involved two alternating conditions, each lasting 30 seconds and repeated five times for a total duration of five minutes. During the "ON" condition, the subject viewed common, concrete nouns ranging from three to seven letters in length and were required to make a word judgment. In each 30-second segment, 12 words were randomly presented in a contiguous fashion, six related to living and six related to nonliving objects, e.g., cow, kettle, lamp, knight, etc. During the "OFF" condition, the subject viewed a blank screen. The subjects were required to make a living/nonliving judgment and to rehearse subvocally their decision for each word. Their decision was made subvocally so as to minimize any potential movement artifact that may correlate with the "ON" condition. Cognitively, this task utilizes the route via the visual input lexicon to establish the visual word form, through to the cognitive system analyzing the meaning of the word, producing a decision which results in articulatory rehearsal but without overt speech. We have shown (Fig. 4) that the angular gyrus, responsible for integrating the visual form of the word with the cognitive system,

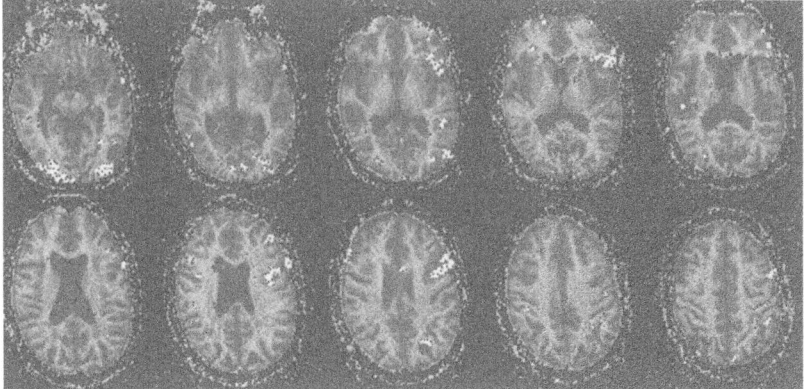

*FIGURE 4.* A brain activation map produced from a single-subject study involving single word reading and subvocal rehearsal. The areas of activation (white pixels) are predominantly located in the subject's language-dominant (left) hemisphere. The data has been presented at a voxelwise probability of false-positive activation of less than 0.001.

Wernicke's area, involved in the comprehension of the word, and Broca's area, implicated in the articulatory loop mechanism and subvocal rehearsal of the decision, were all activated in such a study (Bullmore *et al.,* 1996b). Since this simple word decision task shows focal activation of the cortical language areas that are confined to the language-dominant hemisphere, the procedure has potential utility as a rapid, noninvasive alternative to more interventional methods for establishing language dominance (Wada and Rasmussen, 1960). A recent paper by Binder *et al.* (1996) showed a highly significant correlation between the intracarotid amobarbital (Wada) test and a similar FMRI study involving a single-word, semantic decision task in 22 epilepsy patients. We again envisage that such FMRI maps will be helpful in defining the boundaries of surgical excision.

## 5.3. Verbal Working Memory

The verbal working memory task again involved the comparison of two alternating conditions. In one, the working memory condition, subjects viewed a random series of letters and indicated by means of a button press when the currently projected letter was the same as that presented two previously (e.g., A-B-G-B). In the control condition, subjects presented with letters appearing in alphabetical order were again asked to respond by means of the button press when the letter X appeared out of sequence (e.g., A-B-C-X). The rate of letter presentation (one every 1.2 sec) and the frequency of target responses were the same in each condition. The two conditions were alternated every 30 seconds for a total experiment time of five minutes. Seven right-handed male volunteers were studied, and the significant areas of brain activation for each subject combined in standard stereotactic space. Considerable intersubject variation was observed, but six of the seven subjects studied showed activation in the dorsolateral prefrontal cortex (DLPFC: left > right). Significant activation was also observed in the anterior and posterior cingulate, left and right angular gyri, Broca's area, and occipital striate cortex (Fig. 5). In the context of this study, the DLPFC activation observed may be attributed to cognitive processing associated with the comparison of a current stimulus with stored information held in the visual or phonological scratch pad (Mellers *et al.,* 1995). Anterior cingulate activity may reflect greater demands on attention in the working memory condition when compared with the control condition. Activation in the posterior cingulate and left parietal regions may be attributed, in part, to visual imagery and the phonological store of the "articulatory loop," respectively. It is, however, important to note that this paradigm produced considerable intersubject variation in both the extent and the location of activation. This may, in part, be due to different subjects using alternative strategies (phonological and/or visual) during performance of the task. With this paradigm we have shown that complex mental operations rely on the coordinated activity of widely distributed brain regions that constitute neuro-

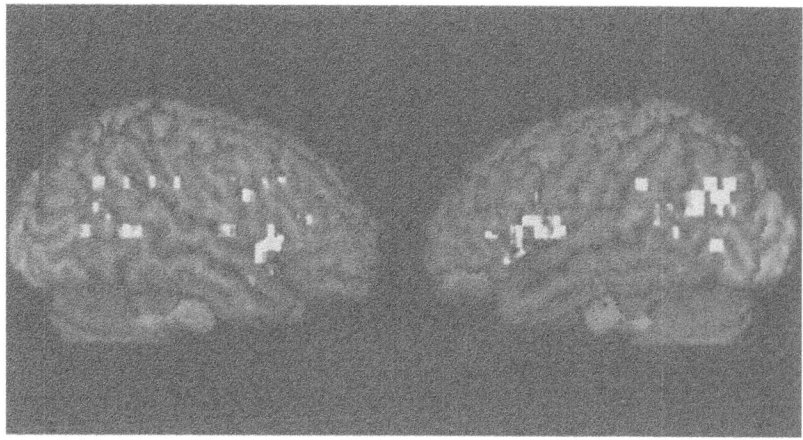

FIGURE 5. A generic brain activation map derived from 7 right-handed male subjects performing a verbal working memory task. The primary areas of activation (white pixels) shown on these surface-rendered images are the left dorsolateral prefrontal cortex and parietal regions.

cognitive networks. We intend to apply such methods to several psychiatric patient groups, e.g., schizophrenics, where such networks are believed to be impaired.

## 5.4. Covert Verbal Fluency

Five right-handed volunteers were studied during covert generation of nouns in response to a letter, e.g., "A," presented every three seconds on audiotape and piped via a pneumatically driven sound system to the subject's headphones for 30-second intervals, after which the letter was replaced by the word "REST" and the subjects were asked to repeat this word silently. All volunteers demonstrated discrete foci of activation in the left frontal region in response to this task. Left inferior frontal gyrus, mid cingulate, and dorsolateral activation were consistently observed. A generic brain activation map from this group of subjects is shown in Fig. 6. The discrete left frontal activations in the dorsolateral and inferior frontal regions suggest functional specialization within the frontal area, with possible inner speech/working memory and internal generation/ initiation being responsible for inferior frontal and dorsolateral prefrontal activations, respectively (Grasby et al., 1995).

We have begun to apply this neuropsychological test to patients with known frontal lobe damage. Monitoring of the effects of psychopharmacological intervention on both performance and functional response to this robust cognitive paradigm is also currently underway.

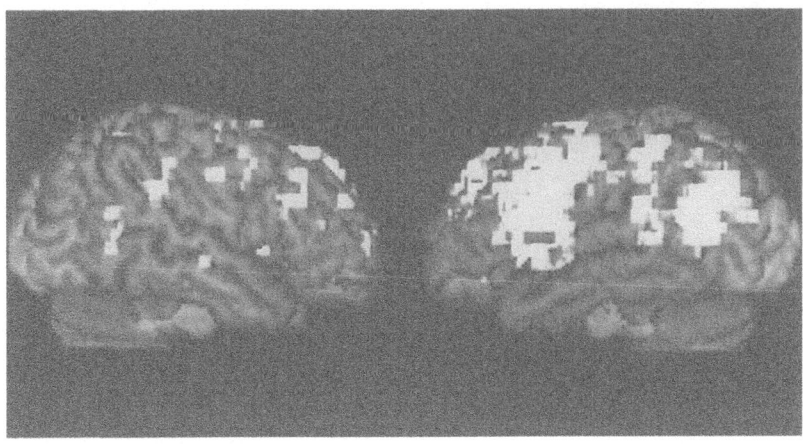

FIGURE 6. A generic brain activation map derived from 5 right-handed subjects performing a covert verbal fluency task. Consistent left, frontal activation was seen for all subjects (shown here as white pixels on a gray-scale surface-rendered image).

## 5.5. Hallucinations

We have previously shown by FMRI that activation of the visual cortex in response to photic stimulation was attenuated during visual hallucinations in a patient suffering from cortical Lewy body dementia (Howard *et al.*, 1995). Here we demonstrate (Fig. 7) an FMRI case study of a man with schizophrenia undergoing two distinct sensory stimuli, once while experiencing and once while free from auditory hallucinations. Sensory stimulation was applied by asynchronous presentation of both photic (flashing LED goggles at 8 Hz for 30 sec "OFF"/30 sec "ON") and auditory (talking book for 39 sec "OFF"/30 sec "ON") stimuli. On collection of 100 EP images per slice and subsequent image registration as before, the data were analyzed to derive the statistical power at both ON–OFF frequencies of sensory stimulation. Activation in response to hearing speech was detected bilaterally in Heschl's and superior temporal gyri in the non-hallucinating state. A normal pattern of activation in response to visual stimulation was also observed in the occipital cortex. During continuous auditory hallucinations, described by the patient as spirits talking to him, there was marked attenuation of activation in response to the same auditory stimulus whereas response to visual activation was unimpaired. We can therefore conclude that hallucinations involve primary, sensory cortical areas which seem to "compete" with exogenous sensory stimuli (David *et al.*, 1996). Further FMRI studies to monitor areas of brain activation during hallucinations directly are in progress (Woodruff *et al.*, 1995).

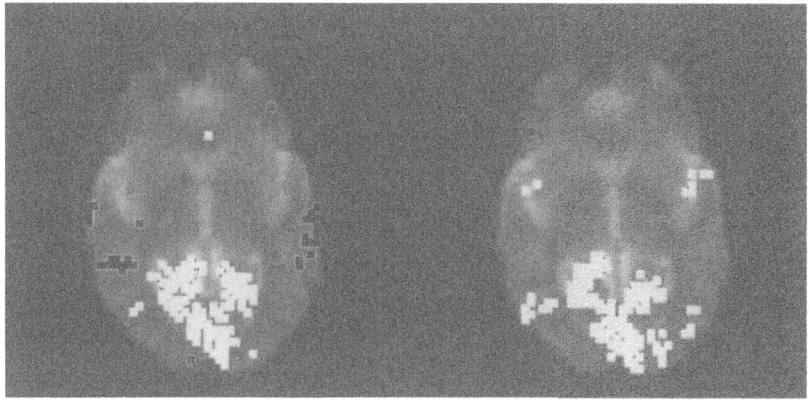

*FIGURE 7.* Two brain activation maps from the same schizophrenic subject undergoing a combined visual and auditory stimulus presentation. Data were collected in the absence (*left*) and presence (*right*) of florid, auditory hallucinations. Both auditory (black) and visual (white) responses were normal in the absence of the hallucinations, whereas the significant response to the same external auditory stimulus was dramatically reduced in the presence of an auditory hallucination.

## 6. CONCLUSION

FMRI has, in five years, proven to be a safe, sensitive, and adaptable tool for the investigation of a wide variety of mental processes. We envisage that many of the paradigms currently in development on normal subjects will, in the near future, be applied to both neurologic and psychiatric patient subgroups.

ACKNOWLEDGMENTS

We would like to thank all our colleagues at the Institute of Psychiatry and Maudsley Hospital for their assistance, support, and enthusiasm during the development of the functional MRI project. Data presented in this chapter were collected in collaboration with Professor Stuart Checkley and Drs. Jonathan Berman, Tony David, Paul Grasby, Rob Howard, John Mellers, Robin Morris, and Peter Woodruff. S.C.R.W. would also like to acknowledge Dr. Gareth Barker and Professor Bob Turner (Institute of Neurology, London) for many fruitful discussions during the initiation of this work. Much of this work would not have been possible without the expertise and assistance of the radiographers and the Neuroimaging group.

## REFERENCES

Bandettini, P. A., Jesmanowicz, A., Wong, E. C., and Hyde, J. S., 1993, Processing strategies for time-course data sets in functional MRI of the brain, *Magn. Reson. Med.* **35:**261–277.

Bandettini, P. A., Wong, E. C., Jesmanowicz, A., and Hyde, J. S., 1995, Functional MR imaging using the BOLD approach, *Diffusion and Perfusion Magnetic Resonance Imaging* **18**:335–349.

Barranco, D., Sutton, L. N., Florin, S., Greenberg, J., Sinnwell, T., Ligeti, L., and McLaughlin, A. C., 1989, Use of $^{19}F$ NMR spectroscopy for measurement of cerebral blood flow: A comparative study using microspheres, *J. Cereb. Blood Flow Metab.* **9**:886–891.

Belliveau, J. W., Kennedy, D. N., McKinstry, R. C., Buchbinder, B. R., Weisskoff, R. M., Cohen, M S., Vevea, J. M., Brady, T. J., and Rosen, B. R., 1991, Functional mapping of the human visual cortex by magnetic resonance imaging, *Science* **254**:716–719.

Binder, J. R., Swanson, S. J., Hammeke, T. A., Morris, G. L., Mueller, W. M., Fischer, M., Benbadid, S., Frost, J. A., Rao, S. M., and Haughton, V. M., 1996, Determination of language dominance using functional MRI: A comparison with the Wada test, *Neurology* **46**:978–984.

Blamire, A. M., Rothman, D. L., and Dixon, T., 1996, Dynamic shim updating: A new approach towards optimized whole brain shimming, *Magn. Reson. Med.* **36**:159–165.

Bullmore, E. T., Brammer, M. J., Williams, S. C. R., Rabe-Hesketh, S., Janot, N., David, A. S., Mellers, J. D. C., Howard, R., and Sham, P, 1996a, Statistical methods of estimation and inference for functional MR image analysis, *Magn. Reson. Med.* **35**:261–277.

Bullmore, E. T., Rabe-Hesketh, S., Morris, R. G., Williams, S. C. R., Gregory, L., Gray, J. A., and Brammer, J. J., 1996b, Functional magnetic resonance image analysis of a large scale neurocognitive network, *Neuroimage* **4**:16–33.

David, A. S., Woodruff, P. W. R., Howard, R. J., Mellers, J. D. C., Brammer, M. J., Bullmore, E. T., Wright, I. C., Andrew, C., and Williams, S. C. R., 1996, Auditory hallucinations inhibit exogenous activation of auditory association cortex, *Neuroreport* **7**:932–936.

Detre, J. A., Subramanium, V. H., Mitchell, M. D., Smith, D. S., Kobayashi, A., Zaman, A., and Leigh, J. S., 1990, Measurement of regional cerebral blood flow in cat brain using intracarotid $^{2}H_2O$ and $^{2}H$ imaging, *Magn. Reson. Med.* **14**:389–395.

Detre, J. A., Leigh, J. S., Williams, D. S., and Koretsky, A. P., 1992, Perfusion imaging, *Magn. Reson. Med.* **23**:37–45.

Edelman, R. R., Siewert, B., Darby, D. G., Thangaraj, V., Nobre, A. C., Mesulam, M. M., and Warach, S, 1994, Qualitative mapping of cerebral blood flow and functional localization with echo planar MR imaging and signal targeting with alternating radio frequency, *Radiology* **192**:513–520.

Eleff, S. M., Schnall, M. D., Ligeti, L., Osbakken, M., Subramanian, V. H., Chance, B., and Leigh, J. S. J., 1988, Concurrent measurements of cerebral blood flow, sodium, lactate, and high-energy phosphate metabolism using $^{19}F$, $^{23}Na$, $^{1}H$, and $^{31}P$ nuclear magnetic resonance spectroscopy, *Magn. Reson. Med.* **7**:412–424.

Ernst, T., and Hennig, J., 1994, Observation of a fast response in functional MR, *Magn. Reson. Med.* **32**:146–149.

Ewing, J. R., Branch, C. A., Fagan, S. C., Helpern, J. A., Simkins, R. T., Butt, S. M., and Welch, K. M., 1990, Fluorocarbon-23 measure of cat cerebral blood by nuclear magnetic resonance, *Stroke* **21**:100–106.

Fisel, C. R., Ackerman, J. L., Buxton, R. B., Garrido, L., Belliveau, J. W., Rosen, B. R., and Brady, T. J., 1991, MR contrast due to microscopically heterogeneous magnetic susceptibility: Numerical simulations and applications to cerebral physiology, *Magn. Reson. Med.* **17**:336–347.

Frahm, J., Merboldt, K. D., Hänicke, W., Kleinschmidt, A., and Boecker, H., 1994, Brain or vein-oxygenation of flow? On signal physiology in functional MRI of human brain activation, *NMR Biomed.* **7**:45–53.

Friston, K. J., Jezzard, P., and Turner, R., 1994, Analysis of functional MRI time series, *Human Brain Mapping* **1**:153–171.

Friston, K. J., Williams, S., Howard, R., Frackowiack, R. S. J., and Turner, R., 1996, Movement-related effects in FMRI time-series, *Magn. Reson. Med.* **35**:346–355.

Frostig, R. D., Lieke, E. E., Ts'o, D. Y., and Grinvald, A., 1990, Cortical functional architecture and local coupling between neuronal activity and the microcirculation revealed by in vivo high-resolution optical imaging of intrinsic signals, *Proc. Natl. Acad. Sci. USA* **87**:6082–6086.

Grasby, P., Williams, S. C. R., Bullmore, E., Brammer, M., Samuel, M., Andrew C., Simmons, A., and Checkley, S., 1995, A functional MRI study of covert verbal fluency in normal volunteers, *in* "Proceedings of the 3rd Meeting of the Society of Magnetic Resonance, Nice, France," Vol. 3, p. 1337.

Hennig, J., Ernst, T., Speck, O., Deuschl, G., and Feifel, E., 1994, Detection of brain activation using oxygenation sensitive functional spectroscopy, *Magn. Reson. Med.* **31**:85–90.

Hill, D. L. G., Simmons, A., Studholme, C., Hawkes, D. J., and Williams, S. C. R., 1995a, Removal of stimulus correlated motion from echo planar FMRI studies, *in* "Proceedings of the 3rd Meeting of the Society of Magnetic Resonance, Nice, France," Vol. 2, p. 840.

Hill, D. L. G., Sahni, V. A. S., Polkey, C. E., Simmons, A., Cox, T. C. S., Edwards, P. J., and Hawkes, D. J., 1995b, Combining anatomical and functional information for epilepsy surgery guidance, *in* "Proceedings of the 3rd Meeting of the Society of Magnetic Resonance, Nice, France," Vol. 2, p. 854.

Howard, R. J., Williams, S. C. R., Bullmore, E., Brammer, M., Mellers, J., Woodruff, P., and David, A. S., 1995, Cortical response to exogenous visual stimulation during visual hallucinations, *Lancet* **345**:70.

Jezzard, P., and Goldstein, S. R., 1994, A head position monitoring device for use in functional MRI studies, *in* "Proceedings of the 2nd Meeting of the Society of Magnetic Resonance, San Francisco," Vol. 2, p. 648.

Kim, S. G., 1995, Quantification of relative cerebral blood flow change by flow sensitive alternating inversion recovery (FAIR) technique: Application to functional mapping, *Magn. Reson. Med.* **34**:293–301.

Kwong, K. K., Belliveau, J. W., Chesler, D. A., Goldberg, I. E., Weisskoff, R. M., Poncelet, B. P., Kennedy, D. N., Hoppel, B. E., Cohen, M. S., Turner, R., Cheng, H-M., Brady, T. J., and Rosen, B. R., 1992, Dynamic magnetic resonance imaging of human brain activity during primary sensory stimulation, *Proc. Natl. Acad. Sci. USA* **89**:5675–5679.

Lai, S., Hopkins, A. L., Haacke, E. M., Li, D., Wasserman, B. A., Buckley, P., Friedman, L., Meltzer, H., Hedera, P., and Friedland, R., 1993, Identification of vascular structures as a major source of signal contrast in high resolution 2D and 3D functional activation imaging of the motor cortex at 1.5T: Preliminary results, *Magn. Reson. Med.* **30**: 387–392.

Lange, N., 1996, Tutorial in biostatistics: Statistical approaches to human brain mapping by functional magnetic resonance imaging, *Stat. Med.* **15**:389–428.

Lange, N., and Zeger, S. L., 1997, Non-linear Fourier time series analysis for human brain mapping by functional magnetic resonance imaging, *J. R. Stat. Soc. Series B* (in press).

Mansfield, P., Coxon, R., and Glover, P., 1994a, Echo-planar imaging of the brain at 3.0 T: First normal volunteer results, *J. Comput. Assist. Tomogr.* **18**:339–343.

Mansfield, P., Glover, P., and Bowtell, R., 1994b, Active acoustic screening: Design principles for quiet gradient coils in MRI, *Meas. Sci. Technol.* **5**:1021–1025.

Mansfield, P., Coxon, R., and Hykin, J., 1995a, Echo-volumar imaging (EVI) of the brain at 3.0 T: First normal volunteer and functional imaging results, *J. Comput. Assist. Tomogr.* **19**:847–852.

Mansfield, P., Chapman, B. L. W., Bowtell, R., Glover, P., Coxon, R., and Harvey, P. R., 1995b, Active acoustic screening: Reduction of noise in gradient coils by Lorentz force balancing, *Magn. Reson. Med.* **33**:276–281.

Mellers, J. D. C., Bullmore, E., Brammer, M., Williams, S. C. R., Andrew, C., Sachs, N., Andrews, C., Cox, T. S., Simmons, A., Woodruff, P., David, A. S., and Howard, R., 1995, Neural correlates of working memory in a visual letter monitoring task: An FMRI study, *Neuroreport* **7**:109–112.

Menon, R. S., Ogawa, S., Hu, X., Strupp, J. P., Anderson, P., and Ugurbil, K., 1995, BOLD-based functional MRI at 4 Tesla includes a capillary bed contribution: Echo planar imaging correlates with previous optical imaging using intrinsic signals, *Magn. Reson. Med.* **33**:453–459.

Ogawa, S., Lee, T. M., Nayak, A. S., and Glynn, P., 1990, Oxygenation-sensitive contrast in magnetic resonance image of rodent brain at high magnetic fields, *Magn. Reson. Med.* **14**:68–78.

Pauling, L., and Coryell, C. D., 1936, The magnetic properties and structure of hemoglobin, oxyhemoglobin and carbonmonoxyhemoglobin, *Proc. Natl. Acad. Sci. USA* **22**:210–216.

Pekar, J., Ligeti, L., Ruttner, Z., Lyon, R. C., Sinnwell, T. M., van Gelderen, P., Fiat, D., Moonen, C. T. W., and McLaughlin, A. C., 1991, In vivo measurement of cerebral oxygen consumption and blood flow using $^{17}O$ magnetic resonance imaging, *Magn. Reson. Med.* **21**:313–319.

Pekar, J., Ligeti, L., Sinnwell, T. M., Moonen, C. T. W., Frank, J. A., and McLaughlin, A. C., 1994, $^{19}F$ magnetic resonance imaging of cerebral blood flow with 0.4 cc resolution, *J. Cereb. Blood Flow Metab.* **14**:656–663.

Press, W. H., Flannery, B. P., Teukolsky, S. A., and Vetterling, W. T., 1992, Numerical Recipes in C-The Art of Scientific Computing, Cambridge University Press, Cambridge, U.K.

Rabe-Hesketh, S., and Bullmore, E. T., Brammer, M. J., 1997, The analysis of functional magnetic resonance images, *Stat. Methods Med. Res.* (in press).

Reese, T. G., Davis, T. L., and Weisskoff, R. M., 1993, Automated shimming at 1.5T using echo planar image frequency maps, *in* "Proceedings of the 12th Annual Meeting of the Society of Magnetic Resonance in Medicine, New York," Vol. 1, p. 354.

Rempp, K. A., Brix, G., Wenz, F., Becker, C. R., Guckel, F., and Lorenz, W. T., 1994, Quantification of regional cerebral blood flow and volume with dynamic susceptibility contrast-enhanced MR imaging, *Radiology* **193**:637–641.

Roy, C., and Sherrington, C., 1890, On the regulation of the blood supply of the brain, *J. Physiol.* **11**:85–108.

Takahashi, T., Takiguchi, K., Itagaki, H., Onodera, Y., Yamamoto, E., and Koizumi, H., 1993, Real time imaging of brain activation during imagination of finger tasks, *in* "Proceedings of the 12th Annual Meeting of the Society of Magnetic Resonance in Medicine, New York," Vol. 3, p. 1415.

Talairach, J., and Tournoux, P., 1993, "A Co-planar Stereotaxic Atlas of the Human Brain," 2nd ed., Thieme Verlag, Stuttgart.

Thulborn, K. R., Waterton, J. C., Matthews, P. M., and Radda, G. K., 1982, Oxygenation dependence of the transverse relaxation time of water protons in whole blood at high field, *Biochim. Biophys. Acta.* **714**:265–270.

Turner, R., and Jezzard, P., 1994, Magnetic resonance studies of brain functional activation using echo planar imaging, *in* "Advances in Functional Neuroimaging," Vol. 1, (R. W. Thatcher, M. Hallett, T. A. Zeffiro, and E. Roy John, eds.), pp. 69–78, Academic Press, New York.

Turner, R., and Keller, P., 1991, Angiography and perfusion measurements by NMR, *Prog. NMR Spectrosc.* **23**:93–133.

Turner, R., LeBihan, D., Moonen, C. T. W., DesPres, D., and Frank, J., 1991, Echo planar time course MRI of cat brain oxygenation changes, *Mang. Reson. Med.* **22**:159–166.

Turner, R., Jezzard, P., Wen, H., Kwong, K. K., Le Bihan, D., Zeffiro, T., and Balaban, R. S., 1993, Functional mapping of the human visual cortex at 4 Tesla and 1.5 Tesla using deoxygenation contrast EPI. *Magn. Reson. Meg.* **29**:277–281.

Tyszka, J. M., Grafton, S. T., Chew, W., Woods, R. P., and Colletti, P. M., 1994, Parcellation of mesial frontal motor areas during ideation and movement using functional magnetic resonance imaging at 1.5 tesla, *Ann. Neurol.* **35**:746–749.

Wada, J., and Rasmussen, T., 1960, Intracarotid injection of sodium amytal for the lateralization of cerebral speech dominance, *J. Neurosurg.* **17**:266–282.

Wehrli, F. W., 1990, Fast scan magnetic resonance: Principles and applications, *Magn. Reson. Q.* 6(3):165–236.

Weisskoff, R. M., and Kiihne, S., 1992, MRI susceptometry: Image-based measurement of absolute susceptibility of MR contrast agents and human blood, *Magn. Reson. Med.* **24:**375–383.

Weisskoff, R. M., Chesler, D., Boxerman, J. L., and Rosen, B. R., 1993, Pitfalls in MR measurement of tissue blood flow with intravascular tracers: Which mean transit time? *Magn. Reson. Med.* **29:**553–558.

Woodruff, P., Brammer, M., Mellers, J., Wright, I., Bullmore, E., and Williams, S. C. R., 1995, Auditory hallucinations and perception of external speech, *Lancet* **346:**1035.

Worsley, K. J., and Friston, K. J., 1995, Analysis of FMRI time series revisited again, *Neuroimage* **2:**173–181.

# MRI AND PROTON MRS IN THE EVALUATION OF MULTIPLE SCLEROSIS

## D. L. ARNOLD, P. M. MATTHEWS, and N. DE STEFANO

## SUMMARY

Magnetic resonance imaging (MRI) has revolutionized our understanding of disease activity in patients with multiple sclerosis. Advanced MR techniques, including magnetization transfer, diffusion, and spectroscopic imaging, now can be used to define specific pathological changes in lesions and offer promise for improved definition of the nature of individual lesions and the dynamics of their evolution. Combined, multimodal MR studies of the brains of patients with multiple sclerosis therefore soon may provide efficient, specific, and quantitative new approaches to assessment of outcome in therapeutic trials.

*D. L. ARNOLD, P. M. MATTHEWS, and N. DE STEFANO* • *Montreal Neurological Institute, McGill University, Montreal, Quebec, Canada H3A 2BH.*

*Magnetic Resonance Spectroscopy and Imaging in Neurochemistry,* Volume 8 of *Advances in Neurochemistry,* edited by Bachelard, Plenum Press, New York, 1997.

## 1. MULTIPLE SCLEROSIS: AN INTRODUCTION

Multiple sclerosis (MS) is a common disease that tends to attack young adults. It is an idiopathic inflammatory disease of the central nervous system (CNS) characterized by neurological impairment referable to multiple lesions disseminated in the CNS that arise independently at different times. Neither the primary trigger for the inflammatory response nor the mechanisms for the sustained immune response are known. It is probable that both a genetic predisposition and exogenous factors are important (W. B. Matthews *et al.*, 1991). An environmental, possibly infectious, etiology is suggested by at least two observations. First, there have been a number of examples of clusters or epidemics of the disease. Second, there are striking differences in the frequency of the disease in different parts of the world, and the risk in migrants depends on the age of migration. People who migrate after adolescence seem to retain the risk of their place of origin, whereas people who migrate early in life acquire the risk of the place they move to.

The natural history of the disease is marked by two general stages. For most patients, the early phase of the disease is characterized by recurrent attacks with neurological deficits that resolve completely or partially, and the disease is classified as "relapsing–remitting" (RR). Eventually, the remissions become less complete, and the disease adopts a more inexorably progressive character (McAlpine *et al.*, 1955). This is referred to as the secondary progressive (SP) phase. Whether the SP phase is simply the result of accumulation of nervous system damage or represents a fundamental alteration in the nature of the disease is a major unresolved question.

Consistent with this clinical presentation, pathological examination of the brains of patients with MS reveals multiple, diffusely scattered lesions of different ages, as inferred from the stage of the inflammatory, repair, and reactive processes (Table 1). Recent lesions show loss of myelin with relatively modest degeneration of oligodendrocytes and large numbers of macrophages, whereas long-standing, severe lesions show profound loss of oligodendroglial cells with gliosis. The lesions (also called "plaques") may vary in diameter from less than a millimeter to several centimeters in size (W. B. Matthews *et al.*, 1991). They have a predilection to accumulate in regions adjacent to the lateral ventricles, where they may become confluent. Sparing of axonal loss in the center of lesions has been a characteristic of these lesions emphasized by pathologists, as it is helpful in discriminating them from those of a more generally destructive process, such as an encephalitis. However, this should not be interpreted as suggesting that axonal damage does not occur. Axons show changes in size, shape, and morphology (Prineas and Connell, 1978; Raine and Cross, 1989; Rodriguez and Scheithauer, 1994), and estimates suggest that there may be loss of as many as 50% of the axons traversing severe, chronic lesions (Prineas and Connell, 1978).

TABLE 1. Pathology of Multiple Sclerosis

| Event | Pathology |
|---|---|
| Initiation of inflammation | Expression of leukocyte adhesion molecules |
| | Class II major histocompatibility complex expression |
| Breakdown of blood–brain barrier | Alteration of extracellular matrix composition |
| Myelin destruction | T-cell-generated lymphokines |
| | Antimyelin antibodies |
| | Macrophage-induced proteolysis and phagocytosis |
| Injury to cells in central nervous system (oligodendroglial cells, axons) | |
| Down-regulation of immune response | |
| Repair | Remyelination |
| | Alteration in abundance and distribution of voltage sensitive (VS) Na channels |
| | Functional plasticity |

Local changes in chemical composition are associated with the inflammation, loss of compact myelin and oligodendroglial cells, secondary axonal damage, recruitment of inflammatory cells, and astrogliosis. Just as classical pathology can define evolution on the basis of an inferred sequence of histological changes in a lesion, chemical pathology defines evolution by concomitant chemical changes. Early changes include an increase in water content and the appearance of immunological markers of inflammation. With inflammation, proteases and other catabolic enzymes are released locally from macrophages and other infiltrating inflammatory cells, lysosomes, and even myelin itself, leading to alterations in myelin lipid and protein composition. With demyelination, there is local accumulation of cholesterol esters (Quarles et al., 1993). These are absent from normal mature brain and arise with esterification of myelin membrane cholesterol by phagocytes. These esters remain in phagocytic vacuoles at the site of the lesion for some time and can be considered to be a marker of relatively recent disease activity.

Nervous system dysfunction in MS has generally been attributed to the consequences of demyelination (Quarles et al., 1993). Myelin, acting both directly as an electrical insulator and indirectly in organization of axonal membrane ion channels, allows the saltatory conduction necessary for normal myelinated nerve conduction. Demyelination leads to conduction block acutely because of failure of propagation of locally generated axonal membrane potentials with exposure of repolarizing potassium channels along previously myelinated axonal segments. Because at least partial remyelination of MS lesions is now well described pathologically (Prineas *et al.*, 1993), one possible explanation for the remission of symptoms is remyelination of previously demyelinated axonal segments. However, there are also other potential explanations. Acute conduction block also can be caused by a humoral immune response directed against axon sodium channels and therefore can be relieved by down-regulation of that response (Waxman, 1995). Although the epitopes against which the immune response is generated are not well characterized, it is clear that a robust humoral immune response does occur within the CNS. Redistribution of sodium channels to allow conduction across demyelinated segments with exposed potassium channels can also occur (McDonald, 1994). In addition, other changes in axonal morphology (e.g., change of axonal diameter) may be important in facilitating functional recovery.

The available experimental evidence supports the possible involvement of multiple mechanisms in recovery by emphasizing that there is not a simple relationship between demyelination and conduction block (Rasminsky, 1984). Nerve injected by antigalactocerebroside antisera develops conduction block associated with minimal paranodal demyelination and edema (Sumner *et al.*, 1982). In contrast, in nerve that is chronically denervated by diphtheria toxin (Rasminsky and Sears, 1972) or in the dystrophic mouse model (Rasminsky *et al.*, 1978), conduction may be preserved across even several millimeters of demyelinated axon.

Treatment of MS is generally directed toward modulating the inflammatory process that is responsible for the nervous system damage (Weiner *et al.*, 1995). A major problem has been the difficulty in assessing disease activity, the overall burden of disease, and the pathological evolution of lesions. *Clinical* disability scales are subject to notoriously high inter-rater variability and are not linear numerical scales but show variable sensitivity to change, depending on the degree of clinical disability (Kurtzke, 1984).

In this chapter we will focus on the use of magnetic resonance (MR) techniques to evaluate the nature and severity of brain injury in MS rather than on the clinical diagnosis, which has been reviewed in several recent publications (Kurtzke, 1994; Morrissey *et al.*, 1993; Mushlin *et al.*, 1993; Offenbacher *et al.*, 1993; Francis *et al.*, 1995). Conventional magnetic resonance imaging (MRI) is sensitive to lesions but gives only limited clues to the stage of pathological evolution of lesions. The relationship between MRI lesions and functional disability (which is the variable of interest for the clinician and patient) is therefore complex (Miller, 1994). However, other MR techniques potentially allow more specific, serial noninvasive characterization of MS pathology through the combined analysis of changes in diffusion, magnetization transfer, and chemical pathological changes (Table 2).

TABLE 2. Chemical Markers of Pathology with Demyelination

| Event | Chemical change |
|---|---|
| Breakdown of blood–brain barrier | Increased water content |
| | Altered water diffusion |
| | Gd enhancement |
| Myelin destruction | Increase in mobile lipids |
| | Increase in choline and *myo*-inositol |
| | Altered magnetization transfer index |
| Cell injury | |
| Glia | Decrease in total creatine |
| Axons | Decrease in *N*-acetylaspartate (NAA) |
| Repair | Decreased water content |
| | Normalization of chemical changes |

## 2. MAGNETIC RESONANCE IMAGING IN THE EVALUATION AND MONITORING OF MS

### 2.1. Conventional MR Imaging of MS Lesions

Conventional MRI studies of MS generally employ a combination of $T_1$-weighted, $T_2$-weighted, and so-called proton density scans. MS lesions appear hypodense relative to normal white matter in $T_1$-weighted images. More striking contrast is obtained with the $T_2$-weighted images, in which the lesions appear hyperintense, but similar hyperintense signal from cerebrospinal fluid (CSF) can impair discrimination of lesions. Proton density images have somewhat lower lesion contrast but allow clearer discrimination between lesions, which appear hyperintense, and CSF, which appears hypointense. More recently, fluid attenuated inversion recovery (FLAIR) images that combine the benefits of $T_2$-weighting with suppression of the strong signal from CSF have become popular.

In all three types of images, lesion contrast arises nonspecifically from edema, demyelination, gliosis, and increased extracellular fluid spaces. It is therefore impossible to distinguish between acute and chronic lesions or between active and inactive chronic lesions on scans taken at a single time point. There is good correspondence between MRI lesions and pathologically defined "plaques," although the former may be larger than the latter, consistent with the sensitivity of MR to tissue edema associated with demyelination.

Lesions in MS are almost always located in white matter, where they are most common in periventricular regions and the centrum semiovale.

*Total Lesions Load in Assessment of the "Burden of Disease"*

The notion of a measure of "burden of disease" as the accumulated extent of pathological damage due to MS is an attractive concept. One measure of disease burden is the total lesion volume on MRI (Paty, 1993). MRI lesion volume offers potential advantages as a measure of disease burden relative to clinical disability in that it is sensitive to all of the lesions in brain (not just those that affect certain clinically obvious functions, such as walking) and it is more easily quantifiable. The utility of total lesion volume as a measure of disease burden has been demonstrated in a number of ways. Several studies have demonstrated that lesion volume on MRI increases by about 10% per year in MS patients (Paty and Li, 1993; Kastrukoff *et al.*, 1990). Cerebral lesion volume at presentation with optic neuritis or myelopathy has demonstrated prognostic significance: patients with normal scans have a much lower probability of progression to MS in 2 to 5 year follow-up. Total lesion volume on MRI also has been shown to be useful as an additional outcome measure in clinical trials. For example, in the β-interferon trial (Kastrukoff *et al.*, 1990), RR patients demonstrated significantly greater accumulation of lesion burden when on placebo that when treated with high-dose β-interferon.

However, a major concern with such applications is that it has been difficult to demonstrate a strong correlation between MRI cerebral lesion load and clinical status in either cross-sectional or longitudinal studies (Filippi *et al.*, 1994; Khoury *et al.*, 1994; Miller, 1994). Despite the consistent changes in MS lesion volume with time for longitudinally followed groups of patients, most clinicians who treat MS patients are all too familiar with the conundrums posed by individual patients showing severe clinical dysfunction with minimal findings on cerebral MRI or, conversely, those with large numbers of lesions and minimal disability. One possibility is that spinal cord lesions may account for a failure to find strong correlations between cerebral lesion load and disability.

The problem may arise from attempting to correlate disability and different sorts of lesion pathology indiscriminately. As noted above, it is clear from classical pathological studies that there is considerable heterogeneity among lesions in brains of MS patients. Consistent with this notion, despite the failure of other groups to find good correlations between lesion load and disability. Filippi *et al.*, (1994) found a good correlation between disability and the change in total cerebral lesion load in *individuals* who were initially monosymptomatic and were then scanned a second time five years after presentation.

Experimental data showing a time-dependent relationship between pathological changes associated with demyelination and conduction block (Rasminsky, 1984) emphasize that lesion volume is unlikely to be the sole determinant of disability also because functional impairment may occur in different ways, depending on the time course of development of pathology. Largely reversible disability in RR MS might arise predominantly from transient conduction impairment with acute edema and demyelination, prior to alterations in expression of axonal membrane K- and Na- channels that allow recovery from conduction block or prior to local remyelination, for example. In contrast, chronic disability in the later SP phase of disease may be more related to irreversible axonal pathology.

## 2.2. MRI Measures of Disease Activity

One of the most important insights into MS provided by MRI is that disease activity is much greater than is apparent clinically. For example, results from an original study of activity by MRI showed that 5 of 7 patients with RR MS who were scanned over a 6-month period had 18 new and 10 enlarging lesions that were all clinically silent (Isaac *et al.*, 1988). Such simple measurement of changes in individual lesion numbers and volumes can be used as a measure of disease activity.

There are a number of potential concerns associated with the use of an MRI measure as an outcome in a clinical trial, however. Methodological concerns are that lesion-volume measurement may be subject to considerable inter-rater or scan-to-scan variability, depending on the pulse sequence used, and that there may be a lesion-volume-dependent bias in assessment of change (Goodkin *et al.*, 1992). Counting of lesions is intrinsically difficult, as the morphology of lesions may be complex and extend over multiple slices in the MRI. Counting also cannot appropriately weight the relative significance of new activity in preexisting lesions relative to that of new lesions. A fundamental biological question also remains, as the relationship between these imaging changes and pathological processes responsible for disability is uncertain (a key point, which will be discussed further below).

An alternative approach to assessment of disease activity is based on use of paramagnetic contrast agents such as gadolinium diethylenetriaminepentaacetic acid (Gd-DTPA). This compound is excluded from the CNS by the blood–brain barrier (BBB), except in regions of inflammation where selectivity of the BBB is reduced. Gd-DTPA shortens the spin–lattice relaxation time ($T_1$) of local tissue water, leading to contrast enhancement on $T_1$-weighted images in acute or active lesions of multiple sclerosis. Serial studies have demonstrated that enhancement can precede development of a lesion visible on unenhanced scans, consistent with the notion that inflammation is a primary event in lesion evolution (Kermode *et al.*, 1990). Previously established lesions may show peripheral enhancement with subsequent changes in shape in size. Enhancement generally persists for 2–4 weeks and can resolve either with or without development of a chronic lesion (Miller *et al.*, 1988). Early work showed that the acute development of clinical symptoms is frequently associated with enhancement of lesions (Grossman *et al.*, 1986). However, the most striking observation in serial Gd-DTPA scanning has been that disease activity assessed as the number of enhancing lesions greatly exceeds clinical activity. The appearance of new lesions or reenhancement of preexisting lesions is 5–10 times more common than clinical relapse (Isaac *et al.*, 1988; Willoughby *et al.*, 1989; Thompson *et al.*, 1991; Koopmans *et al.*, 1989; McDonald, 1994) in patients with MS.

Initial concerns that MRI activity may not be directly relevant to clinical activity have been assuaged in large measure as the relationship between MRI enhancement and clinical characteristics has become more clear in recent years. Frequent serial study of individuals has demonstrated that periods of increased MRI activity are more likely to be associated with increases in clinical disability (Smith *et al.*, 1993). Comparisons of different patient groups showed lower activity in single contrast-enhanced MRIs for patients with so-called "benign" or primary progressive MS relative to those with RR or SP disease (Thompson *et al.*, 1991; Smith *et al.*, 1993). In a cross-sectional study of RR MS patients, a strong correlation between the mean number of enhancing lesions over a 3-month period and clinical disability was shown recently (Stone *et al.*, 1995). Nonetheless, several previous investigators have failed to find a strong association between attack rate and outcome (Weinshenker and Ebers, 1987; Patzold and Pocklington, 1982; Broman *et al.*, 1981; Kurtzke *et al.*, 1977).

One interpretation of this apparent conflict is that there can be significant variation in the extent of pathological progression of lesions defined by MRI and in their functional consequences for the axons traversing them. If there is pathological progression of lesions through distinct stages, then all lesions must begin with a similar early inflammatory stage marked by BBB breakdown. However, as not all enhancing lesions leave permanent imaging changes, and as lesions that persist as chronic lesions and those that disappear without trace can exist together, rates of new inflammation may exceed those of demyelination, axonal atrophy, and chronic gliosis. For optimum utility in definition of effects of therapeutic interventions, "disease activity" therefore may need to be defined more precisely in terms of stage or type of pathological change, particularly if it is clearly demonstrated both that a significant proportion of lesions in earlier stages of pathological evolution do not progress to later stages and that the most profound functional consequences arise from later stages of lesion evolution.

MRI lesion volume or activity measures are attractive as end points in clinical trials because they report on lesions that may be clinically inapparent and they can be quantified more precisely than clinical measures of disease progression. However, in any given patient, the indirect and uncertain relationship that MRI lesion volumes or the degree of enhancement bear to changes in clinical disability limit meaningful interpretation of imaging results (Fig. 1). We there-

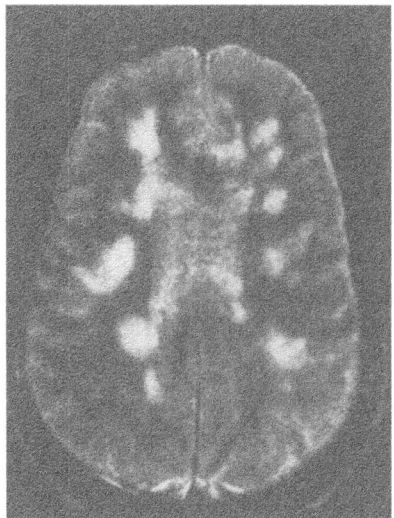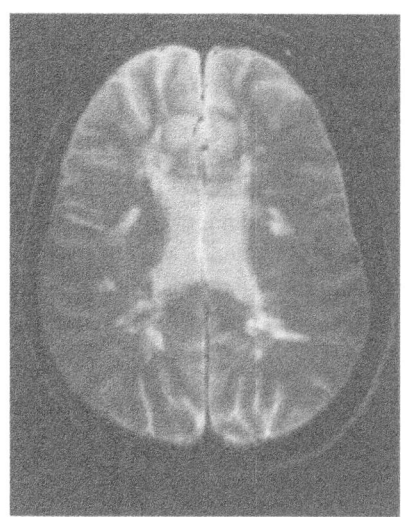

FIGURE 1. Two conventional MRI examinations of the same patient with multiple sclerosis. The image on the left *preceded* that on the right. The substantial decrease in apparent lesion volume suggests that much of this was due to edema.

fore believe that developments of MR that give improved, pathologically specific information will prove important in clinical assessment of MS.

## 2.3. Advanced MR Imaging Techniques for Improved Characterization of MS Pathology *in Vivo*

New MR techniques may offer more specific pathological information than that contained in conventional qualitative $T_2$-weighted spin-echo MRI. For example, quantitative $T_2$ measurements can show changes in normal-appearing white matter that are not apparent by inspection (Miller *et al.*, 1989). Magnetization transfer (MT) sequences show changes in signal due to altered macromolecular interactions of water and may be able to discriminate myelin loss from edema (Gass *et al.*, 1994). Diffusion imaging may be able to define axonal degeneration based on changes in diffusion anisotropy (Christiansen *et al.*, 1993). Finally, as will be discussed below, magnetic resonance spectroscopy (MRS) can provide chemical information that appears to define specific aspects of demyelination and axonal pathology. Multimodal statistical analysis of data from conventional MRI and these additional techniques may provide a more complete, spatially resolved description of the dynamics of MS pathology *in vivo* that could significantly enhance the power and specificity of clinical therapeutic trials in MS.

## 3. APPLICATIONS OF MAGNETIC RESONANCE SPECTROSCOPY TO THE EVALUATION AND MONITORING OF MS

Magnetic resonance spectroscopy is a promising technique for the evaluation and monitoring of MS as it can define simultaneously a number of chemical correlates of pathological change. It also provides a readily quantifiable index of axonal dysfunction or volume loss from lesions that may be important for understanding the pathogenisis of neurological impairments and disabilities in MS. In this section, we describe the techniques of MRS and its use to provide pathologically specific information on the natural history of MS.

### 3.1. Brain MRS in MS

Water-suppressed, localized proton MR spectra of normal human brain at "long" echo times (TE = 136 or 272 msec) reveal four major resonances (Fig. 2): one at 3.2 ppm (Cho) that arises from tetramethylamines (mainly from choline-containing phospholipids), one at 3.0 ppm (Cr) that arises primarily from creatine (or phosphocreatine), one at 2.0 ppm that arises from $N$-acetyl groups [mainly $N$-acetylaspartate (NAA)] and one at 1.3 ppm that arises from the methyl resonance of lactate (LA) or, in certain pathological conditions, or macromolecules. The resonances of these latter compounds (Wolinsky *et al.*, 1990; Davie *et al.*, 1994) and several other compounds (including inositol) and unidentified "marker peaks" (Richards, 1991; Boulet *et al.*, 1992) are better observed using much shorter echo times because of short $T_2$s and $J$-modulation effects.

### 3.2. NAA as an Index of Neuronal Dysfunction or Loss

NAA can be used as a neuronal marker as it is found uniquely in neurons and their processes in the normal mature brain (Simmons *et al.*, 1991; Moffett *et al.*, 1991; Birken and Oldendorf, 1989). So-called O-2A progenitor cells isolated from neonatal rodent brain, which can be induced to differentiate into oligodendroglial cells and astrocytes *in vitro*, also synthesize NAA in culture (Urenjak *et al.*, 1993), but their abundance in the mature rodent brain is low (McLaurin and Yong, 1995). While oligodendroglial cell precursors have been identified in the mature human brain, cells with characteristics of O-2A cells have not been found. Thus, at present it appears that changes in brain NAA reflect neuronal pathology exclusively.

In human brain spectra *in vivo*, NAA is reduced in situations in which neuronal loss is recognized as a major aspect of the pathology, e.g., neuronal degenerative disorders (Weinshenker *et al.*, 1991), stroke (Duijn *et al.*, 1992; Graham *et al.*, 1992), and glial tumors (Arnold *et al.*, 1990b). A striking finding

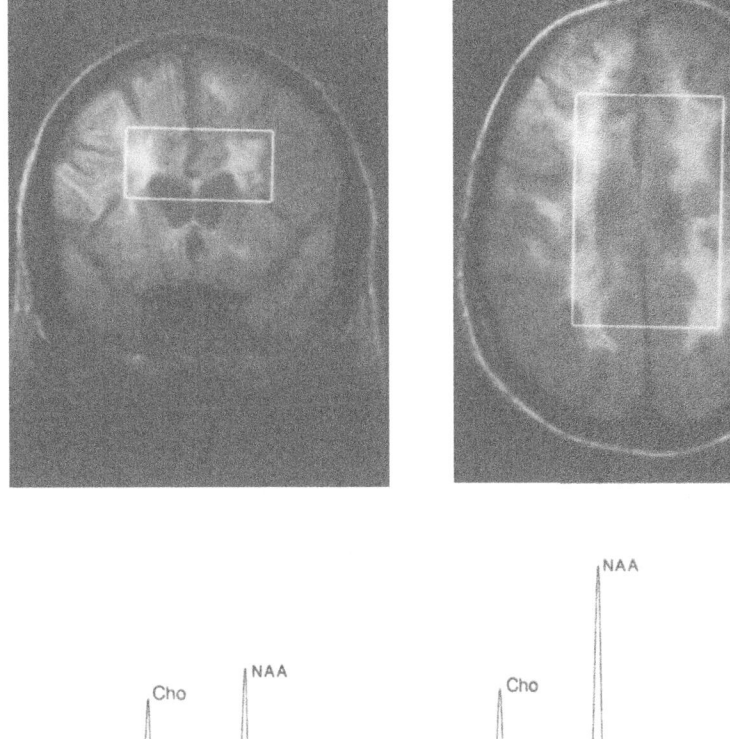

*FIGURE 2.* Scout conventional MRIs used to plan the acquisition of a proton spectrum from a large central volume of interest centered on the corpus callosum. The proton spectrum (TR = 2000 msec, TE = 272 msec) from that volume (*left*) shows a significant decrease in *N*-acetylaspartate (NAA). A normal spectrum from the homologous volume in a normal subject is shown on the right.

in the initial studies of MS was that there is a (variable) loss of brain NAA in patients with MS, with smaller decreases in patients with lower levels of clinical disability. This and subsequent work has emphasized anew that axonal changes occur in MS lesions.

When decreases in the relative NAA signal arise from neuronal or axonal degeneration, irreversible changes are expected. However, we have observed transient decreases in NAA (Fig. 3) in a number of conditions, emphasizing that

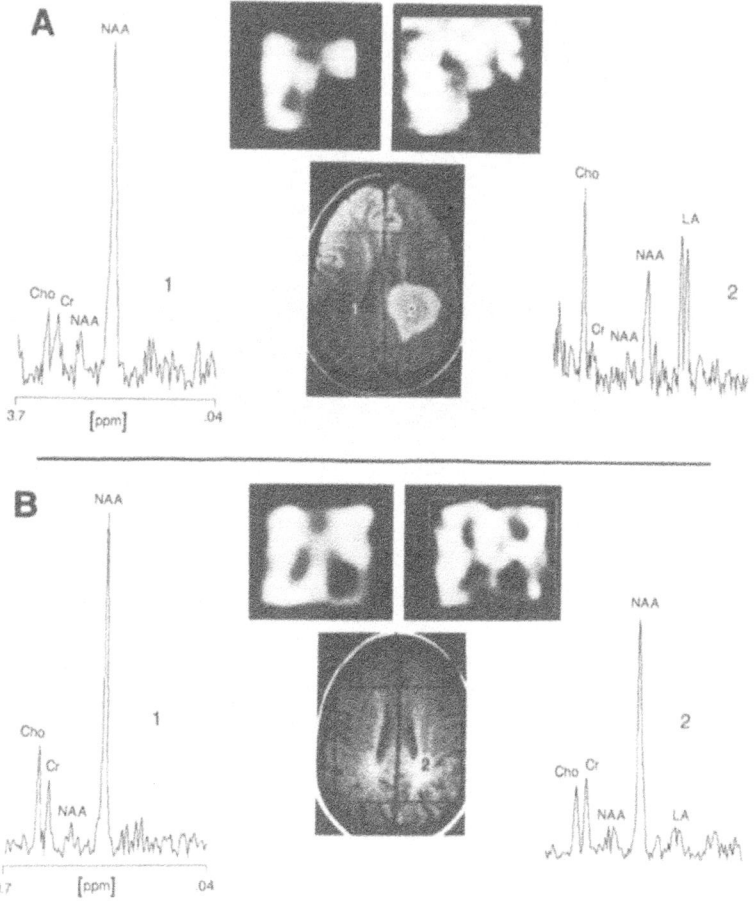

FIGURE 3. Conventional MRI, proton MR spectroscopic image, and sample spectra of brain from a patient with a single large demyelinating plaque during the acute phase of the attack (A) and 13 months later (B). The exam performed one week after the onset of symptoms (panel A) shows focal abnormalities in metabolite images from the NAA resonance at 2 ppm (*upper left*) and 2.6 ppm (*upper right*). These colocalize with the lesion on conventional MRI (shown directly below). Sample spectra from the center of the lesion (A1) and the homologous contralateral voxel (A2) are shown. Note the elevation of Cho, decrease of Cr, and increase in LA, in the spectrum from the lesion. Panel B shows the evolution of the changes in the conventional MRI and the MR spectroscopic image. The sample spectrum (B1 and B2) show substantial normalization of NAA, Cho, Cr, and LA.

neuronal dysfunction or reversible relative volume changes can also lead to decreased NAA (De Stefano *et al.*, 1995b; Davie *et al.*, 1994). The ability to quantify neuronal loss or sublethal damage is one of the most interesting potential applications of MRS in cerebral disorders and will be discussed in more detail below.

### 3.3. Possible Chemical Correlates of Acute Inflammatory Changes and Demyelination

Changes in the resonance intensity of Cho likely result mainly from increases in the steady-state levels of phosphocholine and glycerol phosphocholine. These choline-containing membrane phospholipids are released during active myelin breakdown.

Total Cr concentration is relatively constant throughout the brain and tends to be relatively resistant to change. For this reason, Cr has been used as an internal standard to correct for artifactual variations in signal intensity over space due to magnetic field and radio-frequency inhomogeneity. However, large changes in Cr can be seen with destructive pathology such as malignant tumors (Bruhn *et al.*, 1989; Arnold *et al.*, 1990b; Segebarth *et al.*, 1990; Frahm *et al.*, 1991; Peeling and Sutherland, 1992; Preul *et al.*, 1993) or large acute demyelinating lesions in MS (De Stefano *et al.*, 1995a). Therefore, while total creatine is a reasonable internal standard in some situations, it is not appropriate in others, and its use should be restricted to cases in which it is known or can be demonstrated that Cr does not change.

LA is the end product of glycolysis and accumulates when oxidative metabolism cannot meet energy requirements. However, lactate is both intra- and extracellular, and large amounts may be accumulated *outside* of actively anaerobic tissue (e.g., in necrotic tissue or fluid-filled cysts). In conditions associated with inflammation, LA accumulation may also reflect metabolism of inflammatory cells, rather than brain parenchyma (Petroff *et al.*, 1992).

"Short echo time" spectra have been able to define changes in concentrations of compounds that have short $T_2$s or signals that are obscured at "long" echo times by $J$-modulation effects. Lipid resonances can be seen to be elevated transiently in some lesions (Fig. 4) and are thought to correlate with demyelination due to changes in myelin lipid mobility and accumulation of cholesterol esters in infiltrating macrophages (Wolinsky *et al.*, 1990; Davie *et al.*, 1994). Resonances from inositol can also be observed from acute or active lesions (De Stefano *et al.*, 1995a).

### 3.4. The Chemical–Pathological Evolution of MS Lesions

Serial studies of individual lesions have begun to define a plausible relationship between changes in these compounds and pathological changes in MS

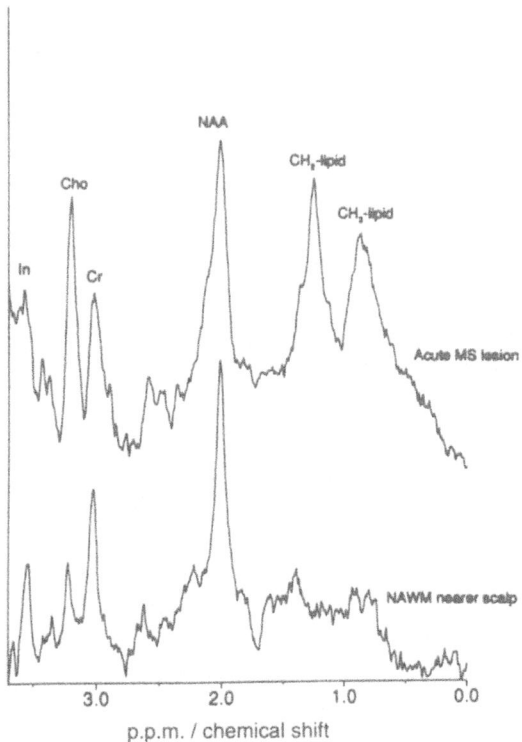

FIGURE 4. Proton spectra obtained at "short" TE, showing an increase in MR visible lipids in an acute plaque compared with normal-appearing white matter (NAWM) between the lesion and the scalp. [From Davie et al., (1994), with permission.]

lesions (Table 2). Proton spectra of acute lesions at long echo times reveal that the Cho and LA resonances increase early in the demyelinating process and that this is followed (at least in many lesions) by a decrease in NAA, presumably from secondary axonal damage or volume loss (P. M. Matthews et al., 1991; Arnold et al., 1990a; Frahm et al., 1989; Larsson et al., 1991; Richards, 1991; Van Hecke et al., 1991; Hanefeld et al., 1991; Miller et al., 1991; Grossman et al., 1992). Identification of lesions in which there is a normal relative concentration of NAA suggests that axonal changes are not found in all lesions, but such changes appear common. Short-echo-time spectra give evidence for transient increases in visible lipids and inositol, lasting between weeks and months (Davie et al., 1993; Posse et al., 1993; Narayana et al., 1992; Larsson et al., 1991; Wolinsky et al., 1990). The available evidence suggests that changes in metabolites result in general from changes in concentration rather than in relaxation times (P. M. Matthews et al., 1991a; Arnold et al., 1992). A likely exception is the lipid resonance, which may become visible largely because of increased mobility with breakdown of the compact myelin. Over a period of weeks there is a progressive reduction of elevated lactate resonance intensities to normal levels. Choline resonance intensity returns to normal over months. The NAA signal intensity may remain decreased or show subsequent partial recovery (Arnold *et al.*, 1992; De Stefano *et al.*, 1995b).

## 3.5. Axonal Dysfunction or Loss in MS Lesions

Changes in axonal metabolism, morphology, or density may be important determinants of functional impairment in MS, despite occurring as only secondary consequences of the pathological process. We believe that the major contribution of MRS to understanding of MS may prove to be quantification of axonal pathology.

Irreversible decreases in NAA within lesions presumably reflect irreversible local axonal volume changes. Pathological studies have shown that axons in chronic plaques shrink by about 30% (Prineas and Connell, 1978). Axonal number or density may be reduced by up to more than 50% (Barnes *et al.*, 1991). Together, the magnitudes of these changes may account well for the maximal decreases of about 70% that we have found in small voxels chosen from the center of lesions to eliminate partial volume averaging with white matter outside of the lesions. In general, less profound decreases in NAA have been reported, however.

Because loss of axon volume or numbers seems to mark a later stage in the evolution of inflammation, it may provide an additional index of disease burden. Early in the course of MS the changes in cerebral NAA are usually relatively focal, and most of the brain has apparently normal levels of NAA (De Stefano *et al.*, 1995b). Axons project through lesions, however, and any axonal interruption that occurs will be associated with anterograde and retrograde axonal (Wallerian) degeneration. A variety of axonal charges have been associated with MS including:

Axonal loss (Wallerian degeneration)
Axonal dystrophy
Interruption of axoplasmic flow
Axonal spheroids
Abortive sprouting (e.g., into Virchow–Robin space)
Increased numbers and degeneration of mitochondria
Sequestration of organelles

Loss of axonal diameter as a result of inflammation and demyelination is likely transmitted beyond the region of acute injury too. Thus, it is not surprising that both single-voxel spectra (Davie *et al.*, 1994; Husted *et al.*, 1994) and MR spectroscopic images (De Stefano *et al.*, 1995a, b) show decreases in NAA extending beyond the borders of plaques defined on MRI. These observations illustrate how, as the disease progresses, and as the multifocal lesions accumulate, axonal dysfunction and loss associated with focal axonal damage could produce a generalized decrease in NAA.

Two approaches can be used to define the generalized decrease of NAA in MS. First, averaging of anatomically segmented conventional MRIs and MR spectroscopic images in a standard anatomical coordinate space makes it possible to generate lesion probability models and average metabolite images. Comparison of these metabolite and MRI lesion probability maps shows loss of NAA beyond regions of high lesion probability into regions of very low lesion probability (Arnold et al., 1993). A second approach that has confirmed this to be a general phenomenon involves independent analysis of the relationship of lesions and metabolic changes, in groups of individual paired MRIs and MR spectroscopic images using spatial statistics (Fu, Wolfson, Worsley, De Stefano, Collins, Narayanan, and Arnold, Unpublished results).

## 3.6. Progression of Axonal Changes in MS

### 3.6.1. Chronic Progressive Charges

The magnitude of the generalized decrease in brain NAA for any patient can be estimated from a single-voxel spectrum centered on the corpus callosum, the main pathway for projections between the two hemispheres (Arnold et al., 1990a). Our initial serial study demonstrated a progressive decrease in the NAA/Cr resonance intensity ratio in spectra from such a central voxel over 18 months (Arnold et al., 1994). This result has been confirmed in a follow-up study with a larger group of patients, although the rate of change was somewhat lower with the larger patient group (unpublished results).

Subgroup analysis of RR and SP patients has shown a consistently slower rate of change of NAA/Cr ratio with time for the SP patients relative to those with RR disease (Arnold et al., 1994, and unpublished observations). The maximum decrease in NAA resonance intensity observed in large central voxels is 30–50%. Together, these results suggest to us that axonal volume changes rather than loss of axons may make the dominant contribution to the generalized NAA changes observed. This is consistent with the relative lack of cortical atrophy until late stages of MS, as widespread Wallerian degeneration would be expected to lead to more prominent secondary changes in the cortical projection neurons. The slower rate of change in SP patients may be hypothesized to result from either a greater relative proportion of disease activity in previously established lesions (in which axonal shrinkage has already occurred) or more widespread axonal shrinkage early in the disease (as the maximum extent of axonal shrinkage may be limited) and a generally longer disease duration in the SP patients. In support of this notion, the extent of the decrease in NAA per unit lesion volume (measured from a large central voxel including the corpus callosum and white matter adjacent to the lateral ventricles) has been found to be higher in SP than in RR patients; that is, there is a greater axonal damage or volume loss per unit volume of lesion in patients with SP MS and greater duration of disease (Matthews, Pioro, Narayanan, De Stefano, Fu, Francis, Antel, Wolfson, and Arnold, 1996).

## 3.6.2. Reversible Axonal Pathology and Functional Impairment

Although serial studies of NAA changes in large central voxels have shown progressive decreases in NAA with time, MR spectroscopic images of single, acute lesions have shown that decreases in individual lesions can be reversible. Recently, we have reported observations in four patients with acute demyelinating lesions (De Stefano et al., 1995b). Initial MR spectroscopic images showed NAA decreases of between 28 and 66% in central lesion voxels. In all patients, serial studies documented recovery of NAA after the acute phase of the disease. Recovery was most rapid in the first few months but thereafter continued for as long as 18 months. The maximum extent of recovery appeared to vary inversely with the magnitude of the initial NAA decrease and was complete for the patient with mildest symptoms and the smallest initial decrease in NAA.

Such recovery of NAA may be related in part to increases in relative axonal volume in lesion voxels. Increases in the diameter of axons that have previously been shrunk might occur with remyelination or clearance of inflammatory factors. As axonal growth is inhibited in the mature CNS, it is unlikely that resprouting and regrowth of degenerated axons could occur to a great enough extent to account for the observed NAA recovery. Alternatively, NAA recovery could reflect reversible metabolic changes in neurons. Recent in vitro studies of a neuronal cell line have demonstrated that decreases in NAA following serum deprivation can be reversed rapidly by reincubation in serum-containing medium (Matthews et al., 1995). In this model system a cell-protein-normalized total NAA value was measured directly by high-pressure liquid chromatography (HPLC), eliminating potential effects of volume changes or cell loss which may occur in vivo.

Axonal metabolic changes secondary to local demyelination may play a direct role in determining functional impairment (McDonald, 1994), or axonal volume or metabolic changes may reflect other local changes important in determining the degree of functional impairment. Whatever the mechanism, preliminary observations suggest that axonal pathology measured by MRS may be closely related to axonal functions. We have recently studied the relationship between functional impairment and NAA in single, large demyelinating lesions responsible for disability in four patients (De Stefano et al., 1995a). The patients were followed for periods from 7 months to $3\frac{1}{2}$ years. In all cases, recovery of initially decreased NAA showed a strong negative correlation with clinical disability.

## 4. CONCLUSIONS

MRI, MRS, and other MR imaging techniques that provide pathologically specific information have much to offer for the improved understanding of the basic mechanism of MS and for investigation of MS patients. By combining different MR modalities for disease assessment, the goal of specific quantitative assessment of the relative amounts and the distribution of myelin destruction, gliosis, and axonal damage might be achieved, allowing a clearer understanding of the mechanisms of functional impairment in MS. Achieving this goal would allow assessment of consequences of therapeutic interventions at various specific stages in the pathological evolution of disease. For example, it would allow evaluation of suppression of immune "triggers" [e.g., by administration of agents such as Cop-1 (Bornstein *et al.,* 1991)], limitation of enlargement or activity in lesions by interfering with early inflammatory mediators (e.g., with $\beta$-interferon), or reduction of the cytotoxic immune response (e.g., using cyclosporin-like agents) or axonal loss (e.g., by neurotrophins). Improvements in the precision with which effects of interventions can be ascertained could lead to effective, rationally designed multimodal therapy for this devastating disease.

## REFERENCES

Arnold, D. L., Matthews, P. M., Francis, G., and Antel, J., 1990a, Proton magnetic resonance spectroscopy of human brain in vivo in the evaluation of multiple sclerosis: Assessment of the load of disease, *Magn. Reson. Med.* **14:**154–159.

Arnold, D. L., Shoubridge, E. A., Villemure, J. G., and Feindel, W., 1990b, Proton and phosphorus magnetic resonance spectroscopy of human astrocytomas *in vivo.* Preliminary observations on tumor grading, *NMR Biomed.* **3:**184–189.

Arnold, D. L., Matthews, P. M., Francis, G. S., O'Connor, J., and Antel, J. P., 1992, Proton magnetic resonance spectroscopic imaging for metabolic characterization of demyelinating plaques, *Ann. Neurol.* **31:**235–241.

Arnold, D. L., Pioro, E., Collins, D. L., Fu, L., Narayanan, S., Francis, G., and Antel, J. P., 1993, Spatial distribution of abnormalities on MRI and MRSI in multiple sclerosis, *Proc. Soc. Magn. Reson. Med.* **1:**282 (abstract).

Arnold, D. L., Riess, G. T., Matthews, P. M., Francis, G. S., Collins, D. L., Wolfson, C., and Antel, J. P., 1994, Use of proton magnetic resonance spectroscopy for monitoring disease progression in multiple sclerosis, *Ann. Neurol.* **36:**76–82.

Barnes, D., Munro, P. M. G., Youl, B. D., Prineas, J. W., and McDonald, W. I., 1991, The longstanding MS lesion; a quantitative MRI and electron microscopic study, *Brain* **114:**1271–1280.

Birken, D. L., and Oldendorf, W. H., 1989, *N*-Acetyl-L-aspartic acid: A literature review of a compound prominent in ¹H-NMR spectroscopic studies of brain, *Neurosci. Biobehav. Rev.* **13:**23–31.

Bornstein, M. B., Miller, A., Slagle, S., Weitzman, M., Drexler, E., Keilson, M., Spada, V., Weiss, W., Appel, S., and Rolak, L., 1991, A placebo-controlled, double-blind, randomized, two-center, pilot trial of Cop 1 in chronic progressive multiple sclerosis, *Neurology* **41**:533–539.

Boulet, L., Karpati, G., and Shoubridge, E. A., 1992, Distribution and threshold expression of the tRNA$^{Lys}$ mutation in skeletal muscle of patients with myoclonic epilepsy and ragged-red fibers (MERRF), *Am. J. Hum. Genet.* **51**:1187–1200.

Broman, T., Andersen, O., and Bergmann, L., 1981, Clinical studies on multiple sclerosis. I. Presentation of an incidence material from Gothenburg, *Acta Neurol. Scand.* **63**:6–33.

Bruhn, H., Frahm, J., Gyngell, M. L., Merboldt, K. D., Hanicke, W., Sauter, R., and Hamburger, C., 1989, Noninvasive differentiation of tumors with use of localized H-1 MR spectroscopy in vivo: Initial experience in patients with cerebral tumors [see comments], *Radiology* **172**:541–548.

Christiansen, P., Gideon, P., Thomsen, C., Stubgaard, M., Henriksen, O., and Larsson, H. B., 1993, Increased water self-diffusion in chronic plaques and in apparently normal white matter in patients with multiple sclerosis, *Acta Neurol. Scand.* **87**:195–199.

Davie, C. A., Hawkins, C. P., Barker, G. J., Brennan, A., Tofts, P. S., Miller, D. H., and McDonald, W. I., 1993, Serial proton MRS in demyelination, *Proc. Soc. Magn. Reson. Med.* **1**:133 (abstract)

Davie, C. A., Hawkins, C. P., Barker, G. J., Brennan, A., Tofts, P. S., Miller, D. H., and McDonald, W. I., 1994, Serial proton magnetic resonance spectroscopy in acute multiple sclerosis lesions, *Brain* **117**:49–58.

De Stefano, N., Matthews, P. M., Antel, J. P., Preul, M., Francis, G., and Arnold, D. L., 1995a, Chemical-pathology of acute demyelinating lesions and its correlation with disability, *Ann. Neurol.* **38**:901–909.

De Stefano, N., Matthews, P. M., and Arnold, D. L., 1995b, Reversible decreases in *N*-acetylaspartate following acute brain injury, *Magn. Reson. Med.* **34**:721–727.

Duijn, J. H., Matson, G. B., Maudsley, A. A., Hugg, J. W., and Weiner, M. W., 1992, Human brain infarction: Proton MR spectroscopy, *Radiology* **183**:711–718.

Filippi, M., Horsfield, M. A., Morrissey, S. P., MacManns, D. G., Rudge, P., McDonald, W. I., and Miller, D. H., 1994, Quantitative brain MRI lesion load predicts the course of clinically isolated syndromes suggestive of multiple sclerosis, *Neurology* **44**:635–641.

Frahm, J., Bruhn, H., Gyngell, M. L., Merboldt, K. D., Hanicke, W., and Sauter, R., 1989, Localized proton NMR spectroscopy in different regions of the human brain *in vivo*. Relaxation times and concentrations of cerebral metabolites, *Magn. Reson. Med.* **11**:47–63.

Frahm, J., Bruhn, H., Hanicke, W., Merboldt, K. D., Mursch, K., and Markakis, E., 1991, Localized proton NMR spectroscopy of brain tumors using short-echo time STEAM sequences, *J. Comput. Assist. Tomogr.* **15**:915–922.

Francis, G. S., Evans, A., and Arnold, D. L., 1995, Neuroimaging in multiple sclerosis, *Neurol. Clin.* **13**:147–171.

Gass, A., Barker, G. J., Kidd, D., Thorpe, J. W., MacManus, D., Brennan, A., Tofts, P. S., Thompson, A. J., McDonald, W. I., and Miller, D. H., 1994, Correlation of magnetization transfer ratio with clinical disability in multiple sclerosis, *Ann. Neurol.* **36**:62–67.

Goodkin, D. E., Ross, J. S., Medendorp, S. V., Konecsni, J., and Rudick, R. A., 1992, Magnetic resonance imaging lesion enlargement in multiple sclerosis. Disease-related activity, chance occurrence, or measurement artifact? *Arch. Neurol.* **49**:261–263.

Graham, G. D., Blamire, A. M., Howseman, A. M., Rothman, D. L., Fayad, P. B., Brass, L. M., Petroff, O. A., Shulman, R. G., and Prichard, J. W., 1992, Proton magnetic resonance spectroscopy of cerebral lactate and other metabolites in stroke patients, *Stroke* **23**:333–340.

Grossman, R. I., Gonzalez-Scarano, F., Atlas, S. W., Galetta, S., and Silberberg, D. H., 1986, Multiple sclerosis: Gadolinium enhancement in MR imaging, *Radiology* **161**:721–725.

Grossman, R. I., Lenkinski, R. E., Ramer, K. N., Gonzalez-Scarano, F., and Cohen, J. A., 1992, MR proton spectroscopy in multiple sclerosis, *Am. J. Neuroradiol.* **13**:1535–1543.

Hanefeld, F., Bauer, H. J., Christen, H. J., Kruse, B., Bruhn, H., and Frahm, J., 1991, Multiple sclerosis in childhood: Report of 15 cases, *Brain Dev.* **13**:410–416.

Husted, C. A., Goodin, D. S., Hugg, J. W., Maudsley, A. A., Tsuruda, J. S., de Bie, S. H., Fein, G., Matson, G. B., and Weiner, M. W., 1994, Biochemical alterations in multiple sclerosis lesions and normal-appearing white matter detected by in vivo $^{31}$P and $^1$H spectroscopic imaging, *Ann. Neurol.* **36**:157–165.

Isaac, C., Li, D., Genton, M., Jardine, C., Grochowski, E., Palmer, M., Kastrukoff, L. F., Oger, J., and Paty, D. W., 1988, Multiple sclerosis: A serial study using MRI in relapsing patients, *Neurology* **38**:1511–1515.

Kastrukoff, L. F., Oger, J. J., Hashimoto, S. A., Sacks, S. L., Li, D. K., Palmer, M. R., Petkau, A. J., Berkowitz, J., and Paty, D. W., 1990, Systemic lymphoblastoid interferon therapy in chronic progressive multiple sclerosis. I. Clinical and MRI evaluation, *Neurology* **40**:479–486.

Kermode, A. G., Thompson, A. J., Tofts, P. S., MacManus, D. G., Kendall, B. E., Kingsley, D. P., Moseley, I. F., Rudge, P., and McDonald, W. I., 1990, Breakdown of the blood–brain barrier precedes symptoms and other MRI signs of new lesions in multiple sclerosis, *Brain* **113**:1477–1489.

Khoury, S. J., Guttmann, C. R., Orav, E. J., Hohol, M. J., Ahn, S. S., Hsu, L., Kikinis, R., Mackin, G. A., Jolesz, F. A., and Weiner, H. L., 1994, Longitudinal MRI in multiple sclerosis: Correlation between disability and lesion burden, *Neurology* **44**:2120–2124.

Koopmans, R. A., Li, D. K., Oger, J. J., Kastrukoff, L. F., Jardine, C., Costley, L., Hall, S., Grochowski, E. W., and Paty, D. W., 1989, Chronic progressive multiple sclerosis: Serial magnetic resonance brain imaging over six months, *Ann. Neurol.* **26**:248–256.

Kurtzke, J. F., 1984, Disability rating scales in multiple sclerosis, *Ann. N. Y. Acad. Sci.* **436**:347–360.

Kurtzke, J. F., 1994, Clinical definition for multiple sclerosis treatment trials [Review], *Ann. Neurol.* **36**(Suppl):S73–S79.

Kurtzke, J. F., Beebe, G. W., Nagler, B., Kurland, L. T., and Auth, T. L., 1977, Studies on the natural history of multiple sclerosis—8. Early prognostic features of the later course of the illness, *J. Chronic Dis.* **30**:819–830.

Larsson, H. B., Christiansen, P., Jensen, M., Frederiksen, J., Heltberg, A., Olesen, J., Henriksen, O., 1991, Localized in vivo proton spectroscopy in the brain of patients with multiple sclerosis, *Magn. Reson. Med.* **22**:23–31.

Matthews, P. M., Francis, G., Antel, J., and Arnold, D. L., 1991, Proton magnetic resonance spectroscopy for metabolic characterization of plaques in multiple sclerosis *Neurology* **41**:1251–1256 [published erratum appears in *Neurology* **41**:1828 (1991)].

Matthews, P. M., Cianfaglia, L., McLaurin, J., Cashman, N., Sherwin, A., Arnold, D. L., and Antel, J., 1995, Demonstration of reversible decreases in *N*-acetylaspartate (NAA) in a neuronal cell line: NAA decreases as a marker of sublethal neuronal dysfunction, *Proc. Soc. Magn. Reson. Med.* (Abstract), **1**:147.

Matthews, W. B., Compston, A., Allen, I. V., and Martyn, C. N., 1991, "McAlpine's Multiple Sclerosis," Churchhill Livingstone, New York.

McAlpine, D., Compston, N. C., and Lumsden, C. E., 1955, Course and prognosis, *in* "Multiple Sclerosis" (D. McAlpine, N. C., Compston, and C. E., Lumsden, eds.), pp. 135–155, Livingstone, Edinburgh.

McDonald, W. I., 1994, Rachelle Fishman-Matthew Moore Lecture. The pathological and clinical dynamics of multiple sclerosis [Review], *J. Neuropathol. Exp. Neurol.* **53**:338–343.

McLaurin, J., and Yong, V. W., 1995, Oligodendrocytes and myelin, *Neurol. Clin.* **13**:23–49.

Miller D. H., 1994, Magnetic resonance in monitoring the treatment of multiple sclerosis [Review], *Ann. Neurol.* **36**(Suppl.):S91–S94.

Miller, D. H., Rudge, P., Johnson, G., Kendall, B. E., MacManus, D. G., Moseley, I. F., Barnes, D., and McDonald, W. I., 1988, Serial gadolinium enhanced magnetic resonance imaging in multiple sclerosis, *Brain* **111**:927–939.

Miller, D. H., Johnson, G., Tofts, P. S., MacManus, D., and McDonald, W. I., 1989, Precise relaxation time measurements of normal-appearing white matter in inflammatory central nervous system disease, *Magn. Reson. Med.* **11**:331–336.

Miller, D. H., Austin, S. J., Connelly, A., Youl, B. D., Gadian, D. G., and McDonald, W. I., 1991, Proton magnetic resonance spectroscopy of an acute and chronic lesion in multiple sclerosis [Letter], *Lancet* **337**:58–59.

Moffett, J. R., Namboodiri, M. A. A., Cangro, C. B., and Neale, J. H., 1991, Immunohistochemical localization of N-acetylaspartate in rat brain, *Neuroreport* **2**:131–134.

Morrissey, S. P., Miller, D. H., Kendall, B. E., Kingsley, D. P., Kelly, M. A., Francis, D. A., MacManus, D. G., and McDonald, W. I., 1993, The significance of brain magnetic resonance imaging abnormalities at presentation with clinically isolated syndromes suggestive of multiple sclerosis. A 5-year follow-up study, *Brain* **116**:135–146.

Mushlin, A. I., Detsky, A. S., Phelps, C. E., O'Connor, P. W., Kido, D. K., Kucharczyk, W., Mooney, C., Tansey, C. M., and Hall, W. J., 1993, The accuracy of magnetic resonance imaging in patients with suspected multiple sclerosis. The Rochester–Toronto Magnetic Resonance Imaging Study Group, *J. Am. Med. Assoc.* **269**:3146–3151.

Narayana, P. A., Wolinsky, J. S., Jackson, E. F., and McCarthy, M., 1992, Proton MR spectroscopy of gadolinium-enhanced multiple sclerosis plaques, *J. Magn. Reson. Imag.* **2**:263–270.

Offenbacher, H., Fazekas, F., Schmidt, R., Freidl, W., Flooh, E., and Payer, F., 1993, Assessment of MRI criteria for a diagnosis of MS, *Neurology* **43**:905–909.

Paty, D. W., 1993, Magnetic resonance in multiple sclerosis [Review], *Curr. Opin. Neurol. Neurosurg.* **6**:202–208.

Paty, D. W., and Li, D. K., 1993, Interferon beta-1b is effective in relapsing–remitting multiple sclerosis. II. MRI analysis results of a multicenter, randomized, double-blind, placebo-controlled trial. UBC MS/MRI Study Group and the IFNB Multiple Sclerosis Study Group [see comments]. Neurology 1993;43(4):662–667.

Patzold, U., and Pocklington, P. R., 1982, Course of multiple sclerosis. First results of a prospective study carried out on 102 MS patients from 1976–1980, *Acta Neurol. Scand.* **65**:248–266.

Peeling, J., and Sutherland, G., 1992, High-resolution ¹H NMR spectroscopy studies of extracts of human cerebral neoplasms, *Magn. Reson. Med.* **24**:123–136.

Petroff, O. A., Graham, G. D., Blamire, A. M., al-Rayess, M., Rothman, D. L., Fayad, P. B., Brass, L. M., Shulman, R. G., Prichard, J. W., 1992, Spectroscopic imaging of stroke in humans: Histopathology correlates of spectral changes, *Neurology* **42**:1349–1354.

Posse, S., Schuknecht, B., Smith M. E., van Zijl, P. C., Herschkowitz, N., and Moonen, C. T., 1993, Short echo time proton MR spectroscopic imaging, *J. Comput. Assist. Tomogr.* **17**:1–14.

Preul, M., Collins, D., Ethier, R., Feindel, W., and Arnold, D. L., 1993, Classification of major intracranial tumor types using ¹H MR spectroscopic imaging and feature space for spectral pattern recognition, *Proc. Soc. Magn. Reson. Med.* **1**:64 (Abstract).

Prineas, J. W., and Connell, F., 1978, The fine structure of chronically active multiple sclerosis plaques, *Neurology* **28**(Suppl. 2):68–75.

Prineas, J. W., Barnard, R. O., Kwon, E. E., Sharer, L. R., and Cho, E. S., 1993, Multiple sclerosis: Remyelination of nascent lesions, *Ann. Neurol.* **33**:137–151.

Quarles, R. H., Morell, P., and McFarlin, D. E., 1993, Diseases involving myelin, *in* "Basic Neurochemistry: Molecular, Cellular, and Medical Aspects" (J. Siegel, B. W. Agranoff, R. W. Albers, and P. B. Molinoff, eds.), pp. 771–792. Raven Press, New York.

Raine, C. S., and Cross, A. H., 1989, Axonal dystrophy as a consequence of long-term demyelination, *Lab. Invest.* **60**:714–725.

Rasminsky, M., 1984, Pathophysiology of demyelination [Review], *Ann. N. Y. Acad. Sci.* **436**:68–85.

Rasminsky, M., and Sears, T. A., 1972, Internodal conduction in undissected demyelinated nerve fibres, *J. Physiol.* **227**:323–350.

Rasminsky, M., Kearney, R. E., Aguayo, A. J., and Bray, G. M., 1978, Conduction of nervous impulses in spinal roots and peripheral nerves of dystrophic mice, *Brain Res.* **143:**71–85.

Richards, T. L., 1991, Proton MR spectroscopy in multiple sclerosis: Value in establishing diagnosis, monitoring progression, and evaluating therapy, *Am. J. Roentgenol.* **157:**1073–1078.

Rodriguez, M., and Scheithauer, B., 1994, Ultrastructure of multiple sclerosis, *Ultrastruct. Pathol.* **18:**3–13.

Segebarth, C. M., Baleriaux, D. F., Luyten, P. R., and den Hollander, J. A., 1990, Detection of metabolic heterogeneity of human intracranial tumors in vivo by ¹H NMR spectroscopic imaging, *Magn. Reson. Med.* **13:**62–76.

Simmons, M. L., Frondoza, C. G., and Coyle, J. T., 1991, Immunocytochemical localization of *N*-acetyl-aspartate with monoclonal antibodies, *Neuroscience* **45:**37–45.

Smith, M. E., Stone, L. A., Albert, P. S., Frank, J. A., Martin, R., Armstrong, M., McFarlin, D. E., and McFarland, H. F., 1993, Clinical worsening in multiple sclerosis is associated with increased frequency and area of gadopentetate dimeglumine-enhancing magnetic resonance imaging lesions, *Ann. Neurol.* **33:**480–489.

Stone, L. A., Smith, M. E., Albert, P. S., Bash, C. N., Maloni, H., Frank, J. A., and McFarland, H. F., 1995, Blood–brain barrier disruption on contrast-enhanced MRI in patients with mild relapsing–remitting multiple sclerosis: Relationship to course, gender, and age, *Neurology* **45:**1112–1126.

Sumner, A. J., Saida, K., Saida, T., Silberberg, D. H., and Asbury, A. K., 1982, Acute conduction block associated with experimental antiserum-mediated demyelination of peripheral nerve, *Ann. Neurol.* **11:**469–477.

Thompson, A. J., Kermode, A. G., Wicks, D., MacManus, D. G., Kendall, B. E., Kingsley, D. P., and McDonald, W. I., 1991, Major differences in the dynamics of primary and secondary progressive multiple sclerosis, *Ann. Neurol.* **29:**53–62.

Urenjak, J., Williams, S. R., Gadian, D. G., and Noble, M., 1993, Proton nuclear magnetic resonance spectroscopy unambiguously identifies different neural cell types, *J. Neurosci.* **13:**981–989.

Van Hecke, P., Marchal, G., Johannik, K., Demaerel, P., Wilms, G., Carton, H., and Baert, A. L., 1991, Human brain proton localized NMR spectroscopy in multiple sclerosis, *Magn. Reson. Med.* **18:**199–206.

Waxman, S. G., 1995, Sodium channel blockade by antibodies: A new mechanism of neurological disease? [editorial; comment], *Ann. Neurol.* **37:**421–423.

Weiner, H. L., Hohol, M. J., Khoury, S. J., Dawson, D. M., and Hafler, D. A., 1995, Multiple sclerosis, *Neurol. Clin.* **13:**173–196.

Weinshenker, B. G., and Ebers, G. C., 1987, The natural history of multiple sclerosis [Review], *J. Can. Sci. Neurol.* **14:**255–261.

Weinshenker, B. G., Rice, G. P. A., Noseworthy, J. H., Carriere, W., Baskerville, J., and Ebers, G. C., 1991, The natural history of multiple sclerosis: A geographically based study, *Brain* **114:**1057–1067.

Willoughby, E. W., Grochowski, E., and Lee, D., 1989, Serial magnetic resonance scanning in multiple sclerosis: A second prospective study in relapsing patients, *Ann. Neurol.* **25:**43–49.

Wolinsky, J. S., Narayana, P. A., and Fenstermacher, M. J., 1990, Proton magnetic resonance spectroscopy in multiple sclerosis, *Neurology* **40:**1764–1769.

# PHOSPHORUS AND PROTON MAGNETIC RESONANCE SPECTROSCOPY OF THE BRAIN OF THE NEWBORN HUMAN INFANT

## ERNEST B. CADY

## SUMMARY

Cerebral damage in premature infants is often related to periventricular hemorrhage, but, in both term and premature infants, intrapartum hypoxia–ischemia ("birth asphyxia") is also a major cause. Hence, there is a need for noninvasive techniques capable of investigating cerebral energy generation and other aspects of neurometabolism. Magnetic resonance spectroscopy (MRS) can safely and noninvasively monitor *in vivo* neurochemistry, and in particular cerebral

ERNEST B. CADY • *Department of Medical Physics and Bio-Engineering, University College London Hospitals NHS Trust, London, WC1E 6JA, United Kingdom.*

*Magnetic Resonance Spectroscopy and Imaging in Neurochemistry,* Volume 8 of *Advances in Neurochemistry,* edited by Bachelard, Plenum Press, New York, 1997.

energy metabolism in the brains of newborn infants. Oxidative phosphorylation and ATP levels can be investigated using phosphorus ($^{31}$P) MRS, and proton ($^1$H) MRS enables the study of anaerobic glycolysis. MRS studies of brain injury in newborn infants commenced in the early 1980s: the first spectra from living human brain were obtained at University College London in 1982. This chapter first summarizes the management during MRS of both normal newborn infants and those undergoing intensive care and then examines MRS localization and quantification techniques. The results from studies of normal developing brain are then described. Finally, detailed sections discuss the roles of MRS in monitoring perinatal hypoxic–ischemic brain injury, in providing prognostic information, and in investigations of putative cerebroprotective methods that may be of clinical benefit.

## 1. INTRODUCTION

Following the development in the early 1980s of magnetic resonance spectroscopy (MRS) systems capable of studies of human limbs and small mammals, this noninvasive modality was soon applied to the investigation of brain injury in newborn infants (Cady et al., 1983). Cerebral damage in premature infants (i.e., born before 37 weeks of gestation) is often related to periventricular hemorrhage (PVH). However, in both term and premature infants, intrapartum hypoxia–ischemia is also a major cause. In ~10% of handicapped children of school age, permanent neurodevelopmental impairments are due to such cerebral injury. During the previous decade, X-ray computerized tomography and, in particular, ultrasound imaging had proven useful for studying the incidence, pathogenesis, and prognostic importance of PVH. However, long-term follow-up studies indicated that hypoxia–ischemia, often a cause of periventricular leukomalacia (PVL), was more important than PVH in the pathogenesis of permanent neurodevelopment impairments in premature infants (Stewart et al., 1983). In infants delivered close to term, acute intrapartum hypoxia–ischemia (often termed "birth asphyxia") was found to be a major cause of cerebral damage. In fact, hypoxia–ischemia was found to be the commonest underlying cause of serious perinatal brain injury. Hence, a real need existed for the development of noninvasive techniques for the investigation of cerebral energy-generating processes, notably oxidative phosphorylation and anaerobic glycolysis, in order to study the pathogenesis and evolution of hypoxic–ischemic injury. Provided existing guidelines are followed, MRS is a safe, noninvasive method for studying neurochemistry, and in particular high-energy metabolism, in the otherwise inaccessible brain of a newborn infant (Cady et al., 1991, 1993; Cady, 1992). Among other applications, the phosphorus ($^{31}$P) nucleus can be employed for investigating oxidative phosphorylation and maintenance of ATP levels, and the proton ($^1$H) for studying anaerobic glycolysis.

The first spectra from living human brain were obtained from a preterm baby at University College London in 1982 (Cady *et al.,* 1983) using a ³¹P pulse–acquire surface-coil acquisition technique in a 1.9-T magnet with a 20-cm bore. Within the next few years, several groups worldwide also commenced MRS investigations of the brains of newborn infants (Younkin *et al.,* 1984; Corbett *et al.,* 1987; Boesch and Martin, 1988). It was soon evident that metabolites involved in energy metabolism [e.g., ATP, phosphocreatine (PCr), and $P_i$] as well as some related to the development and maintenance of cellular membranes [e.g., phosphomonoesters (PME) and phosphodiesters (PDE)] were easily detected. Furthermore, intracellular pH ($pH_i$) could also be estimated using the $P_i$ chemical shift (see chapter 2). Obvious abnormalities (e.g., reduced $[PCr]/[P_i]$) were found in ³¹P brain spectra acquired from infants with hypoxic–ischemic injury (Cady *et al.,* 1983; Younkin *et al.,* 1984). These discoveries were soon followed by the elucidation of developmental changes in normal infants (Hamilton *et al.,* 1986) and later on in older children also (Boesch *et al.,* 1989; van der Knaap *et al.,* 1990).

With the advent of efficient MRS localization methods and the increased availability of high-field, whole-body, magnetic resonance imaging (MRI) systems, ¹H MRS has now been applied to the study of neonatal brain at several centers worldwide (van der Knaap *et al.,* 1990; Huppi *et al.,* 1991; Kreis *et al.,* 1993; Peden *et al.,* 1993; Groenendaal *et al.,* 1994; Cady *et al.,* 1994a, 1995a; Penrice *et al.,* 1995). Using ¹H MRS, resonances from lactate (Lac; a product of anaerobic glycolysis), various amino acids (e.g., Glu and Gln), and other cerebral metabolites such as choline-containing compounds (Cho; related to membrane metabolism) and *N*-acetylaspartate (NAA; a putative neuronal marker) can be detected. ¹H studies provide signals of larger amplitude than those of ³¹P: because of the high ¹H MRS sensitivity and several equivalent nuclei contributing to most resonances, smaller localized volumes of interest (VOI), e.g., a few milliliters, and shorter acquisition times, e.g., a few minutes, can be used.

The aims of this chapter are to summarize the patient management and data acquisition techniques used for MRS of the brain in newborn infants, to describe the results from studies of normal developing brain using both ³¹P and ¹H MRS, and then to examine the role of MRS particularly in the investigation of both perinatal hypoxic–ischemic brain injury and cerebroprotection.

## 2. MANAGEMENT OF THE NEWBORN DURING SPECTROSCOPY STUDIES

MRS studies of newborn infants are safe provided that current guidelines are strictly adhered to (e.g., Department of Health, United Kingdom, 1993). However, requirements for physiological support and monitoring, especially for infants requiring intensive care, are necessarily much greater than for adults undergoing comparable investigations.

## 2.1. Patient Handling

Newborn patients, many of whom may be very ill, should be conveyed to and from the MR suite and studied using a special transport-incubator. This consists of a nonmagnetic pod enclosing the baby and suitable for insertion into the magnet and also includes all the essential life-support and monitoring equipment that would be provided in the intensive care unit.

In studies of newborns, unlike of older children, mild sedation to restrict movement may not be permissible, especially for investigations of normal infants. In many cases, babies can be studied while naturally asleep after feeding. In studies of sick infants, involuntary movement may be a problem (e.g., due to seizures). However, for clinical reasons these infants are often sedated with phenobarbitone or other preparations. Gentle head restraint can be achieved by using, for example, an evacuatable polyethylene bag filled with expanded-polystyrene beads. Patient handling apparatus must be carefully designed to account for the large range of head size of premature and term infants. It is of great importance that data acquisition methods are such that useful information can be obtained even if some movement occurs.

## 2.2. Physiological Monitoring and Maintenance

Standards for monitoring and maintaining sick infants are always improving. Today, equipment should include facilities for environmental temperature control, mechanical ventilation (humidified), and monitoring of heart rate (electrocardiogram or Doppler ultrasound), blood pressure, arterial oxygen saturation (pulse oximetry), end-tidal $CO_2$, transcutaneous $P_{O2}$ and $P_{CO2}$, and body temperature (skin and/or rectal sensors) (Chu et al., 1986; Boesch and Martin, 1988). Intravenous infusions may also have to be provided using controlled drip-feed devices.

It should be noted that, in addition to the problem of radio-frequency (RF) interference being detected by the spectrometer (particularly if microprocessor-controlled monitors are used), the brief RF and magnetic-gradient pulses necessary for MRS can produce artifactual signals and safety risks unless care is taken (Shellock and Slimp, 1989).

## 3. DATA ACQUISITION METHODS

Investigations of the brains of newborn infants often produce better quality spectra if dedicated probe heads are used rather than those intended primarily for adult use. Furthermore, pulse sequences should be resilient to head movement (especially important for collecting data from unsedated normal infants).

## 3.1. Radio-Frequency Probes

Surface coils (i.e., simple loops of wire tuned by a capacitance) with diameters from 5 to 7 cm can be used to obtain ³¹P spectra of good quality from the cerebral cortex of the newborn infant without further localization techniques. Very little contamination comes from superficial scalp tissues since skin and fat contain negligible ³¹P, cranial musculature is immature, and infant bone is poorly mineralized and its ³¹P nuclei are relatively immobile. However, without reduction of adipose-tissue signals, surface coils are of very limited application for ¹H MRS. Some degree of localization of ³¹P spectra can be obtained simply by positioning the surface coil on the scalp close to the brain region of interest. However, surface-coil spectra are acquired only from tissue within about a coil radius, and the sensitive volume is poorly defined. To investigate specific tissue volumes in deeper structures, sophisticated localization techniques are necessary. For full flexibility in terms of positioning the VOI, the RF coil must provide a uniform RF field throughout the head. These "volume coils" are often of the "birdcage" or "slotted-tube resonator" (STR) design and, when manufacturer-supplied, are usually for adult applications. For studies of infants, there are advantages in signal-to-noise ratio (SNR) and shimming (optimizing static magnetic field homogeneity and consequently also spectrum resolution) to be gained by using smaller coils (Cady, 1995). In order to obtain scout images for VOI positioning, ¹H tuning is essential. A double-tuned (¹H and ³¹P) STR dedicated to the examination of infant brain has been developed (Gruetter et al., 1990). However, a Helmholtz coil design has also been used for neonatal MRS (Cady, 1995). This design is easy to "home build": in order improve efficiency, it can be balanced-matched by inductive coupling and series tuning (Decorps et al., 1985), and it can be tuned for both ¹H and ³¹P (Cady, 1995). A further advantage is that the mouthpiece required for mechanical ventilation can be sited between the loops. A Helmholtz coil can therefore be comparatively small, thereby maximizing SNR.

## 3.2. Localization Pulse Sequences

To investigate focal brain pathology and regional metabolic variations in normal and sick infants, and to use ¹H MRS, localization techniques are essential. Recent methods all use pulsed magnetic field gradients in combination with selective RF pulses to obtain spectra from a well-defined VOI (i.e., with very little contamination from extra-VOI signal) positioned by reference to scout images. The VOI can be sited anywhere within the head; however, due to low ³¹P MRS sensitivity and only one equivalent nucleus per resonance (as compared to often three or more per resonance in [1]H MRS), [31]P VOIs must be larger and [31]P acquisition times longer than for [1]H. In these multipulse localization methods, RF power deposition is higher than for "single-pulse" surface-coil methods, and consequently greater attention to safety is essential.

With the general availability of whole-body imaging systems designed primarily for [1]H applications, recent studies of newborn infants have tended to use [1]H MRS. Both the point-resolved spectroscopy (PRESS) and stimulated-echo amplitude mode (STEAM) techniques (see, e.g., Moonen et al., 1989) have found general use for single-VOI [1]H studies of infant brain (van der Knaap et al., 1990; Huppi et al., Kreis et al., 1993; Groenendaal et al., 1994; Cady et al., 1994a, 1995a; Penrice et al., 1995).

For [31]P studies, image-selected in vivo spectroscopy (ISIS; Ordidge et al., 1986) has been used most often (van der Knaap et al., 1990; Buchli et al., 1994). However, with the development of shielded gradient systems (thereby reducing eddy currents), very short echo times (TE) can now be used in both PRESS and STEAM. Furthermore, it has recently been reported that ATP [31]P resonances have longer spin–spin relaxation times $(T_2)$ than previously realized (60–90 msec) and that many other [31]P metabolites (e.g., PCr) have even longer $T_2$s (Jung et al., 1993). These factors have enabled [31]P PRESS studies of infant brain (see Fig. 1A) (Cady, 1995; Cady et al., 1995b). If TE is short enough (e.g., 10 msec), the effects of phase modulation of the γ- and α-nucleotide triphosphate (NTP) resonances are not severe. Loss of signal by $T_2$ relaxation is also small if TE is short. The effects of phase modulation and $T_2$ relaxation on the NTP multiplets can be accounted for during data analysis. [31]P PRESS has advantages in terms of rapid setup of the [31]P study following [1]H MRS ([31]P RF pulse amplitudes can be derived from those for [1]H), and the [31]P VOI can be shimmed using [1]H PRESS to improve spectrum resolution. The [1]H water signal from the same VOI can also be used for quantitation similarly to its use in [1]H MRS.

MR spectroscopic imaging (e.g., Duyn and Moonen, 1993) has not yet been as widely applied in neonatal studies as the single-voxel methods.

## 3.3. Quantitation Techniques

The area of a peak in a spectrum is related to the metabolite concentration. However, peak area also depends on other factors including repetition time (TR), spin–lattice relaxation time $(T_1)$, TE and $T_2$ in spin-echo spectra, the number of equivalent nuclei, VOI volume, the number of summed signals, tissue conductivity (which reduces probe efficiency), and the receiver amplification or attenuation required to digitize the signal adequately. Although peak-area ratios may be useful for comparing normal and pathological tissues, changes can be caused by alterations in $T_1$ or $T_2$ as well as in metabolite concentration. Changes in the latter may be masked by altered relaxation (Wilkinson et al., 1994; Cady et al., 1994b).

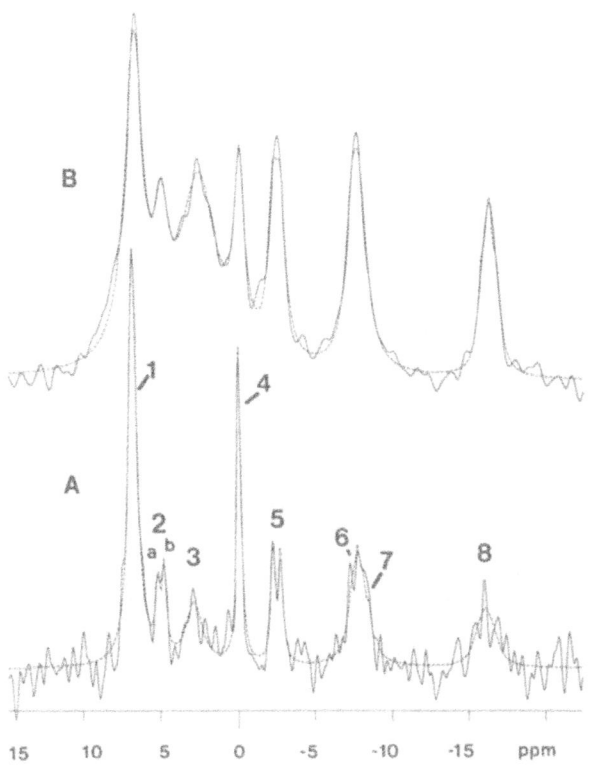

FIGURE 1. <sup>31</sup>P PRESS (A) and pulse-acquire surface-coil (B) spectra acquired from the brains of term newborn infants suspected of perinatal hypoxic–ischemic cerebral injury. In both spectra the mildness of the cerebral injury is indicated by the normal $P_i$, PCr, and NTP levels. The localized (PRESS) spectrum has far better resolution than that acquired by surface coil. In spectrum B, the very broad underlying hump due to relatively immobile <sup>31</sup>P in bone and membrane phospholipids has been removed by postacquisition processing. In spectrum A, both the phospholipid hump and the broad, cross-linked, PDE signal are negligible because their signals dephased during the pulse sequence owing to their short $T_2$s; remaining PDE signal is mainly due to mobile PDEs. Acquisition conditions: (A) 2.4 T, TE 10 msec, TR 12 sec, 160 echoes summed, cubic 125-ml central VOI; (B) 2.4 T, TR 2.256 sec, 256 summed free-induction decays, 6.5-cm-diameter surface coil adjacent to the temporoparietal cortex, 90° coil-center flip angle. Peak identifications: 1, PME; 2, $P_i$ (a, mainly extracellular; b, intracellular); 3, PDE; 4, PCr; 5, 6, and 8, $\gamma$-, $\alpha$-, and $\beta$-NTP, respectively; 7, NAD + NADH and uridine diphosphosugars (these resonances appear as a shoulder on the right side of peak 6). The dashed lines are the results of fitting Lorentzian profiles to the resonances by gradient-search $\chi^2$ minimization. (Spectra acquired in collaboration with J. Penrice, A. Lorek, R. Aldridge, M. Wylezinska, J. Wyatt, and E. O. R. Reynolds.)

## 3.3.1. Relaxometry

To determine metabolite concentrations, relaxation measurements are often essential. Changes in $T_1$ or $T_2$ may also be clinically significant (Prielmeier *et al.*, 1994; Cady *et al.*, 1994b). To measure the $T_2$s of mobile metabolites, it is necessary to collect two (or more) spin-echo spectra with different TEs but with a minimum TE (e.g., $> \sim 75$ msec in $^1$H MRS) such that broad, underlying, short-$T_2$ components are negligible. The metabolite $T_2$ can then be derived from the relationship $A_\tau = A_0 \exp{(-\tau/T_2)}$, where $A_\tau$ and $A_0$ are the peak areas at TE $= \tau$ and TE $= 0$ msec, respectively. In clinical situations, sufficient study time is rarely available to collect fully relaxed localized spectra (i.e., with TR $> 5T_1$). If $T_1$ measurements are necessary, in order to correct for inadequate magnetization recovery, they are best made using the rapid technique utilizing two TRs with TR$_1$ = TR$_2$/2 (Fenstermacher and Narayana, 1990; Kamada *et al.*, 1994): $T_1$ = (TR$_1$ − TR$_2$)/ln($A_{TR2}/A_{TR1}$ − 1). Provided selective RF pulses in the localization sequence flip the magnetization by their nominal 90° or 180° (i.e., the resultant magnetization is entirely in a plane orthogonal to its equilibrium direction), a peak area ($A_{TR}$) can then be corrected for incomplete recovery using the relationship $A_{TR} = A_\infty[1 − \exp(-TR/T_1)]$, where $A_\infty$ is the required fully relaxed peak area. Reasonably accurate $T_1$ estimates can also be made by comparing peak areas from an effectively fully relaxed spectrum (TR $> 5T_1$) with those from a spectrum acquired with, say, TR = $T_1$, but this may take longer.

## 3.3.2. Metabolite Concentrations

$^1$H metabolite concentrations (i.e., in millimoles per kilogram of wet tissue or millimoles per liter of wet tissue) are quite difficult to estimate, largely owing to the use of echo localization methods and the consequent dependence of peak area on $T_2$ in addition to $T_1$, and, in the case of multiplets, on phase modulation also. However, many measurements have now been made on infant brain with the use of either external concentration references (Kreis *et al.*, 1993) or brain water as an internal reference (Christiansen *et al.*, 1993; Toft *et al.*, 1994; Cady, 1995; Cady *et al.*, 1995a). In the latter method, the brain-water signal is measured with TEs from ~25 msec up to, e.g., 2000 msec. The brain-water signal has a double-exponential TE dependence (see Fig. 2a). The rapidly decaying component ($T_2 \sim 200$ msec) originates from extra- and intracellular water whereas that decaying more slowly is due to cerebrospinal fluid (CSF). The brain-tissue water signal at TE = 0 msec can be obtained by fitting a decaying double-exponential function to the data. By comparing this with the metabolite signal (corrected for $T_1$ and $T_2$) and assuming a brain-water fraction (Dobbing and Sands, 1973), the metabolite concentration can be estimated. If metabolite spectra are obtained using short TEs (e.g., 25–75 msec), then these signals also have a double-

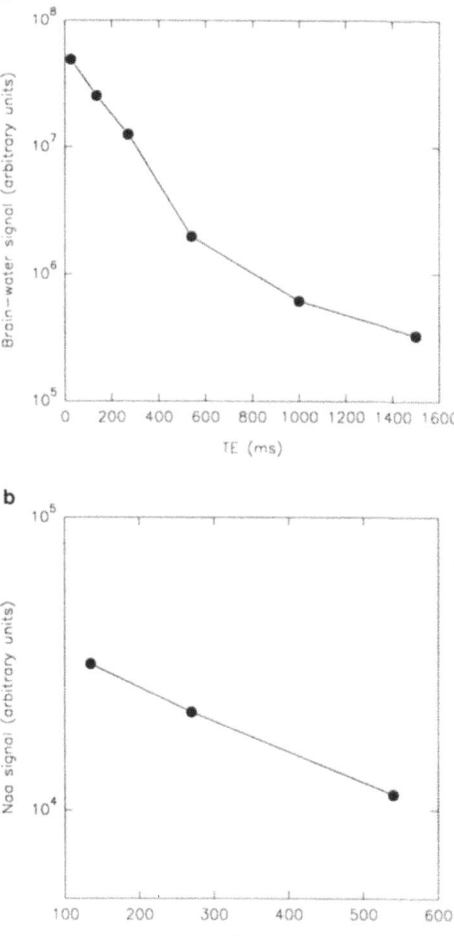

*FIGURE* 2. TE dependences of brain-water (a) and NAA (b) PRESS signals in the thalamic region of a newborn infant. The amplitude scales (vertical) are both logarithmic. The brain-water signal dependence is biexponential and appears to originate from at least two monoexponentially decaying fractions: a major intracellular/extracellular component with a relatively short $T_2$ and a smaller CSF component with a much longer $T_2$. The NAA signal (b) appears to relax purely monoexponentially, presumably due to the exclusively intracellular nature of this metabolite. (Data acquired in collaboration as in caption to Fig. 1.)

exponential TE dependence: the more slowly relaxing components originate from mobile intracellular metabolites. At long enough TEs, these mobile metabolites relax monoexponentially (see Fig. 2b). The fast-relaxing components ($T_2$s of ~20 msec) are mainly due to macromolecules (e.g., proteins) and produce broad underlying spectral features.

In $^{31}$P MRS, metabolite concentrations in developing brain have been estimated by a surface-coil method with brain water as an internal reference (Cady, 1990, 1991), ISIS with a load-matched external reference (Buchli *et al.*, 1994), and PRESS (Cady, 1995; Cady *et al.*, 1995b). The same brain-water internal concentration reference method as used for $^{1}$H MRS and $^{31}$P surface-coil MRS can be applied to $^{31}$P PRESS. However, in addition to measuring the relative $^{31}$P and $^{1}$H sensitivities of the spectrometer receiver, the "trimming" (adjustment of gradient refocusing pulses to counter dephasing across the VOI) must be optimized so as to detect maximum signal from both nuclei.

## 4. NORMAL BRAIN DEVELOPMENT IN PREMATURE AND TERM INFANTS

A great deal of information concerning the relative and molar concentrations and relaxation rates of the metabolites detected by $^{31}$P and $^{1}$H MRS in the brains of newborn infants has now been acquired at several centers worldwide with good corroboration of results. The brain is developing rapidly during the last three months of gestation (e.g., neuronal migration, myelination, axonal proliferation). It is therefore essential to establish normal age-related values for peak-area ratios, concentrations, etc., so that abnormalities can be identified and spectra correctly interpreted. Age-dependent changes in peak-area and concentration ratios and molar concentrations of $^{31}$P and $^{1}$H metabolites have now been defined in several studies. Attention has focused on the rapid changes taking place in the brains of preterm and term infants studied during the first days of life.

### 4.1. $^{31}$P Spectroscopy

Whereas the main resonances in $^{31}$P spectra from skeletal muscle had already been firmly identified, the first spectra obtained from human brain (that of the newborn infant) initially presented problems of resonance assignment. However, the peaks originating mainly from substances directly involved in energy metabolism were easily assigned.

#### 4.1.1. Energy Metabolism.

Resonances attributable to important substances involved in energy metabolism—γ-, α-, and β-NTP, PCr, and $P_i$—were easily detected in the brains of newborn infants (see Fig. 1). Cerebral NTP consists mainly of magnesium-complexed ATP, with contributions from GTP, UTP, and CTP making up about a quarter of the NTP pool (Chapman *et al.*, 1981). The γ- and α-NTP doublet peaks have minor contributions from nucleotide diphosphates; the NAD and NADH quartets and uridine diphosphosugars (UDPS) coresonate close to α-NTP (Glonek *et al.*, 1982) and are often unresolved from this latter doublet in clinical spectra.

Within physiological ranges, the resonant frequency of PCr is independent of both pH (only altering significantly at very acid pHs) and metal-ion (e.g., Mg) concentrations; this resonance is almost universally used as a chemical-shift reference (at 0 ppm).

With surface-coil and ISIS techniques, only a single $P_i$ resonance is detected, as shown in Fig. 1B (peak 1). Its chemical shift relative to PCr ($\delta_{pi}$) depends strongly on the relative proportions of its acid and base forms ($pK_a \simeq$ 6.8) and only slightly on metal-ion concentration. Hence, *in vitro* titrations of $\delta_{pi}$ against pH can be carried out in a medium modeling the cerebral cytosol, and brain intracellular pH ($pH_i$) can then be estimated from the Henderson–Hasselbalch equation (Petroff *et al.*, 1985):

$$pH_i = 6.77 + \log_{10}[(\delta_{pi} - 3.29)/(5.68 - \delta_{pi})] \tag{1}$$

This allows the estimation of $pH_i$ to an accuracy of about 0.1 units (see chapter 2). At high field strengths ($\geq 4.7$ T) and at 2.4 T using ³¹P PRESS (see Fig. 1A), the $P_i$ peak is often resolvable into extra- and intracellular components (Portman and Ning, 1990; Robbins *et al.*, 1990), implying compartments with different pHs. [It is also possible that the extracellular component (peak 2a in Fig. 1A) contains a small contribution from mitochondrial $P_i$ (Garlick *et al.*, 1992).]

### 4.1.2. Membrane Metabolism

The remaining resonances initially posed identification problems. Eventually, it was found that these peaks originated from phosphomonoesters (PME) and phosphodiesters (PDE)—substances related to membrane development. Both the PME peak and the moderately broad, unresolved PDE peak, detected in single-pulse surface-coil spectra (see Fig. 1B), were surprisingly large. Furthermore, surface-coil spectra were superimposed on a very broad hump unless postacquisition processing was used to remove this feature (as in Fig. 1B). Chromatographic and ³¹P and ¹H MRS studies of brain extracts from newborn mammals have identified phosphoethanolamine (PEt) as the major constituent of the PME peak (Pettegrew *et al.*, 1986). However, this peak also includes contributions from phosphocholine (PCh) and several other substances of lower concentration, for example, phosphoserine and phosphoinositol (Glonek *et al.*,

1982). These PMEs are anabolic precursors of phospholipids and are also necessary for myelin synthesis.

The resonance at ~2.9 ppm is quite broad in surface-coil spectra (see Fig. 1B; peak 3) and originates mainly from cross-linked PDEs in membrane-bilayer phospholipids but also includes a small pool of mobile PDEs: the phospholipid breakdown products glycerophosphorylethanolamine (GPE) and glycerophosphorylcholine (GPC). The membrane-bilayer phospholipids exhibit chemical-shift anisotropy, leading to reduced SNR at higher field strengths, e.g., 4.7–7 T (Bates *et al.*, 1989a; Murphy *et al.*, 1989). The very broad underlying signal, removed by postacquisition processing in Fig. 1B, is due to more rigidly bound $^{31}P$ nuclei mainly in membrane phospholipids, myelin, and cranial bone. The cross-linked PDE resonance has a $T_2$ of ~3 msec (Jung *et al.*, 1993). Both this peak and the very broad underlying hump due to immobile $^{31}P$ are almost undetectable in $^{31}P$ PRESS spectra owing to rapid $T_2$ relaxation (see Fig. 1A). This greatly facilitates accurate quantification of the remaining resonances from important mobile molecular species with long $T_2$s, e.g., $P_i$ and PCr.

### 4.1.3. Developmental Changes

$^{31}P$ surface-coil data from the temporoparietal cortex of 30 infants of appropriate weight for gestational age (AGA) studied at a median postnatal age of 4 days are presented in Fig. 3 and Table 1 (Azzopardi *et al.*, 1989a). As gestational plus postnatal age (GPA) increases, [PCr]/[$P_i$] and [PDE]/[total mobile phosphate ($P_{tot}$)] also increase, whereas [PME]/[$P_{tot}$] falls. Normal, small-for-gestational-age (SGA; i.e., birth weight below the 3rd centile) infants had values similar to those of AGA infants. MRS investigations in infants and older subjects using the ISIS technique (see Fig. 4) have shown that these changes, and increases in [PCr]/[NTP], continue during the first few months after birth but then occur at a reduced rate until at about 4 years of age adult values are approached (Boesch *et al.*, 1989; van der Knaap *et al.*, 1990).

[PCr]/[$P_i$] is directly related to both the phosphorylation potential and the free-energy change of ATP hydrolysis. In the newborn human, cerebral [PCr]/[$P_i$] is similar to that in the newborn of altricial mammalian species such as the rat (Tofts and Wray, 1985), but this ratio is smaller than in precocial newborn mammals, e.g., lambs (Hope *et al.*, 1987). The increase in [PCr]/[$P_i$] with development may be related to an increasing energy-reserve requirement and possibly also to less reliance on oxidative phosphorylation in the immature brain. [The latter possibility may relate to the detection of significant amounts of Lac in normal newborn brain; see Section 4.2.1 and Cady *et al.*, (1994a, 1995a) and Penrice *et al.*, (1995).]

In newborn infants, PME has a high amplitude due to an initially high [PEt] (Okumura *et al.*, 1960; Agrawal and Himwich, 1970; Burri *et al.*, 1988). This

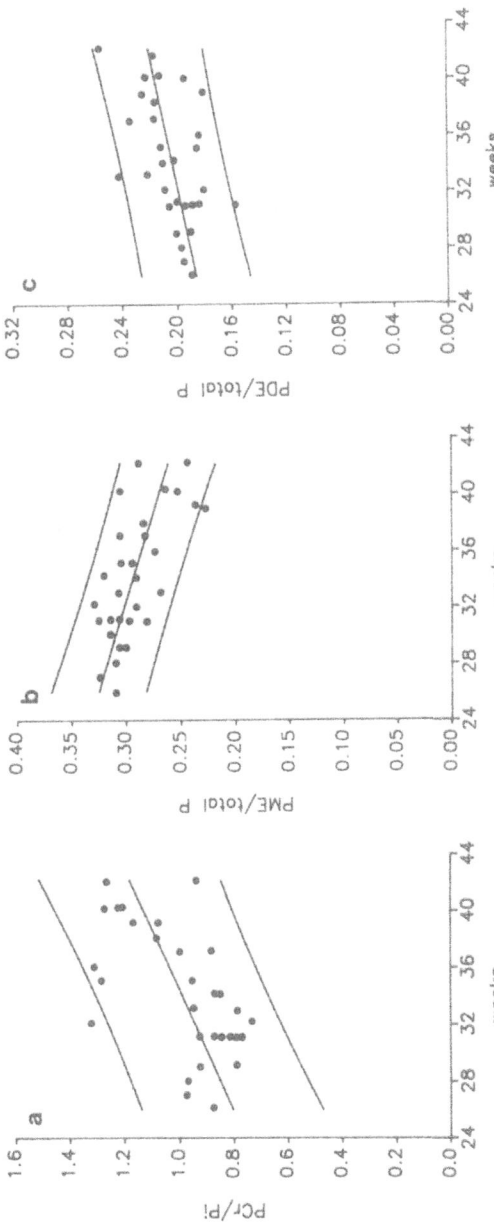

*FIGURE 3.* Relations between cerebral [PCr]/[P$_i$] (a), [PME]/[P$_{tot}$] (b), [PDE]/[P$_{tot}$] (c) and GPA in 30 newborn infants of appropriate weight for gestational age. Linear regression lines and 95% confidence limits are shown. [Reprinted by permission of International Pediatric Research Foundation Inc. from Azzopardi *et al.*, (1989a).]

TABLE 1. [31]P Cerebral Metabolite Concentration Ratios and $pH_i$
at Gestational plus Postnatal Ages (GPAs) of 28 and 42 Weeks
in 30 Normal Infants of Appropriate Weight for Gestational
Age[a]

|  | Mean linear regression value $\pm$ 95% confidence interval at GPA of: | | |
|  | 28 wk | 42 wk | P |
| --- | --- | --- | --- |
| $[PCr]/[P_i]$ | $0.85 \pm 0.33$ | $1.18 \pm 0.33$ | <0.001 |
| $[NTP]/[P_{tot}]$ | $0.09 \pm 0.03$ | $0.10 \pm 0.03$ | NS[b] |
| $[PCr]/[P_{tot}]$ | $0.09 \pm 0.02$ | $0.11 \pm 0.02$ | <0.01 |
| $[P_i]/[P_{tot}]$ | $0.10 \pm 0.02$ | $0.09 \pm 0.02$ | <0.05 |
| $[PME]/[P_{tot}]$ | $0.32 \pm 0.04$ | $0.26 \pm 0.04$ | <0.001 |
| $[PDE]/[P_{tot}]$ | $0.19 \pm 0.04$ | $0.22 \pm 0.04$ | <0.01 |
| $[PCr]/[NTP]$ | $0.79 \pm 0.42$ | $1.09 \pm 0.42$ | NS |
| $[P_i]/[NTP]$ | $1.13 \pm 0.50$ | $0.95 \pm 0.50$ | NS |
| $[PME]/[NTP]$ | $3.49 \pm 1.23$ | $2.79 \pm 1.25$ | <0.05 |
| $[PDE]/[NTP]$ | $2.11 \pm 1.03$ | $2.34 \pm 1.05$ | NS |
| $[Hump]/[NTP]$[c] | $20.5 \pm 20.5$ | $34.7 \pm 21.6$ | <0.05 |
| $pH_i$[d] | $7.14 \pm 0.28$ | $7.09 \pm 0.28$ | NS |

[a]Reprinted, with permission of International Pediatric Research Foundation Inc., from
Azzopardi *et al.* (1989a). $P$ = significance of linear regression correlation coefficient.
[b]NS, Not significant.
[c]$n$ = 24; the spectra from the remaining infants were not well resolved.
[d]$n$ = 23; the spectra from the remaining infants were not well resolved.

substance (produced by the phosphorylation of ethanolamine by ATP) may be
important as a phospholipid precursor related to the high rate of synthesis of
membranes and myelin (e.g., during axonal proliferation) in early life. The decrease in [PME] appears to be related to the increase in [PDE]; as [PEt] declines,
[GPE] increases (Burri *et al.*, 1988).

During development, $pH_i$ ( $\sim$7.10) remains constant (Azzopardi *et al.*,
1989a; van der Knaap *et al.*, 1990).

*4.1.3.1. Relaxation* Among the [31]P resonances, PME alone shows a developmental change in relaxation; $T_1$ decreases from $\sim$1.8 sec at birth to $\sim$1.2 sec in
adulthood (Buchli *et al.*, 1994). The decline in $T_1$ probably relates to alterations
in the relative amounts of the peak constituents, notably the reduced [PEt].

*4.1.3.2. Metabolite Concentrations* Estimates of [31]P metabolite concentrations in the brain of the newborn have been obtained by surface-coil (Cady, 1990,
1991, 1992), ISIS (Buchli *et al.*, 1994), and, more recently, PRESS (Cady, 1995;
Cady *et al.*, 1995b) techniques. However, the results show some disagreement
(see Table 2). With both the ISIS and PRESS methods, [PDE] is much lower than

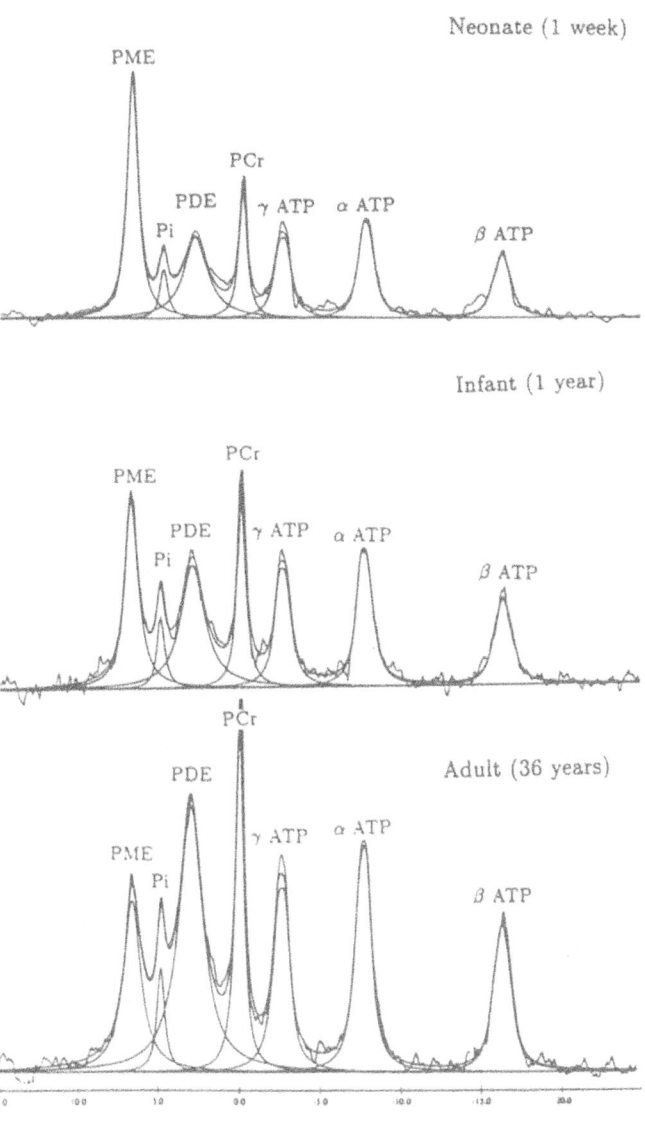

**FIGURE 4.** Fully relaxed 31P ISIS spectra from normal human brain at the ages indicated. Also shown are the calculated model spectra and individual model components generated by fitting decaying exponentials to the time-domain signal. Because each resonance had a similar width in neonatal, infant, and adult spectra, and data were acquired fully relaxed, the spectra could be scaled so that peak heights in the figure provide direct comparison of metabolite concentrations at different stages of cerebral development. Acquisition conditions: 2.4 T (neonates, infants, and 7 adults) and 1.5 T (21 adults); TR 12 sec; 64 averages; 125-ml (neonates), 216-ml (infants), and 343-ml (adults) VOIs. [Reprinted by permission of International Pediatric Research Foundation Inc. from Buchli *et al.* (1994).]

TABLE 2. Concentrations of $^{31}P$ Metabolites in the Brains of Newborn Infants Estimated by Surface-Coil, ISIS, and PRESS Techniques

| | Concentration [mean (SD)] | | |
|---|---|---|---|
| Metabolite | Surface-coil[a] (mmol/kg wet) | ISIS[b] (mmol/l wet) | PRESS[c] (mmol/kg wet) |
| PME | 9.3 (1.5) | 4.5 (0.7) | 5.3 (0.9) |
| $P_i$ | 2.4 (0.6) | 0.6 (0.1) | 1.3 (0.3) |
| PDE | 25.4 (11.0) | 3.2 (0.8) | 2.2 (0.6) |
| PCr | 2.7 (1.1) | 1.4 (0.2) | 2.7 (0.3) |
| NTP | 3.3 (0.6) | 1.6 (0.2) | 2.2 (0.3) |
| $P_{tot}$ | 51.1 (11.5) | 14.9 (2.3) | 19.5 (2.7) |

[a]Gestational plus postnatal age (GPA), 35–37 weeks; data from Cady (1992).
[b]GPA, 39–44 weeks; data from Buchli *et al.* (1994).
[c]GPA, 34–39 weeks; data from Cady *et al.* (1995b).

the surface-coil estimate. In both ISIS and PRESS, because the cross-linked membrane-bilayer PDE fraction has a very short $T_2$ (Jung *et al.,* 1993), [PDE] estimates will be low due to relaxation during the delays and RF pulses of the localization sequences. (In the surface-coil approach, the pulse duration was much shorter than $T_2$ and relaxation was negligible.) The surface-coil and PRESS [NTP] estimates were similar to those from neonatal rat brain (Duffy *et al.,* 1975; Vannucci and Vannucci, 1978; Yager *et al.,* 1992) and canine brain (Vannucci and Duffy, 1977) obtained using invasive techniques especially if NTPs other than ATP are included. However, interspecies variability may be important. A report on rabbit brain (Cohen and Lin, 1962) showed that [ATP] increased with development, and neonatal concentrations in this species were comparable to those obtained by ISIS. [MRS studies of human brain also tend to support an increase in [NTP] with development (Cady *et al.,* 1990; Buchli *et al.,* 1994).] The differences in concentrations estimated by the various techniques may be due to several factors. Contamination of surface-coil spectra by extracerebral tissue may have been one problem; in addition, these spectra were acquired from a different part of the brain than the ISIS and PRESS results.

## 4.2. $^{1}H$ Spectroscopy

For the aforementioned reasons of general availability of high-field imaging spectrometers designed primarily for $^{1}H$ applications and the superior SNR of $^{1}H$ studies, investigations using this nucleus are beginning to provide a vast wealth of information about the newborn human brain.

## 4.2.1. Normal Newborn Brain

If long TEs are used (e.g., 135 msec, 270 msec, or greater), well-resolved PRESS or STEAM ¹H spectra with adequate SNR can be acquired from the brains of newborn infants in spite of the lower molar concentrations of many metabolites, such as NAA (Tallan, 1957; Bates *et al.,* 1989b; Huppi *et al.,* 1991; Kreis *et al.,* 1993). With very short TEs (e.g., <25 msec), spectra become more complicated: $T_2$ relaxation and phase modulation are less important, but many more resonances are detected, leading to analysis difficulties. Further problems associated with the analysis of short-TE spectra are caused by the existence of several broad underlying resonances from both fatty acids (perhaps due to extra-voxel contamination and free fatty acids) and macromolecules (e.g., proteins). Figure 5 compares *in vivo* spectra from thalamic and occipitoparietal regions of the brain of a normal newborn infant. The most prominent *in vivo* peaks originate from Cho, total creatine (Cr; including PCr), NAA, and a clearly detectable doublet resonance due mainly to Lac methyl protons. The *in vivo* Cho resonance includes contributions from, among other substances, GPC, PCh, and taurine (Cerdan *et al.,* 1985). *In vivo,* the NAA resonance at ~2.01 ppm includes contributions from Glu, Gln, and GABA. The biochemical pathways and functions of ³¹P metabolites are largely known. However, many ¹H resonances other than those due to Lac (an end product of anaerobic glycolysis), Cr, and Cho (a precursor for both acetylcholine and phosphatidylcholine) still present interpretational problems (Miller, 1991; Ross, 1991). For example, NAA is now widely looked upon as an intraneuronal marker. However, the biochemical function of NAA remains uncertain. [This substance has also been found in oligodendrocyte progenitor cells (Urenjak *et al.,* 1993).]

The *in vivo* thalamic spectrum (Fig. 5A) has a relatively low Lac signal when compared with that from the occipitoparietal region (Fig. 5B). Quantitation studies indicate that [Lac] is similar in the two regions: [Cho], [Cr], and [NAA] are lower in the occipitoparietal tissue (Cady *et al.,* 1995a). These observations may relate to different stages of development and varying proportions of glia and neurons in the two brain regions.

*¹H Relaxometry* The echo natures of the PRESS and STEAM techniques enable—and quantitation of metabolite concentrations necessitates—the measurement of cerebral metabolite $T_2$s and, if used as an internal reference, also the brain-water $T_2$. Furthermore, recent studies of peritumor edema and acute edematous ischemic stroke have shown reduced metabolite $T_2$s (Kamada *et al.,* 1994) whereas investigations of human immunodeficiency virus (HIV) infection have indicated increases in the $T_2$ of NAA (Wilkinson *et al.,* 1994). Therefore, relaxation times may be of interest, especially following recent discoveries of relationships between these and both deoxyhemoglobin (Hb) and NTP levels. The $T_2$ of cerebral water decreases as blood Hb levels increase during hypoxia

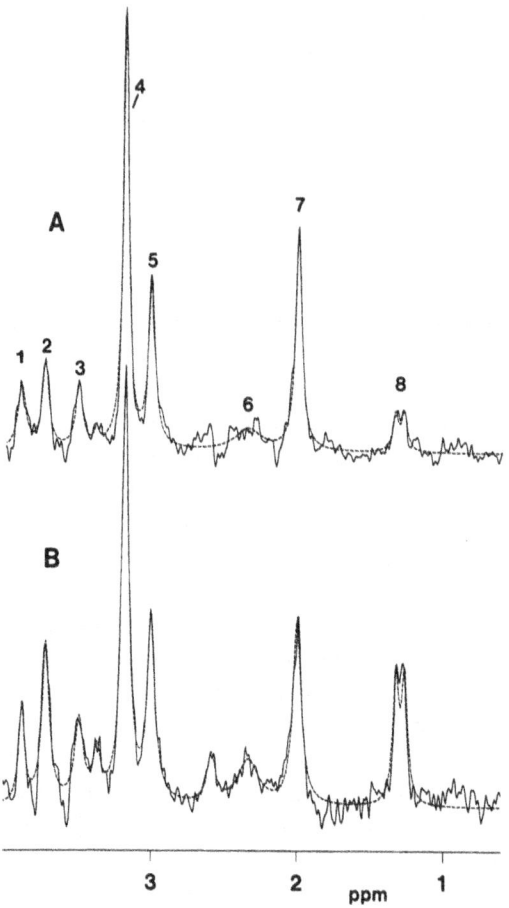

FIGURE 5. *In vivo* ¹H PRESS spectra from thalamic (A) and occipitoparietal (B) regions of normal newborn infants of GPA 34 weeks. In spectrum B, relative to the other peaks, Lac is much larger, and NAA slightly smaller, than in spectrum A. Acquisition conditions: 2.4 T, TE 270 msec, TR 2 sec, 128 (A) and 256 (B) summed echoes, 8-ml cubic VOI. Resonance identifications: 1 and 5, Cr; 2 and 6, Glu and Gln; 3, *myo*-inositol and Gly; 4, Cho; 7, NAA; 8, Lac. The dashed lines are the results of fitting Lorentzian profiles to the resonances by gradient-search $\chi^2$ minimization. (Spectra acquired in collaboration as in the caption to Fig. 1.)

(Prielmeier *et al.*, 1994). This is presumably related to the paramagnetic nature of Hb. Investigations of newborn piglet brain during delayed ("secondary") energy failure following initial metabolic recovery from acute hypoxia–ischemia have suggested large increases in intracellular metabolite $T_2$s as NTP decreases (Cady *et al.*, 1994b). The latter changes are probably related to failure of the ATP-dependent $Na^+/K^+$ pump, which normally extrudes intracellular $Na^+$ in ex-

TABLE 3. Cerebral $^1$H Metabolite and Water $T_2$ Relaxation
in the Brains of Newborn Infants

| | $T_2$ [mean (SD)] (msec) | | |
|---|---|---|---|
| | STEAM[a] | PRESS[b,c] | PRESS[c,d] |
| myo-Inositol | 301[e] (66) | — | — |
| Cho | 431[f] (89) | 328 (88) | 417 (89) |
| Cr | 228 (29) | 209 (36) | 192 (38) |
| NAA | 524[f] (188) | 316 (78) | 473 (140) |
| Lac | — | 232[g] (54) | 260[h] (138) |
| Water | 158–177 | 173[i] (16) | 219[j,k] (28) |

[a]Occipitoparietal and occipital cortex; gestational plus postnatal age (GPA), 42 ± 3 weeks; $n = 6$, except where otherwise noted; data from Kreis et al. (1993).
[b]Thalamic region; GPA, 30–42 weeks; $n = 13$, except where otherwise noted.
[c]Data acquired by the author in collaboration with J. Penrice, A. Lorek, R. Aldridge, M. Wylezinska, J. Wyatt, and E. O. R. Reynolds.
[d]Occipitoparietal region; GPA, 34–40 weeks; $n = 9$, except where otherwise noted.
[e]$n = 4$.
[f]$n = 5$.
[g]$n = 12$.
[h]$n = 8$.
[i]$n = 10$.
[j]$n = 7$.
[k]Statistically significant difference ($P < 0.001$) from value for thalamic region (unpaired $t$-test).

change for extracellular K$^+$ (Flynn et al., 1989). In cells depleted of ATP, the Na$^+$/K$^+$ pump fails; a net Na$^+$ influx ensues, accompanied by increased intracellular water, thereby increasing the mobility of metabolite molecules. This increased mobility appears to manifest itself as raised metabolite $T_2$s.

Experimental protocols whereby, for example, $T_2$s can be measured in vivo have been described in Section 3.3 (see also Fig. 2). Table 3 gives metabolite and brain-water $T_2$s in newborn brain. The in vivo Cho and NAA peaks contain contributions from several other substances: differences in $T_2$s measured by STEAM and PRESS may be due to disparate pulse-sequence-dependent phase modulation (Ernst and Hennig, 1991). In the occipitoparietal region, brain-water $T_2$ is higher than in central gray matter adjacent to the thalami. This is presumably related to greater extracellular water in the former unmyelinated, white-matter region.

### 4.2.2. Developmental Changes

Both the relative peak areas (van der Knaap et al., 1990; Kreis et al., 1993) and concentrations (Kreis et al., 1993) of many $^1$H metabolites change during development. In particular, focusing on perinatal changes, the Lac/NAA peak-

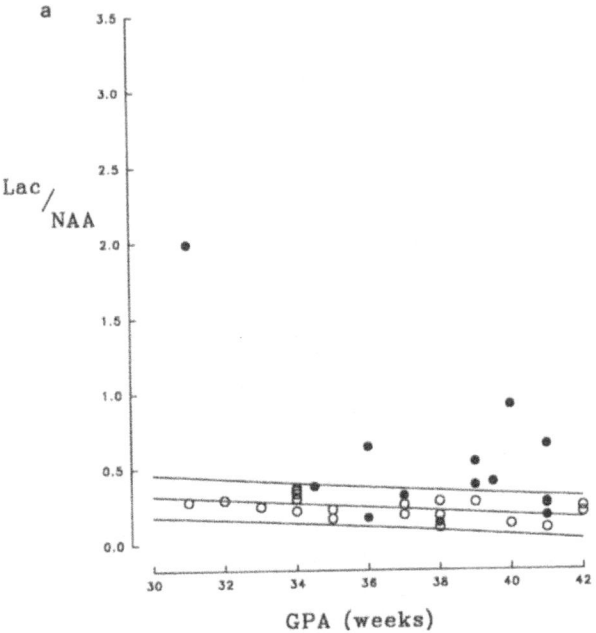

FIGURE 6. Relationships between Lac/NAA peak-area ratios and GPA in thalamic (a) and occipitoparietal (b) regions in normal newborn infants (○) and in infants with suspected hypoxic–ischemic injury (●). In the occipitoparietal region, Lac/NAA is higher in premature infants and falls more rapidly with increasing GPA than in the thalamic tissue. The occipitoparietal region may be less mature than the thalamic, and the proportions of glia and neurons may be different. Linear regression lines and 95% confidence limits are shown for the normal infants. Infants with suspected cerebral injury often have Lac/NAA above 95% confidence limits. Acquisition conditions as for Fig. 5. (Data acquired in collaboration as in caption to Fig. 1.)

area ratio decreases as GPA increases (see Fig. 6), whereas the NAA/Cho peak-area ratio increases. These changes were reported to be more rapid in occipitoparietal than in thalamic regions (Penrice et al., 1995). The relatively high Lac/NAA ratio in the brains of premature infants may reflect a greater reliance on anaerobic glycolysis, and this may be related to the lower [PCr]/[P$_i$] seen in similar babies (Azzopardi et al., 1989a). The relatively low NAA/Cho ratio seen in preterm infants and the increase with GPA are supported by studies showing that [NAA] rises during mammalian brain development (Tallan, 1957; Bates et al., 1989b). Because NAA may be almost exclusively neuronal (Urenjak et al., 1993) and cellularity decreases in regions other than the cerebellum (Dobbing and Sands, 1973), NAA concentration increase probably reflects neuronal maturation. Quantitative MRS studies of human brain have also shown increases with

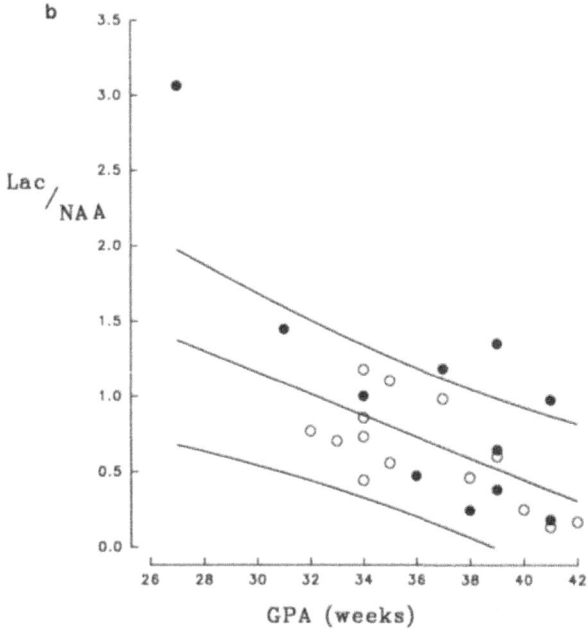

*Figure 6.* Continued

age for both [NAA] and [Cr] whereas decreases with age were found for both [Cho] and [*myo*-inositol] (Huppi *et al.*, 1991; Kreis *et al.*, 1993; Toft *et al.*, 1994). The Cho peak contains signals from PCh, GPC, and taurine among other substances (Cerdan *et al.*, 1985; Bates *et al.*, 1989b); developmental changes in the amplitude of this peak may relate in part to alterations in the concentrations of constituents contributing to the PME and PDE peaks detected by ³¹P spectroscopy. Table 4 gives ¹H metabolite concentrations estimated in various parts of the neonatal human brain. Regional differences are becoming apparent that may be due to different stages of development. For example, high thalamic [Cho] may be related to active myelination in this predominantly gray-matter region at birth.

## 5. CEREBRAL PATHOLOGY

³¹P studies have proven especially valuable for investigating hypoxic–ischemic brain injury in newborn infants. A major portion of the available data has been acquired from infants suffering the consequences of birth asphyxia. However, ³¹P spectra with similar abnormal characteristics have been acquired from infants with PVL and other types of hypoxic–ischemic injury. Comparison with

TABLE 4. Cerebral ¹H Metabolite Concentrations in the Brains of
Newborn Infants

| Metabolite | Concentration [mean (SD)] (mmol/kg wet wt) | | |
| --- | --- | --- | --- |
| | STEAM[a] | PRESS[b] | PRESS[c] |
| myo-Inositol | 10.1 (3.6) | — | — |
| Cho | 2.4 (0.4) | 4.6 (0.8) | 1.8 (0.6)[d] |
| Cr | 6.3 (1.1) | 10.5 (2.0) | 5.8 (1.5)[d] |
| NAA | 4.8 (1.8) | 9.0 (0.7) | 3.4 (1.1)[d] |
| Lac | — | 2.7[e] (0.6) | 3.3[f] (1.3) |

[a]Occipitoparietal and occipital cortex; gestational plus postnatal age (GPA), 38.7 ± 2.3 weeks (all <42 weeks); $n = 11$; data from Kreis et al. (1993).

[b]Thalamic region; GPA, 29–41 weeks; $n = 16$, except where otherwise noted; data from Cady et al. (1996).

[c]Occipitoparietal region; GPA, 29–40 weeks; $n = 10$, except where otherwise noted; data from Cady et al. (1996).

[d]Statistically significant difference ($P < 0.0001$) from value for thalamic region (unpaired t-test).

[e]Statistically significant difference ($P < 0.005$) from value for thalamic region (unpaired t-test).

[f]$n = 15$.

[g]$n = 9$.

age-matched normal controls has provided good correlation between the degree of spectroscopic abnormality seen within a few days of birth and later neuro-developmental outcome. ¹H MRS data from normal controls of various ages are now also adequate, in both quality and quantity, for the interpretation of pathological spectra. This has enabled the use of ¹H MRS for the monitoring of increased glycolytic pathway activity (increased Lac) due to impaired oxidative phosphorylation, observation of neuronal degeneration (decreased NAA), and the detection of metabolic defects.

## 5.1. Perinatal Asphyxia

Before the first ³¹P MRS studies of severely birth-asphyxiated infants with cerebral hypoxic–ischemic injury were carried out, it was falsely anticipated that signs indicative of the acute intrapartum insult would be observed. In conditions where oxygen supply to the brain was sufficiently reduced, or the mechanisms for oxygen consumption were damaged, oxidative phosphorylation would fail, leading to reduced [PCr] and increased [$P_i$] (Siesjö, 1978). If impairment of energy generation continued, alternative methods of energy production (e.g., anaerobic glycolysis) would eventually prove inadequate. [ATP] would then fall, accompanied by serious failure of the energy-consuming transport processes for which ATP is essential, e.g., the Na⁺/K⁺ pump (Flynn et al., 1989). It was

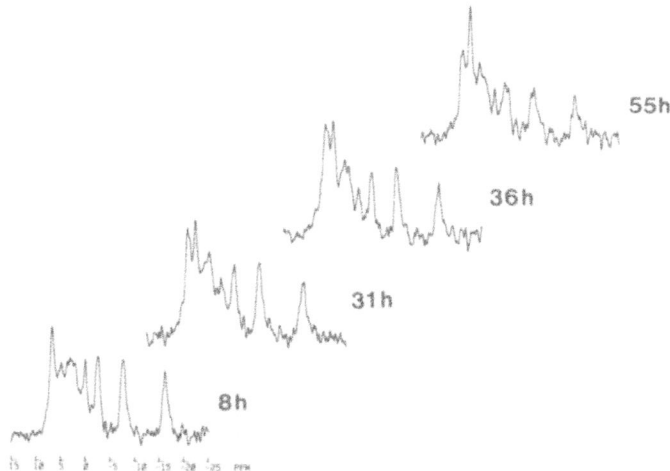

FIGURE 7. $^{31}$P surface-coil spectra from the cerebral cortex of a birth-asphyxiated infant born at 37 weeks' gestation. The postnatal ages at the studies are indicated. Peak assignments are as for Fig. 1. At age 8 hr, [PCr]/[P$_i$] was 0.99, [NTP]/[P$_{tot}$] was 0.09, and pH$_i$ was 7.06: pH$_i$ increased to a maximum of 7.28 at age 36 hr. Minimum observed [PCr]/[P$_i$] was 0.32 at age 55 hr when [NTP]/[P$_{tot}$] was 0.04 and pH$_i$ was 6.99. The infant died aged 60 hr. Acquisition conditions: 1.9 T; TR 2.256 sec; 256 averages; 7-cm-diameter surface coil adjacent to the temporoparietal cortex; 90° coil-center flip angle. [Reprinted by permission of International Pediatric Research Foundation Inc. from Azzopardi, *et al.* (1989b).]

hypothesized that studies of the brains of infants with hypoxic–ischemic injury would show reduced [PCr]/[P$_i$] and, in extreme circumstances, low [ATP] also. It was therefore surprising to find that cerebral $^{31}$P spectra from severely birth-asphyxiated babies often appeared normal on the first day of life (Azzopardi *et al.*, 1989b). However, in spite of normal values for arterial oxygen saturation, blood pressure, and blood glucose, over the next few days [PCr]/[P$_i$] was often low and minimum at age 2–4 days (see Fig. 7 and Table 5). In the most severely affected infants, [NTP] then fell, with an almost certain fatal outcome. During acute experimental hypoxia–ischemia (e.g., Hope *et al.*, 1987), profound intracellular acidosis occurs. However, in birth-asphyxiated infants, pH$_i$ tended to be slightly alkaline. In those who survived, [PCr]/[P$_i$] and [NTP]/[P$_{tot}$] often returned to normal values within a few weeks. However, the overall $^{31}$P signal was often of reduced amplitude, presumably indicating tissue loss. The lowest observed [PCr]/[P$_i$] measured during the first week after birth was strongly related to long-term outcome (Azzopardi *et al.*, 1989b; Roth *et al.*, 1992; see Section 5.3. Due to the time scale of the delayed failure of cerebral energy generation, spectra should be obtained on several occasions during the first few days in order to assess accurately the severity of injury. The metabolic changes observed by

TABLE 5. Age-Dependent Standard Deviation Scores (SDS)[a] for $^{31}$P Metabolite Concentration Ratios and pH$_i$ in Newborn Infants with Suspected Hypoxic–Ischemic Brain Injury[b]

| | SDS ± SD | | | |
| --- | --- | --- | --- | --- |
| | All infants ($n = 61$) | Birth asphyxia ($n = 40$) | Postnatal asphyxia ($n = 5$) | Increased cerebral echodensities ($n = 16$) |
| [PCr]/[P$_i$] | −2.14 ± 2.10[c] | −2.14 ± 2.19[c] | −0.03 ± 0.76 | −2.04 ± 1.78[c] |
| [NTP]/[P$_{tot}$] | −0.98 ± 1.86[d] | −1.17 ± 1.82[d] | −0.01 ± 1.03 | −0.83 ± 1.99 |
| [PCr]/[P$_{tot}$] | −1.72 ± 2.58[c] | −1.79 ± 2.59[c] | 0.33 ± 1.04 | −2.19 ± 2.52[c] |
| [P$_i$]/[P$_{tot}$] | 5.40 ± 8.77[c] | 5.97 ± 8.91[c] | 0.69 ± 0.99 | 5.51 ± 9.06[d] |
| [PME]/[P$_{tot}$] | 0.42 ± 1.69 | 0.63 ± 1.58 | −0.79 ± 1.26 | 0.30 ± 1.83 |
| [PDE]/[P$_{tot}$] | 0.44 ± 1.43 | −0.60 ± 1.44[e] | 0.39 ± 0.50 | −0.32 ± 1.47 |
| [PCr]/[NTP] | 0.52 ± 2.98 | 0.65 ± 3.14 | 0.15 ± 0.66 | 0.31 ± 2.91 |
| [P$_i$]/[NTP] | 7.21 ± 16.41[c] | 8.96 ± 17.94[c] | 0.35 ± 1.06 | 4.94 ± 13.27[d] |
| [PME]/[NTP] | 1.94 ± 4.40[e] | 2.49 ± 4.71[d] | −0.26 ± 1.11 | 1.25 ± 3.72 |
| [PDE]/[NTP] | 1.02 ± 3.84 | 1.43 ± 4.48 | 0.17 ± 0.77 | 0.23 ± 1.81 |
| pH$_i$ | 0.05 ± 1.47[f] | 0.23 ± 0.97[g] | −0.28 ± 0.22 | 0.00 ± 2.21[h] |

[a]Versus normal infants (see Table 1) when PCr/P$_i$ was minimum.
[b]Reprinted, with permission of International Pediatric Research Foundation Inc., from Azzopardi et al. (1989b).
[c]$P < 0.001$.
[d]$P < 0.01$.
[e]$P < 0.05$.
[f]$n = 55$.
[g]$n = 36$.
[h]$n = 14$.

MRS are presumably due to an acute cerebral hypoxic–ischemic episode during labor, leading to "primary" energy failure. These changes reversed on resuscitation after delivery, and the cerebral energy state returned to, or very close to, normal. The later changes in $^{31}$P spectra (i.e., falling [PCr]/[P$_i$], eventually falling [NTP]/[P$_{tot}$], and mildly alkaline pH$_i$) have been attributed to a "secondary" phase of energy generation failure (SEF) initiated directly or indirectly by the primary insult.

Several mechanisms have been proposed to explain the progression from primary energy failure to SEF, which is now known to be strongly associated with neuronal apoptosis (Mehmet et al., 1994). Near-infrared spectroscopy studies in infants developing SEF have shown that both the flow and volume of cerebral blood are increased (Wyatt et al., 1990); hence, inadequate oxygen or substrate supply is probably not causal. Overwhelming entry of calcium into cells and consequent damage to the mitochondrial electron-transport chain may be caused in response to acute hypoxia–ischemia by the release at the synapses of excitatory neurotransmitters (Glu in particular), thereby stimulating $N$-methyl-D-aspartate and other receptors. Further possibilities include those mechanisms involving prostanoids, nitric oxide, free radicals, immune mechanisms, phagocytes, growth factors, and impaired protein synthesis (Palmer et al., 1990; Tan et al., 1993; Thordstein et al., 1993). The delay between birth and the onset of SEF has suggested that suitable therapies may be beneficially applied either during delivery or in the first few hours after.

### 5.1.1. Models of Secondary Energy Failure

Since the development of SEF is strongly related to unfavorable outcome and also because of the social cost of supporting infants with neurodevelopmental impairments, significant international effort is being applied to investigate both causal mechanisms and means of interrupting them. The aim is to provide clinically acceptable cerebroprotective treatments for birth-asphyxiated infants. This research is also of wider importance because similar mechanisms of cerebral damage may be involved in, for example, stroke in adults.

Recently, the development of SEF has been modeled in both the newborn piglet (Lorek et al., 1994) and the immature rat (e.g., Yager et al., 1992); the former model has produced $^{31}$P MRS results closely resembling those from birth-asphyxiated infants (see Fig. 8). In the piglet model, the primary insult consisted of reversible bilateral common carotid occlusion and temporary hypoxia. The extent of SEF (quantified as the minimum [PCr]/[P$_i$] measured 24–48 hr after the primary insult) correlated with primary-insult severity (quantified as the time integral of NTP depletion during the primary insult) (see Fig. 9). This correlation is important because it enables biological variability in the response to the insult to be accounted for when relating outcome to insult severity.

### 5.1.2. Cerebroprotection

These models are suitable for testing putative cerebroprotective therapies. Amelioration of the SEF caused by a severe insult should provide an indication of the efficacy of a proposed therapy. Evidence is accumulating from various studies that cerebroprotection is possible. For example, mild cerebral hypothermia would be safe and relatively simple to apply clinically. Results obtained using the piglet (see Figs. 9 and 10) and other models indicate that this form of therapy may be of value (Thoresen et al., 1995; Laptook et al., 1994).

### 5.2.3. ¹H Studies

Results from ¹H studies of infants with hypoxic–ischemic brain injury are now available from several groups of researchers, and patterns of abnormality are

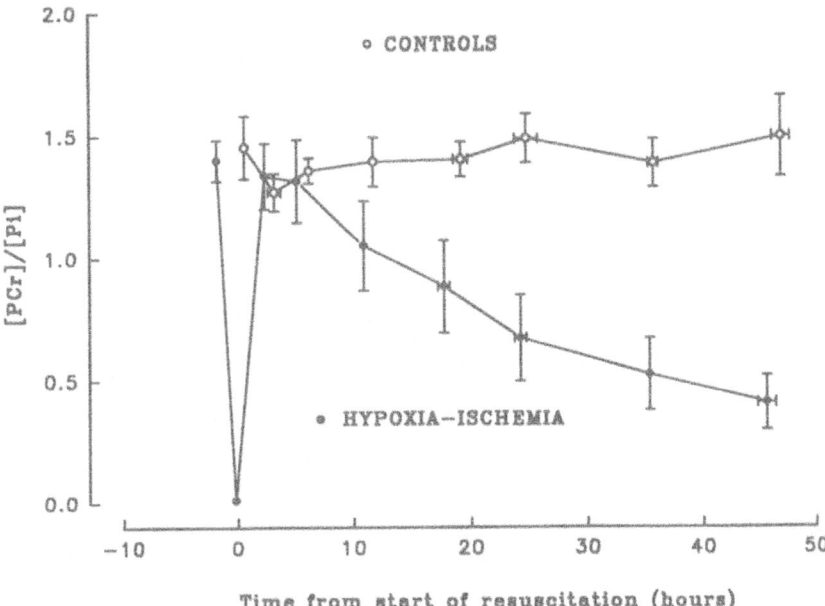

FIGURE 8. A model of secondary energy failure: [PCr]/[P$_i$] (mean values ± SEM) in sham-operated controls (○; $n = 6$) and in newborn piglets subjected to a temporary cerebral hypoxic–ischemic
insult (●; $n = 12$). In the controls, [PCr]/[P$_i$] is unchanged during 48 hr of continuous study. In the
experimental piglets, [PCr]/[P$_i$] falls profoundly during the insult but regains its baseline value within
a few hours of resuscitation. However, over the next two days secondary energy failure develops,
indicated by a gradual decline in [PCr]/[P$_i$] unaccompanied by intracellular acidosis. [Reprinted by
permission of International Pediatric Research Foundation Inc. from Lorek *et al.* (1994).]

becoming apparent. Although Lac is detected in normal infants and its relative
level appears to be higher in less developed tissue (see Fig. 5), increased [Lac],
often quantified by Lac/NAA (see Fig. 6) or Lac/Cho peak-area ratios, is often
seen following severe birth asphyxia (see Figs. 6 and 11) and appears to be
related to cerebral damage (Groenendaal *et al.*, 1994; Cady *et al.*, 1994a; Penrice
*et al.*, 1995). The increase in [Lac] is presumably due to failure of oxidative
phosphorylation and consequent reliance on anaerobic glycolysis. The Lac/NAA
peak-area ratio appears to be a sensitive, early marker of cerebral injury; in
addition to raised [Lac], this ratio may also be eventually increased due, in part,
to reduced [NAA] (presumably caused by neuronal loss). Low [NAA] has been
seen in newborn infants with a variety of nonfocal pathologies (Kreis *et al.*,
1993), and also in older children with generalized demyelination (Grodd *et al.*,
1991). The NAA/Cho peak-area ratio, which increases with development in

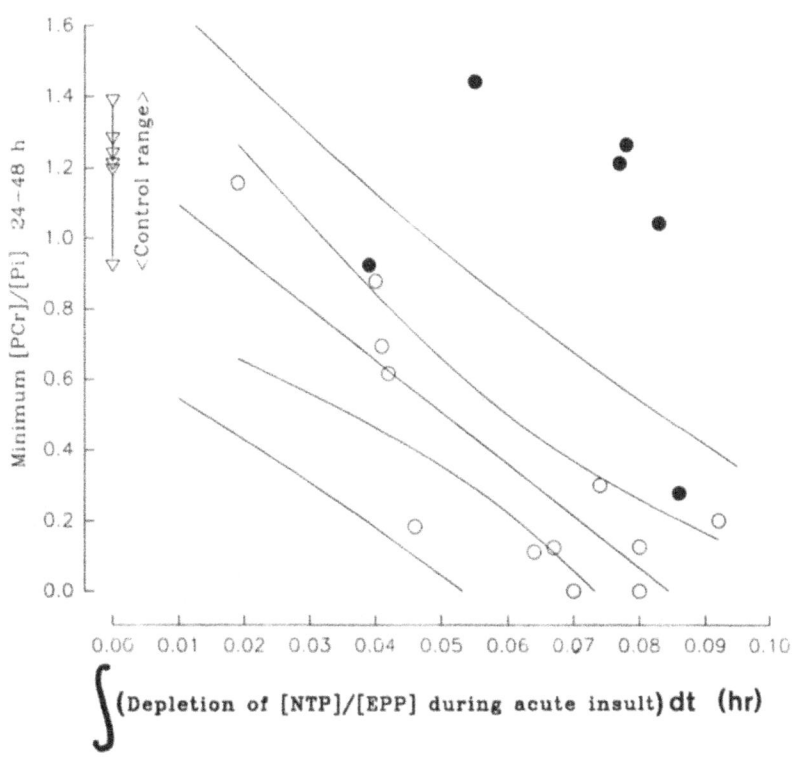

$$\int (\text{Depletion of [NTP]/[EPP] during acute insult}) \, dt \quad (\text{hr})$$

**FIGURE 9.** Relationship between severity of the cerebral hypoxic–ischemic insult (quantified as the time integral of [NTP]/[total exchangeable phosphate (EPP)] depletion during the insult) and the extent of the consequent SEF (quantified as the minimum [PCr]/[$P_i$] measured at 24–48 hr). [EPP] is defined as [$P_i$] + [PCr] + 3[NTP]. Data are presented for 12 normothermic newborn piglets ($\bigcirc$), 6 piglets subjected to mild hypothermia (rectal temperature 35°C for 12 hr; normal 38.5°C) immediately after the insult ($\bullet$), and 6 control piglets ($\nabla$). The linear regression line (central) and its 95% confidence limits (inner pair of lines) and the 95% confidence limits for the normothermic piglets (outer pair of lines) are shown. Two of the six hypothermic piglets fall within the 95% confidence limits for the normothermic piglets. However, only one of these had a major insult. The remaining four piglets with minimum [PCr]/[$P_i$] at 24–48 hr lying above the 95% confidence limits for the normothermic piglets provide evidence that mild hypothermia may be cerebroprotective. (Data acquired in collaboration with J. Penrice, A. Lorek, A. D. Edwards, V. Kirkbride, H. Owen-Reece, G. Brown, C. Cooper, M. Thoresen, Y. Takei, S. Roth, D. Peebles, R. Aldridge, M. Wylezinska, J. Wyatt, and E. O. R. Reynolds.)

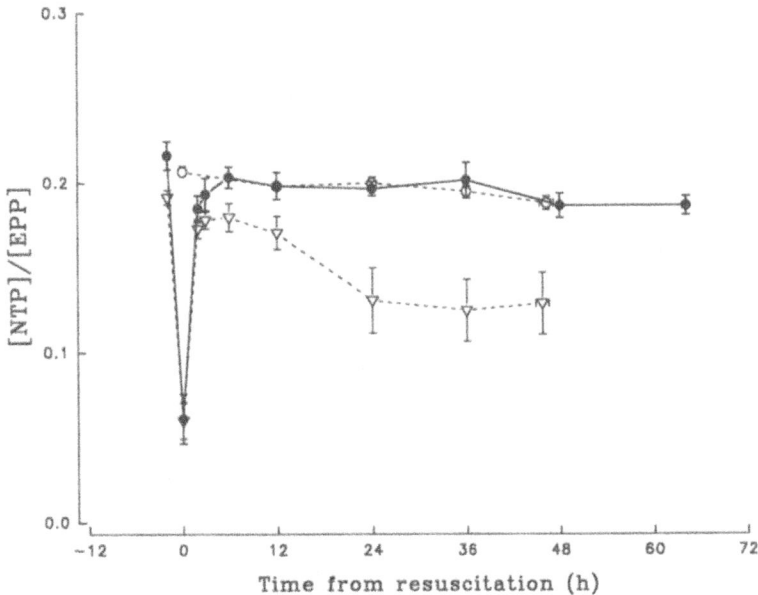

FIGURE 10. [NTP]/[EPP] during and following an acute cerebral hypoxic–ischemic insult in normothermic ($\nabla$; $n = 12$) and hypothermic ($\bullet$; $n = 6$) newborn piglets (mean values $\pm$ SEM). Results for sham-operated controls are also shown ($\bigcirc$; $n = 6$). The data displayed are from the same experiments as in Figs. 8 and 9. The cerebroprotective value of mild hypothermia is further demonstrated both by the complete recovery of [NTP]/[EPP] immediately after cessation of the acute insult and by the maintenance of normal [NTP]/[EPP] over the ensuing 48–64 hr. [Reprinted by permission of International Pediatric Research Foundation Inc. from Thoresen *et al.,* (1995).]

normal infants, may also fall following hypoxic–ischemic damage (Peden *et al.,* 1993; Groenendaal *et al.,* 1994).

## 5.2. Metabolic Disorders

MRS has also been applied to the study of inborn errors of metabolism (see Chapter 12). PCr and NTP were almost undetectable in an infant with propionic acidemia and also in one with arginosuccinic aciduria (Hope *et al.,* 1986). Abnormalities have been found in older children with various inborn errors of metabolism; for example, $^1$H spectra showed relatively increased NAA in Canavan's disease and relatively increased Lac in Leigh, Schilder, and Cockayne disease and also in neuroaxonal dystrophy (Grodd *et al.,* 1991). Recently, $^1$H spectra revealing complete absence of cerebral Cr were acquired from an infant. (Stöckler *et al.,* 1994; see Chapter 12). Cr is produced in the liver and pancreas and was absent due to an inborn error of the enzyme guanidinoacetate methyltransferase.

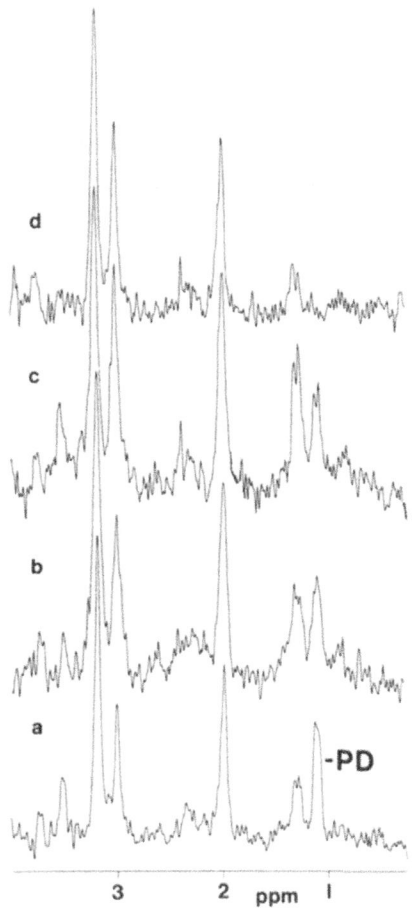

FIGURE 11. ¹H PRESS spectra acquired from the thalamic region of a severely birth-asphyxiated term infant. The spectra were acquired at ages 2 (a), 3 (b), 7 (c), and 14 (d) days. Acquisition conditions and resonance identifications as for Fig. 5. The lactate peak at ~1.3 ppm is large when compared with that in spectra acquired from normal infants (see Fig. 5A). The unusual peak at ~1.1 ppm (PD) is due to propan-1,2-diol (the injection medium for phenobarbitone and phenytoin preparations, commonly given to control neonatal seizures). [Reprinted by permission of Williams and Wilkins from Cady *et al.* (1994c).]

An important aspect of this case was that orally administered Cr appeared in subsequent spectra, and this therapy resulted in an improvement in the infant's condition.

## 5.3. Prognosis

Relationships between $^{31}$P MRS results obtained during the first week after birth from 61 infants suspected of cerebral hypoxic–ischemic injury and neuro-developmental outcome at age 1 year have been investigated (Azzopardi *et al.,* 1989b). Many of these infants were suffering the sequelae of birth asphyxia. The chance of survival and the severity of neurodevelopmental impairments at age 1 year were related directly to the maximum extent of cerebral energy failure seen within the first few days after delivery, quantified as the minimum observed [PCr]/[P$_i$] (see Table 5 and Fig. 12A). Furthermore, fatal outcome was almost inevitable if [NTP]/[P$_{tot}$] fell (see Fig. 12B).

In an extension of the same study, a further group of birth-asphyxiated infants were recruited, so that $^{31}$P MRS results were available from a total of 52 such infants (Roth *et al.,* 1992). These infants were in very poor condition at birth: mean arterial base excess in cord blood or shortly after delivery was -20 mmol/l; 38 infants had seizures; and 26 required mechanical ventilation. The relation between minimum recorded [PCr]/[P$_i$] in the first days after delivery and eventual neurodevelopmental outcome is shown in Fig. 13, confirming that SEF conveyed a bad prognosis. The sensitivity, specificity, and positive predictive value for fatal outcome or multiple disabling impairments of values of [PCr]/[P$_i$] greater than 2 SD below the normal age-matched mean were 72%, 92%, and 91%, respectively, in this study. [Similar results have been obtained by others (e.g., Moorcraft *et al.,* 1991).] A relationship between minimum [PCr]/[P$_i$] and head-circumference growth rate was also found (see Fig. 14). In spite of the fact that the birth-asphyxiated infants had normal head circumferences at birth, subsequent head growth was retarded in those infants whose [PCr]/[P$_i$] values were lowest.

$^{31}$P spectra from the brains of infants with increased ultrasound echoden-sities (usually due to either hypoxic–ischemic injury or hemorrhage) also often show reduced [PCr]/[P$_i$] and sometimes also low [NTP]/[P$_{tot}$] (Hamilton *et al.,* 1986; Azzopardi *et al.,* 1989b); see Table 5. $^{31}$P spectroscopy is useful for separating those infants with echodensities who will have a relatively good outcome from those for whom the outcome will probably be unfavorable.

The number of long-term clinical follow-ups is not yet sufficient to enable judgment of the prognostic value of $^1$H MRS. However, observation of raised Lac and of reduced NAA may be predictive of unfavorable outcome. It has been shown that NAA/Cho peak-area ratios were significantly lower both shortly after

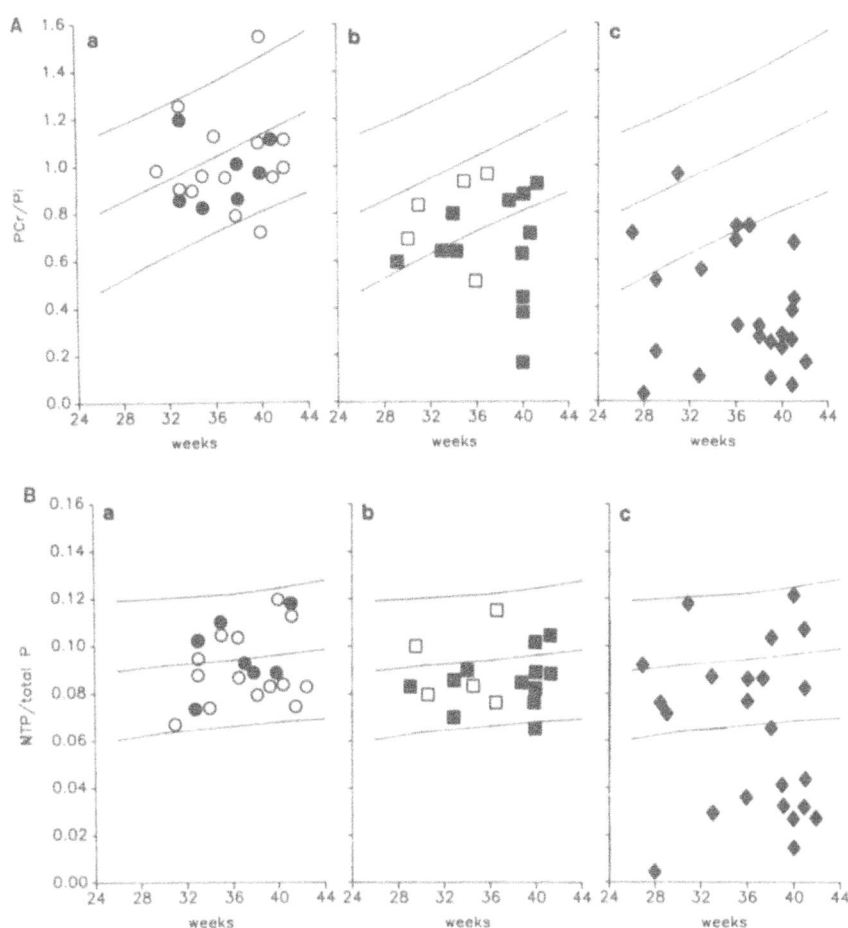

*FIGURE 12.* Relationships between minimum PCr/P$_i$ (A) and NTP/P$_{tot}$ (B) in the brains of newborn infants with suspected perinatal hypoxic–ischemic injury measured during the first week after birth and survival and neurodevelopmental outcome at age 1 year: (a) Normal progress (○) and minor impairments (●); (b) major neuromotor impairments (□) and multiple major impairments (■); (c) fatal outcome (◆). The regression lines and 95% confidence limits for normal values versus GPA are given. (It should be noted that these figures include data from 16 infants studied solely because of increased ultrasound echodensities attributable in many cases to periventricular leukomalacia or intraparenchymal hemorrhage.) [Reprinted by permission of International Pediatric Research Foundation Inc. from Azzopardi, *et al.* (1989b).]

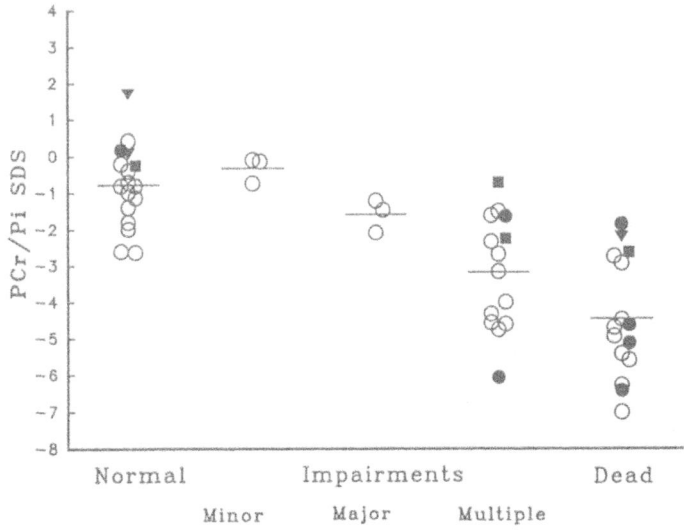

*FIGURE 13.* Minimum observed age-related PCr/P$_i$ standard deviation score (SDS) in birth-asphyxiated newborn infants and neurodevelopmental outcome at age 1 year. ○, Term AGA; ●, term SGA; ▼, preterm AGA; □, preterm SGA. [Reprinted by permission of MacKeith Press from Roth *et al.* (1992).]

delivery and at age 3 months in birth-asphyxiated infants who eventually displayed abnormal neuromotor development (Groenendaal *et al.*, 1994).

So far, insufficient data have been collected to evaluate the prognostic value of metabolite relaxation times and concentrations. However, work is progressing rapidly in this area.

## 6. CONCLUSIONS

The very first *in vivo* MRS studies of human brain were carried out on newborn infants. During recent years, developmental changes in $^{31}$P and $^1$H brain spectra acquired from normal premature and term infants have been investigated and defined. For example, $^{31}$P MRS has shown that [PCr]/[P$_i$] rises with age, indicating increased phosphorylation potential, whereas [PME]/[PDE] falls. $^1$H spectroscopy has indicated a relatively high level of Lac in prematurity and in less developed parts of the brain. These observations may indicate a greater reliance on anaerobic glycolysis, rather than oxidative phosphorylation, in immature brain. Furthermore, the relative level of NAA increases with age in newborn

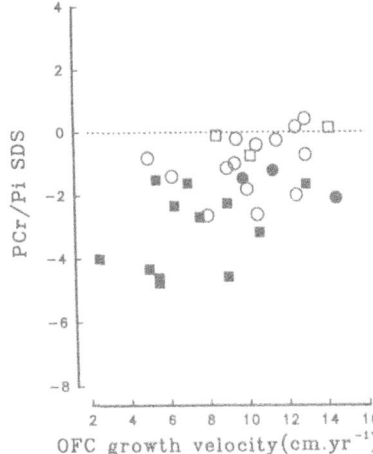

*FIGURE 14.* Relation between PCr/P$_i$ standard deviation score (SDS) and occipitofrontal head circumference (OFC) growth velocity in 32 birth-asphyxiated newborn infants. ○, Normal progress; □, minor impairment; ●, major impairment; ■, multiple impairments. [Reprinted by permission of MacKeith Press from Roth *et al.* (1992).]

infants, and this observation is supported by the known increases in [NAA] that occur during mammalian brain development.

$^{31}$P and $^{1}$H MRS have proven useful for investigating the cerebral metabolic consequences of birth asphyxia both in human infants and, more recently, in an animal model. $^{31}$P spectra are often normal soon after delivery. However, in severe cases, first [PCr]/[P$_i$] and then [NTP]/[P$_{tot}$] gradually fall, both reaching their lowest values at 2–4 days of age. Unlike during acute insults in animal models, in which profound acidosis is observed, these changes are accompanied by a slight alkaline pH$_i$ shift. The probability of long-term neurodevelopmental impairments or fatal outcome is strongly related to the extent of the depletion of both [PCr]/[P$_i$] and [NTP]/[P$_{tot}$]. These reductions in the concentrations of metabolites vital for cerebral energy generation have been termed "secondary" cerebral energy failure. This is based on the hypothesis that the reductions were initiated by a reversed primary energy failure occurring during labor. Although $^{1}$H spectra show early abnormalities, the available data are insufficient to say anything definitive about the time dependence of the changes.

Animal models of SEF utilizing MRS indices (e.g., [PCr]/[P$_i$] to assess insult severity and eventual outcome can be used to test the efficacies of cerebroprotective therapies. In hypoxic–ischemic injury, it is important to acquire diagnostic and prognostic information as soon after birth as possible, thereby identifying those infants at risk of severe cerebral injury and hopefully enabling the future application of such cerebroprotection.

The prognostic information given by $^{31}$P MRS is valuable for directing the clinical care of infants suspected of hypoxic–ischemic brain injury. The prognostic value of $^{1}$H MRS, in particular of increases in [Lac] and falls in [NAA], will only be known when sufficient data have been collected and enough long-term neurodevelopmental assessments have been made. However, there are strong indications that $^1$H MRS will also be of prognostic value. The ability of $^1$H MRS to examine small anatomical regions and the general availability of imaging spectrometers with $^1$H MRS capability are likely to make this modality a very powerful clinical tool.

## REFERENCES

Agrawal, H. C., and Himwich, W. A., 1970, Amino acids, proteins and monoamines of developing brain, in "Developmental Neurobiology" (W. A. Himwich, ed., pp. 298–299, C. C. Thomas, Springfield, Illinois.

Azzopardi, D., Wyatt, J. S., Hamilton, P. A., Cady, E. B., Delpy, D. T., Hope, P. L., and Reynolds, E. O. R., 1989a, Phosphorus metabolites and intracellular pH in the brains of normal and small-for-gestational-age infants investigated by magnetic resonance spectroscopy, *Pediatr. Res.* **25:**440–444.

Azzopardi, D., Wyatt, J. S., Cady, E. B., Delpy, D. T., Baudin, J., Stewart, A. L., Hope, P. L., Hamilton, P. A., and Reynolds, E. O. R., 1989b, Prognosis of newborn infants with hypoxic/ischaemic brain injury assessed by phosphorus magnetic resonance spectroscopy, *Pediatr. Res.* **25:**445–451.

Bates, T. E., Williams, S. R., and Gadian, D. G., 1989a, Phosphodiesters in the liver: The effect of field strength on the $^{31}$P signal, *Magn. Reson. Med.* **12:**145–150.

Bates, T. E., Williams, S. R., Gadian, D. G., Bell, J. D., Small, R. K., and Iles, R. A., 1989b, $^1$H NMR study of cerebral development in the rat, *NMR Biomed.* **2:**225–229.

Boesch, C., and Martin, E., 1988, Combined application of MR imaging and spectroscopy in neonates and children: Installation and operation of a 2.35-T system in a clinical setting, *Radiology* **168:**481–488.

Boesch, C., Gruetter, R., Martin, E., Duc, G., and Wuthrich, K., 1989, Variations in the in vivo P-31 MR spectra of the developing human brain during postnatal life, *Radiology* **172:**197–199.

Buchli, R., Martin, E., Boesiger, P., and Rumpel, H., 1994, Developmental changes of phosphorus metabolite concentrations in the human brain: A $^{31}$P magnetic resonance spectroscopy study in vivo. *Pediatr. Res.* **35:**431–435.

Burri, R., Lazeyras, F., Aue, W. P., Straehli, P., Bigler, P., Althaus, U., and Herschkowitz, N., 1988, Correlation between $^{31}$P NMR phosphomonoester and biochemically determined phosphorylethanolamine and phosphatidylethanolamine during development of the rat brain, *Dev. Neurosci.* **10:**213–221.

Cady, E. B., 1990, Absolute quantitation of phosphorus metabolites in the cerebral cortex of the newborn human infant and in the forearm muscles of young adults using a double tuned surface coil, *J. Magn. Reson.* **87:**433–446.

Cady, E. B., 1991, A reappraisal of the absolute concentrations of phosphorylated metabolites in the human neonatal cerebral cortex obtained by fitting Lorentzian curves to the $^{31}$P NMR spectrum, *J. Magn. Reson.* **91:**637–643.

Cady, E. B., 1992, MRS of the newborn human infant, in "Magnetic Resonance Spectroscopy in Biology and Medicine" (J. D. de Certaines, W. M. M. J. Bovee, and F. Podo, eds.), pp. 437–477, Pergamon Press, Oxford.

Cady, E. B., 1995, Quantitative combined phosphorus and proton PRESS of the brains of newborn human infants, *Magn. Reson. Med.* **33:**557–563.

Cady, E. B., Costello, A. M. de L., Dawson, M. J., Delpy, D. T., Hope, P. L., Reynolds, E. O. R.,

Tofts, P. S., and Wilkie, D. R., 1983, Non-invasive investigation of cerebral metabolism in newborn infants by phosphorus nuclear magnetic resonance spectroscopy, *Lancet* **i:**1059–1062.

Cady, E. B., Roth, S., Azzopardi, D., and Reynolds, E. O. R., 1990, Developmental changes in the absolute concentrations of phosphorylated metabolites in the cerebral cortex of the human infant—results from frequency domain spectrum analysis, *in* "Proceedings of the 9th Annual Meeting of the Society of Magnetic Resonance in Medicine, New York," p. 1003.

Cady, E. B., Hennig, J., and Martin, E., 1991, Magnetic resonance spectroscopy, *in* "Imaging Techniques of the CNS of Neonates" (J. Haddad, D. Christmann, and J. Messer, eds.), pp. 117–146, Springer, Berlin.

Cady, E. B., Boesch, C., and Martin, E., 1993, Magnetic resonance spectroscopy, *in* "Perinatal Asphyxia" (J. Haddad and E. Saliba, eds.), pp. 166–194, Springer, Berlin.

Cady, E. B., Lorek, A., Penrice, J., Wylezinska, M., Cooper, C. E., Brown, G., Owen-Reece, H., Kirkbride, V., Wyatt, J. S., and Reynolds, E. O. R., 1994b, Brain-metabolite transverse relaxation times in magnetic resonance spectroscopy increase as adenosine triphosphate depletes during secondary energy failure following acute hypoxia–ischaemia in the newborn piglet, *Neurosci. Lett.* **182:**201–204.

Cady, E. B., Lorek, A., Penrice, J., Reynolds, E. O. R., Iles, R. A., Burns, S. P., Coutts, G. A., and Cowan, F. M., 1994c, Detection of propan-1,2-diol in neonatal brain by in vivo proton magnetic resonance spectroscopy, *Magn. Reson. Med.* **32:**764–767.

Cady, E. B., Penrice, J., Amess, P. N., Lorek, A., Wylezinska, M., Aldridge, R. F., Franconi, F., Wyatt, J. S., and Reynolds, E. O. R., 1996, Lactate, N-acetylaspartate, choline and creatine concentrations, and spin-spin relaxation in thalamic and occipito-parietal regions of developing human brain, *Magn. Reson. Med.* **36:**878–886.

Cady, E. B., Penrice, J., Lorek, A., Amess, P. N., and Wylezinska, M., 1995b, Quantitative phosphorus PRESS of newborn human brain using internal water as reference standard, *in* "Proceedings of the 3rd Annual Meeting of the Society of Magnetic Resonance, Nice," p. 260.

Cerdan, S., Parilla, R., Santoro, J., and Rico, M., 1985, ¹H NMR detection of cerebral *myo*-inositol, *FEBS Lett.* **187:**167–172.

Chapman, A. G., Westerberg, E., and Siesjö, B. K., 1981, The metabolism of purine and pyrimidine nucleotides in rat cortex during insulin-induced hypoglycemia and recovery, *J. Neurochem.* **36:**179–189.

Christiansen, P., Henriksen, O., Stubgaard, M., Gideon, P., and Larsson, H. B. W., 1993, In vivo quantification of brain metabolites by ¹H-MRS using water as an internal standard, *Magn. Reson. Imaging* **11:**107–118.

Chu, A., Delpy, D. T., and Thalayasingam, S., 1986, A transport and life support system for newborn infants during NMR spectroscopy, *in* "Fetal and Neonatal Physiological Measurements" (P. Rolfe, ed.), pp. 409–415, Butterworths, London.

Cohen, M. M., and Lin, S., 1962, Acid soluble phosphates in the developing rabbit brain, *J. Neurochem.* **9:**345–352.

Corbett, R. J. T., Laptook, A. R., and Nunnally, R. L., 1987, The use of the chemical shift of the phosphomonoester P-31 magnetic resonance peak for the determination of intracellular pH in the brains of neonates, *Neurology* **37:**1771–1779.

Decorps, M., Blondet, P., Reutenauer, H., and Albrand, J. P., 1985, An inductively coupled, series-turned NMR probe, *J. Magn. Reson.* **65:**100–109.

Department of Health, United Kingdom, 1993, "Guidelines for Magnetic Resonance Diagnostic Equipment in Clinical Use", Her Majesty's Stationery Office, Norwich.

Dobbing, J., and Sands, J., 1973, Quantitative growth and development of human brain, *Arch. Dis. Child.* **48:**757–767.

Duffy, T. E., Kohle, S. J., and Vannucci, R. C., 1975, Carbohydrate and energy metabolism in perinatal rat brain: Relation to survival in anoxia, *J. Neurochem.* **24:**271–276.

Duyn, J. H., and Moonen, C. T. W., 1993, Fast proton spectroscopic imaging of human brain using multiple spin-echoes, *Magn. Reson. Med.* **30**:409–414.

Ernst, T., and Hennig, J., 1991, Coupling effects in volume selective $^1$H spectroscopy of major brain metabolites, *Magn. Reson. Med.* **21**:82–96.

Fenstermacher, M. J., and Narayana, P. A., 1990, Serial proton magnetic resonance spectroscopy of ischemic brain injury in humans, *Radiology* **25**:1034–1039.

Flynn, C. J., Farooqui, A. A., and Horrocks, L. A., 1989, Ischemia and hypoxia, *in* "Basic Neurochemistry: Molecular, Cellular, and Medical Aspects" (G. J. Siegel, B. Agranoff, R. W., Albers, and P. Molinoff, eds), pp. 783–795, Raven Press, New York.

Garlick, P. B., Soboll, S., and Bullock, G. R., 1992, Evidence that mitochondrial phosphate is visible in $^{31}$P NMR spectra of isolated perfused rat hearts, *NMR Biomed.* **5**:29–36.

Glonek, T., Kopp, S. J., Kot, E., Pettegrew, J. W., Harrison, W. H., and Cohen, M. M., 1982, P-31 nuclear magnetic resonance analysis of brain: The perchloric acid extract spectrum, *J. Neurochem.* **39**:1210–1219.

Grodd, W., Krägeloh-Mann, I., Klose, U., and Sauter, R., 1991, Metabolic and destructive brain disorders in children: Findings with localized proton MR spectroscopy, *Radiology* **181**:173–181.

Groenendaal, F., Veenhoven, R. H., van der Grond, J., Jansen, G. H., Witkamp, T. D., and de Vries, L. S., 1994, Cerebral lactate and *N*-acetyl-aspartate/choline ratios in asphyxiated full-term neonates demonstrated in vivo using proton magnetic resonance spectroscopy, *Pediatr. Res.* **35**:148–151.

Gruetter, R., Boesch, C., Muri, M., Martin, E., and Wuthrich, K., 1990, A simple design for a double-tunable probe head for imaging and spectroscopy at high fields, *Magn. Reson. Med.* **15**:128–134.

Hamilton, P. A., Hope, P. L., Cady, E. B., Delpy, D. T., Wyatt, J. S., and Reynolds, E. O. R., 1986, Impaired energy metabolism in brains of newborn infants with increased cerebral echodensities, *Lancet* **i**:1242–1246.

Hope, P. L., Costello, A. M. de L., Cady, E. B., Delpy, D. T., Tofts, P. S., Chu, A., Reynolds, E. O. R., and Wilkie, D. R., 1986, Cerebral metabolism in newborn infants studied by phosphorus nuclear magnetic resonance spectroscopy, *in* "Neonatal Physiological Measurements" (P. Rolfe, ed.), pp. 382–389, Butterworths, London.

Hope, P. L., Cady, E. B., Chu, A., Delpy, D. T., Gardiner, R. M., and Reynolds, E. O. R., 1987, Brain metabolism and intracellular pH during ischaemia and hypoxia. An in-vivo $^{31}$P and $^1$H nuclear magnetic resonance study in the lamb, *J. Neurochem.* **49**:75–82.

Huppi, P. S., Posse, S., Lazeyras, F., Burri, R., Bossi, E., and Herschkowitz, N., 1991, Magnetic resonance in preterm and term newborns: $^1$H-spectroscopy in developing human brain, *Pediatr. Res.* **30**:574–578.

Jung, W. I., Widmaier, S., Bunse, M., Seeger, U., Straubinger, K., Schick, F., Kuper, K., Dietze, G., and Lutz, O., 1993, $^{31}$P transverse relaxation times of ATP in human brain *in vivo, Magn. Reson. Med.* **30**:741–743.

Kamada, K., Houkin, K., Hida, K., Matsuzawa, H., Iwasaki, Y., Abe, H., and Nakada, T., 1994, Localised proton spectroscopy of focal brain pathology in humans: Significant effects of edema on spin–spin relaxation time, *Magn. Reson. Med.* **31**:537–540.

Kreis, R., Ernst, T., and Ross, B. D., 1993, Development of the human brain: *In vivo* quantification of metabolite and water content with proton magnetic resonance spectroscopy, *Magn. Reson. Med.* **30**:424–437.

Laptook, A. R., Corbett, R. J., Sterett, R., Burns, D. K., Tollefsbol, G., and Garcia, D., 1994, Modest hypothermia provides partial neuroprotection for ischemic neonatal brain, *Pediatr. Res.* **35**:436–442.

Lorek, A., Takei, Y., Cady, E. B., Wyatt, J. S., Penrice, J., Edwards, A. D., Peebles, D., Wylezinska,

M., Owen-Reece, H., Kirkbride, V., Cooper, C., Aldridge, R. F., Roth, S. C., Brown, G., Delpy, D. T., and Reynolds, E. O. R., 1994, Delayed ('secondary') cerebral energy failure following acute hypoxia–ischaemia in the newborn piglet: Continuous 48-hour studies by phosphorus magnetic resonance spectroscopy, *Pediatr. Res.* **36:**699–706.

Mehmet, H., Yue, X., Squier, M. V., Lorek, A., Cady, E. B., Penrice, J., Sarraf, C., Wylezinska, M., Kirkbride, V., Cooper, C. E., Brown, G. C., Wyatt, J. S., Reynolds, E. O. R., and Edwards, A. D., 1994, Increased apoptosis in the cingulate gyrus of newborn piglets following transient hypoxia–ischaemia is related to the degree of high energy phosphate depletion during the insult, *Neurosci. Lett.* **181:**121–125.

Miller, B. L., 1991, A review of chemical issues in ¹H NMR spectroscopy: N-acetylaspartate, creatine, and choline, *NMR Biomed.* **4:**47–52.

Moonen, C. T. W., von Kienlin, M., van Zijl, P. C. M., Cohen, J., Gillen, J., Daly, P., and Wolf, G., 1989, Comparison of single-shot localisation methods (STEAM and PRESS) for in vivo proton NMR spectroscopy, *NMR Biomed.* **2:**201–208.

Moorcraft, J., Bolas, N. M., Ives, N. K., Ouwerkerk, R., Smyth, J., Rajagopalan, B., Hope, P. L., and Radda, G. K., 1991, Global and depth resolved phosphorus magnetic resonance spectroscopy to predict outcome after birth asphyxia, *Arch. Dis. Child.* **66:**1119–1123.

Murphy, E. J., Rajagopalan, B., Brindle, K. M., and Radda, G. K., 1989, Phospholipid bilayer contribution to ³¹P NMR spectra in vivo, *Magn. Reson. Med.* **12:**282–289.

Okumura, N., Otsuki, S., and Kameyama, A., 1960, Studies of free amino acids in human brain, *J. Biochem.* **47:**315–320.

Ordidge, R. J., Connelly, A., and Lohman, J. A. B., 1986, Image-selected in vivo spectroscopy (ISIS): A new technique for spatially selective NMR spectroscopy, *J. Magn. Reson.* **66:**283–294.

Palmer, C., Brucklacher, R. M., Christensen, M. A., and Vannucci, R. C., 1990, Carbohydrate and energy metabolism during the evolution of hypoxic–ischemic brain damage in the immature rat, *J. Cereb. Blood Flow Metab.* **10:**227–235.

Peden, C. J., Rutherford, M. A., Sargentoni, J., Cox, I. J., Bryant, D. J., and Dubowitz, L. M. S., 1993, Proton spectroscopy of the neonatal brain following hypoxic–ischaemic injury, *Dev. Med. Child Neurol.* **35:**502–510.

Penrice, J., Cady, E. B., Lorek, A., Amess, P. N., Wylezinska, M., Aldridge, R. F., Wyatt, J. S., and Reynolds, E. O. R., 1995, Cerebral metabolite abnormalities after perinatal hypoxia–ischaemia detected by ¹H MRS of the brains of preterm and term infants, *in* "Proceedings of the 3rd Annual Meeting of the Society of Magnetic Resonance, Nice," p. 384.

Penrice, J., Cady, E. B., Lorek, A., Wylezinska, M., Amess, P. N., Aldridge, R. F., Stewart, A. L., Wyatt, J. S., and Reynolds, E. O. R., 1996, Proton magnetic resonance spectroscopy of the brain in normal preterm and term infants, and early changes following perinatal hypoxia-ischaemia, *Pediatr. Res.* **40:**6–14.

Petroff, O. A. C., Prichard, J. W., Behar, K. L., Alger, J. R., den Hollander, J. A., and Shulman, R. G., 1985, Cerebral intracellular pH by ³¹P nuclear magnetic resonance spectroscopy, *Neurology* **35:**781–788.

Pettegrew, J. W., Kopp, S. J., Dadok, J., Minshew, N. J., Feliksik, J. M., Glonek, T., and Cohen, M. M., 1986, Chemical characterization of a prominent phosphomonoester resonance from mammalian brain. ³¹P and ¹H NMR analysis at 4.7 and 14.1 Tesla, *J. Magn. Reson.* **67:**443–450.

Portman, M. A., and Ning, X. H., 1990, Developmental adaptations in cytosolic phosphate content and pH regulation in the sheep heart in vivo, *J. Clin. Invest.* **86:**1823–1828.

Prielmeier, F., Nagatomo, Y., and Frahm, J., 1994, Cerebral blood oxygenation in rat brain during hypoxic hypoxia. Quantitative MRI of effective transverse relaxation rates, *Magn. Reson. Med.* **31:**678–681.

Robbins, R. C., Balaban, R. S., and Swain, J. A., 1990, Intermittent hypothermic asanguineous cerebral perfusion (cardioplegia) protects the brain during prolonged circulatory arrest, *J. Thorac. Cardiovasc, Surg.* **99**:878–884.

Ross, B. D., 1991, Biochemical considerations in ¹H spectroscopy. Glutamate and glutamine; *myo*-inositol and related metabolites, *NMR Biomed.* **4**:59–63.

Roth, S. C., Azzopardi, D., Edwards, A. D., Baudin, J., Cady, E. B., Townsend, J., Delpy, D. T., Stewart, A. L., Wyatt, J. S., and Reynolds, E. O. R., 1992, Relation between cerebral oxidative metabolism following birth asphyxia, and neurodevelopmental outcome and brain growth at one year, *Dev. Med. Child Neurol.* **34**:285–295.

Shellock, F. G., and Slimp, G. L., 1989, Severe burn of the finger caused by using a pulse oximeter during MR imaging, *Am. J. Radiol.* **153**:1105.

Siesjö, B. K., 1978, "Brain Energy Metabolism," Wiley, New York.

Stewart, A. L., Thorburn, R. J., Hope, P. L., Goldsmith, M., Lipscomb, A. P., and Reynolds, E. O. R., 1983, Ultrasound appearance of the brain in very preterm infants and neurodevelopmental outcome at 18 months of age, *Arch. Dis. Child.* **58**:598–604.

Stöckler, S., Holzbach, U., Hanefeld, F., Marquardt, I., Helms, G., Requardt, M., Hänicke, W., and Frahm, J., 1994, Creatine deficiency in the brain: A new treatable inborn error of metabolism, *Pediatr. Res.* **36**:409–413.

Tallan, H. H., 1957, Studies on the distribution of *N*-acetylaspartic acid in brain, *J. Biol. Chem.* **223**:41–45.

Tan, W. K. N., Williams, C. E., Gunn, A. J., Mallard, C., and Gluckman, P. D., 1993, Pretreatment with monosialoganglioside GM1 protects the brain of fetal sheep against hypoxic–ischaemic injury without causing systemic compromise, *Pediatr. Res.* **34**:18–22.

Thordstein, M., Bagenholm, R., Thiringer, K., and Kjellmer, I., 1993, Scavengers of free oxygen radicals in combination with magnesium ameliorate perinatal hypoxic–ischaemic damage in the rat, *Pediatr. Res.* **34**:23–26.

Thoresen, M., Penrice, J., Lorek, A., Cady, E. B., Wylezinska, M., Kirkbride, V., Cooper, C. E., Brown, G. C., Edwards, A. D., Wyatt, J. S., and Reynolds, E. O. R., 1995, Mild hypothermia after severe transient hypoxia–ischaemia ameliorates delayed cerebral energy failure in the newborn piglet, *Pediatr. Res.* **37**:667–670.

Toft, P. B., Christiansen, P., Pryds, O., Lou, H. C., and Henriksen, O., 1994, $T_1$, $T_2$, and concentrations of brain metabolites in neonates and adolescents estimated with H-1 MR spectroscopy, *J. Magn. Reson. Imaging* **4**:1–5.

Tofts, P., and Wray, S., 1985, Changes in brain phosphorus metabolites during the post-natal development of the rat, *J. Physiol. (London)* **359**:417–429.

Urenjak, J., Williams, S. R., Gadian, D. G., and Noble, M., 1993, Proton nuclear magnetic resonance spectroscopy unambiguously identifies different neural cell types, *J. Neurosci.* **13**:981–989.

van der Knaap, M. S., van der Grond, J., van Rijen, P. C., Faber, J. A. J., Valk, J., and Willemse, K., 1990, Age-dependent changes in localized proton and phosphorus MR spectroscopy of the brain, *Radiology* **176**:509–515.

Vannucci, R.C., and Duffy, T. E., 1977, Cerebral metabolism in newborn dogs during reversible asphyxia, *Ann. Neurol.* **1**:528–534.

Vannucci, R. C., and Vannucci, S. J., 1978, Cerebral carbohydrate metabolism during hypoglycemia and anoxia in newborn rats, *Ann. Neurol.* **4**:73–79.

Wilkinson, I. D., Paley, M., Chong, W. K., Sweeney, B. J., Shepherd, J. K., Kendall, B. E., Hall-Craggs, M. A., and Harrison, M. J. G., 1994, Proton spectroscopy in HIV infection: Relaxation times of cerebral metabolites, *Magn. Reson. Imaging* **12**:951–957.

Wyatt, J. S., Cope, M., Delpy, D. T., Richardson, C. E., Edwards, A. D., Wray, S., and Reynolds, E. O. R., 1990, Quantitation of cerebral blood volume in human infants by near-infrared spectroscopy, *J. Appl. Physiol.* **68**:1086–1091.

Yager, J. Y., Brucklacher, R. M., and Vannucci, R. C., 1992, Cerebral energy metabolism during hypoxia–ischaemia and early recovery in immature rats, *Am. J. Physiol.* **262:**H672–H677.

Younkin, D. P., Delivoria-Papadopoulos, M., Leonard, J. C., Subramanian, V. H., Eleff, S., Leigh, J. S., and Chance, B., 1984, Unique aspects of human newborn cerebral metabolism evaluated with 31-P NMR spectroscopy, *Ann. Neurol.* **16:**581–586.

# LOCALIZED PROTON MAGNETIC RESONANCE SPECTROSCOPY OF BRAIN DISORDERS IN CHILDHOOD

## JENS FRAHM and FOLKER HANEFELD

## SUMMARY

Localized proton magnetic resonance spectroscopy is used to study cerebral metabolic disorders in children. The development of single-voxel techniques allows quantitative calculation of the absolute concentrations in different regions of the brain of metabolites that are readily observed in $^1H$ spectra. These include $N$-acetylaspartate (NAA), lactate, amino acids, inositols, choline-containing compounds, and total creatine. The studies are illustrated by the results on numerous childhood disorders: the leukodystrophies, other white-matter diseases, gray-matter diseases, mitochondrial diseases, and miscellaneous deficiency disorders.

JENS FRAHM • Biomedical NMR Research, Inc., at the Max-Planck-Institute for Biophysical Chemistry, D-37070 Göttingen, Germany. FOLKER HANEFELD • Department of Neuropediatrics, University of Göttingen. D-37075 Göttingen, Germany.

Magnetic Resonance Spectroscopy and Imaging in Neurochemistry, Volume 8 of Advances in Neurochemistry, edited by Bachelard, Plenum Press, New York, 1997.

## 1. INTRODUCTION

Beyond structural and functional investigations of the central nervous system by X-ray computed tomography and positron-emission tomography, advances in magnetic resonance have opened new and truly noninvasive avenues into the functioning human brain. In particular, state-of-the-art magnetic resonance imaging (MRI) provides access to high-resolution anatomy and brain function, while magnetic resonance spectroscopy (MRS) offers unique insights into cerebral metabolism.

The primary aim of this chapter is to demonstrate the usefulness and potential of localized ¹H MRS and its application to brain disorders in childhood. To emphasize practical access to *in vivo* neurochemistry, Section 2 describes a simple and robust single-voxel technique that facilitates transformation of spectral data into neurochemical quantities such as absolute metabolite concentrations. Pertinent applications in the field of neuropediatrics are presented in Sections 3–9 and underline the remarkable progress during recent years. Major achievements are illustrated by selected examples from our experience with more than 400 MRS examinations of over 300 children (Table 1).

The focus of this chapter is on neurometabolic, neurodegenerative, and neuroinflammatory disorders. Although this selection is driven by our personal interest, it also reflects preliminary observations that ¹H MRS seems to give less information in cases of unclassified mental retardation, psychosis, generalized epilepsies, and other conditions without definite biochemical or structural abnormalities.

A comprehensive coverage of *in vivo* MRS or even ¹H MRS of the brain is beyond the scope of this chapter. The use of MRS for the understanding of disease has been discussed in a large number of articles, e.g., by Bottomley *et al.* (1985), Radda (1986), Prichard and Shulman (1986), Frahm *et al.* (1989), Howe *et al.* (1993), and Kemp and Radda (1994). Applications to brain disorders in childhood have been described by Grodd *et al.* (1991), van der Knaap *et al.* (1992), Tzika *et al.* (1993a, b), and Ross and Michaelis (1994). The clinical, genetic, biochemical, and radiological characteristics of neurological disorders in childhood have been summarized by Aicardi (1992) and Barkovich (1995).

## 2. LOCALIZED PROTON MAGNETIC RESONANCE SPECTROSCOPY

The purpose of this section is to outline the methodology used for quantitative localized ¹H MRS and its application to the study of brain disorders in childhood. A discussion of basic localization strategies is followed by a description of the sti nulated-echo acquisition mode (STEAM) technique. Experimental

TABLE 1. Brain Disorders in Childhood Studied by Localized ¹H MRS

| Classification | Disease/syndrome | Number of patients |
|---|---|---|
| Leukodystrophies | | |
|   Known origin | Metachromatic leukodystrophy | 8 |
| | Globoid cell leukodystrophy | 2 |
| | Pelizaeus–Merzbacher disease | 4 |
| | Canavan's disease | 1 |
| | Adrenoleukodystrophy | 26 |
|   Unknown origin | Alexander's disease | 2 |
| | Congenital muscular dystrophy plus leukodys-trophy | 6 |
| | Myelinopathia centralis diffusa | 6 |
| | Cystic leukoencephalopathy | 3 |
| | Unclassified leukodystrophies | 26 |
| Other white-matter disorders | L-2-Hydroxyglutaric aciduria | 1 |
| | Succinate dehydrogenase deficiency | 1 |
| | Multiple sclerosis | 24 |
| Gangliosidoses | $G_{M_2}$ gangliosidosis | 1 |
| | Neuronal ceroid lipofuscinosis | 7 |
| Mitochondrial disorders | Leigh syndrome | 15 |
| | Mitochondrial encephalopathy, lactic acidosis, and stroke-like episodes (MELAS) | 1 |
| | Myoclonic epilepsy with ragged red fibers (MERRF) | 2 |
| | Kearns–Sayre Syndrome | 3 |
| | Others | 17 |
| Mucopolysaccharidoses | Type VI | 1 |
| Other lysosomal disorders | Niemann–Pick type C | 1 |
| | Mannosidosis | 1 |
| Peroxisomal disorders | Cerebro-hepato-renal syndrome | 3 |
| Amino aciduria | Nonketotic hyperglycinemia | 5 |
| | Phenylketonuria | 4 |
| | Isovaleric acidemia | 1 |
| Purine/pyrimidine disorders | Lesh–Nyhan syndrome | 2 |
| | Phosphoribosylpyrophosphate synthetase defect | 2 |
| Other metabolic disorders | Carbohydrate-deficient glycoprotein syndrome | 5 |
| | Creatine deficiency | 1 |
| Central nervous system malfor-mations | Hemimegalencephaly | 4 |
| | Double cortex | 1 |
| | Pachygyria | 1 |
| Heredodegenerative disorders | Huntington chorea | 2 |
| | Olivopontocerebellar atrophy | 1 |
| Miscellaneous disorders | Rett syndrome | 21 |
| | Hyperekplexia | 2 |
| | Anorexia nervosa | 13 |
| | Hallervorden–Spatz disease | 2 |
| | Epilepsy | 9 |
| | Brain tumors | 9 |
| | Extrapyramidal disorders | 8 |
| | Cerebrovascular disorders | 2 |
| | Unclassified diseases | 64 |

conditions are presented that allow quantification of the concentrations of metabolites identified in ¹H STEAM spectra of human brain. Regional variability, age dependencies, and other physiological influences are briefly discussed prior to a description of the requirements for a typical patient examination.

## 2.1. Single-Voxel Spectroscopy: A Rationale

Meaningful insights into the chemistry of the intact human brain require spatial discrimination of MRS responses. Two different concepts have been proposed that either emphasize spatial (imaging) or biochemical resolution (spectroscopy). *Chemical shift imaging* techniques employ one to three gradient pulses to encode spatial information along respective gradient axes into the phase of a time-domain signal from the entire object. This signal is acquired in the absence of a magnetic field gradient and thus carries the chemical shift information. Multidimensional image reconstruction then leads to metabolite maps that represent the spatial intensity distributions of individual metabolite resonances at selected chemical shift frequencies. Alternatively, *single-voxel spectroscopy* techniques attempt to restrict the radio-frequency (RF) excitation to a predefined volume of interest (VOI). They acquire the uncompromised though spatially localized frequency spectrum directly as a time-domain signal from the VOI, and, therefore no further processing is required except for one-dimensional Fourier transformation.

To complement high-resolution morphological information from MRI with detailed neurochemical information from MRS, it seems advisable to employ localized single-voxel spectra, which, in general, reveal a much more detailed and quantifiable metabolic picture of selected foci than metabolite maps. More specifically, the advantages of the single-voxel approach are due to the excellent spectral resolution and water suppression that are achievable by optimizing the magnetic field homogeneity over only a small VOI of 1–18 ml. Moreover, the relatively short measurement times of 1–10 min are clearly beneficial for a reliable quantification of metabolite concentrations as they allow the use of long repetition times (TR). For example, selection of TR $\geq$ 6000 msec ensures full relaxation of proton spin systems between successive RF excitations and thus avoids the need for corrections due to $T_1$ saturation when transforming resonance areas into metabolite concentrations. This is of particular importance for studies of disease states of the brain because potential changes in $T_1$ relaxation are usually unknown.

In general, of course, the choice of a technique should only depend on the question to be answered. While details about localization techniques are necessary to properly apply such methods and to facilitate the evaluation and interpretation of the resulting data, MRS findings in metabolic research and medical decision making must be independent of the technique chosen. The best way to accomplish this goal is to convert *relative* spectral findings, often expressed as peak heights or area ratios, into *absolute* spectral quantities, i.e., absolute resonance areas, and to convert these values into metabolite concentrations by calibrating *in vivo* resonance areas with those of model metabolite solutions.

## 2.2. Localization Using Stimulated Echoes

In accordance with the aforementioned considerations, most investigations have been performed using fully relaxed single-voxel ¹H MRS. The two techniques that may be employed for single-shot gradient localization are PRESS (Bottomley, 1984, 1987; Ordidge *et al.*, 1985) and STEAM sequences (Frahm *et al.*, 1984, 1987). In the work described here, we have exclusively used STEAM sequences because this technique provides optimum conditions for MRS acquisition at short echo times (TE) that avoid gross corrections for $T_2$ relaxation and/or spin-coupling modulation. For a comparison of localization strategies, see Frahm and Hänicke (1994, 1996).

In contrast to the double-spin-echo 90°–180°–180° PRESS technique, STEAM localization sequences comprise three frequency-selective 90° RF pulses that are applied in the presence of orthogonal gradients. As shown in Fig. 1, the resulting stimulated-echo (STE) signal (Hahn, 1950) represents magnetizations that originate from a VOI defined by the intersection of three perpendicular sections. During the course of the STEAM sequence, transverse magnetizations excited by the first RF pulse are transformed into longitudinal magnetizations by application of the second pulse. Corresponding components are subject to $T_1$ relaxation during the middle interval TM. They refocus as a stimulated echo at TE/2 after application of the third pulse. To minimize signal losses due to $T_2$ relaxation, spin-coupling modulation, and multiple-quantum interferences (Ernst and Hennig, 1991), the echo time TE should be chosen as short as possible. In our studies, the STEAM intervals were TE = 20 msec and TM = 30 msec. Data acquisition covered the second half of the STE signal (1024 msec) with a receiver bandwidth of ±1000 Hz.

Under these conditions, the STEAM sequence is best suited for ¹H MRS, although ³¹P MRS of human brain at even shorter TE and TM values has been successfully demonstrated (Merboldt *et al.*, 1990). For proton applications, the sequence must be preceded by one to three chemical-shift-selective (CHESS) water suppression pulses plus associated spoiler gradients (Haase *et al.*, 1985). In practice, three successive CHESS pulses of 60 Hz bandwidth (i.e., 0.7 ppm at 2.0 T) attenuate the water proton resonance by about a factor of 1000 while not affecting chemical shifts outside the 4.1–5.3 ppm region of the proton spectrum. Technical details of the water-suppressed STEAM sequence have been described by Frahm *et al.* (1990).

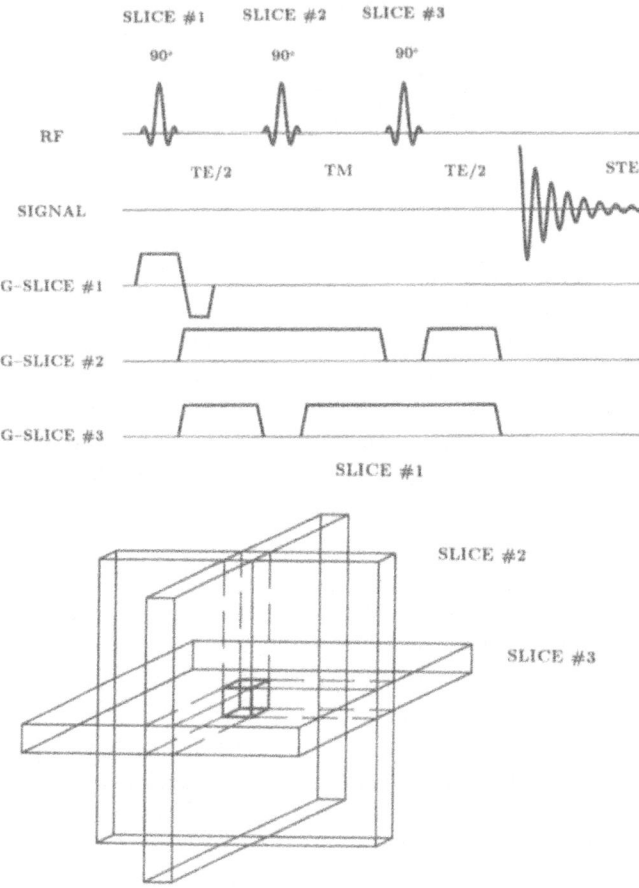

FIGURE 1. Single-voxel spectroscopy using a short-echo-time stimulated-echo acquisition mode (STEAM) localization technique. The STEAM sequence (*top*) defines the volume of interest (VOI) by the intersection of three cross sections (*bottom*) that are excited by frequency-selective 90° radio-frequency (RF) pulses in the presence of orthogonal magnetic field gradients (slices #1 to #3). The stimulated-echo (STE) signal from the selected VOI is attenuated by $T_1$ relaxation during the middle interval TM and $T_2$ relaxation during the two echo intervals TE/2.

## 2.3. Identification of Cerebral Metabolites

Although metabolite concentrations represent the obvious biochemical parameters to be derived from *in vivo* MRS, the accessible information may be extended to rate constants and fluxes provided suitable resonances are detectable and relevant metabolite pools are changing in response to physiological stimuli, pharmacological treatment, or pathological conditions. In all cases, prerequisites for a reliable interpretation of MRS data are a detailed understanding of the spectra (proton, phosphorus, or carbon-13) and objective methods for spectral evaluation and quantification (Bottomley, 1991).

Figure 2 shows *in vivo* ¹H MR spectra of human gray and white matter obtained by fully relaxed (TR = 6000 msec) short-echo-time (TE/TM = 20/30 msec) STEAM spectroscopy. Spectral processing involved zero filling of the original 2K complex time-domain data points to 4K, Gaussian filtering (center zero, half-width 317 msec), Fourier transformation, and zero- and first-order phase correction without spectral baseline manipulation or resolution enhancement. Resonance assignments were guided by biochemical data from biopsies and autopsies (Perry *et al.*, 1971, 1981) and high-resolution high-field ¹H MR spectra of perchloric acid extracts from animal (Behar *et al.*, 1983; Arus *et al.*, 1985) and human brain (Petroff *et al.*, 1989; Peeling and Sutherland, 1993). They were confirmed by recording ¹H MR spectra of individual metabolites using identical experimental conditions as for *in vivo* human studies (Michaelis *et al.*, 1991; Frahm *et al.*, 1991a; Gyngell *et al.*, 1991; Michaelis *et al.*, 1993a). Such model spectra not only eliminate complications due to spectral overlap and allow assessment of the influence of strong spin–spin coupling at the relatively low field strength of 2.0 T used here but also serve as concentration-calibrated references for a user-independent quantification of absolute concentrations.

So far, the list of major detectable metabolites includes *N*-acetylaspartate (NAA), *N*-acetylaspartylglutamate (NAAG), glutamate (Glu), glutamine (Gln), creatine and phosphocreatine (Cr), choline-containing compounds (Cho), and *myo*-inositol (*myo*-Ins). In addition, the spectra may include resonances from glucose (Glc), lactate (Lac), aspartate (Asp), alanine (Ala), taurine (Tau), *scyllo*-inositol (*scyllo*-Ins), γ-aminobutyrate (GABA), guanidinoacetate (G), and cytosolic protein residues (Kauppinen *et al.*, 1992), although reliable assessments of many of these compounds are only possible for elevated concentrations in disease states of the brain. In the absence of a chemical shift standard, the *in vivo* ¹H MR spectra are commonly referenced to the *N*-acetyl resonance of NAA at 2.01 ppm. If NAA is depleted in a brain disorder, then Cr (3.04 ppm), Cho (3.22 ppm), or even Lac (1.33 ppm, center of doublet) may be taken as a reference.

For an understanding of neurochemical and cellular metabolite alterations, it is most important to note that NAA is exclusively located in neurons [see, e.g., the review by Birken and Oldendorf (1989)]. Conversely, the brain osmolyte *myo*-Ins turns out to be glia-specific as it has only been found in perchloric acid extracts of primary glia as well as of C6 and F98 glioma cell lines, but not in neurons (Brand *et al.*, 1993). Thus, taking NAA and *myo*-Ins as markers for neuronal and glial tissue components, respectively, changes in their concentrations reflect metabolic alterations in neuronal and glial pools. In fact, the *irreversible* reduction or loss of NAA observed in diseases ranging from cerebral stroke

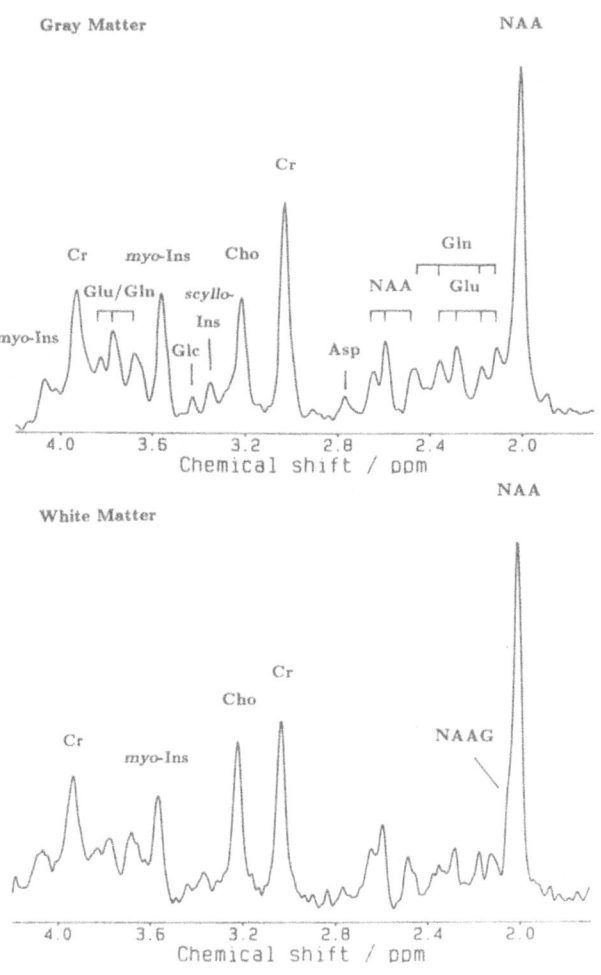

*FIGURE 2.* Cerebral metabolites detected in fully relaxed (TR = 6000 msec) short-echo-time (TE = 20 msec) ¹H MR spectra (STEAM, TM = 30 msec, 64 accumulations) of parietal gray (18 ml) and white matter (12 ml) of a young healthy adult. Spectral assignments include *N*-acetylaspartate (NAA), *N*-acetylaspartylglutamate (NAAG), glutamate (Glu), glutamine (Gln), aspartate (Asp), creatine and phosphocreatine (Cr), choline-containing compounds (Cho), *myo*-inositol (*myo*-Ins), *scyllo*-inositol (*scyllo*-Ins), and glucose (Glc). All spectra in this chapter are scaled in proportion to VOI size and plotted with identical standards.

to brain tumors suggests that spectroscopically detected NAA levels are in direct proportion to the amount of viable neuroaxonal tissue present in the investigated VOI. While putative increases of NAA resulting from regenerative nerve processes should be detectable in principle, occasional reports of *reversible* NAA levels in follow-up MRS studies either were not based on absolute concentrations or may be explained by tissue replacement of a focal lesion during recovery.

The methyl (3.04 ppm) and methylene (3.93 ppm) resonance signals of creatine and phosphocreatine (PCr) are indistinguishable at the spectral resolution achievable *in vivo*. Thus, the ¹H MRS detected total Cr level is not affected by altered energy demands that are regulated via the creatine kinase reaction and phosphate exchange between PCr and γ-adenosine triphosphate (ATP). An independent evaluation of PCr levels and other aspects of cellular bioenergetics may be obtained by ³¹P MRS.

The most likely candidates for the Cho signal are phosphorylcholine (PCh), glycerophosphorylcholine (GPC), and choline plasmalogen, while the concentrations of acetylcholine and choline are at least one order of magnitude lower. Proton-decoupled ³¹P MRS of human brain (Merboldt *et al.*,1990) as well as more recent findings based on quantitative fitting of proton data with the use of complete metabolite spectra suggests that the predominant contribution to the cerebral Cho level results from GPC (P. J. W. Pouwels, unpublished results).

In normal brain, contamination of the 3.56-ppm resonance from *myo*-Ins by the methylene group of glycine (Gly) at 3.55 ppm is unlikely owing to the low Gly concentration. Pathologically enhanced Gly levels should be distinguishable from *myo*-Ins as the Gly singlet does not contribute to the other resonances of *myo*-Ins, e.g., at 4.06 ppm, if unaffected by water suppression. The contributions from inositol phosphates (secondary messengers) or the head groups of inositolphospholipids (membrane constituents) to the *myo*-Ins spectrum can also be neglected owing to low concentrations, slightly different proton spectra, or partial immobilization.

## 2.4. Quantification of Cerebral Metabolite Concentrations

Spectral evaluation and quantification of metabolite concentrations is accomplished with use of the fully automated spectral evaluation program LCModel (Provencher, 1993). The approach takes advantage of a nearly model-free constrained regularization method (Provencher, 1982) and incorporates prior knowledge by fitting a library of metabolite reference spectra to the *in vivo* time-domain data. Because RF coil loading and corresponding receiver sensitivity differ for objects of different electrical conductivity (e.g., due to different head sizes), such effects have to be accounted for by correcting resonance areas with the inverse voltage amplitude required for a rectangular 90° reference RF pulse (Hoult and Richards, 1976). The procedure results in absolute metabolite concentrations that are largely independent of instrumental inadequacies and requires only a minimum number of assumptions (e.g., no lineshape and baseline assumptions are necessary). The use of model metabolite solutions may be referred to as an *absolute reference* method (Michaelis *et al.*, 1993b). The alternative choice of an *internal reference* such as water has the potential drawback of uncertain alterations of tissue water content in disease states of the brain. No corrections

TABLE 2. Absolute Tissue Concentrations (Mean ± SD) of Major Cerebral
Metabolites in Parietal Gray and White Matter, Basal Ganglia, and Cerebellum
of Young Healthy Adults (Mean Age 27 ± 4 Years)

| Metabolite[a] | Concentration (mM) | | | |
|---|---|---|---|---|
| | Gray matter (n = 60) | White matter (n = 72) | Basal ganglia (n = 11) | Cerebellum (n = 8) |
| NAA+NAAG | 7.7 ± 1.3 | 8.9 ± 1.3 | 9.0 ± 1.9 | 8.8 ± 1.6 |
| NAAG | 0.3 ± 0.3 | 1.3 ± 0.8 | 0.4 ± 0.5 | 0.5 ± 0.6 |
| PCr+Cr | 5.8 ± 0.9 | 5.0 ± 0.7 | 6.1 ± 1.3 | 7.4 ± 1.5 |
| Cholines | 1.0 ± 0.2 | 1.4 ± 0.3 | 1.7 ± 0.4 | 2.1 ± 0.7 |
| myo-Inositol | 3.9 ± 0.7 | 3.2 ± 0.8 | 3.3 ± 0.7 | 6.3 ± 3.3 |
| Glutamate | 7.3 ± 1.5 | 5.2 ± 1.2 | 7.7 ± 2.3 | 6.2 ± 2.8 |
| Glutamine | 3.3 ± 1.2 | 1.8 ± 0.9 | 3.9 ± 1.8 | 2.7 ± 1.7 |

[a]Abbreviations: NAA, N-acetylaspartate; NAAG, N-acetylaspartylglutamate; PCr, phosphocreatine; Cr, creatine.

were applied for partial-volume effects with cerebrospinal fluid (CSF). However, in the presence of atrophy, true metabolic alterations may be easily identified by taking ratios of metabolite concentrations.

Table 2 summarizes the concentrations of major metabolites in standardized VOIs of parietal gray and white matter, basal ganglia (including thalamus), and cerebellum (hemispheric center) for young healthy adults. Although gross regional differences are also detectable by visual inspection of the spectra shown in Fig. 3, concentration changes may be obscured by differences in magnetic field homogeneity, i.e., resonance linewidth. It should be noted, however, that all spectra presented in this chapter are scaled in proportion to VOI size and plotted with identical standards.

The quantitative data in Table 2 demonstrate that the combined pool of N-acetyl compounds constituted by NAA and NAAG does not vary beyond one standard deviation in the investigated regions. Recently, a more detailed study of metabolite levels within cortical gray matter revealed a gradient of increasing NAA (and decreasing Cho) from frontal and parietal cortex to occipital cortex (Pouwels et al., 1995). A possible explanation involves differences in cellular composition such as increased glial contributions in frontal brain and stronger neuronal populations in occipital brain (visual cortex), but putative links between cortical anatomy, neurochemistry, and brain function remain to be elucidated.

Relatively high levels of NAAG well above detectability (15% of total N-acetyl concentration) were only found in white matter. The concentration of Cr increases in the order white matter < gray matter < basal ganglia < cerebellum. It should be noted that, unless otherwise specified, the abbreviations NAA and Cr

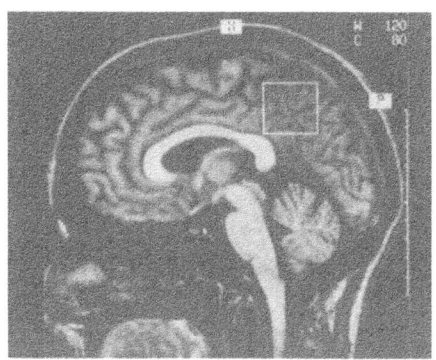

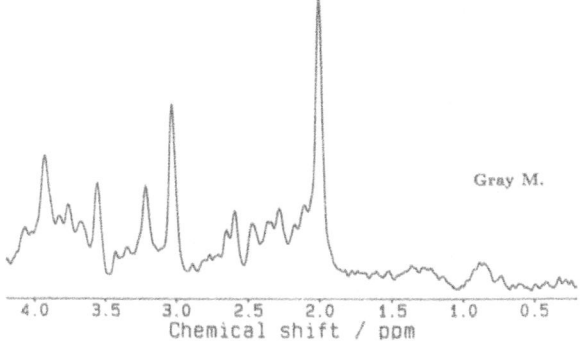

Gray M.

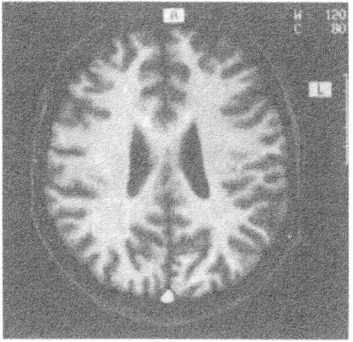

FIGURE 3. Regional variability of the metabolite pattern in the human brain (young adults) as detected by ¹H MRS (STEAM, TR/TE/TM = 6000/20/30 msec, 64 accumulations) of gray matter (18 ml), white matter (12 ml), basal ganglia (8 ml), and cerebellum (8 ml). The VOI locations are indicated on $T_1$-weighted images (RF spoiled 3-D FLASH, TR/TE = 15/6 msec, 20° flip angle, 4-mm partitions).

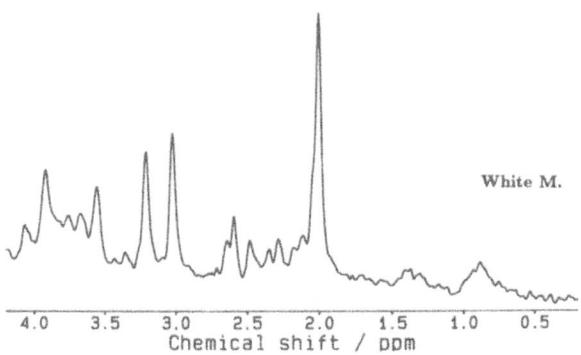

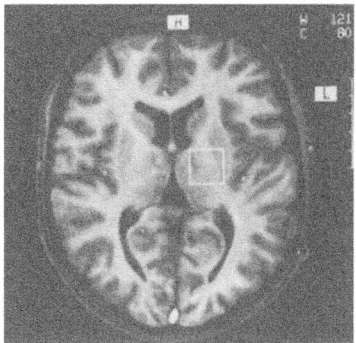

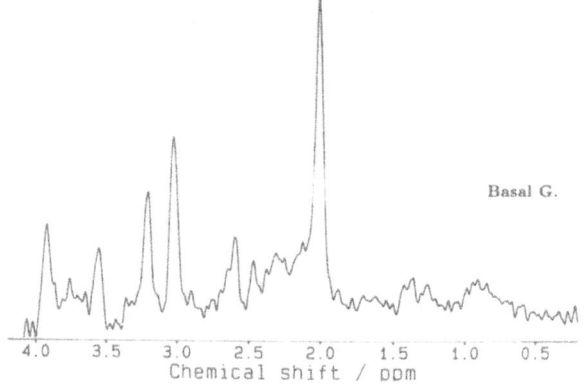

*Figure 3.* Continued

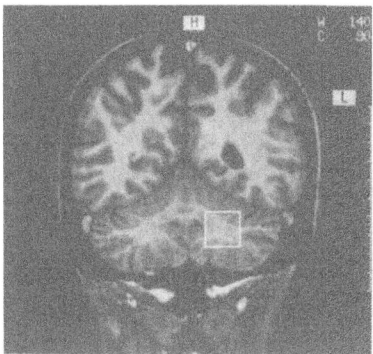

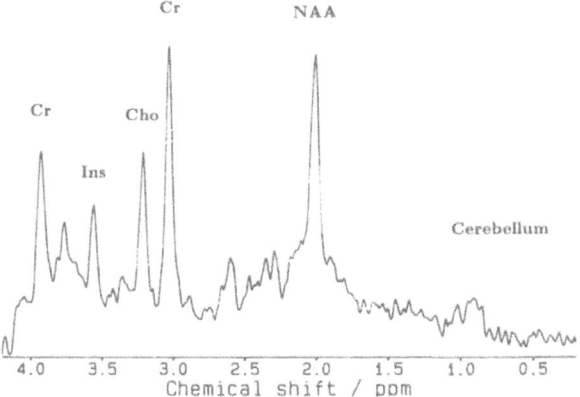

*Figure 3.* Continued

are used throughout this chapter to denote *total N*-acetyl levels, i.e., NAA and NAAG, and *total* Cr, i.e., Cr and PCr.

Regional Cho levels roughly parallel the degree of tissue myelination and therefore provide a marker of normal and disturbed myelination. Unfortunately, however, elevated Cho concentrations may arise either from enhanced membrane turnover and elevated PCh during cell proliferation (e.g., in early infancy, gliosis, and primary brain tumors) or from the accumulation of breakdown products such as GPC from myelin or other cell membranes (e.g., in leukodystrophies). Related findings apply to the elevation of free *myo*-Ins in various brain disorders including leukodystrophies, Alzheimer's disease (Miller *et al.*, 1993), and gliomas (Frahm *et al.*, 1991b; Bruhn *et al.*, 1992a). As a precursor of phospholipid membrane constituents, *myo*-Ins also is linked to processes involved in the formation and breakdown of myelin or, more generally, in the growth or degeneration of glial cells. Since, at least in early disease stages, such processes are likely to occur simultaneously, [1]H MRS observations of altered *myo*-Ins and Cho levels in brain are also compatible with merely a change in the phospholipid composition of myelin and/or other glial cell membranes rather than complete structural disintegration or cellular loss. The latter processes are likely to reflect final stages of disease progression and are expected to yield a global reduction of metabolite concentrations.

A decrease of brain *myo*-Ins levels has been reported in patients with impaired liver function (e.g., liver cirrhosis and hepatic encephalopathy) and high blood ammonia levels by Michaelis *et al.* (1991) and Kreis *et al.* (1990, 1992). The same authors found a concomitant elevation of Gln that was most pronounced in gray matter, where relatively high concentrations of glutamine synthetase in astrocytes produce Gln from brain Glu and blood ammonia.

The range of Lac concentrations in parietal gray matter largely reflects intra- and intersubject variability of physiological states during the MRS examination. Pathologically enhanced concentrations of brain Lac are observed in disorders associated with increased energy demand and/or impaired ability for oxidative phosphorylation. Examples range from acute stroke (Bruhn *et al.*, 1989a) and rapidly growing malignant tumors (Bruhn *et al.*, 1989b; Frahm *et al.*, 1991b; Bruhn *et al.*, 1992a) to mitochondrial myopathies. Similarly, also Glc levels of $0.7 \pm 0.4$ mM in parietal gray matter reflect variations in physiology and are subject to changes due to a variety of factors ranging from functional to nutritional state. When controlled for plasma Glc levels, preliminary data on long-term diabetics revealed marked differences in brain-to-plasma Glc ratios that may be explained by alternations of Glc transport kinetics across the blood–brain barrier (Bruhn *et al.*, 1991; Kreis and Ross, 1992). Additional metabolite concentrations not included in Table 2 are $2.0 \pm 1.0$ mM for Tau, $1.8 \pm 0.6$ mM for Asp, and $1.0 \pm 0.6$ mM for GABA in parietal gray matter.

Apart from basic function-related differences in cellular composition (and metabolism) of various brain regions, metabolite profiles may be affected by maturation and aging, acute and chronic changes in nutrition (Frahm *et al.*, 1995), functional state (Prichard *et al.*, 1992; Merboldt *et al.*, 1992; Frahm *et al.*, 1996), and medication. Figure 4 illustrates spectral variations in gray and white matter of healthy subjects at the age of 4 months, 2 years, and 12 years. Although current data bases are still insufficient to support a detailed understanding of age-related changes during maturation, and despite the fact that preliminary reports are based on *in vivo* and *in vitro* data acquired under different technical conditions (van der Knaap *et al.*, 1990; Peden *et al.*, 1990; Hüppi *et al.*, 1991, 1995; Kreis *et al.*, 1993; Toft *et al.*, 1994; Hashimoto *et al.*, 1995), some consistent findings have emerged that are in agreement with our own experimental experience:

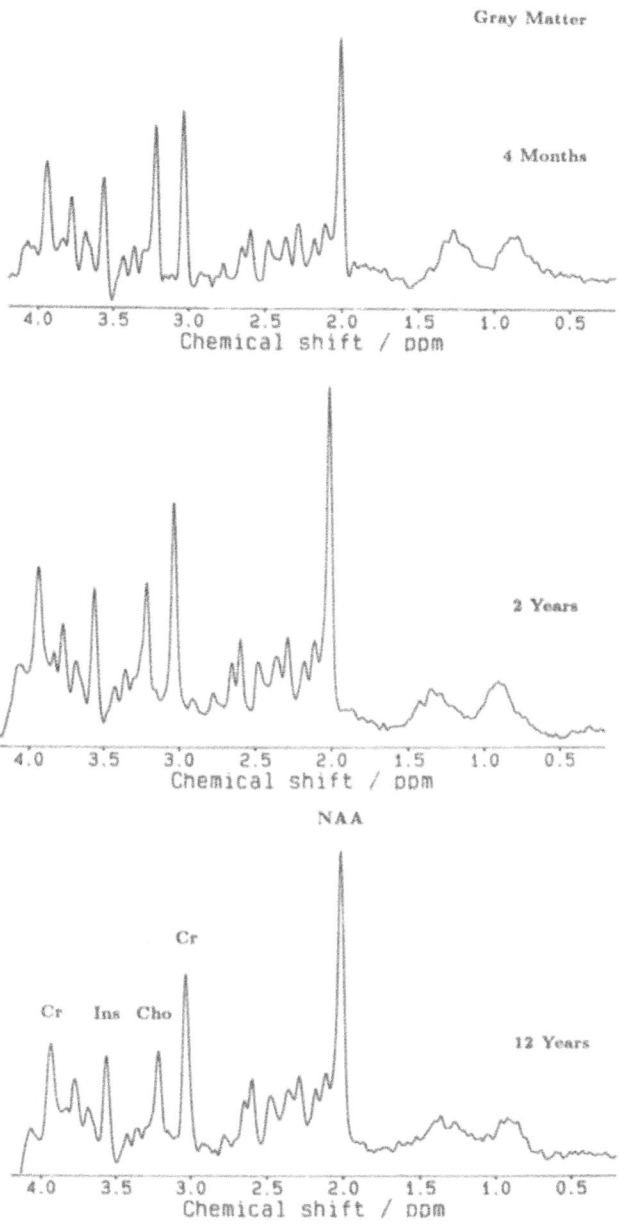

FIGURE 4. Age dependence of the metabolite pattern in the human brain (healthy subjects) as detected by ¹H MRS (STEAM, TR/TE/TM = 3000/20/30 msec, 128 accumulations) of gray (12–18 ml) and white matter (5.1–12 ml) at the age of 4 months (*top*), 2 years (*middle*), and 12 years (*bottom*). It should be noted that true changes in concentration, i.e., resonance area, may not be reflected by corresponding changes in resonance *peak height*, if spectra are measured with different magnetic field homogeneities, i.e., resonance linewidths.

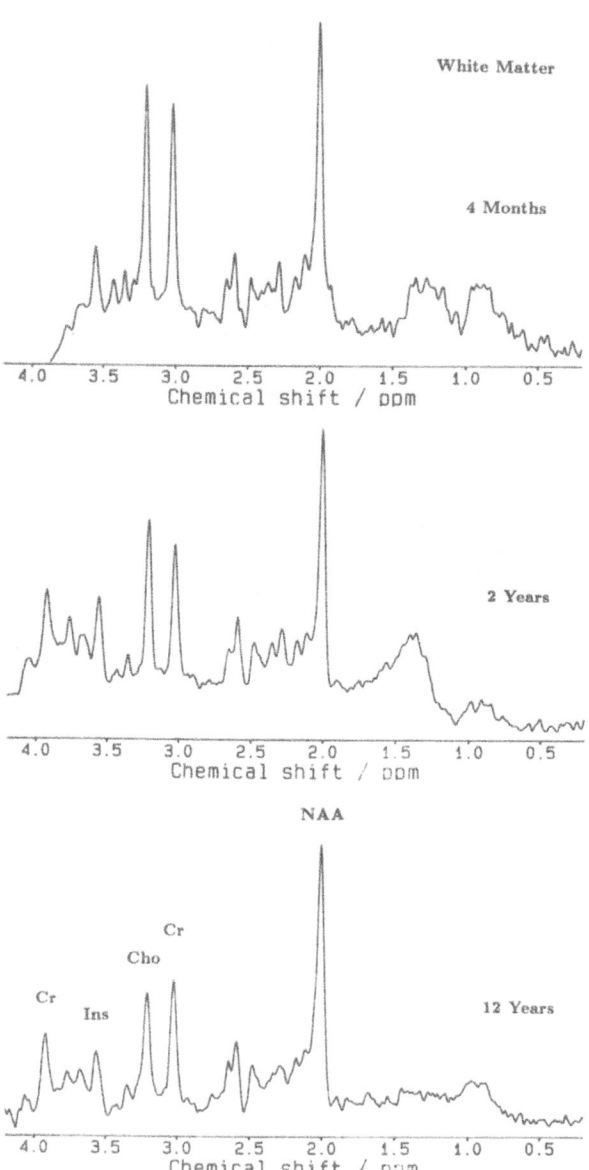

FIGURE 4. Continued

- NAA increases from low values immediately after birth to (almost) adult levels at the age of 3 years, probably in parallel with myelination.
- Cho concentrations decrease over the same time period.
- *myo*-Ins concentrations rapidly fall from very high values after birth to adult levels within the first 3–4 months of life.

All changes occur in both gray and white matter and are also seen in other brain areas, although no systematic investigations are available. For NAA and Cho, normative time courses correlate best with gestational age, while *myo*-Ins decreases with postnatal age (Kreis *et al.,* 1993). The concentration of Cr seems to be fairly stable at the adult level. An interesting observation is the absence of NAAG in white matter of young children. Detectable levels are not reached before adolescence, as indicated by a broadening of the NAA resonance (left shoulder) in the white-matter spectrum of the 12-year-old subject (Fig. 4, bottom right). In agreement with conventional analyses, our data further demonstrate that the concentration of Tau (underlying Glc and *scyllo*-Ins resonances) is elevated in young children in comparison to adult controls.

## 2.5. Patient Examinations

All MRI and ¹H MRS studies were carried out at 2.0 T (Siemens Magnetom SP4000) using the standard circularly polarized head coil. Examinations of young children of less than about 8 years of age are performed under mild oral sedation using either chloral hydrate or benzodiazepines. Cardiovascular and pulmonory function are monitored throughout the investigation. Since most children sleep during the examination, the combined MRI and MRS study typically lasts 1–1.5 hr and comprises the following:

- $T_1$-weighted three dimensional (3-D) gradient-echo MRI, eventually complemented by $T_2$-weighted fast-scan MRI (Frahm *et al.,* 1991c)
- Single-voxel ¹H MRS of parietal gray and white matter and 2–4 additional brain areas reflecting disease-specific requirements

In these studies the measuring time of individual spectra is kept to 6.5 min using 64 accumulations with TR = 6000 msec or 128 accumulations with TR = 3000 msec in some earlier studies. More than 90% of attempted examinations have been successful. Written informed consent is obtained from the parents prior to the examination.

In general, meaningful applications of quantitative ¹H MRS to studies of brain disorders require not only control of data acquisition and evaluation but also careful selection of patients and diseases to avoid unnecessary and costly investigations. In addition to neurological, neurophysiological, and biochemical investigations, conventional computerized tomography (CT) or MRI should precede MRS. Biochemical screening should include the determination of amino acids, organic acids, ammonia, catecholamines, and lactate in urine, blood, and, if possible, CSF. Such prior information improves MRS examinations by guiding both the selection of VOI locations and the analysis of suspected metabolic/spectroscopic abnormalities.

As far as diseases are concerned, at least three different categories may be identified:

1. *Diseases with abnormal MRI* that lead to focal or generalized lesions in patients with or without clinical symptoms. The main candidates in this category are gray- and white-matter diseases as well as basal ganglia and cerebellar disorders, e.g., leukodystrophies, mitochondrial disorders, infections, epilepsies with structural abnormalities, amino and organic acidopathias, and brain tumors.

2. *Diseases with normal MRI* but distinct neurological and/or biochemical abnormalities. Neurological abnormalities include disturbances of consciousness (acute and chronic encephalopathies), disturbances of muscle tonus (severe hypotonia), movement disorders (extrapyramidal, particularly dystonic syndromes), seizures of suspected metabolic origin, and dementias. Biochemical abnormalities are lactic acidosis, hyperammonemia, and organic and amino acidurias.

3. *Miscellaneous disease conditions,* mainly neurological disorders of unknown origin with normal MRI and unsuspicious biochemistry but suspected metabolic abnormalities (e.g., neurotransmitter disorders).

## 3. LEUKODYSTROPHIES

Leukodystrophies comprise a group of genetic diseases that affect brain myelin and, in a few disorders, also peripheral myelin. While onset is usually in early childhood, the spectrum ranges from congenital types to adult-onset types. Leukodystrophies represent the classical type of white-matter disorders with motor disturbances, visual loss, and deafness as early symptoms. Associated dysfunction of gray matter such as dementia and seizures is usually secondary and occurs later during the evolution of the disease. Numerous reviews on leukodystrophies cover the molecular biology and genetic basis (Rosenberg *et al.*, 1993) as well as clinical aspects (Aicardi, 1993; Kolodny, 1993) and MRI appearances (Valk and van der Knaap, 1989; Kendall, 1993).

White-matter abnormalities in leukodystrophy are often characterized as demyelination, dysmyelination, and hypomyelination, defined as follows:

• *Demyelination* refers to a breakdown of structurally and biochemically normal myelin (e.g., in multiple sclerosis).

- *Dysmyelination* denotes a breakdown of structurally and biochemically abnormal and/or unstable myelin (e.g., in adrenoleukodystrophy).
- *Hypomyelination* indicates disturbance and delay in the formation of normal myelin (e.g., in Pelizaeus–Merzbacher disease).

Although the clinical course and neuropathology of most leukodystrophies were described almost a century ago, the biochemical and genetic defects were only clarified during the last decades. Despite all diagnostic efforts and progress in modern neuroradiological, biochemical, and genetic methods, a substantial proportion of leukodystrophies in childhood remain unclassified. Even in those cases in which the metabolic defect is known and the gene has been mapped, e.g., in metachromatic and globoid cell leukodystrophy, the pathogenesis of brain abnormalities is still unclear. The following description is based on our experience in more than 100 cases of leukodystrophies including 84 in whom MRS was performed during initial diagnostic procedures and follow-up of the disease.

## 3.1. Metachromatic Leukodystrophy

Metachromatic leukodystrophy (MLD) was first described by Scholz (1925) as a familial type of diffuse sclerosis. Austin *et al.* (1963) and Mehl and Jatzkewitz (1965) discovered a deficiency of arylsulfatase A that causes a deposition of typical metachromatic material in central and peripheral myelin. The gene has been mapped to chromosome 22 (Polten *et al.*, 1991), and a pseudodeficiency gene was identified. Three clinical forms of MLD can be distinguished according to age of onset: a late infantile, a juvenile, and an adult type. The incidence of MLD is estimated to be 1:40,000. Demyelination as shown on MRI starts most frequently in frontal white matter and capsula interna and is accompanied by peripheral neuropathy in almost all cases. It leads to ataxia, spasticity, and severe motor and mental handicap (Kolodny, 1989).

¹H MRS of cerebral metabolic alterations in MLD reveals variable degrees of NAA reduction and a generalized and pronounced increase in brain *myo*-Ins as conspicuous abnormalities (Kruse *et al.*, 1993). As shown in Fig. 5 for a 2-year-old patient with late infantile MLD, *myo*-Ins concentrations of gray (7.5 mM) and white matter (9.9 mM) are 2- to 3-fold above normal levels (of. Table 2). A parallel enhancement is seen for *scyllo*-Ins in white matter (0.5 mM; normal 0.2 ± 0.1 mM). Its concentration is known to be tightly coupled to that of *myo*-Ins via *myo*-inosose-2 as a single intermediate (Sherman *et al.*, 1968). The observation of a specific metabolic disturbance in glial cells is paralleled by significant increases of Cho (2.3 mM, + 65%) and Cr (7.3 mM, + 45%) in white matter.

As noted previously (Kruse *et al.*, 1993), the MRS findings are in line with a demyelination of white matter and (subsequent) degeneration of neuroaxonal tissue. Although enhanced levels of Ins and Cho may reflect accumulation of

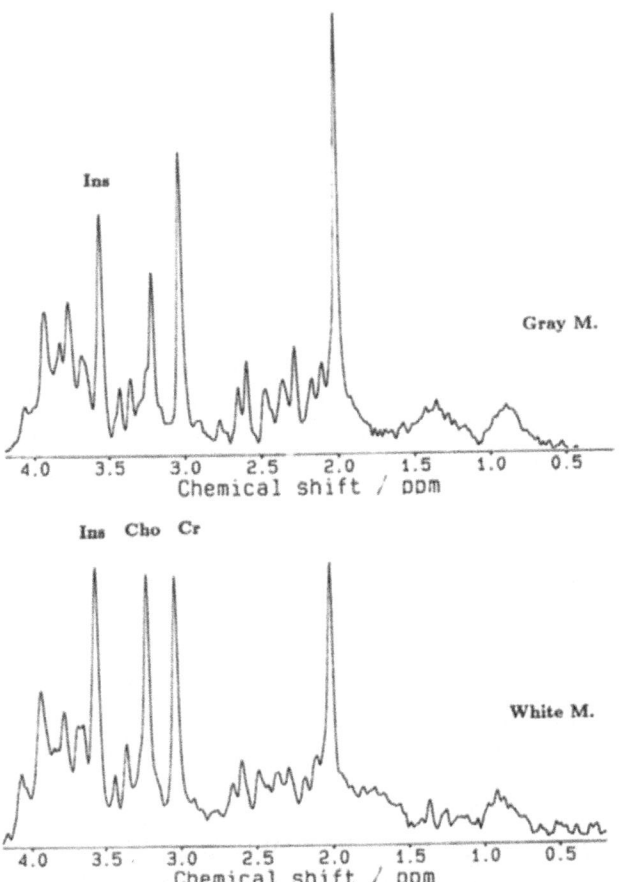

FIGURE 5. Metabolic alternations in a 2-year-old patient with metachromatic leukodystrophy as detected by [1]H MRS (STEAM, TR/TE/TM = 6000/20/30 msec, 64 accumulations) of gray (18 ml) and white matter (12 ml). Most prominent findings are a generalized elevation of *myo-* and *scyllo-*Ins as well as increased levels of Cho and Cr in white matter.

myelin breakdown products, they are more likely accounted for by changes in glial membrane composition that involve free *myo-*Ins and PCh contributions to the Cho level as major constituents for a synthesis of phospholipid membranes in general and the formation of myelin in particular. In fact, since other explanations such as dietary considerations or renal dysfunction could be excluded in our patients and since autopsy studies revealed a strong elevation of phosphatidylinositol relative to other phospholipids in isolated myelin from a patient with MLD (Norton and Poduslo, 1982), it may be hypothesized that [1]H MRS primarily reflects lipid alterations in demyelinating areas of MLD patients. Thus, elevation of *myo-*Ins (as well as of Cho and Cr) may be of more fundamental importance for the pathogenesis of demyelination as well as for related neuro-degenerative processes and neurological symptoms than the accumulation of sulfatides in MLD and, as such, clearly deserves further consideration. Related observations have been made in other leukodystrophies.

## 3.2. Globoid Cell Leukodystrophy (Krabbe's Disease)

Globoid cell leukodystrophy (GLD) was first described by Krabbe (1916) and is now recognized as an inherited autosomal recessive disorder. Patients develop symptoms from the age of 2 to 6 months onward and die within one or two years mostly in a decerebral state. The clinical stages of Krabbe's disease have been delineated by Hagberg *et al.* (1970). A late-onset form was found by Crome *et al.* (1973). The brain shows extensive lack of myelin as well as proliferation of glial cells in affected areas with mononuclear epitheloid cells and clusters of large multinucleated globoid cells. Involvement of the peripheral nervous system is much less pronounced than in MLD. The biochemical defect is a deficiency of galactocerebroside β-galactosidase (Suzuki *et al.*, 1971) and mapped on chromosome 14. Many symptoms are related to the increase of psychosin in brain tissue as the result of the enzyme deficiency. GLD is a rare disease; its exact incidence is unknown.

MRI reveals a diffuse demyelination of the cerebral hemispheres, brain stem, and cerebellum, while arcuate fibers are often spared (Baram *et al.*, 1986; Sasaki *et al.*, 1991; Percy *et al.*, 1994). The $T_1$-weighted image of an 8-month-old patient in Fig. 6 shows pronounced hypodensities indicating structural disintegration in periventricular white matter. Analysis of a spectrum from left parietal white matter depicts elevated concentrations of *myo*-Ins (8.7 mM), Cho (2.6 mM), and Cr (6.5 mM) that resemble the findings in MLD. However, reduction of NAA (3.5 mM) in affected white matter is even more pronounced than in MLD as is the occurrence of elevated Lac (3.2 mM). These MRS findings suggest that there is a similar disturbance of myelination in MLD and GLD but that the latter produces a more severe neuroaxonal degeneration and/or an even more rapid disease progression.

## 3.3. Pelizaeus–Merzbacher Disease

The classical X-linked type of Pelizaeus–Merzbacher disease (PMD) was described by Pelizaeus (1885) and Merzbacher (1910). Seitelberger (1970) added the congenital variant in his description of three affected brothers. Hypotonus, ataxia, spasticity, and nystagmus are early symptoms. Classical PMD does not belong to the lysosomal disorders. It is caused by a genetic defect of one of the major myelin proteins, proteolipidprotein (PLP), which results in a disturbance of myelin formation and maintenance. PLP has also a role in the differentiation of

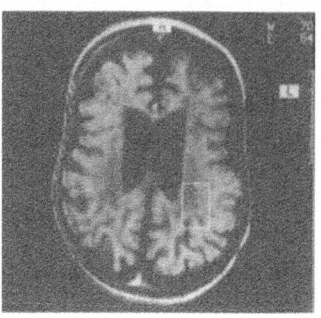

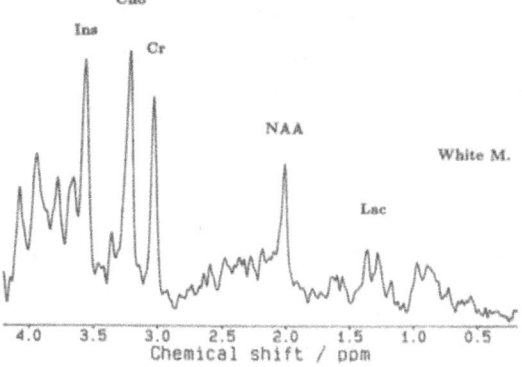

FIGURE 6. Morphological and metabolic alterations in an 8-month-old patient with Krabbe's disease as detected by $T_1$-weighted MRI (RF spoiled 3-D FLASH, TR/TE = 15/6 msec, 20° flip angle, 4-mm partitions) and ¹H MRS (STEAM, TR/TE/TM = 6000/20/30 msec, 64 accumulations). Affected white matter (6.4 ml) shows reduced NAA as well as elevated *myo*-Ins, Cho, Cr, and Lac.

oligodendrocytes. Various point mutations in its gene have been detected in many but not all patients (Doll *et al.*, 1992; Boespflug-Tanguy *et al.*, 1994). The neuropathology in classical PMD is described as patchy dys- or hypomyelination, whereas the Seitelberg type exhibits almost total absence of myelin. The peripheral nervous system is spared.

MRI of PMD shows diffuse white-matter abnormalities. The degree of signal intensity reversal in gray and white matter reflects the stage of the disease. Figure 7 shows residual myelination in basal ganglia in a 12-year-old patient. ¹H MRS of white matter revealed increased *myo*-Ins (5.0 mM) and Cr (6.6 mM) but mildly *decreased* Cho (1.2 mM). NAA levels were normal in both gray and white matter. In general, the altered white-matter spectra closely resembled that of normal gray matter (not shown). In comparison to MLD and GLD, the neuro-

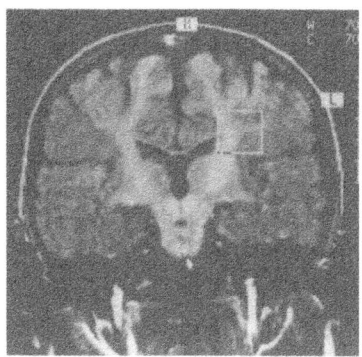

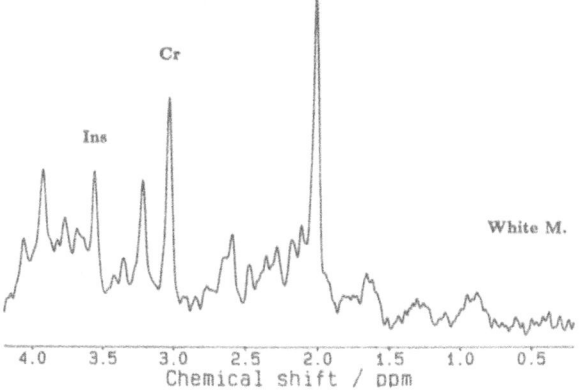

*FIGURE 7.* Morphological and metabolic alterations in a 12-year-old patient with Pelizaeus–Merzbacher disease as detected by $T_1$-weighted MRI (RF spoiled 3-D FLASH, TR/TE = 15/6 msec, 20° flip angle, 4-mm partitions) and ¹H MRS (STEAM, TR/TE/TM = 6000/20/30 msec, 64 accumulations) of white matter (8 ml). Mild reduction of Cho is accompanied by elevated *myo*-Ins and Cr.

chemical alterations associated with classical PMD yield an independent and characteristic pattern of abnormal myelination that differs by the absence of Cho elevation and a lack of neuroaxonal degeneration.

## 3.4. Canavan's Disease

Spongiform encephalopathy was first described in a case report by Canavan (1931) and later extensively studied by van Bogaert and Bertrand (1967). Among the three clinical variants of this autosomal recessive disorder, the infantile type of Canavan's disease (CD) is most frequently seen. Its clinical characteristics are macrocephaly associated with progressive spasticity, blindness, and motor and mental retardation. The metabolic defect is a deficiency of aspartoacylase, the enzyme responsible for NAA breakdown (Matalon *et al.*, 1988). Cloning of the

human aspartoacylase cDNA was accomplished and a common missense mutation in CD was reported by Kaul *et al.* (1993).

Neuropathology of CD is characterized by rarefication, vacuolation, and
ultimately breakdown of neural tissue. Cavitations occur at the junction of white
matter and cortex. Vacuoles are caused by the separation of the lamellae of
myelin. Astrocytic cell membranes are also affected, and large pale astrocytic
nuclei devoid of visible cytoplasm represent a characteristic feature of the cerebral cortex (so-called Alzheimer-II cells). MRI of the brain reveals diffuse white-
matter degeneration with partial sparing of the corpus callosum. This is well
appreciated in the coronal image of a 4-year-old patient with CD shown in Fig. 8.

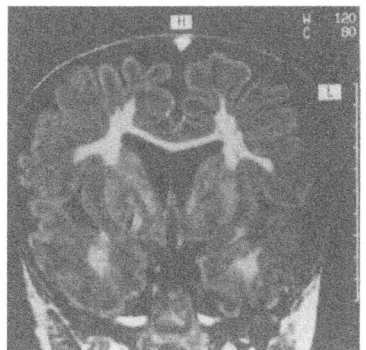

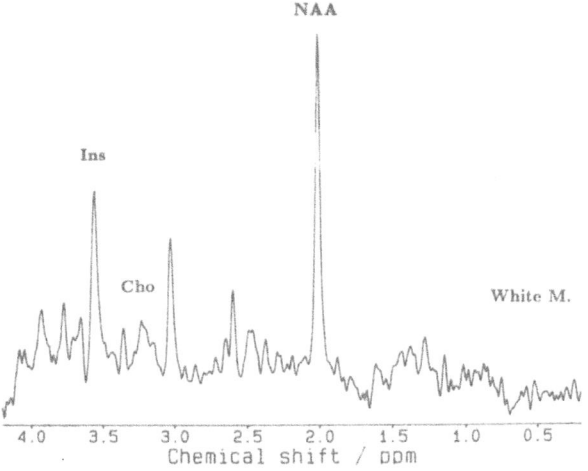

*FIGURE 8.* Morphological and metabolic alterations in a 4-year-old patient with Canavan's disease
as detected by $T_1$-weighted MRI (RF spoiled 3-D FLASH, TR/TE = 15/6 msec, 20° flip angle, 4-mm
partitions) and ¹H MRS (STEAM, TR/TE/TM = 6000/20/30 msec, 64 accumulations) of white
matter (4.1 ml), gray matter (18 ml), and basal ganglia (4.1 ml). The disorder is characterized by
generalized increases of NAA and *myo*-Ins as well as marked reductions of Cho.

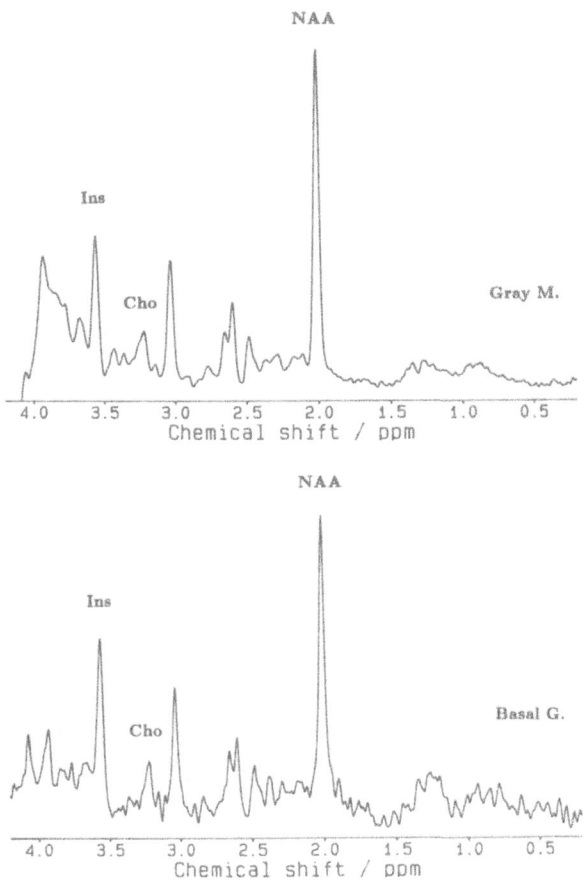

*FIGURE 8.* Continued

So far, several MRS case reports of CD have appeared (Grodd *et al.*, 1990; Austin *et al.*, 1991; Barker *et al.*, 1991). Experimental limitations in these studies were mainly associated with the range of identifiable metabolites (e.g., due to long-TE rather than short-TE conditions) and neurochemical data evaluation (e.g., use of short TR values and peak ratios rather than absolute metabolite concentrations). As demonstrated in Fig. 8, quantitative ¹H MRS resulted in a pattern of findings common to all areas investigated. Elevation of NAA was mild in white matter and marked in gray matter (11.5 mM, +50%), consistent with aspartoacylase deficiency and different from the findings in all other brain disorders investigated so far. With Cr levels within normal ranges, the concentrations of Cho and *myo*-Ins were markedly decreased and increased, respectively. For example, in white matter Cho (0.6 mM) was reduced by 60% while *myo*-Ins (9.7 mM) was enhanced 3-fold.

Similar to MRS findings in other forms of leukodystrophy, the strong enhancement of free *myo*-Ins in white (and gray) matter seems to indicate alterations of glial metabolism that are closely associated with disturbances of myelination. Because oligodendrocytes are involved in the formation of myelin, it may be hypothezised that they represent the primary location of *myo*-Ins. Accordingly, the observation of low Cho levels may be due to degeneration of astrocytes in this disease. High Cho concentrations have been found in astrocytomas and related to cell proliferation. An alternative explanation of the "decoupling" of Cho and *myo*-Ins changes in white matter involves a specific disease-related change in myelin composition. Further open questions are the link between elevated NAA and demyelination in CD and the discrepancy between the only mild to moderate NAA increases in brain and the strongly increased levels in plasma (and urine).

## 3.5. Alexander's Disease

The other macrocephalic encephalopathy identified as a clinical and probably also genetic entity (Ochi *et al.*, 1991) is Alexander's disease (AD). The description of brain pathology by Alexander (1949) referred to a "hydrocephalic infant" with progressive fibrinoid degeneration of fibrillary astrocytes and the presence of so-called Rosenthal fibers. Head enlargement occurs gradually during the first years of life. Etiology is unknown. Although the disease is rare and usually sporadic, familial cases as well as late-onset types have been reported.

In general, MRI findings are in line with a progressive white-matter disease affecting frontal lobes primarily (Arend *et al.*, 1991; Schuster *et al.*, 1991). Brain stem involvement is common, but the peripheral nervous system is spared. Ataxia, spasticity, swallowing difficulties, blindness, and severe psychomotor retardation develop early during AD. Seizures are rare and typically occur during unspecific viral infections or following mild head trauma.

Figure 9 shows the case of a 2-year-old patient with AD and severe disturbances of cortical white matter detected by MRI. [1]H MRS revealed a marked decrease of NAA (2.1 mM) and Cr (3.4 mM) but elevated Cho (1.8 mM) and Lac (3.8 mM). Interestingly, these abnormalities clearly resemble the metabolite pattern of astrocytomas. The white-matter findings in AD are therefore compatible with astrocytic degeneration severely affecting the viability of neuroaxonal tissue compartments.

The neurochemical behavior seen in cortical and subcortical gray matter clearly differs from that in white matter. While 30–50% reductions of NAA most likely indicate neuronal damage, the concentrations of *myo*-Ins (7.4 mM), Cho (1.7 mM), and Cr (6.8 mM) are enhanced in gray matter and even further in basal ganglia (8.6 mM, 2.7 mM, and 8.5 mM, respectively). This pattern parallels metabolic abnormalities previously found in affected white matter of patients

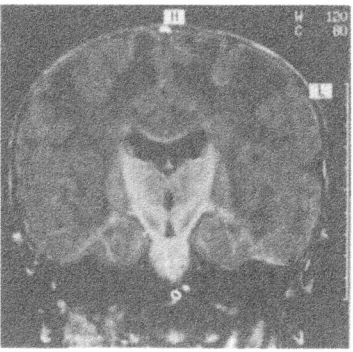

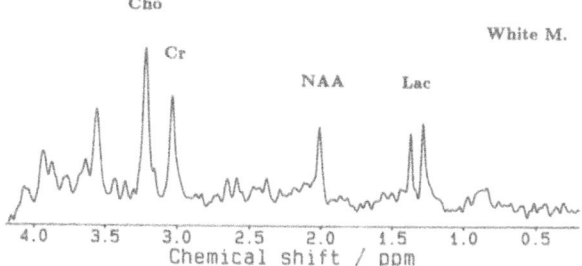

*FIGURE 9.* Morphological and metabolic alterations in a 2-year-old patient with Alexander's disease as detected by $T_1$-weighted MRI (RF spoiled 3-D FLASH, TR/TE = 15/6 msec, 20° flip angle, 4-mm partitions) and ¹H MRS (STEAM, TR/TE/TM = 6000/20/30 msec, 64 accumulations) of white matter (12 ml), gray matter (18 ml), and basal ganglia (6.4 ml). While white-matter disturbances include reduced NAA and Cr as well as elevated Cho and Lac, cortical and subcortical gray matter shows decreased NAA and elevated *myo*-Ins, Cho, and Cr.

with MLD and GLD and therefore suggests alterations in glial phospholipid metabolism rather than structural disintegration of glial cells.

## 3.6. Adrenoleukodystrophy

Adrenoleukodystrophy (ALD) is an X-linked disorder occurring at an incidence of 1:15,000 in males. Different clinical phenotypes are known. The cerebral form (40%) manifests itself between 5 and 12 years of age. Clinical symptoms are similar to those seen in other white-matter diseases and are dominated by ataxia, spasticity, deafness, visual problems, personality changes, and mental deterioration. In most cases, associated adrenal insufficiency (Addison's disease) emerges as the first—and in rare cases as the only—clinical expression of ALD.

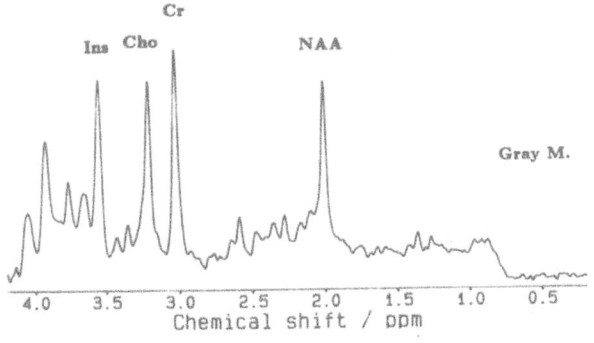

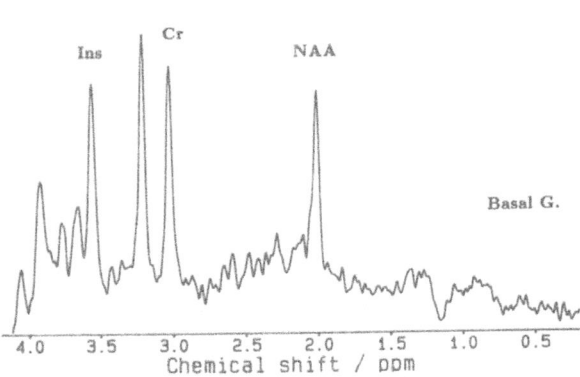

FIGURE 9. Continued

Adrenomyeloneuropathy (AMN) represents the adult phenotype. Both forms may occur in the same family. The biochemical defect is a deficiency of VLCFA synthetase in peroxisomes, which leads to an increase of very-long-chain fatty acids (VLCFA) in fibroblasts and plasma. The ALD gene on the X chromosome (Xq28) encodes a peroxisomal membrane protein (ALDP) of the "ATP binding cassette" family (Mosser *et al.*, 1993).

Typically, pathology reveals most severe demyelination in the occipital,

FIGURE 10. Morphological and metabolic alterations in an 8-year-old patient with adrenoleuko-dystrophy as detected by $T_1$-weighted MRI (RF spoiled 3-D FLASH, TR/TE = 15/6 msec, 20° flip angle, 4-mm partitions) and ¹H MRS (STEAM, TR/TE/TM = 6000/20/30 msec, 64 accumulations) of right frontoparietal white matter (8 ml) as well as of left and right parieto-occipital white matter (6.4 ml). While frontal white matter shows reduced NAA and enhanced levels of *myo*-Ins and Cho, left parieto-occipital white-matter disturbances include a complete loss of NAA as well as elevated *myo*-Ins and Cho. The most affected white matter in right parieto-occipital brain shows a complete loss of NAA as well as reduced Cho and Cr but still elevated *myo*-Ins.

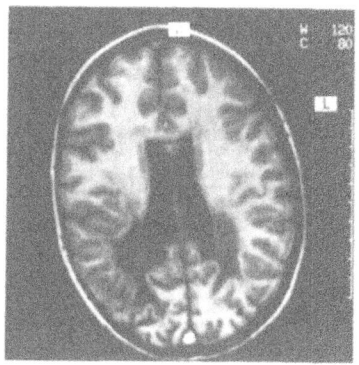

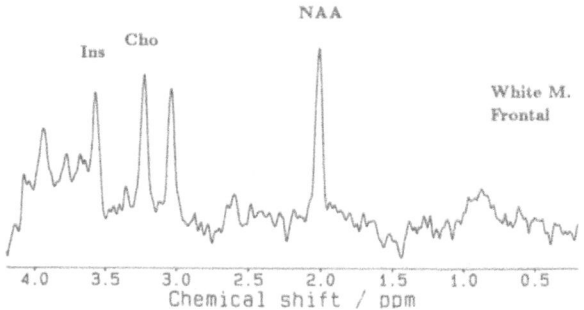

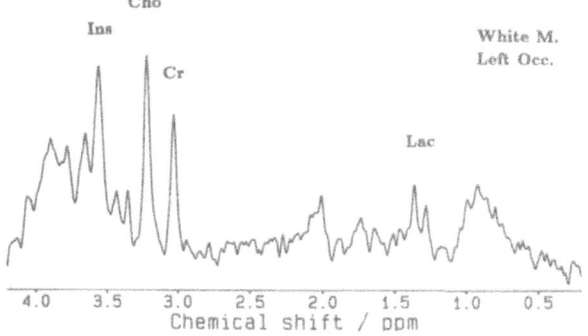

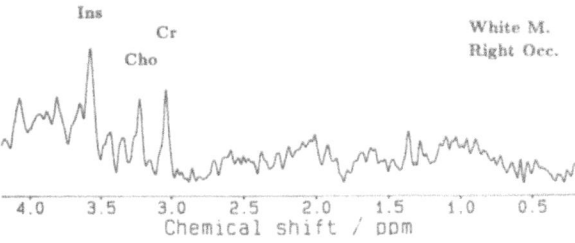

parietal, and temporal lobes, although some of our patients presented with an initial affection of frontal lobes. The lesions are histologically characterized by three zones, respectively showing (i) destruction of myelin and proliferation of sudanophilic macrophages, (ii) myelinated and demyelinated axons with an associated inflammatory response, and (iii) dense gliosis with loss of oligodendrocytes, myelin, and axons.

Inflammatory responses with proliferation of macrophages are particularly prominent at the edges of the lesions, and perivascular lymphocytic cuffing is marked. MRI shows related signal alterations that form a characteristic pattern of periventricular dysmyelination. This is demonstrated in Fig. 10 for the case of an 8-year-old patient. While $^1$H MRS revealed no changes in parietal gray matter (not shown), disease progression in white matter is evidenced by metabolic alterations of increasing severity in the three regions investigated. Despite the fact that frontal white matter appears unsuspicious on MRI, it shows reduced NAA (6.2 mM) and elevated myo-Ins (5.6 mM) and Cho (1.7 mM). These trends are accentuated in affected left parieto-occipital white matter, yielding a severe reduction of NAA (2.0 mM) as well as strongly elevated myo-Ins (6.4 mM) and Cho (2.2 mM). Increase of brain Lac (3.6 mM) in such regions is a frequent finding in ALD. Finally, the lesion in right parieto-occipital white matter exhibits a generalized reduction of metabolites with residual concentrations of Cho (0.6 mM) and Cr (2.3 mM) but still elevated myo-Ins (5.5 mM).

The metabolic findings in white matter of ALD parallel observations in other leukodystrophies, e.g., MLD and GLD. In this case, the simultaneous occurrence of glial and neuroaxonal changes of different severity suggests a characterization of disease progression according to the following stages:

- Normal-appearing white matter (on MRI) may be affected by moderate neuroaxonal damage (or loss) as well as signs of demyelination that are in line with myelin phospholipid changes in glia.
- Affected white matter (on MRI) shows severe neuroaxonal loss and elevation of Lac as well as signs of severe demyelination (glial changes).
- Most severely affected white matter (lesions on MRI) shows a complete neuroaxonal loss as well as structural disintegration of glial cells.

A previous study using proton chemical shift imaging indicated a correlation of the Cho/NAA resonance ratio with clinical course (Kruse et al., 1994a). Our data not only confirm this suggestion, with use of the Cho/NAA *concentration* ratio, but extend the correlation to the myo-Ins/NAA concentration ratio. The latter parameter allows a more specific assessment of the degree of neuroaxonal damage and of activity of the demyelinating process in respective neuronal and glial compartments (Pouwels et al., unpublished results). For the case shown in Fig. 10, the myo-Ins/NAA concentration ratio increases from 0.9 (normal-appearing white matter) to 3.2 (affected) and 5.6 (severely affected) as compared to 0.36 for normal controls. The hypothesis that the *myo*-Ins/NAA ratio is sensitive to true disease progression is further substantiated by follow-up studies in selected patients over several years (Pouwels *et al.*, unpublished results). In particular, continuous examinations of asymptomatic ALD patients at early stages prove useful, because an increase in *myo*-Ins/NAA clearly precedes a subsequent clinical deterioration. Tight patient monitoring by ¹H MRS may thus guide the selection of candidates for bone marrow transplantation (Aubourg *et al.*, 1990).

## 3.7. Unclassified Leukodystrophies

Among 35 children with unclassified white-matter disease, two subgroups with identical clinical characteristics and MRS abnormalities could be identified. A first disease entity has been described by Hanefeld *et al.* (1993) and termed myelinopathia centralis diffusa (MCD). Subsequent to a normal development in early infancy up to the age of 2 to 3 years, children with MCD develop ataxia and spasticity. They become wheelchair-bound between 4 and 6 years of age. Optic atrophy and dementia are late features. Six children, including one pair of siblings (brother and sister), have been studied so far. Head circumference was within normal limits. MRI revealed diffuse homogeneous hypointensity of white matter ($T_1$-weighting) identical to the signal from CSF in the ventricles. U-fibers were also affected. Figure 11 gives two examples, showing an early phase (top) with residual preserved white matter and a late phase (middle) characterized by cellular disintegration of white matter and a complete loss of neuroaxonal and glial tissue. This is also evidenced by ¹H MRS, showing a significant reduction (top) and final loss (middle) of all cellular metabolites except those found in CSF (Glc at 3.1 mM, Lac at 1.7 mM).

A second disease entity among patients with diffuse white-matter demyelination was first described by van der Knaap *et al.* (1995) and may be characterized as spongy or cystic leukoencephalopathy. In this group, all children are macrocephalic from birth. After normal development, focal seizures start around 3 years of age. Ataxia and spasticity develop later than in the normocephalic patients with MCD. In addition to diffuse white-matter changes, MRI is characterized by cystic lesions in the temporal and high parietal regions. As demonstrated for a 6-year-old patient in Fig. 11 (bottom), these findings are clearly distinct from the pattern seen in other macrocephalic white-matter diseases (e.g., Alexander's disease). ¹H MRS shows a severe reduction of all metabolites but no elevation of Lac. Major differences with respect to MCD patients are the macrocephaly, a slower disease progression, and the cystic white-matter abnormalities. Whether final stages entail a complete replacement of white matter by CSF and a metabolic profile equivalent to that of late MCD remains to be seen.

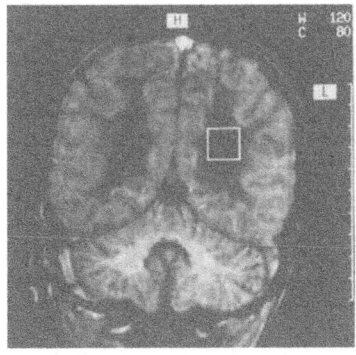

White M.

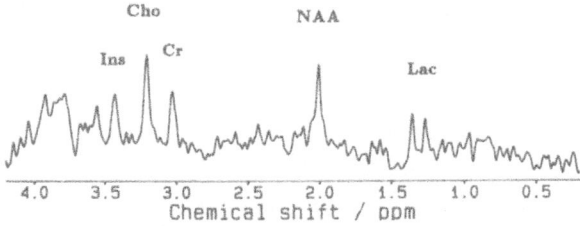

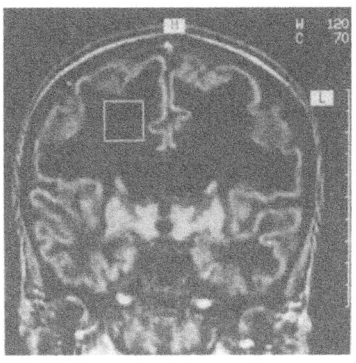

FIGURE 11. Morphological and metabolic alterations in three patients with unclassified leuko-dystrophy as detected by $T_1$-weighted MRI (RF spoiled 3-D FLASH, TR/TE = 15/6 msec, 20° flip angle, 4-mm partitions) and ¹H MRS (STEAM, TR/TE/TM = 6000/20/30 msec, 64 accumulations) of white matter (5.1–12 ml). *First part:* 3-year-old patient with myelinopathia centralis diffusa (MCD) as described by Hanefeld *et al.* (1993). (early phase). *Second part:* 12-year-old patient with MCD (late phase). *Third part:* 6-year-old patient with spongy or cystic leukoencephalopathy as described by van der Knaap *et al.* (1995). Both disorders are characterized by a loss of white matter and pertinent metabolites.

White M.

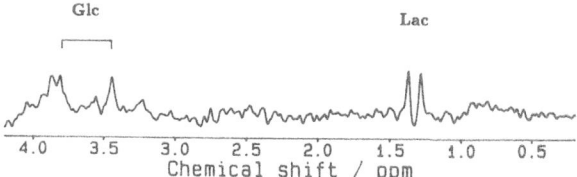

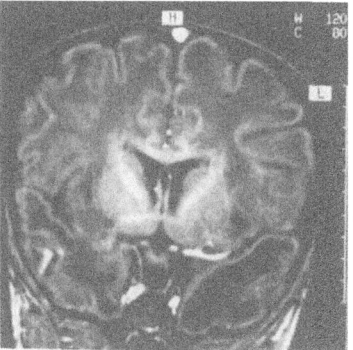

White M.

NAA

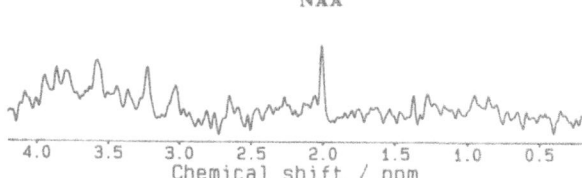

*Figure 11.* Continued

## 4. OTHER WHITE-MATTER DISEASES

In recent years the more widespread availability of MRI has led to the recognition of an increasing number of disorders with white-matter abnormalities. It would be wrong to classify these disorders among the leukodystrophies, although the clinical manifestations and the predominance of white-matter changes often suggest such a diagnosis. The following examples should illustrate this problem.

## 4.1. L-2-Hydroxyglutaric Aciduria

L-2-Hydroxyglutaric (L2OHglu) aciduria is a rare inherited neurometabolic disease that has only recently been described. (Barth *et al.,* 1992). Cerebellar ataxia, mental retardation, and seizures in combination with extrapyramidal and pyramidal symptoms are the main clinical characteristics. Biochemical findings include increased levels of L-2-hydroxyglutaric acid in urine, plasma, and CSF.

MRI reveals enlargement of internal and external CSF spaces as well as patchy white-matter lesions in subcortical regions and adjacent to the frontal and occipital horn. A typical example is shown in Fig. 12 for a 16-year-old patient. $^1$H MRS of the periventricular white matter lesion revealed reduced levels of NAA (3.9 mM, −40%), Cr (3.5 mM, −20%), and Cho (0.9 mM, −25%), as well as elevation of *myo*-Ins (5.9 mM) by a factor of 2. Because these concentrations are not corrected for the use of a TR of 3000 msec, the relative deviations are given

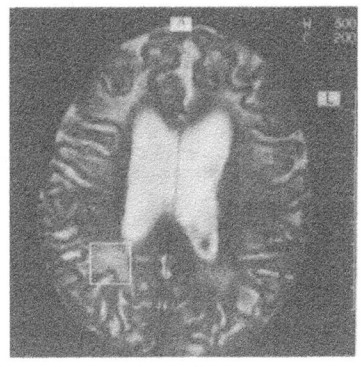

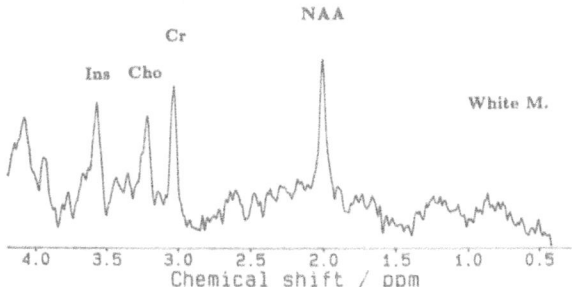

*FIGURE 12.* Morphological and metabolic alterations in a 16-year-old patient with L-2-hydroxyglutaric aciduria as detected by $T_2$-weighted MRI (CE-FAST, TR/TE = 14/−6 msec, 40° flip angle, 4-mm thickness) and $^1$H MRS (STEAM, TR/TE/TM = 3000/20/30 msec, 128 accumulations) of white matter (8 ml). The periventricular lesion in right parietal white matter shows a decrease of NAA, Cr, and Cho as well as elevated *myo*-Ins.

with respect to corresponding control values. A reduced concentration of NAA (4.2 mM, −25%) was also found in gray matter. The observations are consistent with a generalized neurodegenerative process in L20Hglu aciduria but also suggest a defect in inositolphospholipid metabolism of glial cells (Hanefeld *et al.*, 1994). A follow-up examination of the same patient at the age of 18 years showed a slight progress of cerebral atrophy, an even more dramatic decrease of NAA (−70%), and a further increase of *myo*-Ins, in close correlation to a further clinical deterioration.

## 4.2. Succinate Dehydrogenase Deficiency

Figure 13 depicts metabolic alterations in a patient with succinate dehydrogenase (SDH) deficiency as detected by ¹H MRS of gray and white matter. The child presented with severe progressive spasticity at the end of the first year of life. MRI revealed marked periventricular white-matter abnormalities that suggested a leukodystrophy. However, all known peroxisomal and lysosomal disorders as well as amino and organic acidopathias and congenital infections could be excluded.

Diagnosis of the biochemical defect was made at the age of 17 months by MRS. While the spectrum of gray matter yielded normal metabolite concentrations except for a comparatively high level of Tau (4.7 mM), white matter was characterized by an extremely high concentration (8.7 mM) of an unknown compound whose MR signal was subsequently identified as originating from the four protons of the two equivalent methylene groups of succinate (Suc) resonating at 2.40 ppm. In addition, white-matter alterations included marked reductions of NAA (2.2 mM), Cr (2.2 mM), Glu, and Gln as well as strongly elevated Lac (7.6 mM). Since *myo*-Ins and Cho levels were within normal ranges, the Suc and Lac findings are consistent with a specific neuroaxonal degeneration, most likely caused by mitochondrial enzyme deficiencies.

The assumption that Suc accumulation is caused by a deficiency of SDH was confirmed biochemically. Muscle tissue showed a partial deficiency of SDH as well as a deficiency of complex II of the respiratory chain. While Suc was found to be 10-fold increased in CSF, the discrepancy between gray and white matter remains unresolved. The child died at the age of 20 months. Neuropathological observations are consistent with the diagnosis of a Leigh syndrome, suggesting a classification as a mitochondrial disease. In fact, similar findings were reported for two sisters with Leigh syndrome that presented as a leukodystrophy (Bourgeois *et al.*, 1992).

## 4.3. Multiple Sclerosis

The characteristics of multiple sclerosis (MS) in childhood have recently been elaborated as a result of a prospective study in Göttingen (Hanefeld *et al.*,

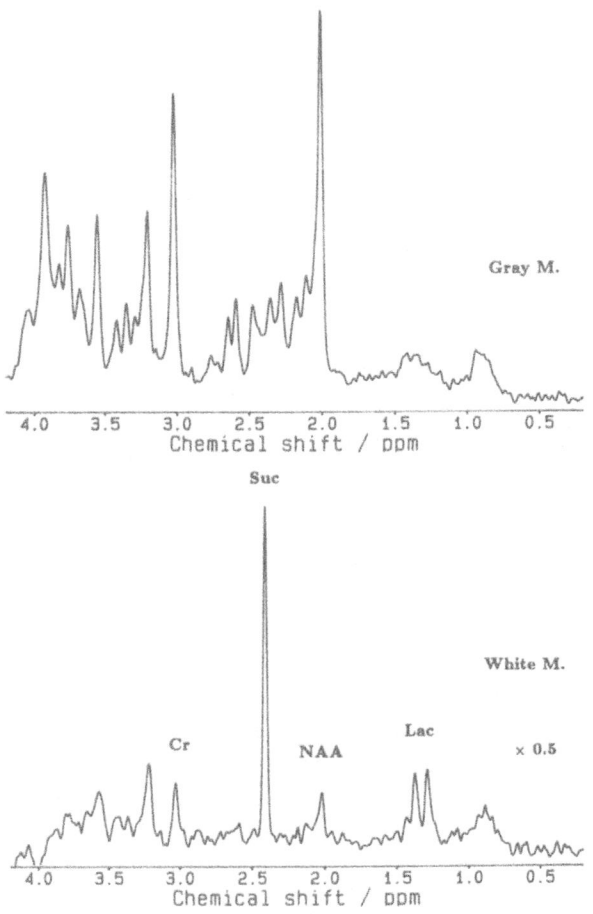

*FIGURE 13.* Metabolic alterations in a 17-month-old patient with succinate dehydrogenase deficiency as detected by ¹H MRS (STEAM, TR/TE/TM = 6000/20/30 msec, 64 accumulations) of gray (18 ml) and white matter (4.1 ml). Note the difference in scale due to the high concentration of succinate (Suc) in white matter.

1991, 1995). Early symptoms include optic neuritis, cranial nerve palsy, bulbar symptoms, ataxia, and seizures in some cases. MRI shows widespread periventricular, cerebellar, and brain stem lesions.

An example of the type of metabolic alterations found in nonenhancing (contrast agent) chronic lesions is shown in Fig. 14. In agreement with a previous report of moderate to severe decreases of NAA and Cr as well as increases of Cho in childhood MS (Bruhn *et al.,* 1992b), the spectrum (TR = 3000 msec) yields reduced NAA (4.7 mM, −30%) and elevated Cho (1.7 mM, +45%).

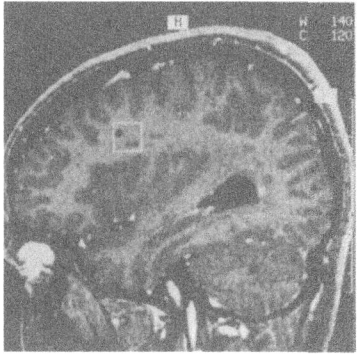

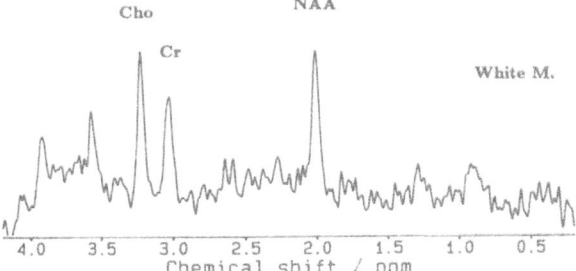

*FIGURE 14.* Morphological and metabolic alterations in a 13-year-old patient with multiple sclerosis as detected by $T_1$-weighted MRI (RF spoiled 3-D FLASH, TR/TE = 15/6 msec, 20° flip angle, 1-mm partitions, post Gd-DTPA) and ¹H MRS (STEAM, TR/TE/TM = 3000/20/30 msec, 128 accumulations) of a plaque in white matter (4.1 ml). The nonenhancing chronic lesion shows reduced NAA and elevated Cho.

Increases of *myo*-Ins were less pronounced than in MLD or GLD. No elevations of Lac, lipid, or protein resonances were observed in the children investigated. An interesting observation was a notable reduction of NAA in cortical gray matter receiving projections from neighboring MS lesions.

These findings reflect neuroaxonal degeneration and tissue replacement by glial cells with altered phospholipid composition. Probably, elevated Cho is due to enhanced concentrations of GPC as a myelin and/or membrane breakdown product. This interpretation is supported by a reduction of the broad phospholipid component in ³¹P MRS of both lesions and normal-appearing white matter in MS subjects (Husted *et al.,* 1994). Focal demyelination may further be associated with secondary neuronal shrinkage or loss, perhaps extending into functionally related cortical gray matter.

## 5. GRAY-MATTER DISEASES

The term "gray matter disease" mainly applies to a group of storage disorders that affect the nervous system. The storage material—in most cases gangliosides or ceroid lipofuscin—accumulates in neurons, which ultimately cease to function and die. The vague and inaccurate expressions "amaurotic familial idiocy" and "cerebromacular degeneration" were used to designate these disorders before biochemical classification became possible. The typical clinical symptomatology of gray-matter disorders results from disturbed neuronal function: seizures, dementia, blindness, and ataxia. Ocular symptoms are the cherry-red spot of the macula, retinitis pigmentosa, and optic atrophy. Storage may also occur in other organs and in neurons of the autonomous nervous system, e.g., Auerbach's plexus. These phenomena can be detected in a rectal biopsy or as vacuoles in peripheral lymphocytes. Both tests are used for diagnosis of ceroid lipofuscinosis. Different clinical forms have been identified according to age of onset or specific enzyme deficiencies in sphingolipid degradation and resulting storage material, e.g., gangliosidoses $G_{M1}$ and $G_{M2}$. An overview has been given by Herschkowitz and Schulte (1984).

### 5.1. Gangliosidoses

The fact that storage of gangliosides in neurons leads to cell death has been known for almost a century. Deficiency of hexosaminidase A ($\alpha-\beta$) and B ($\beta-\beta$) are the biochemical characteristics of a variant described by Sandhoff et al. (1989). The region for the $\alpha$ unit has been mapped on chromosome 15 and for the $\beta$ unit on chromosome 5. As typical signs of a gray-matter disorder, children with Sandhoff's disease develop seizures early in life as well as blindness and severe mental defects. Fundoscopy reveals a characteristic cherry-red spot and later optic atrophy.

MRI of a 13-month-old patient (Fig. 15) showed marked disturbances that are characterized by almost identical signal intensities from gray and white matter, cortical atrophy, and enlargement of ventricular spaces (Uyama et al., 1992). $^1$H MRS not only demonstrated a reduction of NAA in gray matter (4.1 mM), white matter (5.0 mM), and basal ganglia (6.3 mM) but also revealed the generalized occurrence of an additional N-acetyl compound (NA) at 2.04 ppm (left shoulder of NAA at 2.01 ppm). Although the resonance of NAAG has been found in the spectrum of normal adult white matter at this position, the presence of other N-acetylated compounds is much more likely in this disorder. In fact, deficiency of $\beta$-hexosaminidase A blocks the degradation of sphingolipids at a stage that results in the accumulation of gangliosides comprising N-acetylgalactosamine (NAcgal) and N-acetylneuraminic acid (NANA). The N-acetyl groups of both compounds yield proton spectra with a strong singlet resonance at 2.04

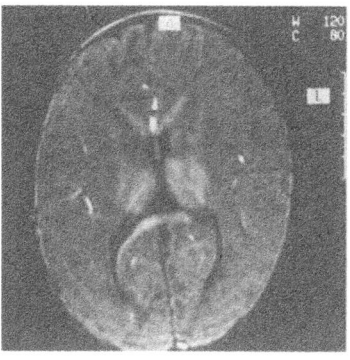

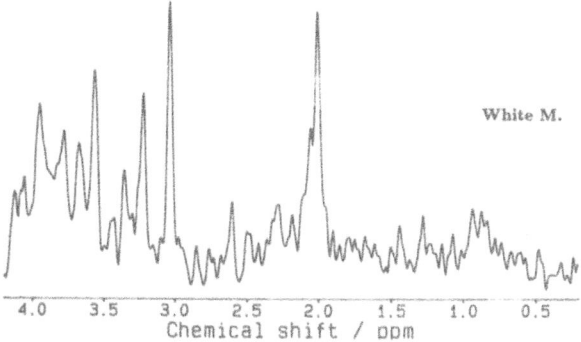

*FIGURE 15.* Morphological and metabolic alterations in a 13-month-old patient with Sandhoff's disease as detected by $T_1$-weighted MRI (RF spoiled 3-D FLASH, TR/TE = 15/6 msec, 20° flip angle, 4-mm partitions) and ¹H MRS (STEAM, TR/TE/TM = 6000/20/30 msec, 64 accumulations) of white matter (4.1 ml), gray matter (12 ml), and basal ganglia (4.1 ml). Most prominent findings are a generalized decrease of NAA and elevation of *myo*-Ins, *scyllo*-Ins, Cr, and *N*-acetyl compounds (NA) at 2.04 ppm, most likely *N*-acetylgalactosamine (NAcgal) and *N*-acetylneuraminic acid (NANA).

ppm (T. Michaelis, unpublished results). Here, the concentrations derived from this resonance were 1.7 mM in gray matter, 2.5 mM in white matter, and 2.0 mM in basal ganglia. Since no related *N*-acetyl disturbances have been seen with any other brain disorder investigated so far, the observation seems to be specific for Sandhoff's disease. Pertinent brain levels may be expected to directly reflect the amount of *N*-acetylated gangliosides stored in this $G_{M2}$ disorder.

The observed regional uniformity of neuroaxonal changes also holds true for glial compartments. Prominent metabolic alterations common to gray and white matter and basal ganglia were increases of *myo*- and *scyllo*-Ins, Cho, and Cr. For example, gray-matter levels were 7.2 mM (+85%) for *myo*-Ins, 1.7 mM (+70%) for Cho, and 7.9 mM (+36%) for Cr. These abnormalities resemble

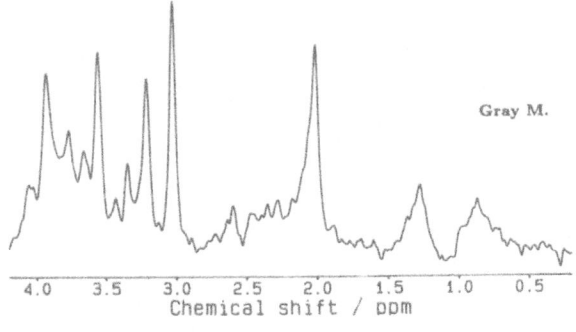

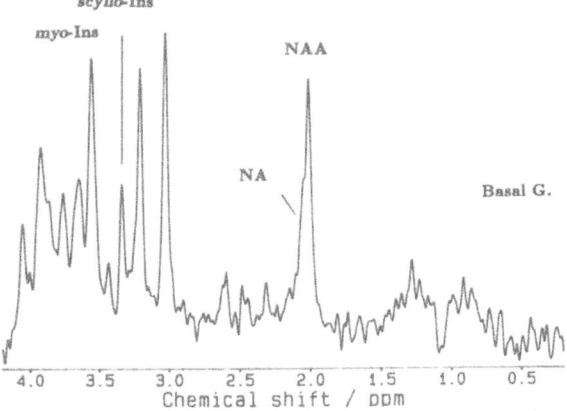

*FIGURE 15.* Continued

neurochemical disturbances in glial membrane metabolism usually seen in affected white matter in MLD and GLD. The results suggest a severe and generalized alteration of glial membrane composition in the brain of patients with Sandhoff's disease and, more generally, shed new light on the pathogenesis of symptoms and course of the gangliosidoses.

## 5.2. Neuronal Ceroid Lipofuscinosis

Neuronal ceroid lipofuscinosis (NCL) is the most frequently occurring inherited disorder affecting neurons in childhood (Claussen *et al.,* 1992). Early and late infantile, juvenile, and adult types have been identified. NCL is characterized by massive intralysosomal accumulation of storage materials due to abnormal protein processing, disturbances in protein and glycoconjugate metabolism, and impaired lysosomal function (Santavuori *et al.,* 1991; Autti *et al.,* 1992). Al-

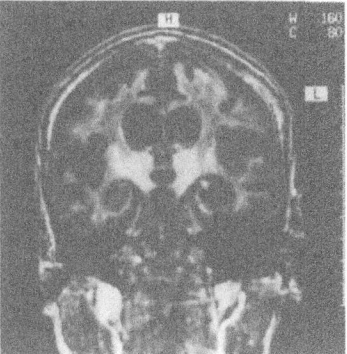

Gray M.

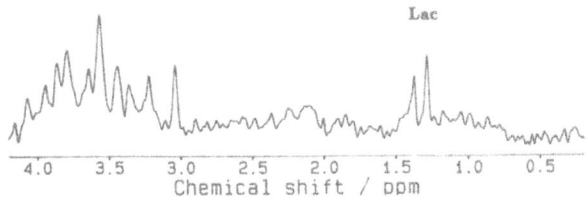

FIGURE 16. Morphological and metabolic alterations in two patients with neuronal ceroid lipofuscinosis (NCL) as detected by $T_1$-weighted MRI (RF spoiled 3-D FLASH, TR/TE = 15/6 msec, 20° flip angle, 4-mm partitions) and ¹H MRS (STEAM, TR/TE/TM = 3000/20/30 msec, 128 accumulations) of gray matter. *First part:* 5-year-old patient with infantile NCL, showing depleted NAA, Glu, Gln, marked reduction of Cr and Cho, and elevated Lac and Glc (8 ml). *Second part:* Disease progression in a patient with juvenile NCL from the age of 9 to 13 years as indicated by a reduction of NAA, Cr, and *myo*-Ins.

though the basic biochemical defect is still unknown, the gene for the early infantile type (Haltia–Santavuori) has been mapped on chromosome 1, and that for the juvenile type (Spielmeyer–Vogt) on chromosome 16 (Järvelä *et al.*, 1992). Myoclonic epilepsy, amaurosis due to retinitis pigmentosa, and progressive mental deterioration are leading symptoms. Neuropathology reveals varying degrees of cerebral and cerebellar atrophy. Neurons are distended by the storage of ceroid lipofuscin and are reduced in number. Cerebellar atrophy is most pronounced in the infantile type, combined with an almost isoelectrical electroencephalogram (EEG).

Depending on the stage and type of the disease, MRI shows severe internal and external cerebral and cerebellar atrophy. Corresponding morphological distortions and metabolic abnormalities in gray matter were readily detected in a 5-year-old patient with infantile NCL (Fig. 16, first part). In this case, ¹H MRS

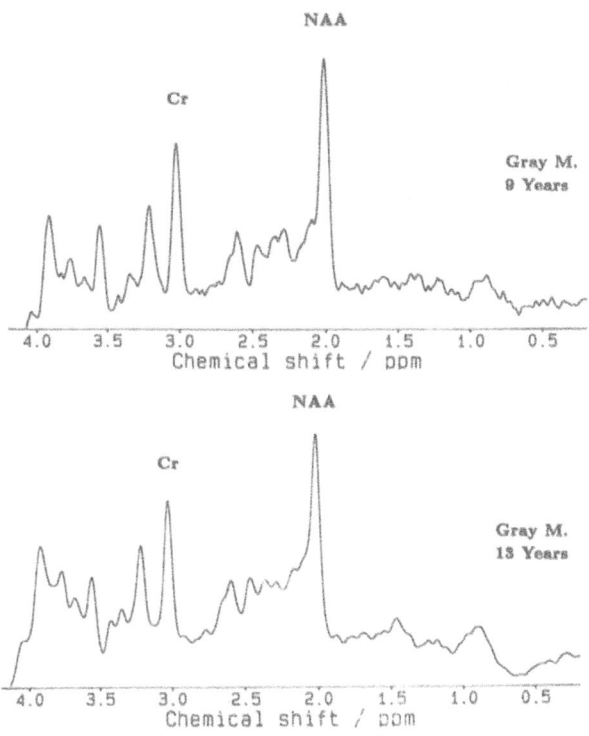

*Figure 16* Continued

(TR = 3000 msec) revealed depleted NAA, Glu, and Gln, markedly reduced Cho (0.2 mM, −70%) and Cr (1.0 mM, −80%), and elevated Lac (1.9 mM) and Glc (3.5 mM). An almost identical metabolite pattern was found in white matter (not shown). These results are consistent with complete structural disintegration of gray and white matter, with a loss of viable neuroaxonal tissue being followed by a disruption of glial cells. In comparison to only moderate reductions of NAA in Sandhoff's disease, loss of neurons in infantile NCL turns out to be much more marked than in gangliosidoses. As in other diseases that lead to a loss of brain tissue, MRS findings of elevated Lac may be due to anaerobic metabolism of infiltrating macrophages or may simply reflect CSF contributions rather than indicating a mitochondrial enzyme deficiency. In fact, once cellular disintegration is completed, the resulting Lac and Glc concentrations may entirely originate from CSF in pertinent spaces.

The second part of Fig. 16 indicates disease progression in NCL by comparing follow-up studies of gray matter in a patient at the age of 9 and 13 years. In this juvenile form, the metabolite pattern was normal at the age of 9 years, but a decrease of NAA (5.7 mM, $-25\%$), Cr (4.2 mM, $-15\%$), and *myo*-Ins (2.5 mM, $-30\%$) was observed at the time of the second examination (TR = 3000 msec), indicating slow progression of neuronal and glial damage. It should be noted that the ¹H MRS results of both patients shown in Fig. 16 are not entirely in agreement with a case report by Confort-Gouny *et al.* (1993) on early infantile NCL. In particular, the reported increase of Ins and Tau may be due to age-related adjustments in early infancy and requires further clarification. In addition, the discussion of a 2.04-ppm resonance from *N*-acetylglucosamine is not substantiated here.

# 6. MITOCHONDRIAL DISEASES

Mitochondrial disorders comprise a group of diseases with structurally and functionally abnormal mitochondria in muscle tissue and other organs (Sengers *et al.*, 1984; Barkovich *et al.*, 1993; Cassedy and Edwards, 1993). Accordingly, mitochondrial myopathies may occur as separate disorders or as multisystem diseases (mitochondrial cytopathies, encephalomyopathies). A large number of mitochondrial myopathies with or without abnormalities in the mitochondrial DNA have been described that refer to disturbances of oxidative energy metabolism at different stages of the Krebs cycle or respiratory chain. Examples of these disorders are Leigh syndrome and the encephalomyopathies described in the following sections.

## 6.1. Leigh Syndrome

Leigh (1951) described an autosomal recessive disorder which he called subacute necrotizing encephalomyelopathy. Symptoms usually commence in infancy as hypotonia or acute life-threatening encephalopathy or later in life as intermittent ataxia or dystonia. An increase of Lac in urine, blood, and CSF points toward a disturbed oxidative energy metabolism. The biochemical basis of Leigh syndrome (LS) has been the subject of many studies. Defects associated with this disorder include deficiencies of pyruvate dehydrogenase (PDH) and carboxylase, cytochrome *c* oxidase (complex IV), NADH-coenzyme-Q reductase (complex I), and biotinidase. Recently, a complex II deficiency has also been noted (Bourgeois *et al.*, 1992). Mutation of mitochondrial DNA at the ATPase gene (NARP mutation) was demonstrated in LS.

Brain pathology reveals bilateral symmetrical areas of rarefication and proliferation with relatively good preservation of neurons. The optic nerves, chiasma, and tracts, basal ganglia, brain stem, corpora quadrigemina, tegmentum, and inferior olives are mainly involved, but cerebellar dentate nuclei and cerebral cortex may also be affected. MRI frequently shows abnormalities in these areas,

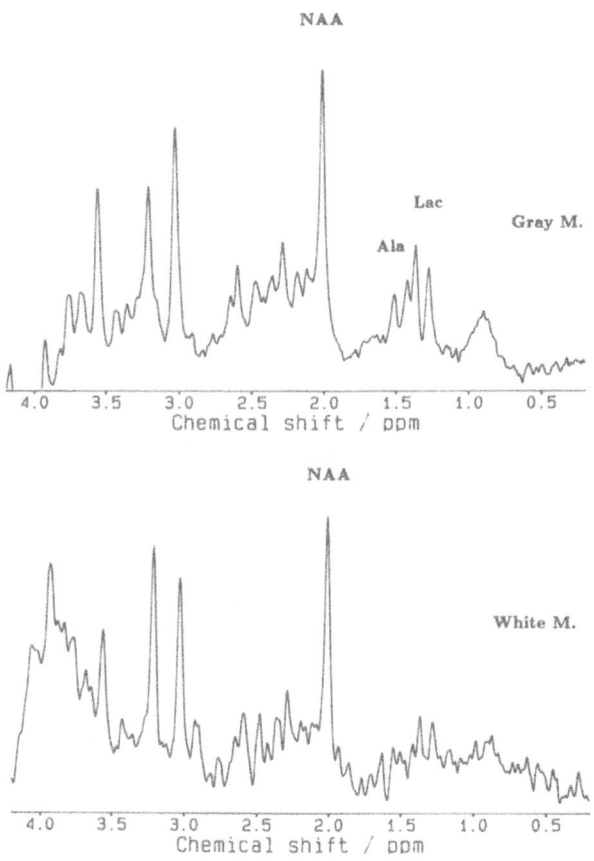

FIGURE 17. Metabolic alterations in three patients with Leigh syndrome as detected by ¹H MRS (STEAM, TR/TE/TM = 3000/20/30 msec, 128 accumulations) of gray and white matter. *First part:* Gray- (12 ml) and white-matter (4.1 ml) spectra of a 5-year-old patient with Leigh syndrome and elevated Lac and alanine (Ala) in gray matter. *Second part:* Basal ganglia (4.1 ml) and white-matter (4.1 ml) spectra of a patient with Leigh syndrome at the age of 12 months and 17 months, respectively, showing reduced NAA and elevated *myo*-Ins (white matter) but no elevated Lac. *Third part:* Gray-matter (9.2 ml) spectra of a patient with Leigh syndrome and high levels of Lac at the age of 4 months (*top*) and 18 months (*bottom*) after dichloroacetate therapy.

particularly in basal ganglia and brain stem, although such findings may vary in intensity in the same patient at different times. Rare cases present with a significant demyelination.

The heterogeneity of enzyme deficiencies in LS is also reflected in the variability of cerebral metabolite alterations detected by ¹H MRS. An example of diverse observations is shown in Fig. 17, comprising data from three different

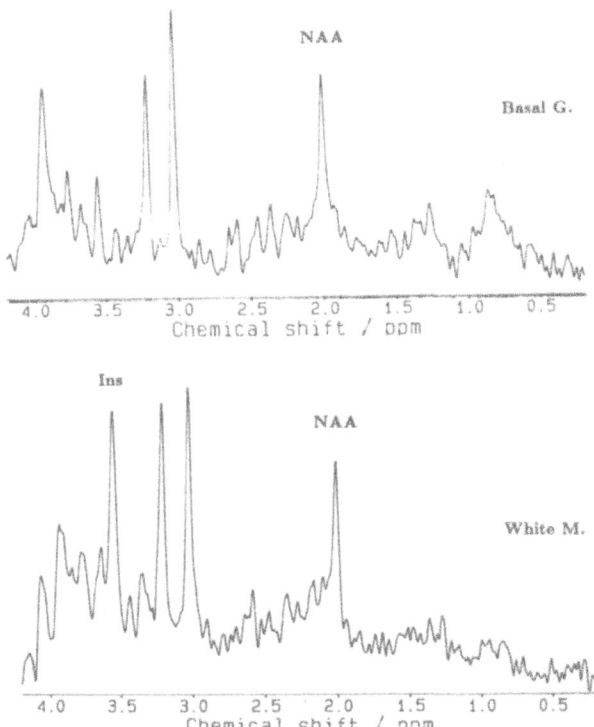

*FIGURE 17.* Continued

subjects (TR = 3000 msec). The gray- and white-matter spectra from a 5-year-old patient are characterized by elevated concentrations of Lac (3.3 mM) and Ala (1.6 mM) in gray matter. NAA was reduced in both gray (4.2 mM, −25%) and white matter (4.2 mM, −40%), while no glial changes were noted. These findings are consistent with a classical PDH deficiency blocking the pathway from cytosolic pyruvate (produced by glycolysis) to mitochondrial acetyl coenzyme A. It is not surprising that the consequences of a restricted capacity for oxidative metabolism are more pronounced in well-perfused cortical gray matter than in white matter.

The second part of Fig. 17 shows basal ganglia and white-matter spectra of another patient at the age of 12 and 17 months. This case clearly differs from the previous case. While reduced levels of NAA in basal ganglia (3.3 mM) and frontal white matter (3.3 mM, −50%) also suggest neuroaxonal degeneration, a 2.5-fold increase of *myo*-Ins (6.4 mM) in white matter in conjunction with Cho and Cr levels at the upper end of the normal range suggest a disturbance of

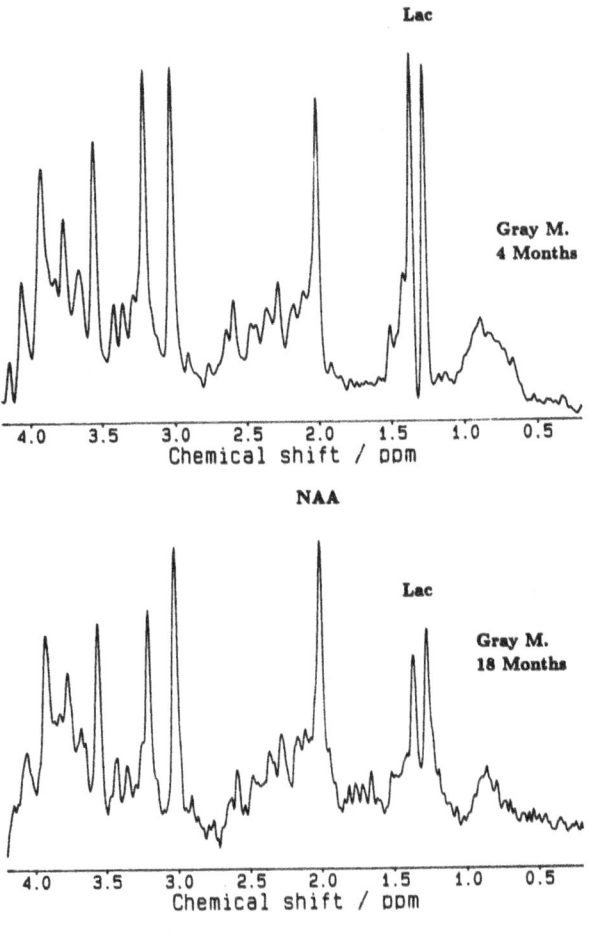

*FIGURE 17.* Continued

myelination similar to that seen in many leukodystrophies. Most remarkably, however, no elevation of brain Lac was detected despite a Lac concentration of 6.7 mM found in CSF.

The third part of Fig. 17 compares gray-matter spectra of a patient with LS and extremely high levels of Lac at the age of 4 months and during treatment with dichloroacetate at 18 months. Analysis of these data indicated a reduction of Lac from 8.1 to 5.1 mM and of Ala from 1.0 to 0.5 mM. While the concurrent decrease of Cho from 1.2 to 0.9 mM and of Tau from 4.5 to 2.3 mM may be considered as an age-related normalization, the NAA concentration (3.1 mM) remained low during this period instead of exhibiting the expected increase during brain maturation.

Summarizing our experience with a total of 15 patients, the only uniform and consistent finding in patients with LS is a generalized reduction of NAA by about 50%. The concentration of Lac, however, shows marked regional variability (Detre *et al.*, 1991). Moreover, since some patients show normal brain Lac levels even when CSF Lac is elevated (cf. Fig. 17, second part), the utility of brain Lac levels as a diagnostic marker for LS, as hypothesized on the basis of 5 patients with elevated Lac in basal ganglia (Krägeloh-Mann *et al.*, 1993), may be questioned (Kruse *et al.*, 1994b). While ¹H MRS determinations of brain Lac and other cerebral metabolites in the individual patient provide additional diagnostic hints, the possibility of "negative" Lac findings, even when serum and/or CSF levels are high, must be emphasized. In other words, elevated brain Lac is not a necessary finding in LS, while normal brain Lac does not exclude its diagnosis.

## 6.2. Encephalomyopathies

The encephalomyopathies belong to the continuously increasing number of known mitochondrial diseases. From the original description of progressive external ophthalmoplegia (von Graefe, 1866), the concept of encephalomyopathies has evolved as a representation of multiorgan diseases that are caused by a mutation or deletion of mitochondrial DNA. At least four different disorders can be distinguished according to their clinical characteristics and mitochondrial defect:

* Mitochondrial encephalopathy, lactic acidosis, and stroke-like episodes (MELAS)
* Myoclonic epilepsy with ragged red fibers (MERRF)
* Kearns–Sayre syndrome (KSS)
* Chronic progressive external ophthalmoplegia (CPEO)

These myopathies present with a wide range of additional organ involvement, e.g., of the endocrine and hematopoietic system. Diagnosis is based on clinical, morphological, biochemical, and genetic abnormalities. Muscle biopsy shows ragged red fibers as an accumulation of abnormally structured and malfunctioning mitochondria, which can also be observed by electron microscopy. Deficiencies of respiratory chain complexes are infrequently found in muscle tissue, but biochemical determinations often demonstrate lactic acidosis and increase of Lac in plasma and CSF. Mitochondrial DNA deletions have been reported in KSS and CPEO, and point mutations in MELAS and MERRF.

MRI of mitochondrial disorders frequently shows involvement of basal ganglia as well as white-matter abnormalities. The stroke-like lesions in MELAS (Gropen *et al.*, 1994; Lee *et al.*, 1994) resemble infarcts of the cortex and adjacent white matter (Matthews *et al.*, 1991a), as shown in Fig. 18, but are not restricted to a specific vascular territory. Postmortem investigation of the 18-

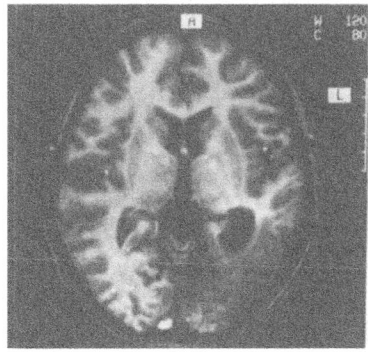

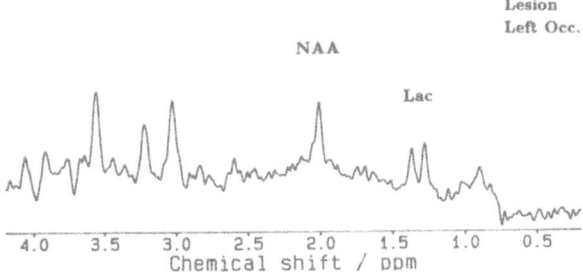

*FIGURE 18.* Morphological and metabolic alterations in an 18-year-old patient with mitochondrial encephalopathy, lactic acidosis, and stroke-like episodes (MELAS) as detected by $T_1$-weighted MRI (RF spoiled 3-D FLASH, TR/TE = 15/6 msec, 20° flip angle, 4-mm partitions) and ¹H MRS (STEAM, TR/TE/TM = 6000/20/30 msec, 64 accumulations) of a lesion (8 ml) and gray (18 ml) and white matter (6.4 ml). The lesion is characterized by decreased levels of NAA, Cr, and Cho as well as elevated Lac.

year-old subject of this MRI investigation revealed distortions in endothelial and smooth muscle cells of small cerebral arteries. Proton MRS of the left occipital lesion showed reduced NAA, Cr, and Cho but elevated *myo*-Ins (4.9 mM) and Lac (3.5 mM). This pattern of metabolite alterations is consistent with findings in affected brain areas of patients after acute stroke (Bruhn *et al.,* 1989a). While contralateral white matter was normal with regard to morphological and metabolic criteria, neighboring midsagittal parietal gray matter revealed a mild reduction of NAA (6.2 mM) and Cho (0.8 mM) but also high Lac (3.4 mM).

Figure 19 shows metabolic alterations in gray and white matter of a 6-year-old patient with MERRF (TR = 3000 msec). The general pattern is similar to that found for MELAS and consistent with neuroaxonal degeneration, possibly leading to complete cellular disruption. Changes of NAA (3.8 mM, −45%) and Lac 3.7 mM) are observed in white matter but become much more pronounced in gray matter, where function is tightly linked to oxidative energy metabolism (Glc

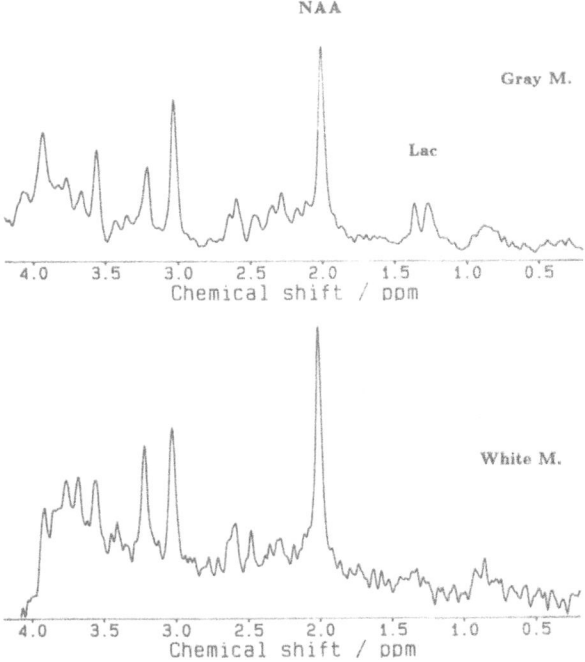

*Figure 18.* Continued

consumption). Corresponding concentrations of NAA (1.5 mM, −75%), Cr (2.0 mM, −55%), and Cho (0.5 mM, −40%) are markedly reduced, while Lac is considerably enhanced (5.5 mM). A $^{31}P$ MRS study of 8 patients with MERRF revealed increased intracellular inorganic phosphate ($P_i$) and decreased $PCr/P_i$ ratios in resting muscle but no alterations of phosphate metabolites and intra-cellular pH in the brain (Matthews *et al.,* 1991b).

Gray- and white-matter spectra of a 23-year-old patient with KSS are shown in Fig. 20. Partial malfunction of mitochondrial respiration is indicated by a generalized though mild elevation of Lac in gray (2.4 mM) and white matter (2.9 mM). An unexpected finding is a concomitant reduction of *myo*-Ins (2.7 and 2.1 mM, respectively). In addition, gray matter showed reduced NAA (6.5 mM) and Cho (0.7 mM), while Cho (1.6 mM) and Cr (6.5 mM) were elevated in white matter. The significance of these observations is not yet clear.

## 7. RETT SYNDROME

The syndrome described by Rett (1966) is now defined by criteria for diagnosis first formulated at a conference in 1985 in Vienna (Hagberg, 1993).

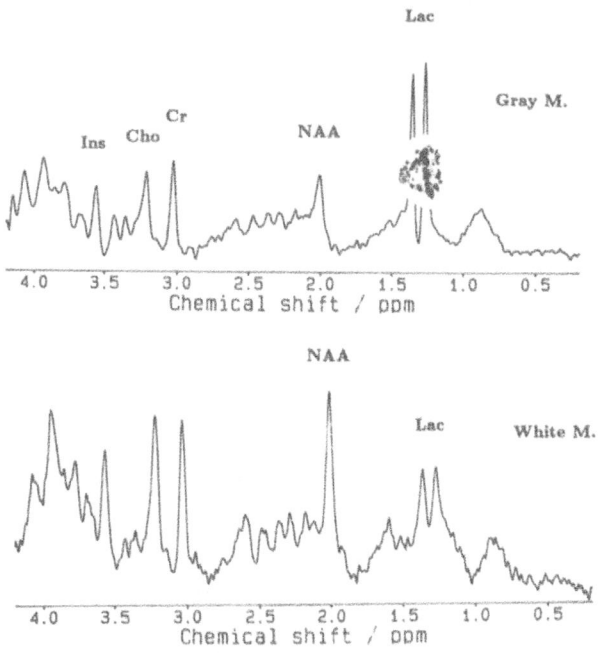

FIGURE 19. Metabolic alterations in a 6-year-old patient with myoclonal epilepsy with ragged red fibers (MERRF) as detected by ¹H MRS (STEAM, TR/TE/TM = 3000/20/30 msec, 128 accumulations) of gray (18 ml) and white matter (8 ml). The reduction of NAA and elevation of Lac is most pronounced in gray matter.

Only girls are affected. They are born into normal families after uneventful pregnancy. Usually between 6 and 8 months of age, movement disorders (ataxia, apraxia), behavioral changes (autistic features), and alterations of head growth (deceleration leading to microcephaly) develop as characteristic symptoms. Psychomotor regression is associated with loss of purposeful use of hands, acquired words, and ambulatory capacity in many children. No biochemical, metabolic, neurophysiological, or morphological marker for the disease is known at present. Neuropathology shows small brains packed with small neurons in normal numbers. Dendritic branching has been found to be decreased or deranged in combination with other synaptic abnormalities. The absence of typical degenerative changes, signs of gliosis, or mitochondrial dysfunction (Nielsen *et al.*, 1993) suggests a developmental disorder.

In a preliminary ¹H MRS study, we hypothesized that slow disease progression with no or only mild metabolic alterations in early infancy might become more pronounced in older children and young adult patients (Hanefeld *et al.*, 1995a). As demonstrated in Table 3, more recent examinations of two larger

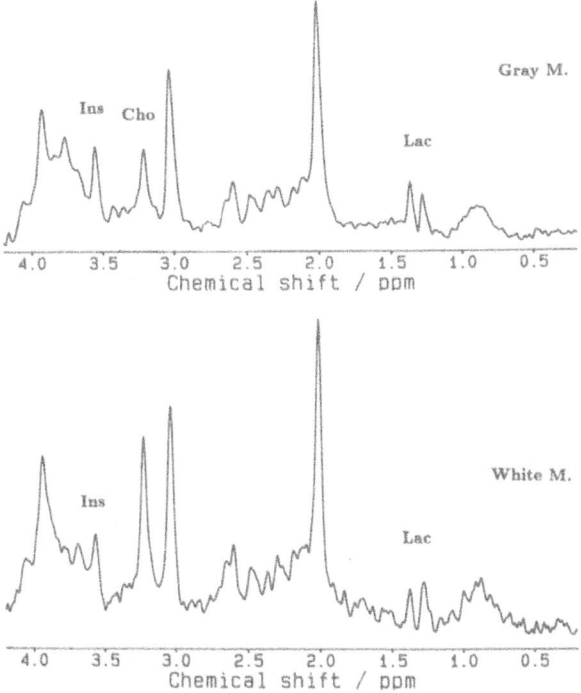

*FIGURE 20.* Metabolic alterations in a 23-year-old patient with Kearns–Sayre syndrome (KSS) as detected by ¹H MRS (STEAM, TR/TE/TM = 6000/20/30 msec, 64 accumulations) of gray (18 ml) and white matter (8 ml). Generalized findings are a mild increase of Lac and a decrease of *myo*-Ins.

*TABLE 3.* Absolute Tissue Concentrations (Mean ± SD) of Major Cerebral Metabolites in Parietal Gray and White Matter of Patients with Rett Syndrome below and above 10 Years of Age

| Metabolite[a] | Concentration (mM) in gray matter at age (years): | | Concentration (mM) in white matter at age (years): | |
|---|---|---|---|---|
| | 4.3 ± 2.0 (n = 9) | 18 ± 7 (n = 7) | 4.1 ± 1.8 (n = 12) | 18 ± 7 (n = 10) |
| NAA+NAAG | 6.8 ± 0.8 | 6.8 ± 1.0 | 7.8 ± 0.8 | 8.1 ± 0.6 |
| PCr+Cr | 5.7 ± 0.8 | 5.7 ± 0.7 | 5.1 ± 0.3 | 5.3 ± 0.3 |
| Cholines | 1.0 ± 0.2 | 1.1 ± 0.1 | 1.5 ± 0.2 | 1.5 ± 0.2 |
| *myo*-Inositol | 3.9 ± 0.6 | 3.4 ± 0.4 | 2.5 ± 0.3 | 2.5 ± 0.6 |
| Glutamate | 7.5 ± 1.1 | 6.7 ± 1.4 | 6.3 ± 1.1 | 6.0 ± 1.4 |
| Glutamine | 4.1 ± 0.9 | 4.4 ± 1.1 | 2.6 ± 0.9 | 3.4 ± 2.4 |

[a] Abbreviations as in Table 2.

patient populations below and above 10 years of age resulted in rather stable values for the mean concentrations of most metabolites. Whether disease progression in individual patients may be monitored, for example, by following decreasing NAA concentrations, remains to be elucidated in a longer follow-up study. A comparison with adult controls yields only mild (10%) reductions of NAA in both groups of patients and a slight trend for the elevation of Gln at the expense of Glu with increasing age. Accordingly, the Glu/Gln ratio decreases from 2.2 (control) to 1.8 (Rett, <10 years) and 1.5 (Rett, >10 years) in gray matter and from 2.9 to 2.4 and 1.8 in white matter, respectively.

## 8. CREATINE DEFICIENCY

Creatine ($\alpha$-methylguanidinoacetate) and creatine phosphate play essential roles in the storage and transmission of phosphate-bound energy. In humans, Cr is synthesized in liver and pancreas using Arg, Gly, and $S$-adenosylmethionine as substrates and arginine:glycine amidinotransferase and guanidinoacetate methyltransferase as enzymes. Blood transport ensures delivery to muscle and brain, which contain high activities of creatine kinase. This enzyme catalyzes phosphorylation and dephosphorylation of Cr/PCr and thus provides a high-energy phosphate buffering system during states of ATP synthesis and utilization.

Using combined $^1$H and $^{31}$P MRS, we were able to

- diagnose the first known case of a generalized and selective Cr deficiency in a 22-month-old boy,
- propose adequate treatment and monitor its efficacy over more than two years, and
- identify the biochemical defect of the disease as a lack of guanidinoacetate methyltransferase activity (Stöckler *et al.*, 1994).

The patient with this new treatable inborn error of metabolism first developed abnormalities at the age of 5 months. They consisted of muscular hypotonia and severe extrapyramidal symptoms of a hemiballistic-dystonic type. Owing to uncoordinated swallowing and frequent vomiting, he required nasogastric tube feeding. Particular metabolic findings included hyperammonemia, orotic aciduria, hyperornithinemia, and extremely low concentrations of creatinine in serum and urine. Oral substitution of creatine monohydrate (400 mg kg$^{-1}$ day$^{-1}$) but not of Arg (300 mg kg$^{-1}$ day$^{-1}$) improved his clinical state considerably including recovery from his extrapyramidal movement disorder and epileptic seizures. His previously pathological EEG normalized, and bilateral hypointensities in globus pallidus ($T_1$-weighted MRI) vanished during treatment.

Figure 21 a–c shows $^1$H MR spectra of gray matter at diagnosis, after 1 month of Arg supplementation, and after 18 months of oral Cr treatment, respectively. During Arg treatment as the initial therapeutic trial, the depletion of Cr in

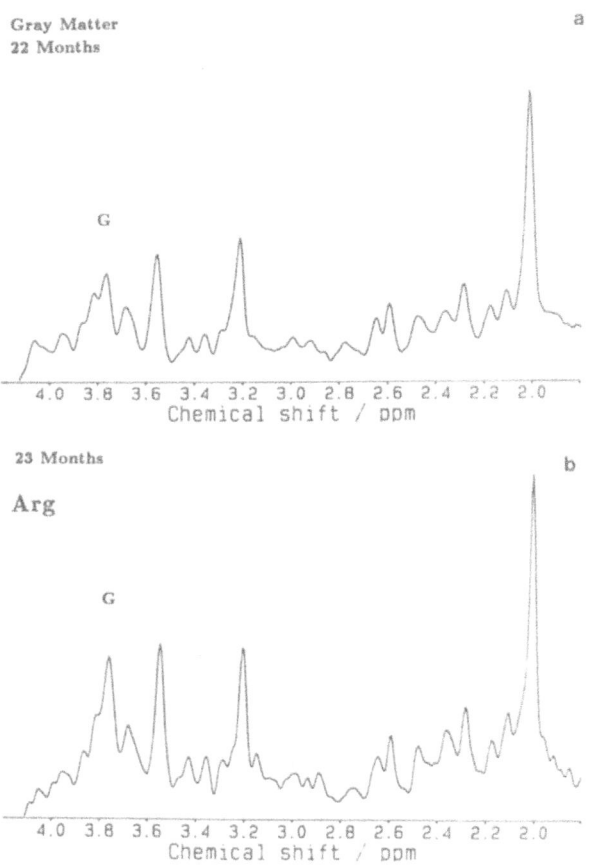

**FIGURE 21.** Metabolic alterations in a 2-year-old patient with Cr deficiency as detected by ¹H and ³¹P MRS. (a)–(c) Localized proton MRS (STEAM, TR/TE/TM = 6000/20/30 msec, 64 accumulations) of gray matter (8 ml) at diagnosis (22 months), during oral Arg substitution (23 months), and after 18 months of oral Cr substitution (41 months) reveals a partial restoration of brain Cr and a reduction of initially elevated guanidinoacetate (G), the immediate precursor of Cr. (d)–(f) Non-localized ³¹P MRS (FID, TR = 22,000 msec, 16 accumulations) of the head during oral Arg substitution (23 months), after 3 months of Arg and 2 months of Cr substitution (25 months), and after 18 months of oral Cr substitution (41 months) shows an almost complete restoration of brain PCr and normalization of phosphorylated guanidinoacetate (GP). Other ³¹P resonances are due to adenosine triphosphate (ATP), inorganic orthophosphate ($P_i$), phosphomonoesters (PME), and phosphodiesters (PDE).

all brain areas investigated was not reversed but was complemented by a substantial increase of guanidinoacetate (resonance G in Fig. 21a,b) as the immediate precursor of Cr. Since this finding indicated that Cr synthesis was blocked at the final step, direct substitution of Cr led to remarkable spectroscopic and clinical improvements. During the course of more than 18 months of Cr treatment, brain

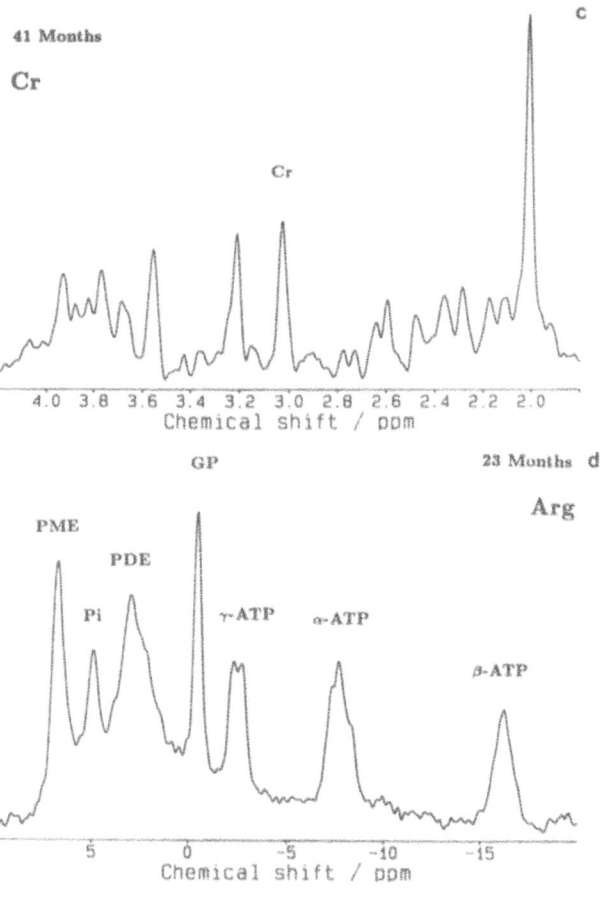

FIGURE 21. Continued

Cr concentrations steadily increased to 65% of adult control levels, while guanidinoacetate levels returned to very low values. Time courses of NAA and Cr
concentrations in gray and white matter are summarized in Table 4. The increase
of NAA over this period represents normal brain maturation.

The deficiency of guanidinoacetate methyltransferase suggested by [1]H MRS
was confirmed biochemically by demonstrating lack of pertinent enzyme activity
in liver tissue. To evaluate whether the newly formed pool of brain Cr was
metabolically active, [31]P MRS (homogeneous phosphorus head coil) was performed subsequent to [1]H MRS but during the same examination (Requardt,
1995). Figure 21d demonstrates that no PCr was detectable during oral Arg
substitution. Instead, [31]P MRS revealed a strong resonance that could be assigned
to guanidinoacetate phosphate (GP). Its distinction from PCr becomes evident in

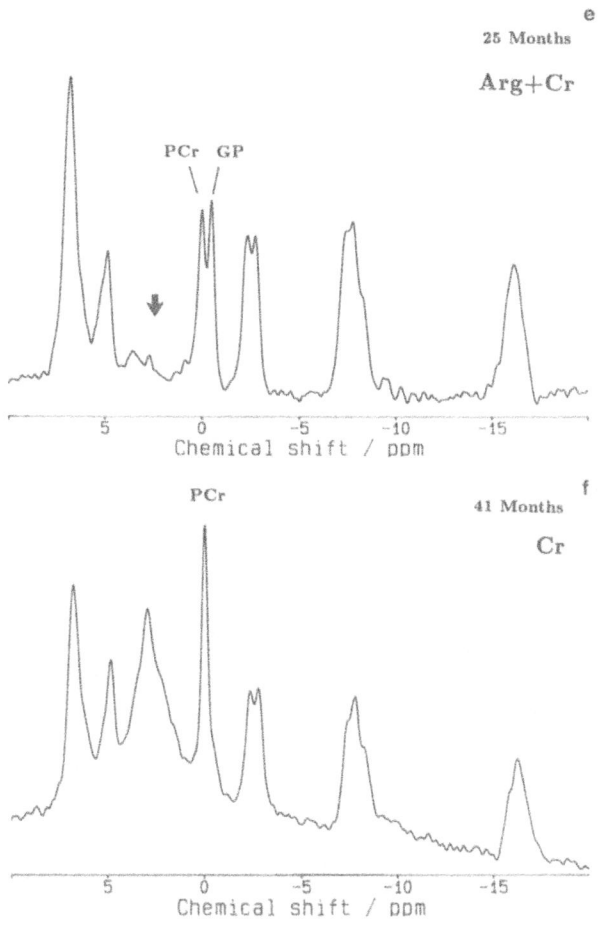

FIGURE 21. Continued

spectra obtained after oral Cr substitution and concomitant PCr increase (Fig. 21e, f). GP nearly vanished over the course of treatment.

Quantitative assessments of high-energy phosphate concentrations were accomplished with use of localized STEAM spectra (TR/TE/TM = 9000/4/5 msec) from a 96-ml VOI ($4 \times 4 \times 6$ cm$^3$) in parietal cortex (Merboldt et al., 1990; Requardt, 1995). The time course of PCr concentrations in Table 4 demonstrates a proportional increase to total Cr during therapy. In all stages, PCr was about 50% of the mean ¹H-MRS-detected concentration of total Cr in gray and white matter. These findings indicate normal Cr phosphorylation, i.e., creatine kinase activity, which is further supported by the fact that control measurements in young adults ($n = 21$) resulted in a PCr level of $2.5 \pm 0.4$ mM, or about 50%

TABLE 4. Absolute Tissue Concentrations (Mean ± SD) of Major Cerebral
Metabolites in Parietal Gray and White Matter of a Patient with Creatine Deficiency
as Detected by ¹H and ³¹P MRS at Diagnosis and during Treatment[a]

| Age (months) | Concentration (mM) | | | | | |
| | Gray matter | | White matter | | Whole brain PCr | Therapy |
| | NAA | Cr | NAA | Cr | | |
|---|---|---|---|---|---|---|
| 22 | 5.7 | 0.0 | 6.1 | 0.1 | — | — |
| 23 | 6.0 | 0.0 | 6.3 | 0.0 | 0.0 | Arg |
| 25 | 6.2 | 2.0 | 6.9 | 1.2 | 0.7 | Arg + Cr |
| 26 | 5.8 | 2.5 | 6.9 | 1.8 | — | Cr |
| 33 | 7.3 | 3.0 | 8.1 | 2.6 | 1.5 | Cr |
| 38 | 8.1 | 4.1 | 7.9 | 3.1 | 2.1 | Cr |
| 41 | 7.4 | 3.8 | 8.7 | 3.7 | 2.0 | Cr |

[a]Abbreviations: NAA, N-acetylaspartate; Cr, creatine; PCr, phosphocreatine; Arg, arginine.

of total Cr (Requardt, 1995). In addition, direct measurements of the kinetics of phosphate exchange between PCr and ATP using a pulsed saturation transfer technique failed to demonstrate any transfer from $\gamma$-ATP to the metabolically inactive GP. On the other hand, Cr substitution led to a significant phosphate transfer from $\gamma$-ATP to the newly formed PCr at a rate similar to that found in controls.

Thus, dynamic MRS studies in this patient not only demonstrated the cerebral uptake of oral Cr by ¹H MRS but further proved its phosphorylation and normal metabolic function, i.e., phosphate exchange with ATP on the proper time scale, by ³¹P MRS. In addition to providing information for the clinical management of the patient, these results may further help to elucidate the questions of Cr "visibility," its compartmentalization into different (neuronal and glial) pools, and the phosphorylation of the respective concentrations. Whether a full restoration of brain Cr (and PCr) levels will be possible still remains to be seen, as time constants of cerebral uptake are surprisingly small. Correspondingly, the delayed onset of clinical symptoms several months after birth may also have been due to a slow depletion of a normal Cr pool supplied by maternal Cr during pregnancy and/or postnatal feeding.

## 9. MISCELLANEOUS DISORDERS

### 9.1. Carbohydrate-Deficient Glycoprotein Syndrome

The carbohydrate-deficient glycoprotein (CDG) syndromes are genetic multisystemic disorders characterized by glycosylation defects in certain glycoproteins, most likely at the level of protein synthesis and processing in endoplasmatic reticulum (Jaeken *et al.*, 1991). The gene for type I has recently been mapped on chromosome 16 (Martinsson *et al.*, 1994). The demonstration of abnormal transferrins is used for diagnosis. Multiorgan involvement includes cerebral and peripheral nervous system, subcutaneous fat, eyes, liver function, coagulation, and the endocrine system depending on the stage of the disease.

Three types of CDG syndrome have been described by Hagberg *et al.* (1993), Ramaeckers *et al.* (1991), and Stibler *et al.* (1993), respectively. Type I represents the most frequent form and typically leads to severe cerebellar atrophy. Type III shows the most severe clinical course with infantile spasms, cerebral atrophy, and severe mental and motor retardation. The skin of these patients exhibits characteristic areas of depigmentation.

Metabolic alterations detected by ¹H MRS of patients with CDG syndromes reflect the above clinical classification (Holzbach *et al.*, 1995). Figure 22 shows gray- and white-matter changes in two different patients with type I and III CDG syndrome, respectively. In type I CDG syndrome (Fig. 22, first part), gray-matter metabolite concentrations were all within normal ranges, in line with only mild cortical atrophy. White matter showed mild reductions of NAA (6.6 mM), Cr (4.2 mM), and Cho (1.2 mM) but no signs of disturbed myelination. In a 2-year-old girl with type III CDG syndrome (Fig. 22, second part), ¹H MRS revealed more severe metabolic alterations in both gray and white matter. Gray-matter spectra yielded reduced NAA (6.0 mM) as well as elevated *myo*-Ins (5.1 mM), Cho (1.15 mM), and Cr (6.6 mM). Similar though more pronounced alterations were seen in white matter, representing a marked reduction of NAA (4.7 mM) and increases in *myo*-Ins (4.1 mM), Cho (1.8 mM), and Cr (6.2 mM). These findings indicate that neuroaxonal degeneration, as evidenced by a 50% loss of NAA, is accompanied by a demyelinating process similar to that in certain leukodystrophies. Histological observations and ¹H MRS suggest that both peripheral and central axons are affected in CDG syndrome.

## 9.2. Cerebro-Hepato-Renal Syndrome of Zellweger

The cerebro-hepato-renal syndrome of Zellweger is the prototype of peroxisomal disorders (Barth *et al.*, 1988). It presents in the newborn with typical facial dysmorphism, high forehead, low and broad nasal bridge, epicanthus, and dysplastic external ears. Dominating neurological symptoms are severe hypotonia, nystagmus, and poor sucking and swallowing as well as generalized seizures often starting during the first days of life, retinal degeneration, and optic atrophy. Hepatomegaly, polycystic kidneys, and skeletal deformities are constant findings. MRI shows gyral abnormalities, polymicroglia, and pachygyria. White-matter abnormalities have been described as sudanophilic leukodystrophy. Absence of demonstrable peroxisomes in liver tissue and loss of peroxisomal en-

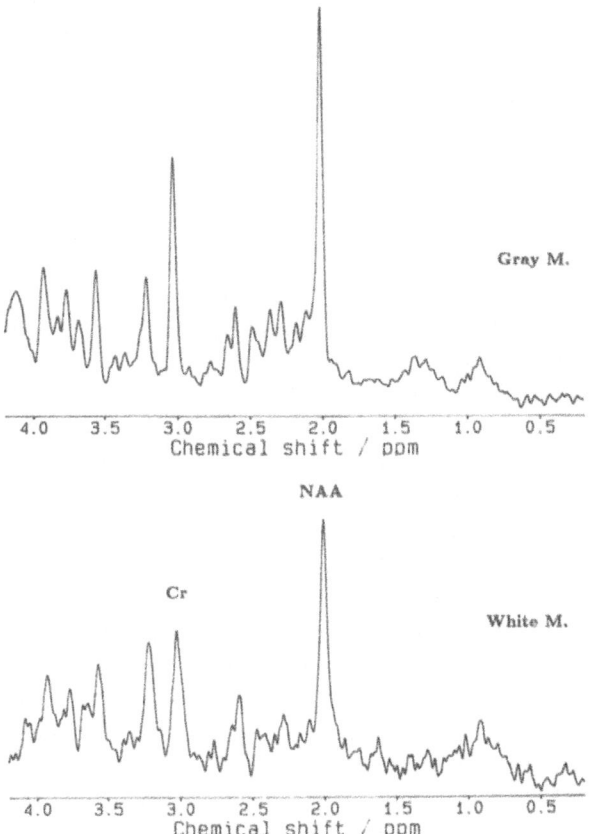

*FIGURE 22.* Metabolic alterations in two patients with carbohydrate-deficient glycoprotein (CDG) syndrome as detected by ¹H MRS (STEAM, TR/TE/TM = 6000/20/30 msec, 64 accumulations) of gray (8–12 ml) and white matter (5.1 ml). *First part:* 8-year-old patient with CDG syndrome type I and mildly decreased NAA and Cr in white matter. *Second part:* 2-year-old patient with CDG syndrome type III, showing elevated *myo*-Ins and Cr in gray matter as well as elevated *myo*-Ins, Cho, and Cr and decreased NAA in white matter.

zyme function lead to an increase of VLCFA and later also of phytanic acid. Most children succumb during the first year of life.

Figure 23 shows ¹H MR spectra (TR = 3000 msec) of gray and white matter of two patients with Zellweger syndrome, aged 3 months (first part) and 12 months (second part). In the younger patient, the most characteristic abnormality is a drastic increase of aliphatic hydrocarbon resonances (0.5–1.5 ppm range) from cytosolic proteins, cholesterol, or—less likely—mobile lipids. The effect is most notable in white matter and probably reflects the accumulation of break-

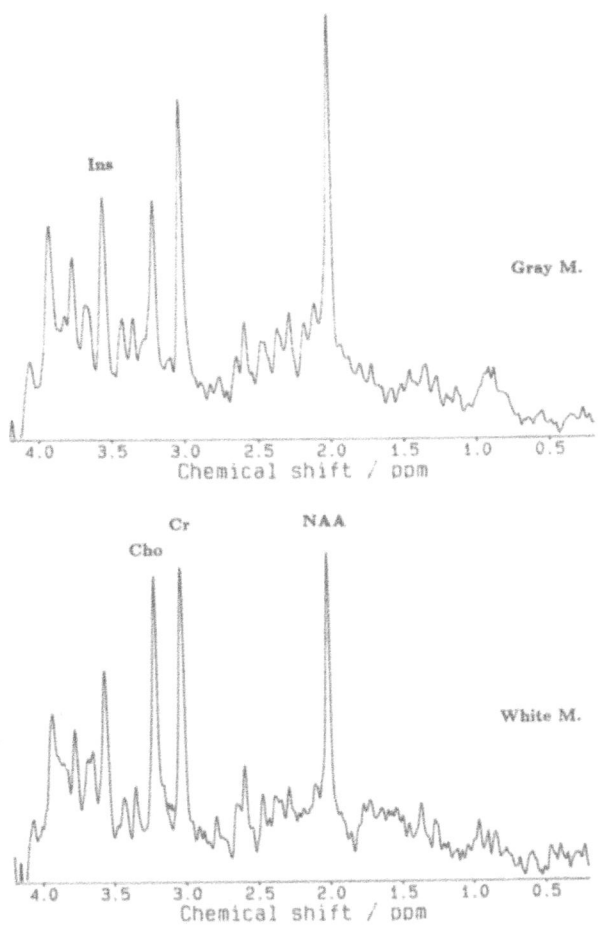

*Figure 22* Continued

down products due to disturbed developmental processes such as impaired formation of myelin. A correlation between the extent to which these resonances are increased and the severity of the clinical status has been suggested (Bruhn *et al.*, 1992c). Although the detailed origin of the hydrocarbon resonances has not yet been identified, they may be due to mobile residues from thymosin β4 that have been reported to contribute to ¹H spectra of brain tissue (Kauppinen *et al.*, 1992).

Because only a limited number of controls are yet available at this age, reliable assessments of other disturbances remain difficult. However, apart from rather low NAA levels in gray (2.9 mM) and white matter (2.3 mM), strong elevation of Gln (8.2 and 9.8 mM, respectively) clearly manifests a major abnor-

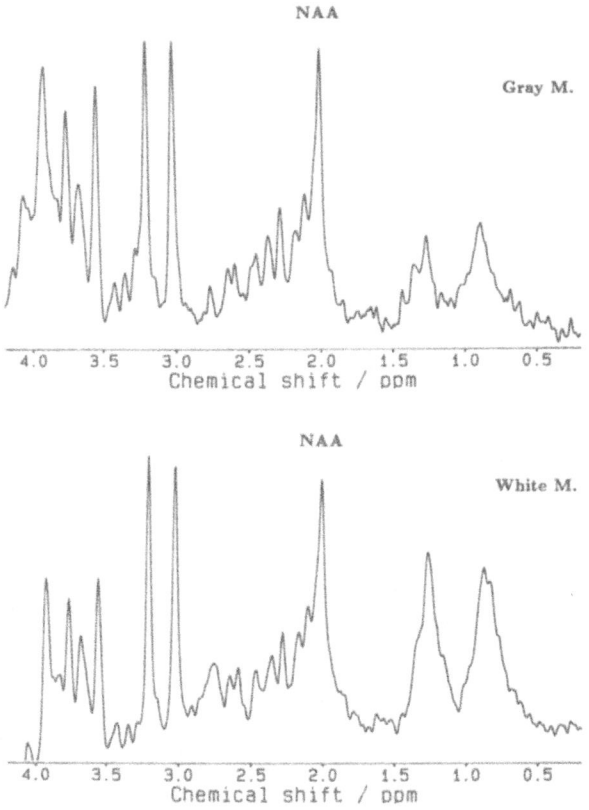

*FIGURE 23.* Metabolic alterations in two patients with cerebro-hepato-renal syndrome (Zellweger) as detected by ¹H MRS (STEAM, TR/TE/TM = 3000/20/30 msec, 128 accumulations) of gray (8 ml) and white matter (8 ml). *First part:* 3-month-old patient showing reduced NAA and elevation of aliphatic hydrocarbon resonances (0.5–1.5 ppm), most likely from mobile residues of cytosolic proteins. *Second part:* 12-month-old patient with strongly elevated Gln in gray matter, indicating impaired liver function.

mality that may be ascribed to impaired liver function as a characteristic feature of this disease with peroxisomal enzyme deficiencies. The resulting cerebral metabolic disturbance is even more pronounced in the 1-year-old patient; Fig. 23 (second part) shows extremely elevated Gln in gray matter (18.1 mM) and high values in white matter (9.1 mM). Enhanced synthesis of brain Gln from Glu and elevated blood ammonia has also been reported in hepatic encephalopathy (Kreis *et al.,* 1990, 1992). It particularly occurs in gray matter as a result of glutamine synthetase activity in astrocytes. The absence of protein resonances may be related to the later age of onset (Bruhn *et al.,* 1992c).

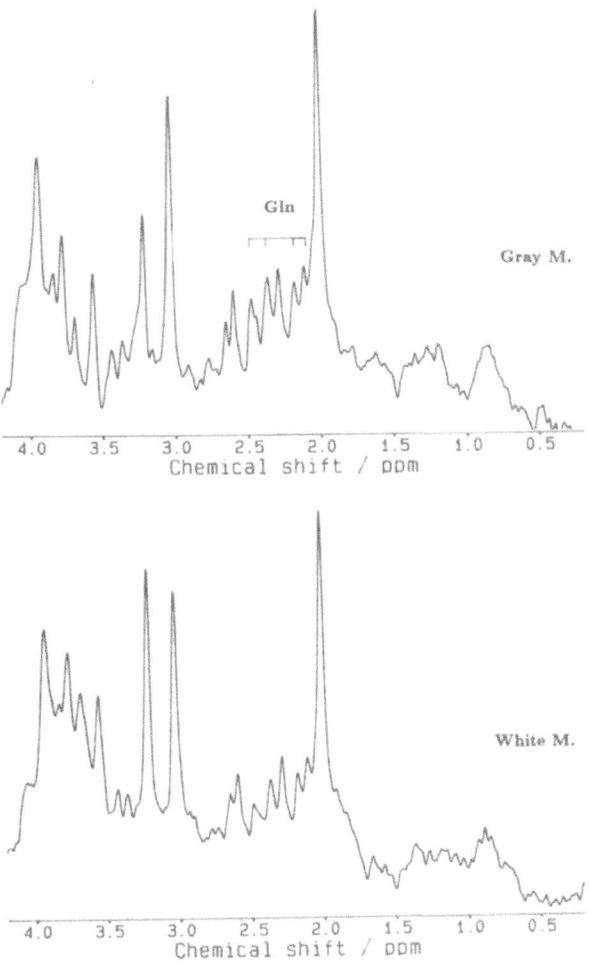

*Figure 23.* Continued

## 9.3. Hemimegalencephaly

Hemimegalencephaly (HME) is now frequently diagnosed during life. It may be idiopathic but is more often associated with vascular malformations and neurocutaneous syndromes (Hallervorden, 1923; Barkovich and Chuang, 1990). Clinical symptoms include mental retardation, hemiparesis, and intractable epileptic seizures. The enlarged cerebral hemisphere usually shows additional anatomical anomalies such as pachygyria, polymicrogyria, and gliosis. MRI may demonstrate the thickened pachygyric cortex and white-matter gliosis.

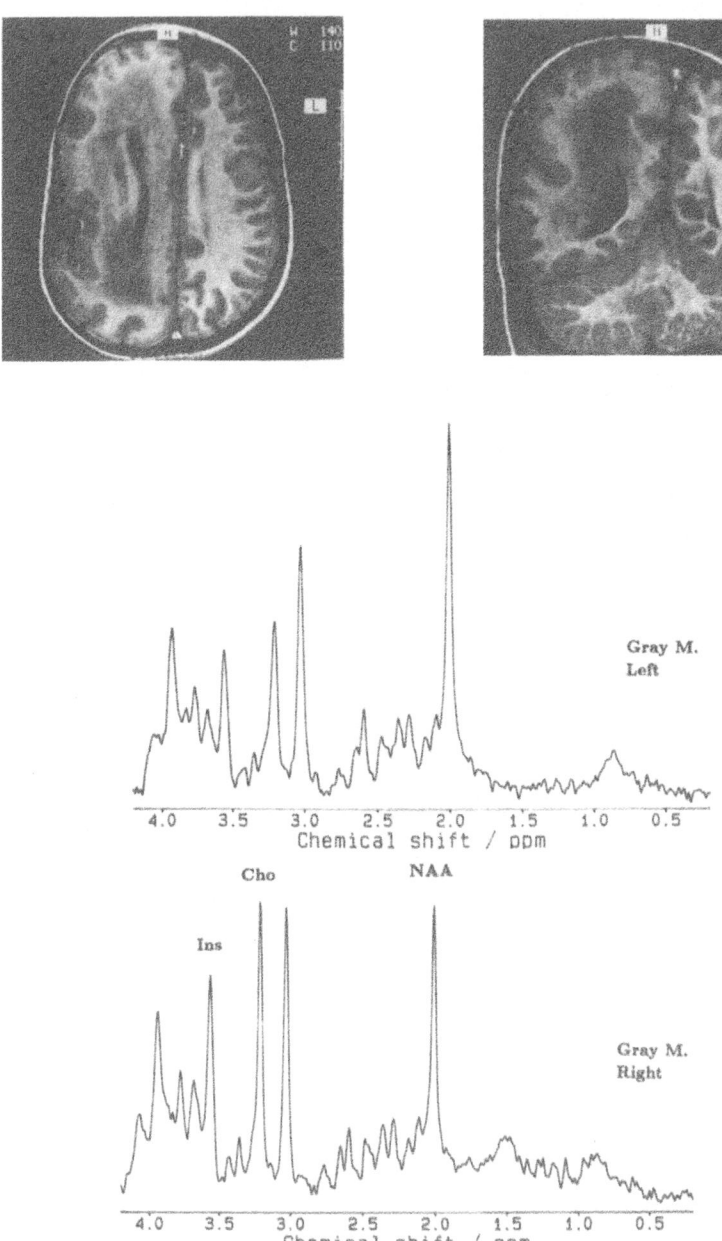

FIGURE 24. Morphological and metabolic alterations in a 13-year-old patient with hemimegalen-cephaly as detected by $T_1$-weighted MRI (RF spoiled 3-D FLASH, TR/TE = 15/6 msec, 20° flip angle, 4-mm partitions) and $^1$H MRS (STEAM, TR/TE/TM = 6000/20/30 msec, 64 accumulations) of gray matter in the insular area (14.4 ml) and parietal white matter (12 ml) in the left and the right hemisphere. The affected hemisphere shows decreased NAA and increased myo-Ins and Cho in gray matter as well as markedly reduced NAA, Glu, and Cr and increased myo-Ins in white matter.

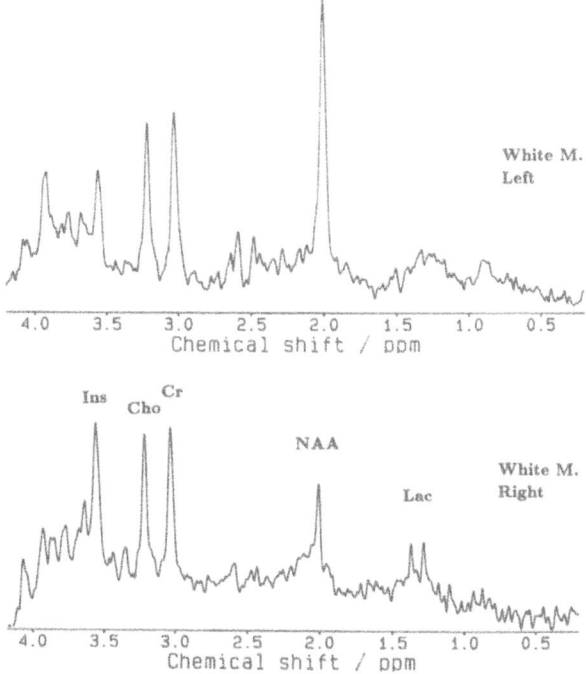

*Figure 24.* Continued

A ¹H MRS study of HME was undertaken to characterize the abnormal half of the brain and to identify neurochemical abnormalities in the apparently healthy hemisphere. Morphological and metabolic alterations in a 13-year-old patient are shown in Fig. 24 (Hanefeld *et al.*, 1995b). Metabolic disturbances in white matter of the enlarged hemisphere include a marked reduction of NAA (2.1 mM), Glu- (1.5 mM), Cr (3.3 mM), and Cho (1.0 mM) as well as elevation of *myo*-Ins (5.0 mM). Affected gray matter in the right insular area showed reduced NAA (4.9 mM) as well as increased *myo*-Ins (5.9 mM), Cho (1.8 mM), and Cr (6.3 mM). These findings may result from neuroaxonal degeneration and further reflect a pattern of disturbed glial metabolism such as found in leukodystrophies. The reduction of all metabolites in white matter except *myo*-Ins suggests a process that leads to complete cellular disintegration starting with neuroaxonal tissue components.

The normal-appearing hemisphere showed mildly decreased NAA (7.1 mM) in contralateral white matter as well as increased Cho (1.5 mM) and mild elevation of *myo*-Ins (4.4 mM) and Cr (6.8 mM) in gray matter. Cerebral Gln was strongly enhanced in all areas investigated, e.g., in affected (10.8 mM) and normal (12.8 mM) gray matter, indicating impaired liver function. Partial in-

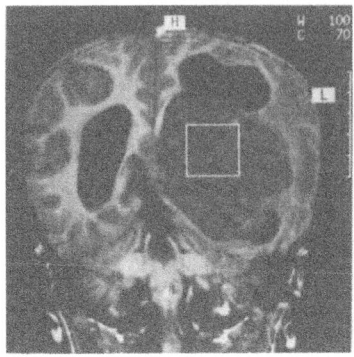

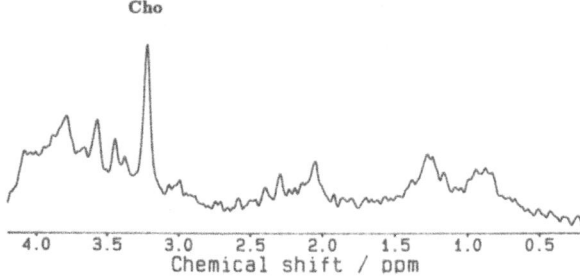

*FIGURE 25.* Morphological and metabolic alterations in two patients with intracranial tumors as detected by $T_1$-weighted MRI (RF spoiled 3-D FLASH, TR/TE/ = 15/6 msec, 20° flip angle, 4-mm partitions) and ¹H MRS (STEAM, TR/TE/TM = 6000/20/30 msec, 64 accumulations) of the neoplasm. *First part:* 20-month-old patient with a plexus papilloma (15.6 ml), showing strongly reduced metabolite levels except for Cho. *Second part:* 2-year-old patient with a cerebellar astrocytoma (4.1 ml), showing marked increases of *myo*-Ins and Cho as well as decreased NAA.

volvement of the contralateral hemisphere was in line with the presence of additional epileptic activity in this patient. The ability to detect metabolic abnormalities in the normal-appearing hemisphere is expected to significantly improve the diagnostic evaluation of HME patients, in particular when hemispherectomy is considered as a therapeutic option due to intractable seizures.

## 9.4. Brain Tumors

Brain tumors are the second most frequent malignancy in childhood. Medulloblastomas, ependymomas, and brain stem gliomas dominate, and infratentorial locations occur in more than 50% (Kucharczyk *et al.,* 1985; Cohen and Duffner, 1994). However, despite methodological progress in CT, MRI, and angiography, accurate diagnosis often poses problems. For example, in a child with cerebellar symptoms differential diagnosis between a neoplasm and gliosis

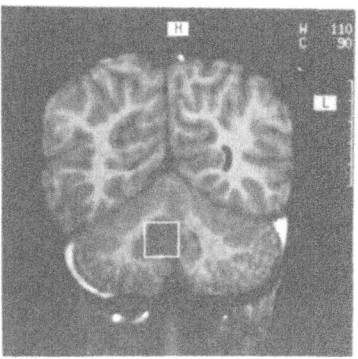

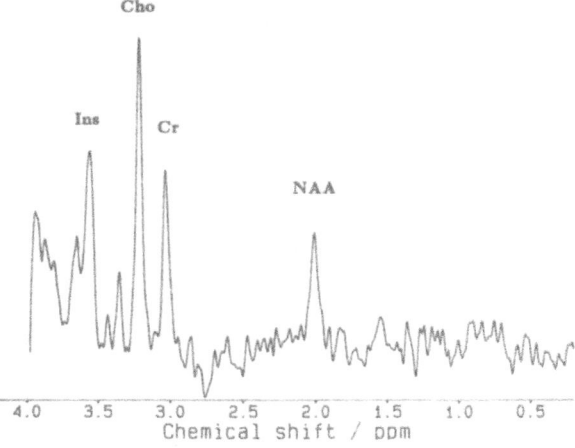

*Figure 25* Continued

may be difficult. Similar arguments hold true for the distinction between a brain tumor and perinatally acquired gliotic scars, chronic inflammation, or hamartomas in supratentorial lesions. The putative role of MRS aims at enhanced specificity in tissue characterization as it provides metabolic profiles that may be linked to the proliferation (or degeneration) of certain cells types.

Two cases in which ¹H MRS has been successfully employed for differential diagnosis are shown in Fig. 25. In both the 20-month-old patient with a plexus papilloma (first part) and the 2-year-old patient with a cerebellar astrocytoma (second part), suggestion of a tumor by MRS, as opposed to, for example, a vascular malformation, was confirmed histologically after surgery. The following combination of findings may be considered diagnostic for a brain tumor:

- A focal lesion on MRI
- Moderate to strong elevation of Cho (1.7 and 3.0 mM in the papilloma and astrocytoma, respectively)
- A partial or complete loss of NAA and Cr

Here, the observed metabolic profiles are clearly in line with previous findings in primary brain tumors both *in vitro* (Kinoshita *et al.*, 1994) and *in vivo* [e.g., see Bruhn *et al.* (1989b), Frahm *et al.* (1991b), and Fulham *et al.* (1992)].

ACKNOWLEDGEMENTS

We gratefully acknowledge the help of Petra Pouwels, Ph.D., Wolfgang Hänicke, Dipl. Math., and Knut Brockmann, M.D., in the preparation of this chapter. Many thanks go to our colleagues in the Biomedizinische NMR Forschungs GmbH am Max-Planck-Institut für biophysikalische Chemie and the Abteilung Neuropädiatrie des Klinikums der Universität Göttingen. We also thank the many colleagues from external hospitals for allowing us to study their patients. Last but not least, our warm thanks go to the parents and their children. Without their cooperation and understanding, the present study would not have been possible.

## REFERENCES

Aicardi, J., 1992, "Diseases of the Nervous System in Childhood," Mac Keith Press, London.

Aicardi, J., 1993, The inherited leukodystrophies: A clinical overview, *J. Inherited Metab. Dis.* **16**:733–743.

Alexander, W. S., 1949, Progressive fibrinoid degeneration of fibrillary astrocytes associated with mental retardation in a hydrocephalic infant, *Brain* **72**:373–381.

Arend, A. O., Leary, P. M., and Rutherfoord, G. S., 1991, Alexander's disease: A case report with brain biopsy, ultrasound, CT scan and MRI findings, *Clin. Neuropathol.* **10**:122–126.

Arus, C., Chang, Y., and Barany, M., 1985, Proton nuclear magnetic resonance spectra of excised rat brain. Assignment of resonances, *Physiol. Chem. Phys. Med. NMR* **17**:23–33.

Aubourg, P., Balancne, S., Jambaque, I., Rocchiccioli, R., Kalifa, G., Naud-Saudreau, C., Rolland, M. O., Debre, M., Chaussain, J. L., Greiscelli, C., Fische, A., and Baugneres P.-F., 1990, Reversal of early neurologic and neuroradiologic manifestations of X-linked adrenoleukodystrophy by bone marrow transplantation, *N. Engl. J. Med.* **322**:1860–1866.

Austin, J. H., Balasubramanian, A. S., Pattabiraman, T. N., Saraswathi, S., Basu, D. K., and Bachhawat, B. K., 1963, A controlled study of enzymatic activities in three human disorders of glycolipid metabolism, *J. Neurochem.* **10**:805–816.

Austin, S. J., Connelly, A., Gadian, D. G., Benton, J. S., and Brett, E. M., 1991, Localized proton NMR spectroscopy in Canavan's disease: A report of two cases, *Magn. Reson. Med.* **19**:439–445.

Autti, T., Raininko, R., Jaunes, J., Nuutila, A., and Santavuori, P., 1992, Jansky–Bielschowsky variant disease: CT, MRI, and SPECT findings, *Pediatr. Neurol.* **8**:121–126.

Baram, T. Z., Goldman, A. M., and Percy, A. K., 1986, Krabbe disease: Specific MRI and CT findings, *Neurology* **36**:111–115.

Barker, P. B., Kumar, A. J., and Naidu, S., 1991, ¹H NMR spectroscopy of Canavan's disease, *in* "Book of Abstracts," 10th Annual Meeting, p. 381, Society of Magnetic Resonance in Medicine, Berkeley, California.

Barkovich, A. J., 1995, "Pediatric Neuroimaging," Raven Press, New York.

Barkovich, A. J., and Chuang, S. H., 1990, Unilateral megalencephaly: Correlation of MR imaging and pathologic characteristics, *Am. J. Neuroradiol.* **11**:523–531.

Barkovich, A. J., Good, W. V., Koch, T. K., and Berg, B. O., 1993, Mitochondrial disorders: Analysis of their clinical and imaging characteristics, *Am. J. Neuroradiol.* **14**:1119–1137.

Barth, P. G., Schüttgens, R. B. H., Wanders, R. J., and Heymans, H. S. A., 1988, The spectrum of peroxisomal disorders, *in* "Child Neurology and Developmental Disabilities. Selected Proceedings of the Forth International Child Neurology Congress" (J. H. French, ed.), Paul H. Brookes, Baltimore, pp. 67–82.

Barth, P. G., Hoffmann, G. F., Jaeken, J., Lehnert, W., Hanefeld, F., van Gennip, A. H., Duran, M., Valk, J., Schutgens, R. B. H., Trefz, F. K., Reimann, G., and Hartung, H. P., 1992, L-2-Hydroxyglutaric acidemia: A novel inherited neurometabolic disease, *Ann. Neurol.* **32**:66–71.

Behar, K. L., den Hollander, J. A., Stromski, M. E., Ogino, T., Shulman, R. G., Petroff, O. A. C., and Prichard, J. W., 1983, High-resolution ¹H nuclear magnetic resonance study of cerebral hypoxia *in vivo, Proc. Natl. Acad. Sci. USA* **80**:4945–4948.

Birken, D. L., and Oldendorf, W. H., 1989, N-Acetyl-L-aspartic acid: A literature review of a compound prominent in ¹H-NMR spectroscopic studies of brain, *Neurosci. Biobehav. Rev.* **13**:23–31.

Boespflug-Tanguy, O., Mimault, C., Melki, J., Cavagna, A., Giraud, G., Dinh, D. P., Dastugue, B., and Dautigny, A., 1994, Genetic homogeneity of Pelizaeus–Merzbacher disease: Tight linkage to the proteolipoprotein locus in 16 affected families, *Am. J. Hum. Genet.* **55**:461–467.

Bottomley, P. A., 1984, U.S. Patent 44 80 228.

Bottomley, P. A., 1987, Spatial localization in NMR-spectroscopy *in vivo, Ann. N. Y. Acad. Sci.* **508**:333–348.

Bottomley, P. A., 1991, The trouble with spectroscopy papers, *Radiology* **181**:344–350.

Bottomley, P. A., Edelstein, W. A., Foster, T. H., and Adams, W. A., 1985, *In vivo* solvent-suppressed localized hydrogen nuclear magnetic resonance spectroscopy: A window to metabolism?, *Proc. Natl. Acad. Sci. USA* **82**:2148–2152.

Bourgeois, M., Goutieres, F., Chretien, D., Rustin, P., Munnich, A., and Aicardi, J., 1992, Deficiency in complex II of the respiratory chain, presenting as a leukodystrophy in two sisters with Leigh syndrome, *Brain Dev.* **14**:404–408.

Brand, A., Richter-Landsberg, C., and Leibfritz, D., 1993, Multinuclear NMR studies on the energy metabolism of glial and neuronal cells, *Dev. Neurosci.* **15**:289–298.

Bruhn, H., Frahm, J., Gyngell, M. L., Merboldt, K. D., Hänicke, W., and Sauter, R., 1989a, Cerebral metabolism in man after acute stroke: New observations using localized proton NMR spectroscopy, *Magn. Reson. Med.* **9**:126–131.

Bruhn, H., Frahm, J., Gyngell, M. L., Merboldt, K. D., Hänicke, W., Sauter, R., and Hamburger, C., 1989b, Noninvasive differentiation of tumors using localized H-1 MR spectroscopy *in vivo:* Initial experience in patients with cerebral tumors, *Radiology* **172**:541–548.

Bruhn, H., Michaelis, T., Merboldt, K. D., Hänicke, W., Gyngell, M. L., and Frahm, J., 1991, Monitoring cerebral glucose in diabetics by proton MRS, *Lancet* **337**:745–746.

Bruhn, H., Michaelis, T., Merboldt, K. D., Hänicke, W., Gyngell, M. L., Hamburger, C., and Frahm, J., 1992a, On the interpretation of proton NMR spectra from brain tumours *in vivo* and *in vitro, NMR Biomed.* **5**:253–258.

Bruhn, H., Frahm, J., Merboldt, K. D., Hänicke, W., Christen, H. J., Kruse, B., Hanefeld, F., and

Bauer, H. J., 1992b, Multiple sclerosis in children. Cerebral metabolic alterations monitored by localized proton magnetic resonance spectroscopy *in vivo, Ann. Neurol.* **32:**140–150.

Bruhn, H., Kruse, B., Korenke, G. C., Hanefeld, F., Hänicke, W., Merboldt, K. D., and Frahm, J., 1992c Proton NMR spectroscopy of cerebral metabolic alterations in infantile peroxisomal disorders, *J. Comput. Assist. Tomogr.* **16:**335–344.

Canavan, M. M., 1931, Schilder's encephalitis periaxialis diffusa. Report of a case in a child aged sixteen and one-half months, *Arch. Neurol. Psychiatry* **25:**299–308.

Cassedy, K. J., and Edwards, M. K., 1993, Metabolic and degenerative diseases of childhood, *Top. Magn. Reson. Imaging* **5:**73–95.

Claussen, M., Heim, P., Knispel, J., Goebel, H. H., and Kohlschütter, A., 1992, Incidence of neuronal ceroid-lipofuscinoses in West Germany: Variation of a method for studying autosomal recessive disorders, *Am. J. Med. Genet.* **42:**536–538.

Cohen, M. E., and Duffner, P. K., 1994, "Brain Tumors in Children. Principles of Diagnosis and Treatment," Raven Press, New York.

Confort-Gouny, S., Chabrol, B., Vion-Dury, J., Mancini, J., and Cozzone, P. J., 1993, MRI and localized proton MRS in early infantile form of neuronal ceroid-lipofuscinosis, *Pediatr. Neurol.* **9:**57–60.

Crome, L., Hanefeld, F., Patrick, D., and Wilson, J., 1973, Late onset globoid cell leukodystrophy, *Brain* **96:**841–848.

Detre, J. A., Wang, Z. Y., Bogdan, A. R., Gusnard, D. A., Bay, C. A., Bingham, P. M., and Zimmerman, R. A., 1991, Regional variation in brain lactate in Leigh syndrome by localized $^1$H magnetic resonance spectroscopy, *Ann. Neurol.* **29:**218–221.

Doll, R., Natowicz, M. R., Schiffman, R., and Smith, F. I., 1992, Molecular diagnostics for myelin proteolipid protein gene mutations in Pelizaeus–Merzbacher disease, *Am. J. Hum. Genet.* **51:**161–169.

Ernst T., and Hennig, J., 1991, Coupling effects in volume-selective $^1$H spectroscopy of major brain metabolites, *Magn. Reson. Med.* **21:**82–96.

Frahm, J., 1993, Nuclear magnetic resonance studies of human brain *in vivo:* Anatomy, function, and metabolism, *Adv. Exp. Med. Biol.* **333:**257–271.

Frahm, J., and Hänicke, W., 1994, Single voxel localized spectroscopy, *in* "Advanced Magnetic Resonance Spectroscopy," Syllabus, 2nd Annual Meeting, pp. 206–213, Society of Magnetic Resonance, Berkeley, California.

Frahm, J., and Hänicke, W., 1996, Single voxel proton NMR. Methods and applications to human subjects, *in* "Encyclopedia of Nuclear Magnetic Resonance" (D. M. Grant and R. K. Harris, eds.), Wiley, Chichester.

Frahm, J., Haase, A., Hänicke, W., Merboldt, K. D., and Matthaei, D., 1984, Verfahren und Einrichtung zur ortsaufgelösten Untersuchung einer Probe mittels magnetischer Resonanz von Spinmomenten, German Patent DE 3445689.

Frahm, J., Merboldt, K. D., and Hänicke, W., 1987, Localized proton spectroscopy using stimulated echoes, *J. Magn. Reson.* **72:**502–508.

Frahm, J., Michaelis, T., Merboldt, K. D., Hänicke, W., Gyngell, M. L., Chien, D., and Bruhn, H., 1989, Localized NMR spectroscopy *in vivo.* Progress and problems, *NMR Biomed.* **2:**188–195.

Frahm, J., Michaelis, T., Merboldt, K. D., Bruhn, H., Gyngell, M. L., and Hänicke, W., 1990, Improvements in localized $^1$H-NMR spectroscopy of human brain. Water suppression, short echo times, and 1 mL resolution, *J. Magn. Reson.* **90:**464–473.

Frahm, J., Michaelis, T., Merboldt, K. D., Hänicke, W., Gyngell, M. L., and Bruhn, H., 1991a, On the *N*-acetyl methyl resonance in localized $^1$H-NMR spectra of human brain *in vivo, NMR Biomed.* **4:**201–204.

Frahm, J., Bruhn, H., Merboldt, K. D., Hänicke, W., Mursch, K., and Markakis, E., 1991b, Localized proton NMR spectroscopy of brain tumors. Methodologic improvements using short-echo time STEAM sequences, *J. Comput. Assist. Tomogr.* **15**:915–922.

Frahm, J., Gyngell, M. L., and Hänicke, W., 1991c, Rapid scan techniques, *in* "Magnetic Resonance Imaging" (D. D. Stark and W. G. Bradley, eds.), pp. 165–203, CV Mosby, St. Louis.

Frahm, J., Bruhn, H., and Hanefeld, F., 1995, Proton NMR studies of human brain metabolism, *in* "Proceedings of the 2nd International Conference on Applications of Magnetic Resonance in Food Science, University of Aveiro, Portugal, 19–21 September 1994" (P. S. Belton, I. Delgadillo, A. M. Gil, and G. A. Webb, eds.), pp. 191–205, Royal Society of Chemistry, Cambridge.

Frahm, J., Krüger, G., Merboldt, K. D., and Kleinschmidt, A., 1996, Dynamic uncoupling and recoupling of perfusion and oxidative metabolism during focal brain activation in man, *Magn. Reson. Med.* **35**:143–148.

Fulham, M. J., Bizzi, A., Dietz, M. J., Shih, H. H., Raman, R., Sobering, G. S., Frank, J. A., Dwyer, A. J., Alger, J. R., and Di Chiro, G., 1992, Mapping of brain tumor metabolites with proton MR spectroscopic imaging: Clinical relevance, *Radiology* **185**:675–686.

Grodd, W., Krägeloh-Mann, I., Petersen, D., Trefz, F. K., and Harzer, K., 1990, *In vivo* assessment of *N*-acetylaspartate in brain in spongy degeneration (Canavan's disease) by proton spectroscopy, *Lancet* **336**:437–438.

Grodd, W., Krägeloh-Mann, I., Klose, U., and Sauter, R., 1991, Metabolic and destructive brain disorders in children: Findings with localized proton MR spectroscopy, *Radiology* **181**:173–181.

Gropen, T. I., Prohovnik, I., Tatemichi, T. K., and Hirano, M., 1994, Cerebral hyperemia in MELAS, *Stroke* **25**:1873–1876.

Gyngell, M. L., Michaelis, T., Bruhn, H., Hänicke, W., Merboldt, K. D., and Frahm, J., 1991, Cerebral glucose is detectable by localized proton NMR spectroscopy in normal rat brain *in vivo*, *Magn. Reson. Med.* **19**:489–495.

Haase, A., Frahm, J., Hänicke, W., and Matthaei, D., 1985, ¹H NMR chemical shift selective (CHESS) imaging, *Phys. Med. Biol.* **30**:341–344.

Hagberg, B., 1993, "Rett Syndrome—Clinical & Biological Aspects", Mac Keith Press, London.

Hagberg, B., Kollberg, H., Sourander, P., and Akerson, H. O., 1970, Infantile globoid cell leucodystrophy (Krabbe's disease): A clinical, morphological, and genetical study of 32 Swedish cases, *Neuropädiatrie* **1**:74–88.

Hagberg, B., Blennow, G., Kristiansson, B., and Stibler, H., 1993, Carbohydrate-deficient glycoprotein syndromes: Peculiar group of new disorders, *Pediatr. Neurol.* **9**:255–262.

Hahn, E. L., 1950, Spin echoes, *Phys. Rev.* **80**:580–592.

Hallervorden, J., 1923, Angeborene Hemihypertrophie der linken Körperhälfte ein-schließlich des Gehirns, *Zentralbl. Gesamte Neurol. Psychiatr.* **33**:518–519.

Hanefeld, F., 1995, Characteristic of childhood multiple sclerosis, *Int. M. S. J.* **1**:91–97.

Hanefeld, F., Bauer, H. J., Christen, H. J., Kruse, B. Bruhn, H., and Frahm, J., 1991, Multiple sclerosis in childhood: Report of 15 cases, *Brain Dev.* **13**:410–416.

Hanefeld, F., Holzbach, U., Kruse, B., Wilichowski, E., Christen, H. J., and J. Frahm, 1993, Diffuse white matter disease in three children: An encephalopathy with unique features on magnetic resonance imaging and proton magnetic resonance spectroscopy, *Neuropediatrics* **24**:244–248.

Hanefeld, F., Kruse, K., Bruhn, H., and Frahm, J., 1994, *In vivo* proton magnetic resonance spectroscopy of the brain in a patient with L-2-hydroxyglutaric aciduria, *Pediatr. Res.* **35**:614–616.

Hanefeld, F., Christen, H. J., Holzbach, U., Kruse, B., Frahm, J., and Hänicke, W., 1995a, Cerebral proton magnetic resonance spectroscopy in Rett syndrome, *Neuropediatrics* **26**:126–127.

Hanefeld, F., Kruse, B., Holzbach, U., Christen, H. J., Merboldt, K. D., Hänicke, W., and Frahm, J., 1995b, Hemimegalencephaly. Localized proton magnetic resonance spectroscopy *in vivo*, *Epilepsia* **36**:1215–1224.

Hashimoto, T., Tayama, M., Miyazaki, M., Fuji, E., Harada, M., Miyoshi, H., Tanouchi, M., and Kuroda, Y., 1995, Developmental brain changes investigated with proton magnetic resonance spectroscopy, *Dev. Med. Child Neurol.* **37**:398–405.

Herschkowitz, N., and Schulte, F. J., 1984, Gangliosidoses and leukodystrophies: A correlative approach in pediatric neurobiology, *Neuropediatris* **15**(Suppl.):1–112.

Holzbach, U., Hanefeld, F., Helms, G., Hänicke, W., and Frahm, J., 1995, Localized proton magnetic resonance spectroscopy of cerebral abnormalities in children with carbohydrate-deficient glycoprotein syndrome, *Acta Paediatr.* **84**:781–786.

Hoult, D. I., and Richards, R. E., 1976, The signal-to-noise ratio of the nuclear magnetic resonance experiment, *J. Magn. Reson.* **24**:71–85.

Howe, F. A., Maxwell, R. J., Saunders, D. E., Brown, M. M., and Griffiths, J. R., 1993, Proton spectroscopy in vivo, *Magn. Reson. Q.* **9**:31–59.

Hüppi, P. S., Posse, S., Lazeyras, F., Burri, R., Bossi, E., and Herschkowitz, N., 1991, Magnetic resonance in preterm and term newborns: [1]H-spectroscopy in developing human brain, *Pediatr. Res.* **30**:574–578.

Hüppi, P. S., Fusch, C., Boesch, C., Burri, R., Bossi, E., Amato, M., and Herschkowitz, N., 1995, Regional metabolic assessment of human brain during development by proton magnetic resonance spectroscopy *in vivo* and by high-performance liquid chromatography/gas chromatography in autopsy tissue, *Pediatr. Res.* **37**:145–150.

Husted, C. A., Matson, G. B., Adams, D. A., Goodin, D. S., and Weiner, M. W., 1994, In vivo detection of myelin phospholipids in multiple sclerosis with phosphorus magnetic resonance spectroscopic imaging, *Ann. Neurol.* **36**:239–241.

Jaeken, J., Stibler, H., and Hagberg B. (eds.), 1991, The carbohydrate-deficient glycoprotein syndrome. A new inherited multisystemic disease with severe nervous system involvement, *Acta Paediatr. Scand.* **375**:1–71.

Järvelä, I., Vesa, J., Santavuori, P., Hellsten, E., and Peltonen, L., 1992, Molecular genetics of neuronal ceroid lipofuscinoses, *Pediatr. Res.* **32**:645–648.

Kaul, R., Gao, G. P., Balamurugan, K., and Matalon, R., 1993, Cloning of the human aspartoacylase cDNA and a common missense mutation in Canavan disease, *Nat. Genet.* **5**:118–123.

Kauppinen, R. A., Nissinen, T., Kärkkäinen, A. M., Pirttila, T. R. M., Palrimo, J., Kokko, H., and Williams, S. R., 1992, Detection of thymosin β4 *in situ* in a guinea pig cerebral cortex preparation using [1]H NMR spectroscopy, *J. Biol. Chem.* **267**:9905–9910.

Kemp, G. J., and Radda, G. K., 1994, Quantitative interpretation of bioenergetic data from [31]P and [1]H magnetic resonance spectroscopic studies of skeletal muscle: An analytical review, *Magn. Reson. Q.* **10**:43–63.

Kendall, B. E., 1993, Inborn errors and demyelination: MRI and the diagnosis of white matter disease, *J. Inherited Metab. Dis.* **16**:771–786.

Kinoshita, Y., Kajiwara, H., Yokota, A., and Koga, Y., 1994, Proton magnetic resonance spectroscopy of brain tumors: An in vitro study, *Neurosurgery* **35**:606–614.

Kolodny, E. H., 1989, Metachromatic leukodystrophy and multiple sulfatase deficiency: Sulfatide lipidosis, *in* "The Metabolic Basis of Inherited Disease" (C. R. Scriver, A. L. Beaudet, W. S. Sly, and D. Valle, eds), pp. 1721–1750, McGraw-Hill, New York.

Kolodny, E. H., 1993, Dysmeylinating and demyelinating conditions in infancy, *Curr. Opin. Neurol. Neurosurg.* **6**:379–386.

Krabbe, K., 1916, A new familial form of diffuse brain-sclerosis, *Brain* **39**:74–114.

Krägeloh-Mann, I., Grodd, W., Schöning, M., Marquard, K., Nägele, I., and Ruitenbeek, W., 1993, Proton spectroscopy in five patients with Leigh's disease and mitochondrial enzyme deficiency, *Dev. Med. Child. Neurol.* **35**:769–776.

Kreis, R., and Ross, 1992, Cerebral metabolic disturbances in patients with subacute and chronic diabetes mellitus: Detection with proton MR spectroscopy, *Radiology* **184**:123–130.

Kreis, R., Farrow, N. A., and Ross, B. D., 1990, Diagnosis of hepatic encephalopathy by proton magnetic resonance spectroscopy, *Lancet* **336**:635–636.

Kreis, R., Ross, B. D., Farrow, N. A., and Ackermann, Z., 1992, Metabolic disorders of the brain in chronic hepatic encephalopathy detected with H-1 MR spectroscopy, *Radiology* **182**:19–27.

Kreis, R., Ernst, T., and Ross, B. D., 1993, Development of the human brain: In vivo quantification of metabolite and water content with proton magnetic resonance spectroscopy, *Magn. Reson. Med.* **30**:424–437.

Kruse, B., Hanefeld, F., Christen, H. J., Bruhn, H., Michaelis, T., Hänicke, W., and Frahm, J., 1993, Alterations of brain metabolites in metachromatic leukodystrophy as detected by localized proton MR spectroscopy *in vivo, J. Neurol.* **241**:68–74.

Kruse, B., Barker, P. B., van Zijl, P. C., Duyn, J. H., Moonen, C. T., and Moser, H. W., 1994a, Multislice proton magnetic resonance spectroscopic imaging in X-linked adrenoleukodystrophy, *Ann. Neurol.* **36**:595–608.

Kruse, B., Hanefeld, F., Holzbach, U., Wilichowski, E., Christen, H. J., Merboldt, K. D., Hänicke, W., and Frahm, J., 1994b, *Dev. Med. Child. Neurol.* **36**:839–840. [Letter to: Krägeloh-Mann, I., Grodd, W., Schöning, M., Marquard, K., Nägele, T., and Ruitenbeek, W., Proton spectroscopy in patients with Leigh's disease and mitochondrial enzyme deficiency, *Dev. Med. Child. Neurol.* **35**:769–776].

Kucharczyk, W., Brant-Zawadzki, M., Sobel, D., Edwards, M. B., Kelly, W. M., Norman, D., and Newton, T. H., 1985, Central nervous system tumors in children: Detection by magnetic resonance imaging, *Radiology* **155**:131–136.

Lee, M. L., Chaou, W. T., Yang, A. D., Jong, Y. J., Tsai, J. L., Pang, C. Y., and Wei, Y. H., 1994, Mitochondrial myopathy, encephalopathy, lactic acidosis and stroke-like episodes (MELAS): Report of a sporadic case and review of the literature, *Acta Paediatr. Sin.* **35**:148–156.

Leigh, D., 1951, Subacute necrotizing encephalomyelopathy in an infant, *J. Neurol. Neurosurg. Psychiatry,* **14**:216–221.

Martinsson, T., Bjursell, C., Stibler, H., Kristiansson, B., Skovby, F., Jaeken, J., Blennow, G., Strömme, P., Hanefeld, F., and Wahlström, J., 1994, Linkage of a locus for carbohydrate-deficient glycoprotein syndrome type I (CDG I) to chromosome 16p and linkage dysequilibrium to microsatellite marker D16S406, *Hum. Mol. Genet.* **3**:2037–2042.

Matalon, R., Michals, K., Sebesta, D., Deanching, M., Gashkoff, P., and Casanova, J., 1988, Aspartoacylase deficiency and *N*-acetylaspartic aciduria in patients with Canavan disease, *Am. J. Med. Genet.* **29**:463–471.

Matthews, P. M., Tampieri, D., Berkovic, S. F., Andermann, F., Silver, K., Chityat, D., and Arnold, D. L., 1991a, Magnetic resonance imaging shows specific abnormalities in the MELAS syndrome, *Neurology* **41**:1043–1046.

Matthews, P. M., Berkovic, S. F., Shoubridge, E. A., Andermann, F., Karpati, G., Carpenter, S., and Arnold, D. L., 1991b, *In vivo* magnetic resonance spectroscopy of brain and muscle in a type of mitochondrial encephalopathy (MERRF), *Ann. Neurol.* **29**:435–438.

Mehl, E., and Jatzkewitz, H., 1965, Evidence for the genetic block in metachromatic leukodystrophy (ML), *Biochem. Biophys. Res. Commun.* **19**:407–411.

Merboldt, K. D., Chien, D., Hänicke, W., Gyngell, M. L., Bruhn, H., and Frahm, J., 1990, Localized ³¹P-NMR spectroscopy of the adult human brain *in vivo* using stimulated echo (STEAM) sequences, *J. Magn. Reson.* **89**:343–361.

Merboldt, K. D., Bruhn, H., Hänicke, W., Michaelis, T., and Frahm, J., 1992, Decrease of glucose in the human visual cortex during photic stimulation, *Magn. Reson. Med.* **25**:187–194.

Merzbacher, L., 1910, Eine eigenartige familiär-hereditäre Erkrankungsform (Aplasia axialis extracorticalis congenita), *Z. Gesamte Neurol. Psychiatr.* **3**:1–138.

Michaelis, T., Merboldt, K. D., Hänicke, W., Gyngell, M. L., Bruhn, H., and Frahm, J., 1991, On the identification of cerebral metabolites in localized ¹H-NMR spectra of human brain *in vivo, NMR Biomed.* **4:**90–98.

Michaelis, T., Helms, G., Merboldt, K. D., Hänicke, W., Bruhn, H., and Frahm, J., 1993a, Identification of *scyllo*-inositol in proton NMR spectra of human brain *in vivo, NMR Biomed.* **6:**105–109.

Michaelis, T., Merboldt, K. D., Bruhn, H., Hänicke, W., and Frahm, J., 1993b, Absolute concentrations of metabolites in the adult human brain *in vivo:* Quantification of localized proton MR spectra, *Radiology* **187:**219–227.

Miller, B. L., Moats, R. A., Shonk, T., Ernst, T., Woolley, S., and Ross, B. D., 1993, Alzheimer disease: Depiction of increased cerebral *myo*-inositol with proton MR spectroscopy, *Radiology* **187:**433–437.

Mosser, J., Douar, A. M., Sarde, C. O., Kioschis, P., Feil, R., Moser, R., Poustka, A. M., Mandel, J. L., and Aubourg, P., 1993, Putative X-linked adrenoleukodystrophy gene shares unexpected homology with ABC transporters, *Nature (London)* **361:**726–730.

Nielsen, J. B., Toft, P. B., Reske-Nielsen, E., Jensen, K. E., Christiansen, P., Thomsen, C., Henriksen, O., and Lou, H. C., 1993, Cerebral magnetic resonance spectroscopy in Rett syndrome. Failure to detect mitochondrial disorder, *Brain Dev.* **15:**107–112.

Norton, W. T., and Poduslo, S. E., 1982, Biochemical studies of metachromatic leukodystrophy in three siblings, *Acta Neuropathol.* **57:**188–196.

Ochi, N., Kobayashi, K., Mehara, M., Nakayama, A., Negoro, T., Shinohara, H., Watanabe, K., Nagatsu, T., and Kato, K., 1991, Increment of α B-crystallin mRNA in the brain of a patient with infantile type Alexander's disease, *Biochem. Biophys. Res. Commun.* **179:**1030–1035.

Ordidge, R. J., Bendall, M. R., Gordon, R. E., and Connelly, A., 1985, Volume selection for in-vivo biological spectroscopy, *in* "Magnetic Resonance in Biology and Medicine" (G. Govil, C. L. Khetrapal, and A. Saran, eds.), p. 387, Tata McGraw-Hill, New Delhi.

Peden, C. J., Cowan, F. M., Bryant, D. J., Sargentoni, J., Cox, I. J., Menon, D. K., Gadian, D. G., Bell, J. D., and Dubowitz, L. M., 1990, Proton MR spectroscopy of the brain in infants, *J. Comput. Assist. Tomogr.* **14:**886–894.

Peeling, J., and Sutherland, G., 1993, ¹H magnetic resonance spectroscopy of extracts of human epileptic neocortex and hippocampus, *Neurology* **43:**589–594.

Pelizaeus, E., 1885, Über eine eigentümliche Form spastischer Lähmungen mit Cerebralerscheinungen auf hereditärer Grundlage, *Arch. Psychiatr.* **16:**698–710.

Percy, A. K., Odrezin, G. T., Knowles, P. D., Rouah, E., and Armstrong, D. D., 1994, Globoid cell leukodystrophy: Comparison of neuropathology with magnetic resonance imaging, *Acta Neuropathol.* **88:**26–32.

Perry, T. L., Hansen, S., Berry, K., Mok, C., and Lesk, D., 1971, Free amino acids and related compounds in biopsies of human brain, *J. Neurochem.* **18:**521–528.

Perry, T. L., Hansen, S., and Gandham, S. S., 1981, Postmortem changes of amino compounds in human and rat brain, *J. Neurochem.* **36:**407–412.

Petroff, O. A. C., Spencer, D. D., Alger, J. R., and Prichard, J. W., 1989, High-field proton magnetic resonance spectroscopy of human cerebrum obtained during surgery for epilepsy, *Neurology* **39:**1197–1202.

Polten, A., Fluharty, A. L., Fluharty, C. B., Kappler, J., von Figura, K., and Gieselmann, V., 1991, Molecular basis of different forms of metachromatic leukodystrophy, *N. Engl. J. Med.* **324:**18–22.

Pouwels, P. J. W., Hänicke, W., and Frahm, J., 1995a, On a concentration gradient of cerebral metabolites in human gray matter as determined by quantitative localized proton MRS, *in* "Book of Abstracts," 3rd Annual Meeting, Society of Magnetic Resonance, Berkeley, California.

Pouwels, P. J. W., Kruse, B., Hanefeld, F., and Frahm, J., 1995b, Quantitative localized proton MRS in adrenoleukodystrophy, *Neuropediatrics.* **26:**341–341.

Prichard, J. W., and Shulman, R. G., 1986, NMR spectroscopy of brain metabolism in vivo, *Annu. Rev. Neurosci.* **9**:61–85.

Prichard, J., Rothman, D., Novotny, E., Petroff, O., Kuwabara, T., Avison, M., Howseman, A., Hanstock, C., and Shulman, R. G., 1992, Lactate rise detected by ¹H NMR in human visual cortex during physiologic stimulation, *Proc. Natl. Acad. Sci. USA* **88**:5829–5831.

Provencher, S. W., 1982, A constrained regularization method for inverting data represented by linear algebraic or integral equations, *Comput. Phys. Commun.* **27**:213–227.

Provencher, S. W., 1993, Estimation of metabolite concentrations from localized *in vivo* NMR spectra, *Magn. Reson. Med.* **30**:672–679.

Radda, G. K., 1986, The use of NMR spectroscopy for the understanding of disease, *Science* **233**:640–645.

Ramaeckers, V. T., Stibler, H., Kint, J., and Jaeken, J., 1991, A new variant of the carbohydrate-deficient glycoprotein syndrome, *J. Inherited Metab. Dis.* **14**:385–388.

Requardt, M., 1995, Dissertation, University of Göttingen, Göttingen.

Rett, A., 1966, Über ein eigenartiges hirnatrophisches Syndrom bei Hyperammonämie im Kindesalter, *Wien. Med. Wochenschr.* **116**:723–726.

Rosenberg, R. N., Prusiner, S. B., DiMauro, S., Barchi, R. L., and Kunkel, L. M. (eds), 1993, "The Molecular Biology and Genetic Basis of Neurological Disease," Butterworth-Heineman, Boston.

Ross, B. D., and Michaelis, T., 1994, Clinical applications of magnetic resonance spectroscopy, *Magn. Reson. Q.* **10**:191–248.

Sandhoff, K., Conzelmann, E., Neufeld, E. F., Kaback, M. M., and Suzuki, K., 1989, The GM2-gangliosidosis, *in* "The Metabolic Basis of Inherited Disease" (C. R. Scriver, A. L. Beaudet, W. S. Sly, and D. Valle, eds.), pp. 1807–1839, McGraw-Hill, New York.

Santavuori, P., Rapola, J., Nuutila, A., Raininko, R., Lappi, M., Launes, J., Herva, R., and Sainio, K., 1991, The spectrum of Jansky–Bielschowsky disease, *Neuropediatrics* **22**:92–96.

Sasaki, M., Sakuragawa, N., Takashima, S., Hanaoka, S., and Arima, M., 1991, MRI and CT findings in Krabbe disease, *Pediatr. Neurol.* **7**:283–288.

Scholz, W., 1925, Klinische, pathologisch-anatomische und erbbiologische Untersuchungen bei familiärer, diffuser Hirnsklerose im Kindesalter, *Z. Gesamte Neurol. Psychiatr.* **99**:651–717.

Schuster, V., Horwitz, A. E., and Kreth, H. W., 1991, Alexander's disease: Cranial MRI and ultrasound findings, *Pediatr. Radiol.* **21**:133–134.

Seitelberger, F., 1970, Pelizaeus–Merzbacher disease, *in* "Handbook of Clinical Neurology, Vol. 10, Leucodystrophies and Poliodystrophies" (P. J. Vinken and G. W. Brown, eds.), pp. 150–202, North-Holland, Amsterdam.

Sengers, R. C. A., Stadhouders, A. M., and Trijbels, J. M. F., 1984, Mitochondrial myopathies. Clinical, morphological, and biochemical aspects, *Eur. J. Pediatr.* **141**:192–207.

Sherman, W. R., Stewart, M. A., Kurien, M. M., and Goodwin, S. L., 1968, The measurement of *myo*-inositol, *myo*-inosose-2 and *scyllo*-inositol in mammalian tissues, *Biochim. Biophys. Acta* **158**:197–205.

Stibler, H., Westerberg, B., Hanefeld, F., and Hagberg, B., 1993, Carbohydrate-deficient glycoprotein (CDG) syndrome—a new variant type III, *Neuropediatrics* **24**:51–52.

Stöckler, S., Holzbach, U., Hanefeld, F., Marquardt, I., Helms, G., Requardt, M., Hänicke, W., and Frahm, J., 1994, Creatine deficiency in the brain: A new treatable inborn error of metabolism, *Pediatr. Res.* **36**:409–413.

Suzuki, K., Suzuki, Y., and Eto, Y., 1971, Deficiency of galactocerebroside β-galactosidase in Krabbe's globoid cell leukodystrophy, *in* "Lipid Storage Diseases: Enzymatic Defect and Clinical Implications" (J. Bersohn and H. J. Grossman, eds.), p. 396, Academic Press, New York.

Toft, P. B., Leth, H., Lou, H. C., Pryds, O., and Henrikson, O., 1994, Metabolite concentrations in the developing brain estimated with water proton MR spectroscopy, *J. Magn. Reson. Imaging* **4:**674–680.

Tzika, A. A., Vigneron, D. B., Ball, W. S., Jr., Dunn, R. S., and Kirks, D. R., 1993a, Localized proton MR spectroscopy of the brain in children, *J. Magn. Reson. Imaging* **3:**719–729.

Tzika, A. A., Ball, W. S., Jr., Vigneron, D. B., Dunn, R. S., and Kirks, D. R., 1993b, Clinical proton MR spectroscopy of neurodegenerative disease in childhood, *Am. J. Neuroradiol* **14:**1267–1281.

Uyama, E., Terasaki, T., Watanabe, S., Naito, M., Owada, M., Araki, S., and Ando, M., 1992, Type 3 GM1 gangliosidosis: Characteristic MRI findings correlated with dystonia, *Acta Neurol. Scand.* **86:**609–615.

Valk, J., and van der Knaap, M. S., 1989, "Magnetic Resonance of Myelin, Myelination, and Myelin Disorders," Springer, Berlin.

van Bogaert, L., and Bertrand, I., 1967, "Spongy Degeneration of the Brain in Infancy," North-Holland, Amsterdam.

van der Knaap, M. S., van der Grond, J., van Rijen, P. C., Faber, J. A., Valk, J., and Willemse, K., 1990, Age-dependent changes in localized proton and phosphorus MR spectroscopy of the brain, *Radiology* **176:**509–515.

van der Knaap, M. S., van der Grond, J., Luyten, P. R., den Hollander, J. A., Nauta, J. P., and Valk, J., 1992, ¹H and ³¹P magnetic resonance spectroscopy of the brain in degenerative cerebral disorders, *Ann. Neurol.* **31:**202–211.

van der Knaap, M. S., Barth, P. G., Stroink, H., van Nieuwenhuizen, O., Arts, W. F. M., Hoogenraad, F., and Valk, J., 1995, Leukoencephalopathy with swelling and a discrepancy mild clinical course in eight children, *Ann. Neurol.* **37:**324–334.

von Graefe, A., 1866, Bemerkungen über doppelseitige Augenmuskellähmungen basilären Ursprunges, *Albrecht von Graefes Arch. Ophthalmol.* **12:**265.

# INDEX

The manufacturer's authorised representative in the EU is Springer
Nature Customer Service Centre GmbH, Europaplatz 3, 69115 Heidelberg,
Germany. If you have any concerns regarding our products, please
contact ProductSafety@springernature.com

Printed and bound by CPI Group (UK) Ltd, Croydon, CR0 4YY
23/04/2026
02095585-0001